Stem Cell Biology and Regenerative Medicine

Series Editor
Kursad Turksen, Ph.D.
kursadturksen@gmail.com

More information about this series at http://www.springer.com/series/7896

Dario O. Fauza • Mahmud Bani
Editors

Fetal Stem Cells in Regenerative Medicine

Principles and Translational Strategies

Springer

Editors
Dario O. Fauza
Department of Surgery
Boston Children's Hospital
and Harvard Medical School
Boston, MA, USA

Mahmud Bani
Department of Cellular and Molecular Medicine
University of Ottawa, Faculty of Medicine
Ottawa, ON, Canada

ISSN 2196-8985 ISSN 2196-8993 (electronic)
Stem Cell Biology and Regenerative Medicine
ISBN 978-1-4939-3481-2 ISBN 978-1-4939-3483-6 (eBook)
DOI 10.1007/978-1-4939-3483-6

Library of Congress Control Number: 2016934211

Printed on acid-free paper

This Springer imprint is published by Springer Nature
The registered company is Springer Science+Business Media LLC New York

To the patients and their families,
who may benefit from the developments
discussed in this book

To our mentors and colleagues,
who share our quest for such developments

To our families, who enrich our lives
through love, care, and understanding
amid our quest

Dario O. Fauza and Mahmud Bani

Foreword

This book brings together two important fields of science and engineering with roots in the latter part of the twentieth century and propels them into the twenty-first century. Fetal biology and fetal therapy are interwoven into the disciplines of regenerative medicine and tissue engineering. The marriage is logical from many standpoints. Developmental biology has undergone explosive growth in new knowledge and understanding. The gene programs and signaling pathways in many ways overlap with the signaling of successful regeneration of tissues. Much of this knowledge can be harnessed into new strategies to improve patient care. Engineering science stands at this nexus in many circumstances.

Fauza and Bani have carefully assembled experts in the key areas of these fields and have put together a thoughtful sequence of chapters which brings the reader through sophisticated science and technology in a coherent and readable way. New populations of stem cells including fetal stem cells, embryonic stem cells, amniotic stem cells, and placental and umbilical cord stem cells are all described and discussed. As well, their potential use in human therapy is a fundamental part of the book.

In short, this book provides a readable summary of an area in science and medicine that has the potential to transform the way we think about improving patient care and a paradigm shift in the way we approach future studies to understand developmental biology and apply that knowledge to improve the human condition.

August 25, 2015 — Joseph P. Vacanti

Preface

Despite significant experimental advances, much promise, and excessive publicity, most cell-based therapies have yet to deliver meaningful impact to patient care. Conspicuous exceptions have been therapies based on amplifying the biological role played by cells in their natural environment. Evident examples are variations of blood transfusions and bone marrow transplantation. These long-established cell-based therapies have had unparalleled impacts in health care, to a large extent due to the fact that the cells involved fulfill the very same roles that they already perform in nature. Therein lies much of the appeal of fetal stem cell-based regenerative medicine, particularly as it applies to the perinatal period, during which the normal biological activities of these cells are regulated within the distinctive environment in which they already operate, aiming at therapeutic benefit. As much as fetal stem cells have shown to possess unique characteristics compared with other stem cells, so do the fetus and neonate when compared with any other age group, converging into a perfect storm that enables unparalleled biomedical discoveries, original therapeutic paradigms, and ultimate translational significance. Although fetal stem cells have been increasingly used in recipients of all ages, this book is focused on their perinatal applications, exploring the exceptionality of their fundamental roles in fetal development, arguably the purest form of regeneration. This relationship lends overt biological validation to the use of these cells in therapeutic strategies within this specific period, confirmed by prolific advances in the field. It also allows for the establishment of select service-based models of on-demand individualized stem cell processing, while validating fetal stem cell banking as clinically relevant.

In light of such tangible biological and therapeutic correlations, it is perhaps surprising that fetal regenerative medicine is still in its infancy, even when compared with its parent field. Therefore, expectedly, much of the nomenclature used has yet to be properly standardized. This is reflected in some of the chapters, which expose terminology overlaps typical of an emerging discipline, while we deliberately avoided attempts to arbitrarily systematize it. Also typical of a burgeoning field is its fluidity. This has led us to favor basic principles and general translational strategies, as opposed to multiple, as of yet unwarranted fragmented chapters devoted to narrower specific applications, conferring a more universal nature to the book. This should appeal to a broader readership of physicians, scientists, and trainees.

We were fortunate to have attracted contributions from esteemed, highly prominent colleagues in their respective areas of expertise, to whom we are greatly obliged. We are also grateful to Michael Griffin at Springer for his patience and precious assistance throughout the preparation of this volume. Special thanks from MB to the late Andree Gruslin, a kind, passionate, and cheerful clinician-scientist who will be always remembered for her devotion to promoting regenerative medicine. A personal, deep expression of gratitude from DOF goes to Kevin and Kate McCarey for their sustained generous support, without which a number of the developments discussed herein would not have taken place.

Boston, MA, USA — Dario O. Fauza, MD, PhD
Ottawa, ON, Canada — Mahmud Bani, PhD

Contents

About the Editors

Dr. Dario Fauza received his M.D. from the University of São Paulo Medical School, in São Paulo, Brazil, where he also received a PhD-equivalent degree. There, he attended an internship and residencies in both general and pediatric surgery. He then moved to the United States, where he completed postdoctoral training in different clinical and research fellowships at Boston Children's Hospital and Harvard Medical School. He is currently an Associate in Surgery and Associate Professor of Surgery at these two institutions, respectively. Dr. Fauza serves as a member of various professional organizations, including the American Academy of Pediatrics, the International Fetal Medicine and Surgery Society, the Society of University Surgeons, and the Royal College of Surgeons of England, where he received a Fellowship *ad eundem*. He has an extensive bibliography as well as a patent portfolio. His research is directed at developing original, more effective ways to repair birth defects, both pre- and postnatally. To that end, he has pioneered and explored a variety of approaches, including fetal tissue engineering and transamniotic stem cell therapy.

Dr. Mahmud Bani received his Ph.D. in neurodevelopment from the Department of Anatomy and Cell Biology at the University of Western Ontario in Ontario, Canada. Following the completion of his doctoral training, he took a postdoctoral fellowship in the Neural Stem Cell Laboratory at the Robarts Research Institute, studying neural cell fate and signaling pathways. In 2002, Dr. Bani joined the National Research Council (NRC) Canada in Ottawa, where he has established his research on the applications of stem cells in the brain. He is currently a Team Leader and Senior Research Officer in the Department of Translational Bioscience at NRC, and an Adjunct Professor in the Department of Cellular and Molecular Medicine at the University of Ottawa. Dr. Bani serves as a member of several scientific organizations and committees, including the Natural Sciences and Engineering Research Council of Canada, Ontario Institute for Regenerative Medicine, and the University of Ottawa Brain and Mind Research Institute. Dr. Bani's primary research interests are cellular and molecular mechanisms of neurogenesis and developing neuroprotective and neuroregenerative strategies for brain injury.

Contributors

Sandra Acosta, PhD Center of Excellence for Aging and Brain Repair, Department of Neurosurgery and Brain Repair, Morsani College of Medicine, University of South Florida College of Medicine, Tampa, FL, USA

Ji Yeon Ahn, PhD Departments of Agricultural Biotechnology, Seoul National University, Seoul, South Korea

Mohammad Z. Albanna, PhD Cook General Biotechnology, LLC, Indianapolis, IN, USA

Mofeedah Al Shammary, MD Gynecology and Newborn Care, Division of Maternal Fetal Medicine, Department of Obstetrics and Gynecology, The Ottawa Hospital, University of Ottawa, Ottawa, ON, Canada

Karen K. Ballen, MD Division of Hematology/Oncology, Massachusetts General Hospital Cancer Center, Boston, MA, USA

Mahmud Bani-Yaghoub, PhD Department of Cellular and Molecular Medicine, University of Ottawa, Faculty of Medicine, Ottawa, ON, Canada

Marina Bastawrous, BS Center of Excellence for Aging and Brain Repair, Department of Neurosurgery and Brain Repair, Morsani College of Medicine, University of South Florida College of Medicine, Tampa, FL, USA

Julie Di Bernardo, PhD Department of Surgery, C.S. Mott Children's Hospital and Von Voigtlander Women's Hospital, University of Michigan, Ann Arbor, MI, USA

Matthew M. Boelig, MD Department of Surgery, Hospital of the University of Pennsylvania, Philadelphia, PA, USA

Cesar V. Borlongan, BS, MS, PhD Center of Excellence for Aging and Brain Repair, Department of Neurosurgery and Brain Repair, Morsani College of Medicine, University of South Florida College of Medicine, Tampa, FL, USA

Joseph A. Brazzo III, MS Department of Surgery, Boston Children's Hospital, Boston, MA, USA

Timothy M. Crombleholme, MD Laboratory for Fetal and Regenerative Biology, Children's Hospital Colorado, Aurora, CO, USA

Anna L. David, PhD, MRCOG Institute for Women's Health, University College London, London, UK

Scott M. Deeney, MD Department of Surgery, Center for Fetal and Regenerative Biology, University of Colorado School of Medicine, Aurora, CO, USA

S. Christopher Derderian, MD Department of Surgery and The Center for Maternal-Fetal Precision Medicine, University of California, San Francisco, San Francisco, CA, USA

Beatrice Dionigi, MD Department of Surgery, Brigham and Women's Hospital, Boston, MA, USA

Dario O. Fauza, MD, PhD Department of Surgery, Boston Children's Hospital, Harvard Medical School, Boston, MA, USA

Christina Feng, MD Department of Surgery, Beth Israel Deaconess Medical Center, Boston, MA, USA

Alan W. Flake, MD, FACS, FAAP Division of Pediatric General, Thoracic and Fetal Surgery, The Children's Hospital of Philadelphia, Philadelphia, PA, USA

David T. Harris, PhD Department Immunobiology, University of Arizona, Tucson, AZ, USA

Julie Haukenfrers, BSc Translational Bioscience, Human Health Therapeutics Portfolio, National Research Council Canada, Ottawa, ON, Canada

Diana Hernandez-Ontiveros, MS, PhD Department of Physiology and Pharmacology, University of South Florida College of Medicine, Tampa, FL, USA

David A. Hess, PhD Krembil Centre for Stem Cell Biology, Robarts Research Institute, London, ON, Canada

Department of Physiology and Pharmacology, University of Western Ontario, London, ON, Canada

Ryan J. Hodges, MBBS, PhD, FRANZCOG The Ritchie Centre, Department of Obstetrics and Gynaecology, Monash University, Monash Medical Centre, Clayton, VIC, Australia

Cerine Jeanty, MD Department of Surgery and The Center for Maternal-Fetal Precision Medicine, University of California, San Francisco, San Francisco, CA, USA

Yuji Kaneko, PhD Center of Excellence for Aging and Brain Repair, Department of Neurosurgery and Brain Repair, Morsani College of Medicine, University of South Florida College of Medicine, Tampa, FL, USA

Shaun M. Kunisaki, MD, MSc Department of Surgery, C.S. Mott Children's Hospital and Von Voigtlander Women's Hospital, University of Michigan, Ann Arbor, MI, USA

Bryan A. Liang, MD, PhD, JD Department of Anesthesiology, San Diego Center for Patient Safety, San Diego School of Medicine, University of California, San Diego, CA, USA

Jeong Mook Lim, DVM, PhD Departments of Agricultural Biotechnology, Seoul National University, Seoul, South Korea

Cancer Research Institute, Seoul National University Hospital, Seoul, South Korea

Research Institutes of Agriculture and Life Sciences, Seoul National University, Seoul, South Korea

Rebecca Lim, PhD The Ritchie Centre, Department of Obstetrics and Gynaecology, Monash University, Monash Medical Centre, Clayton, VIC, Australia

Qing Yan Liu, PhD Translational Bioscience, Human Health Therapeutics Portfolio, National Research Council Canada, Ottawa, ON, Canada

Dao Ly, BSc Translational Bioscience, Human Health Therapeutics Portfolio, National Research Council Canada, Ottawa, ON, Canada

Tippi C. MacKenzie, MD Department of Surgery and The Center for Maternal-Fetal Precision Medicine, University of California, San Francisco, San Francisco, CA, USA

Timothy Ken Mackey, MAS, PhD Department of Anesthesiology, San Diego Center for Patient Safety, School of Medicine, University of California, San Diego, San Diego, CA, USA

Felipe Mangoni Moretti, MD Division of Maternal Fetal Medicine, Department of Obstetrics, Gynecology and Newborn Care, The Ottawa Hospital, University of Ottawa, Ottawa, ON, Canada

Dema Najem, MSc Translational Bioscience, Human Health Therapeutics Portfolio, National Research Council Canada, Ottawa, ON, Canada

Heiner Niemann, DVM, PhD Department of Biotechnology, Institute of Farm Animal Genetics (FLI), Mariensee, Neustadt, Germany

Monika Nowak-Imialek, DVM Department of Biotechnology, Institute of Farm Animal Genetics (FLI), Mariensee, Neustadt, Germany

Mibel M. Pabón, MS, PhD Center of Excellence for Aging and Brain Repair, Department of Neurosurgery and Brain Repair, Morsani College of Medicine, University of South Florida College of Medicine, Tampa, FL, USA

Paolina Pantcheva, BS, MS Center of Excellence for Aging and Brain Repair, Department of Neurosurgery and Brain Repair, Morsani College of Medicine, University of South Florida College of Medicine, Tampa, FL, USA

Ike de la Peña, PhD Pharmaceutical and Administrative Sciences, Loma Linda University, Loma Linda, CA, USA

Mariusz Z. Ratajczak, MD, PhD Department of Medicine, James Graham Brown Cancer Center, University of Louisville, Louisville, KY, USA

Janina Ratajczak, MD, PhD, DSci Department of Medicine, James Graham Brown Cancer Center, University of Louisville, Louisville, KY, USA

Maria Ribecco-Lutkiewicz, PhD Translational Bioscience, Human Health Therapeutics Portfolio, National Research Council Canada, Ottawa, ON, Canada

Gabriela Schneider, PhD Department of Medicine, James Graham Brown Cancer Center, University of Louisville, Louisville, KY, USA

Panicos Shangaris, MBBCh, BAO, MSc Institute for Women's Health, University College London, London, UK

Stephen E. Sherman, BSc Department of Physiology and Pharmacology, Western University, London, ON, Canada

Kazutaka Shinozuka, PhD Neurosurgery and Brain Repair, University of South Florida Morsani College of Medicine, Tampa, FL, USA

Center of Excellence for Aging and Brain Repair, Department of Neurosurgery and Brain Repair, Morsani College of Medicine, University of South Florida College of Medicine, Tampa, FL, USA

Marianna Sikorska, PhD Center of Excellence for Aging and Brain Repair, Department of Neurosurgery and Brain Repair, Morsani College of Medicine, University of South Florida College of Medicine, Tampa, FL, USA

Brandon Smith, BSc, MSc Translational Bioscience, Human Health Therapeutics Portfolio, National Research Council Canada, Ottawa, ON, Canada

Caroline Sodja, MSc Translational Bioscience, Human Health Therapeutics Portfolio, National Research Council Canada, Ottawa, ON, Canada

Danica Stanimirovic, MD, PhD Translational Bioscience, Human Health Therapeutics Portfolio, National Research Council Canada, Ottawa, ON, Canada

Department of Cellular and Molecular Medicine, Faculty of Medicine, University of Ottawa, Ottawa, ON, Canada

Meaghan Staples, BS Center of Excellence for Aging and Brain Repair, Department of Neurosurgery and Brain Repair, Morsani College of Medicine, University of South Florida College of Medicine, Tampa, FL, USA

Naoki Tajiri, PhD Center of Excellence for Aging and Brain Repair, Department of Neurosurgery and Brain Repair, Morsani College of Medicine, University of South Florida College of Medicine, Tampa, FL, USA

Jean Tan, BSc The Ritchie Centre, Department of Obstetrics and Gynaecology, Monash University, Monash Medical Centre, Clayton, VIC, Australia

Euan M. Wallace, MBChB, MD The Ritchie Centre, Department of Obstetrics and Gynaecology, Monash University, Monash Medical Centre, Clayton, VIC, Australia

Erik J. Woods, PhD Cook General Biotechnology, LLC, Indianapolis, IN, USA

Part I
Fetal Cell Biology

Chapter 1
Historical Perspectives

Scott M. Deeney and Timothy M. Crombleholme

Introduction

Stem cells are a source of intense scientific inquiry due to their unique properties. Research into the nature of these cells has deepened our understanding of cell and molecular biology and has led to the development of many important therapies. Human stem cells have been harvested from embryos and have been induced from adult tissues, but stem cells from embryonic and adult sources have demonstrated certain inherent limitations.

In recent decades, the human fetus has been increasingly viewed as a distinct entity worthy of study, aided largely by advancements in prenatal diagnostic and imaging techniques as well as by the development of the concept of the fetus as a patient. Fetal interventions are already being performed as an early way of preventing or treating progressive diseases. The ability of fetal skin to heal without scarring has been noted and is just one example of the unique properties possessed by fetal tissue [1]. Further investigation into the properties of fetal tissue has revealed the existence of stem cells which are notably distinct from embryonic and adult stem cells. These fetal stem cells have recently been described as a separate category of stem cells with its own characteristics, some of which show promise in bypassing the limitations inherent in embryonic and adult stem cell research.

S.M. Deeney, M.D.
Department of Surgery, Laboratory for Fetal and Regenerative Biology, University of Colorado School of Medicine, 12700 East 19th Ave. Mailstop 8618, Aurora, CO 80045, USA
e-mail: Scott.deeney@ucdenver.edu

T.M. Crombleholme, M.D. (✉)
Laboratory for Fetal and Regenerative Biology, Children's Hospital Colorado, 13123 East 16th Avenue, B 323, Aurora, CO 80045, USA
e-mail: Timothy.crombleholme@childrenscolorado.org

D.O. Fauza, M. Bani (eds.), *Fetal Stem Cells in Regenerative Medicine*, Stem Cell Biology and Regenerative Medicine,
DOI 10.1007/978-1-4939-3483-6_1

To properly understand fetal stem cells and their potential therapeutic applications, the nature of stem cells in general must first be appreciated. In addition, familiarity with how the field of stem cells has progressed from the early 1900s to the flurry of interest today is helpful in placing the trajectory of current investigation in its appropriate historical context. While a significant focus of current research has been on embryonic and induced pluripotent adult stem cells (iPC), fetal stem cells have recently emerged as a distinct group with characteristics that are intermediate to these two groups. Knowledge of how fetal stem cells compare to their embryonic and adult counterparts is important in understanding how they may be used in therapeutic applications.

What Is a Stem Cell?

A stem cell is a distinct type of cell which possesses the unique set of characteristics which include clonality, proliferative capacity, and plasticity. During mitosis, stem cells divide asymmetrically to maintain **clonality**. One daughter cell retains the original characteristics of the stem cell population to retain clonal self-renewal, while the other takes a step toward differentiation down a specialized cell lineage. Stem cells also are known to have an impressive **proliferative capacity**, with some lineages sustaining up to 250 cell culture passages and beyond without loss of their original characteristics. They perform this feat in part by expressing elevated levels of telomerase which sustains the ability to maintain telomere length [2]. Stem cells are undifferentiated cells, and depending on the specific stem cell population, may possess the ability to differentiate along multiple types of cell lineages. This differentiation capability is known as **plasticity**. The various levels of plasticity include totipotency, pluripotency, multipotency, or unipotency (Fig. 1.1). *Totipotent* stem cells have the ability to differentiate along all three germ layers: the ectoderm, mesoderm, and endoderm, and they can also differentiate into the extraembryonic tissues of the trophoblast. Similarly, *pluripotent* stem cells can differentiate along ectoderm, mesoderm, and endoderm germ lines, but they are unable to form the extraembryonic structures. *Multipotent* stem cells can differentiate into more than one specific cell type, within one or two—but not all three—germ lines. For example, one hypothetical multipotent stem cell could differentiate into either a neuron or an astrocyte, but not a chondrocyte or keratinocyte. *Unipotent* stem cells, the most restricted type, are able to maintain a continuous line of one specific type of cell. For example, epithelial progenitor cells live in the epidermis and provide the source for continuous epithelial cell turnover.

Some researchers have made efforts to define stem cells based on specific molecular markers. Defining stem cells as cells with clonality, proliferative capacity, and plasticity utilizes the phenotypic characteristics of the cells, but this definition requires observing cell behavior over time, making it difficult to isolate stem cells in early primary culture. A molecular definition is a useful approach to help isolate stem cells from the surrounding tissues. Unfortunately, a universal molecular definition has yet to be elucidated. That being said, several commonly accepted markers

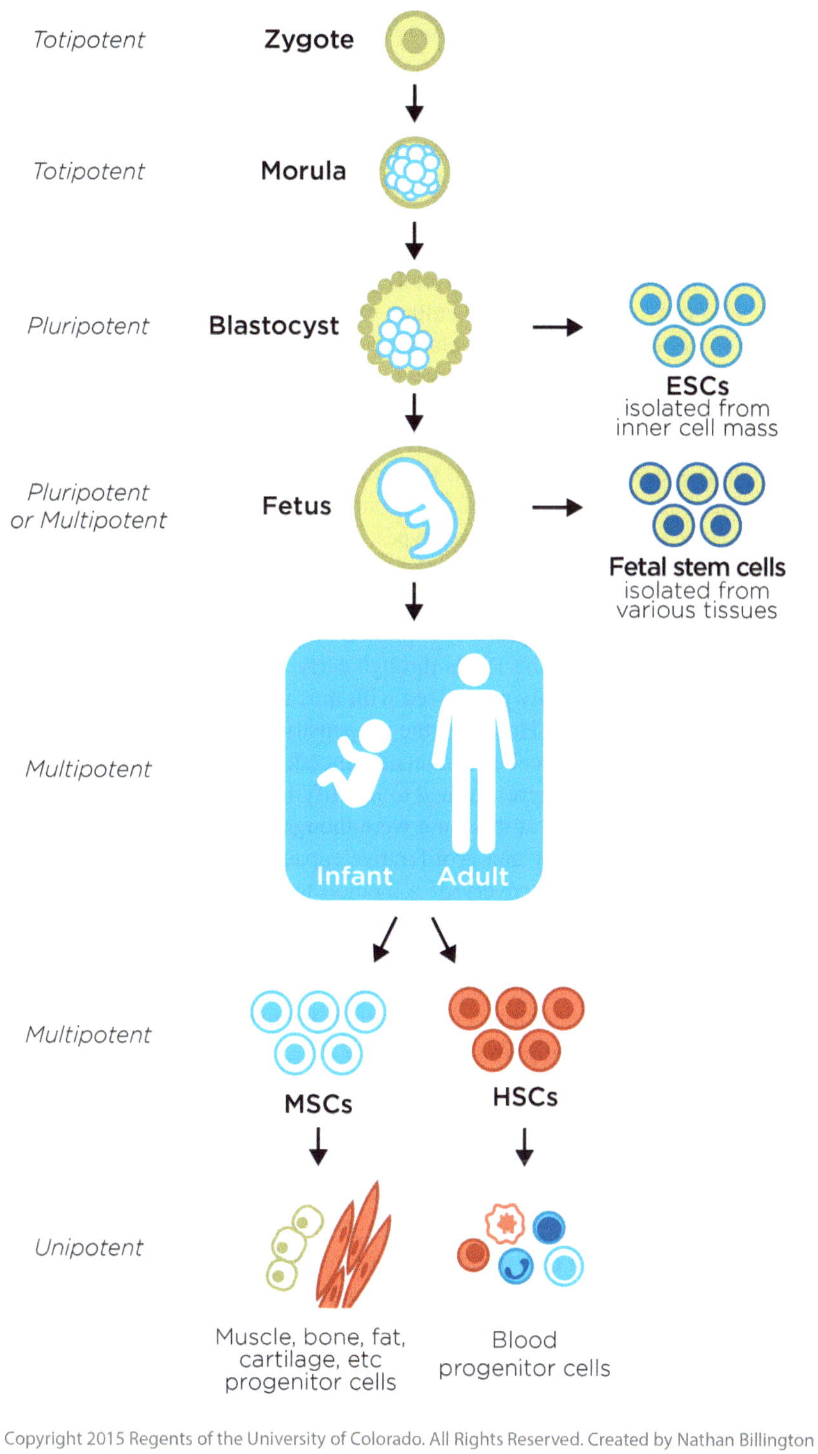

Fig. 1.1 Stem cell plasticity through the lifespan. *ESC* embryonic stem cell, *HSC* hematopoietic stem cell, *MSC* mesenchymal stem cell. Copyright 2015 Regents of the University of Colorado. All Rights Reserved. Created by Nathan Billington

of pluripotency have been identified, including Oct-3 and Oct-4, TRA-1-60 and TRA-1-81, SSEA-3 and SSEA-4, Sox2, Rex-1, and Nanog.

Two broad categories of highly investigated lines of multipotent stem cells are the hematopoietic stem cells (HSC) and mesenchymal stem cells (MSC). Hematopoietic stem cells maintain the various cell lines of the blood and immune system. They are usually identified by expression of the cell surface markers CD34, CD45 and c-kit. Mesenchymal stem cells, alternatively known as mesenchymal stromal cells, can, depending on the culture conditions, give rise to progenitor cells along the mesodermal germ line including adipocytes, chondroblasts, and osteoblasts [3]. However, as research has shown, HSCs and MSCs are more plastic than once thought, and under specific *in vitro* or *in vivo* conditions they can be induced to differentiate along other lineages as well.

Stem cells are located throughout the body's various tissues and organs. They also are found along the lifespan of the organism, from embryonic stages through adulthood. Generally, there are three types of age-specific stem cells described: embryonic, fetal, and adult stem cells [4]. Clinical research on stem cells began with adult bone marrow transplantation in the middle of the twentieth century when their ability to repopulate host marrow was exploited in a broad range of blood disorders and malignancies. Later, in the 1990s through early 2000s, advances in embryonic stem cell research were highly publicized with much excitement generated regarding their pluripotent nature. However, the propensity to form teratomas as well as the intense ethical and political controversies impeded progress in embryonic stem cell research. As a result, interest turned to identifying other sources of pluripotent stem cells. Adult stem cells at this time were thought to be less useful given their much more limited plasticity and proliferative capacities. Interest was renewed in adult stem cells after the discovery of a method to induce pluripotent cells from terminally differentiated cells. More recently, various types of fetal stem cells have been described as a third source of stem cells. They have proven to possess characteristics intermediate between embryonic and adult stem cells. The field of fetal stem cell research is still very much in its infancy, but exciting prospects for clinical applications are already being investigated.

A Brief History of the Field of Stem Cell Science

The healing potential of stem cells has been exploited for over a 100 years, beginning with fetal tissue, although the specific role of fetal stem cells was not understood at the time. In 1910, Davis described using amniotic membranes for skin transplants [5]. In 1913, amniotic membranes were further described for use in augmenting burn wound healing and in skin grafting [6, 7]. Amniotic membrane application to healing conjunctival injuries was later described in the 1940s [8]. The healing properties of fetal adrenal glands were investigated in 1922, when Hurst performed the first fetal tissue transplant, placing a human fetal adrenal graft into a patient with Addison's disease [9]. However, the source of the unique healing properties of fetal tissues, namely fetal stem cells, would not be recognized for many decades.

In 1957, Thomas first described performing an adult bone marrow transplantation in a patient with leukemia following chemotherapy and radiation in an attempt to repopulate the patient's bone marrow [10]. His accomplishment would begin a flurry of investigations of other possible therapeutic uses for bone marrow transplantation, all with varying levels of success. The development of bone marrow transplantation stimulated interest in identifying the unique properties of the cellular components of bone marrow. Years of painstaking work eventually lead to the characterization of hematopoietic and mesenchymal stem cells in bone marrow. Early descriptions of HSCs were published by McCulloch and Till in the 1960s regarding what they described at the time as "colony forming unit-spleen cells" [4]. The first definitive isolation and identification of human HSCs was published a decade later in 1975 [11]. Around the same time, the work of Owen and Friedenstein in the 1970s and 1980s led to the discovery of MSCs in bone marrow [12, 13]. These marrow cells were multipotent and could form daughter cells with the capacity to differentiate into each line of the host's original bone marrow population. This work provided our earliest understanding of adult stem cells.

Bone marrow transplantation grew in scope as a therapeutic option for many conditions and diseases, but in the process certain limitations arose. Graft versus host disease restricted the pool of potential donors to those with similar human leukocyte antigen (HLA) haplotypes. The relative lack of suitable donor matches led researchers to search for an alternative source of these adult bone marrow stem cells. Knudtzon, in 1974, described the presence of hematopoietic cells in umbilical cord blood (UCB) [14]. These hematopoietic cells were proven to be an acceptable replacement for adult bone marrow when in 1988 the first successful UCB transplant was performed, treating a young boy with Fanconi's anemia [15]. The ability to use UCB as an alternate source of HSCs spurred interest in collecting and storing UCB at birth for later use. This prompted the establishment of the first umbilical cord blood bank in 1994 for use in unrelated recipients [16]. One limiting factor of cord blood was the relatively small volume of cells compared to samples obtainable from adult bone marrow donors. This initially restricted UCB transplantation to pediatric recipients, but later experience proved that UCB could also effectively engraft in adult recipients, resulting in acceptable clinical outcomes [17].

The era of totipotent and pluripotent embryonic stem cells (ESC) originated in the 1950s and 1960s when Pierce, Stevens and Kleinsmith proposed the existence of ESCs while studying the properties of mouse teratomas [18–20]. Embryonic stem cells were first isolated from mouse embryos by Martin in 1981 [21], and human ESCs were first isolated in 1998 by Thomson and colleagues [22]. Embryonic stem cells were obtained by removing cells from the inner cell mass of the blastocyst, called the embryoblast. The ESCs possessed seemingly unlimited replication potential, while still maintaining their original characteristics. Also notably, these inner cell mass-derived cells were pluripotent, and as such could differentiate into all three germ layers of the embryo [23]. If the blastomers were isolated even earlier in development, they maintained totipotent potential, with the ability to form the tissues of the extraembryonic trophoblast as well as the embryo [24]. The discovery of these pluripotent and totipotent cells led researchers to imagine vast possibilities in terms of their therapeutic potential. However, the method of ESC procurement necessarily resulted in the death of the embryo. In human subjects, this was cause for significant ethical concern. Early research into the therapeutic potential of these ESCs was also hindered by their tendency to form teratomas *in vivo*. The malignant potential of these cells was certainly an unwelcome side-effect. These ethical and practical limitations of ESCs led researchers to search for stem cells with similar levels of plasticity from alternate sources.

Adult stem cells by this time had been discovered in multiple tissues of the human body, including bone marrow, blood, adipose, skin, and liver [24]. Initial expectations of their therapeutic utility had been only modest due to the adult cells' limited plasticity and proliferative capacities compared to their ESC counterparts. However, the search for ESC replacements prompted researchers to take a second look at adult stem cells for ways to increase their "*stemness*"—their clonality, proliferative capacity, and plasticity. In 1996, Dolly the cloned lamb had been born [25], breathing new life into the potential plasticity of adult stem cells. The success of this highly publicized story was built upon research begun half a century earlier, when in 1958 somatic cell nuclear transfer was used by Gurdon and colleagues to transform adult somatic cells into pluripotent stem cells [26]. Dolly the lamb provided proof of concept that a terminally differentiated adult cell could be de-differentiated to a totipotent stem cell using nuclear transfer cloning technique. This nuclear transfer technique, however, was highly inefficient [27] and some scientists called into question the reproducibility of the experiment [28]. Also, potential human applications were limited because human cloning was not palatable to society at large. It wasn't until 10 years after Dolly when adult stem cells realistically became potential replacements for ESCs. In 2006, Takahashi and Yamanaka published a landmark paper describing a method to induce terminally differentiated cells in mice to become pluripotent. They transfected cells with viral vectors containing the OKSM factors (Oct3/4, Klf4, Sox2, and c-Myc) [29]. However, while the inclusion of c-Myc was helpful in mouse models, it proved to have an adverse effect on human pluripotent stem cell models [30]. In 2007, Thomson and colleagues described an alternate approach using OSNL (Oct4, Sox2, Nanog, and Lin28) factors, notably without c-Myc [31]. This was also the first published trial of induced pluripotent cells in human adult cells.

After initial descriptions of the use of amniotic membranes and fetal tissues in the early half of the twentieth century, interest in therapeutic applications of fetal tissues did not resurface again until the 1990s with the use of umbilical cord blood. Interest also was augmented after the discovery of ESCs in the search for an ethical alternative. While studies in the adult stem cell populations were proceeding, the search for pluripotent fetal tissue stem cells began.

In theory, fetal tissues would be an ideal source of pluripotent cells. Fetal-derived tissue had already been readily available. The extra-embryonic products of conception were simply discarded after birth as standard medical practice. In addition, amniocentesis, performed as early as the 1930s [4], and chorionic villus sampling became standard practice for prenatal diagnosis and were widely accepted as ethical procedures. If pluripotent stem cells could be isolated from the amniotic fluid or from the placental tissues obtainable during chorionic villus sampling or after birth, then these tissues could prove to be ethically acceptable alternatives to ESCs. Furthermore, these fetal stem cells could in theory be stored for the potential use by the same individual later in life as a source of complete immune-matched pluripotent stem cells. In the case of amniocentesis and chorionic villus sampling, the ability to obtain fetal-derived stem cells early in utero introduces the possibility of directly autotransplanting stem cells for therapeutic purposes or creating stem cell-derived tissue engineered organs for use in the same fetus prior to birth or in the immediate postnatal period [32].

In 1993, Torricelli first identified the presence of HSCs in amniotic fluid [33]. Streubel, in 1996, discovered MSCs in amniotic fluid and determined that they could be induced to differentiate into myocytes [34]. Early studies on amniotic fluid and umbilical cord blood-derived stem cells proved they were at least multipotent, if not pluripotent, and a growing body of evidence indicated that many other fetal stem cell populations existed as well. Since the 1990s, fetal stem cells have been isolated from multiple fetal and extra-fetal sources and their therapeutic potential is currently being investigated. Fetal stem cells have emerged as an intermediate between the totipotent/pluripotent nature of embryonic stem cells and the more limited, but ethically less objectionable, nature of adult stem cells.

Fetal Stem Cells: What Are They?

There does not exist in the literature, to our knowledge, an accepted standard definition of a fetal stem cell. Embryonic stem cells are described as originating from the embryoblast, but they have also been isolated later in gestation from prenatal organs while still being described as embryonic stem cells [35]. Adult stem cells originate from adult tissue, but also from pediatric tissue. In contrast, fetal stem cells originate from the fetal and extra-fetal products of conception and span the timeline between the other two described cell types. To avoid conflating the age extremes of embryonic and adult stem cells with fetal stem cells, for the purposes of this discussion we propose a working definition of fetal stem cells: *human fetal stem cells are cells from the end of the 8th week of conception until birth, obtained from tissue of fetal origin including the fetus itself as well as the embryo-derived extra-fetal products of conception, and possessing the ability to maintain clonality, high proliferative capacity, and plasticity.*

Investigations into their properties have shown that fetal stem cells are similar to, but distinct from, embryonic stem cells and adult stem cells, yet they display characteristics of each. There is significant heterogeneity even between populations of fetal stem cells themselves depending on the tissue of origin. Comparisons and contrasts can be made in terms of plasticity, proliferative capacity, immunogenicity, tumorigenicity, paracrine effects, morphology, stability, engraftment, accessibility, and safety.

Plasticity

Depending on the isolated population, fetal stem cells have been shown to possess a pluripotent nature, with the potential to differentiate into all three germ cell layers. For stem cells in general, there exist several hypotheses about how this conversion from one type of cell into another can occur, including differentiation, de-differentiation, transdifferentiation, and fusion. *Differentiation* is simply the transformation of an undifferentiated cell into a differentiated cell. *De-differentiation* refers to partially or fully differentiated cells reverting back to undifferentiated cells after which differentiation to another cell type takes place. The process of de-differentiation is likely what occurs in amniotic epithelial cells, which are differentiated epithelial cells that maintain the ability to differentiate into all three germ layers [36]. *Transdifferentiation* refers to a differentiated cell directly converting into another differentiated cell type. Transdifferentiation, first proposed in the early 1970s [37], is unlikely as an actual mechanism of cell transformation, given the lack of any evidence that this occurs [38]. *Fusion* occurs when an undifferentiated stem cell fuses with a differentiated cell and takes on its properties. Cell fusion has been demonstrated to occur, for example, in the mechanism of HSC differentiation to hepatocytes [39].

While adult stem cells are relatively restricted in their differentiation potential, and are only pluripotent when induced via somatic cell nuclear transfer or viral gene transfection in a highly inefficient manner, fetal stem cells are much easier to induce to pluripotent states [27, 40]. Fetal stem cells have been shown to differentiate into cell types of all three germ layers by simply placing them in a favorable culture medium. They are induced to express even more markers of pluripotency and form teratomas when valproic acid is added to the medium, ostensibly making them even closer in nature to embryonic stem cells [41]. It has been suggested that the relative reprogramming ease of fetal stem cells may derive from the similarity of the epigenetic state of fetal stem cells and ESCs [27].

Proliferative Capacity

Stem cells differ in their proliferative capacities. Embryonic stem cells can proliferate indefinitely *in vitro* [4]. Depending on the cell line, fetal stem cells have varying proliferative capacities. In general, they proliferate faster and through more passages in culture compared to adult cells [2, 42]. For example, adult MSCs have been found to become genetically unstable after 20 population doublings [43], whereas stem cells isolated from the amniotic fluid are stable after more than 250 doublings [44]. Amniotic epithelial cells have even been described as having a proliferative capacity rivaling that of embryonic stem cells [45]. A possible mechanism for this advanced proliferative capacity is the increased telomerase expression observed in fetal stem cells compared to adult stem cells [2].

Fetal-derived stem cells have been shown to proliferate faster *in vitro* than similar adult stem cells [46]. There are also differences in mitotic rates among fetal stem cell populations. For example, amniotic fluid-derived MSCs have been found to proliferate even faster *in vitro* than both fetal and adult derived MSCs obtained from subcutaneous connective tissue [46]. The ability to expand more rapidly over more culture passages makes fetal stem cells an attractive source of stem cells which can be expanded to levels required for therapeutic applications [27].

Some studies have demonstrated that, within fetal stem cell lines, the "stemness" of the cell population is dependent upon the gestational age of the fetus from which they were harvested. Jones and colleagues demonstrated that, compared to term chorionic mesenchymal stromal cells (CMSC), first trimester CMSCs were smaller, proliferated faster, and expressed more stem cell markers [47]. Also, Portmann-Lanz and colleagues similarly found that, compared to third trimester, first trimester amniotic and chorionic mesenchymal stromal cells grew faster and survived over more passages [48].

Immunogenicity

Early studies of amniotic membrane grafting revealed that it does not provoke immune rejection [49]. Later studies of other fetal tissues similarly revealed that they do not evoke the same immune response that adult tissues do. For example, umbilical cord blood used in place of adult bone marrow for transplantation has been noted to have a lower risk of graft versus host disease [50]. In general, fetal MSCs express low levels of HLA class 1 and do not express HLA class 2 [24, 42, 45, 48]. The particular level of HLA expression is specific to the fetal cell's age and tissue of origin [42]. For example, amniotic epithelial cells, which do not elicit an immune response in early isolates, do begin expressing HLA class 1 and a low level of HLA class 2 in later culture passages [49] and also after induced differentiation down hepatic and pancreatic lineages [45]. Studies on placental-derived stem cells revealed they don't express co-stimulatory molecules like CD40 and CD80 [24, 51]. This may be one mechanism of their anergic effect on T-cells [52]. Amniotic epithelial stem cells also have been shown to secrete immunomodulating cytokines including IL-6, IL-11 [24] and MIF [53]. They have been shown to inhibit lymphocyte proliferation *in vitro* and suppress IL-1 expression [36]. Studies of therapeutic uses of fetal stem cells in mice have been performed both with and without immunosuppression with varying levels of immunological tolerance.

Although both fetal and embryonic stem cells possess very low levels of immunogenicity, their use in allogenic transplantation still elicits concern about their potential for immunorejection or graft versus host disease [54]. If stem cell transplantation could be performed without need for immunosuppression, then the side effects of immunosuppression could be avoided. Induced pluripotent cells created from differentiated adult tissue of the intended recipient have been investigated as a way of creating immune-tolerated stem cells for therapeutic purposes. Induced pluripotent fetal cells have also recently been investigated for the same reason [27, 41], and with their reduced immunogenicity compared to adult cells, they may prove more successful in achieving the goal of immunosuppression-free stem cell transplantation.

Tumorigenicity

Embryonic stem cells are known to form embryoid bodies *in vitro* and teratomas *in vivo* when engrafted into a host [36]. No line of fetal stem cells has shown the same tendency for tumorigenicity *in vivo*, possibly adding a layer of safety for their therapeutic use. For example, experimentation with ESCs and human amniotic epithelial cells led to the formation of embryoid bodies *in vitro*, but amniotic epithelial cells did not show the same tendency to form teratomas *in vivo* [45]. Similarly, amniotic fluid stem cells did not form teratomas when tested by De Coppi and colleagues [44]. However, as described earlier, fetal stem cells are able to be induced into a more embryonic-like-state. When amniotic fluid stem cells were cultured in a medium containing valproic acid, they were induced to form teratomas *in vivo* similar to embryonic stem cells [41].

Paracrine Effects

Stem cells are known to produce and secrete angiogenic and trophic growth factors which aid in tissue repair and regeneration, including VEGF, epidermal growth factor, and M-CSF [24, 42, 55]. This paracrine activity enhances their own ability to engraft and proliferate, as well as enhances the regeneration of the surrounding tissues [42]. As stated earlier, fetal stem cells have also been known to secrete immunomodulatory cytokines [24].

It appears that much of the benefit of stem cell transplantation comes not from structural incorporation in a tissue, but rather from their paracrine effects. For instance, engraftment of amniotic epithelial cells or umbilical cord HSCs have been found to aid in spinal cord repair, but the evidence does not support their restorative function being from the stem cells replacing the native neurons [56, 57]. Instead, the evidence is highly suggestive of a paracrine effect of the transplanted stem cells on modulating the native tissue's response to injury [36]. These paracrine effects are not unique to fetal stem cells, as they have been readily demonstrated in adult stem cells as well.

Morphology

Umbilical cord blood multilineage stem cells express a leukocyte-like morphology upon isolation, and then take on a more fibroblast-like appearance after establishment in culture about 7 days later [43]. Likewise, amniotic and chorionic stem cells take on a fibroblast-like appearance during *in vitro* culture, and adhere to the plastic plates, similar in behavior to that expressed by adult bone marrow MSCs [27, 42, 58]. Some heterogeneity also occurs in the morphologic expression of fetal MSCs. Transmission electron microscopy of amniotic MSCs reveals characteristics of both mesenchymal and epithelial cells, but this was not seen on visualization of chorionic MSCs which appeared more primitive in phenotype in terms of the organization of their cytoplasm and organelles [49].

Stability

In vitro, fetal stem cells have also been noted to be more resistant to hypoxia than adult stem cells [42]. They tolerate refrigeration and cryopreservation better as well [42, 43]. Embryonic stem cells are known to spontaneously differentiate in culture, and differentiate into an assortment of tissue types. Fetal stem cells have not demonstrated this same tendency and therefore are thought to be more stable in culture than ESCs [43]. In addition, the phenotype of amniotic and chorionic stem cells has been known to remain more stable in culture than adult stem cells after several passages [27].

Engraftment

It appears that, at least under certain conditions, fetal stem cells have an easier time engrafting in a host than adult stem cells. In comparing the ability of fetal liver stem cells and adult bone marrow stem cells to repopulate irradiated marrow in mice, Harrison and colleagues demonstrated that the fetal stem cells exhibit a long-term but not short-term engraftment survival advantage [59]. In a subsequent study, Harrison demonstrated the engraftment advantage of fetal stem cells was more pronounced when engrafted into a fetus versus an adult host. When fetal liver stem cells and adult bone marrow cells were transplanted into a *fetal* SCID mouse host, the fetal liver stem cells demonstrated an engraftment advantage over adult bone marrow cells. Yet when the same two cell types were transplanted into an *adult* SCID mouse host, fetal and adult stem cells engrafted with about the same efficiency [60]. This study suggested that fetuses may be more receptive to stem cell transplants, and that fetal stem cells engraft better than adult stem cells, at least in fetal applications.

Accessibility

Another advantage of fetal stem cells over both adult and embryonic stem cells is their accessibility. Adult bone marrow cells are obtainable, but the process is invasive and painful, with relatively low yield [36]. The availability of embryonic stem cells is hindered by ethical and political hurdles, and the volume obtainable from each embryo is quite small. Obtaining fetal stem cells from the tissues of the fetus itself similarly presents ethical and technical challenges, since ethical objections are raised when collecting tissue from electively aborted fetuses, and genetic and environmental impurity tends to impede collection from spontaneously aborted fetuses [42]. However, fetal tissue stem cells obtained from umbilical cord blood banks, prenatal diagnostic procedures like amniocentesis and chorionic villus sampling, and postpartum placenta and amnion collection, are easier to obtain and raise fewer ethical objections. High yields of fetal stem cells can also be obtained: a term amniotic membrane can yield between 50 and 70 million epithelial cells [36], and umbilical cord blood contains 4 % CD34 positive HSCs, compared to 1 % in adult bone marrow [61].

Safety

As stated previously, fetal stem cells may be safer than embryonic stem cells for therapeutic use due to presenting less risk of teratoma formation in the recipient and having more stability maintaining differentiated lines *in vitro*. Fetal stem cells may be safer than adult stem cells due to a lower risk of transmitting infections via latent viruses like cytomegalovirus [43]. They also have a lower likelihood of possessing silent genetic mutations [27]. Their ability to be induced to pluripotency with relative ease, and without viral vectors, reduces the likelihood of causing genetic mutations by the induction process itself.

Fetal Stem Cells: Where Do They Come From?

Fetal stem cells are a heterogeneous group of cells with varying characteristics in terms of their relative plasticity, proliferative capacities, phenotypic features and cell markers, all depending on their source and gestational age [27]. These stem cells can be isolated from tissue from the fetus itself, or from extra-fetal tissues like umbilical cord blood, Wharton's jelly, amniotic fluid, placenta, amniotic membrane, and even maternal circulation. Each source contains its own unique types of fetal stem cells (Fig. 1.2; Table 1.1).

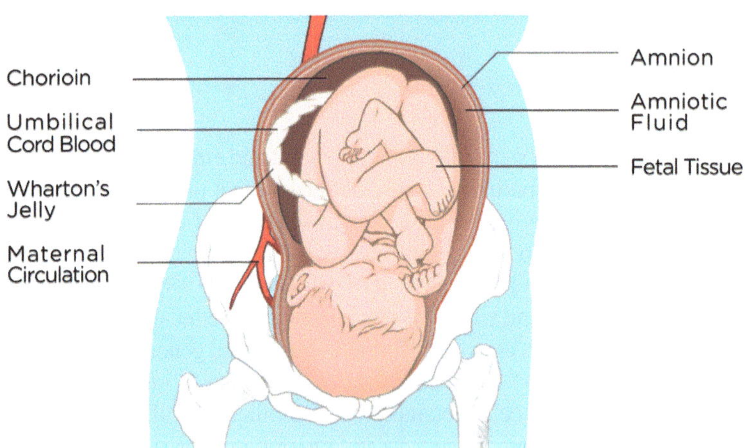

Fig. 1.2 Fetal stem cell sources. Fetal stem cells may be isolated from fetal tissue, umbilical cord blood, Wharton's jelly, amniotic fluid, the amniotic membrane, the chorion of the placenta, and the maternal circulation

Table 1.1 Fetal stem cell types organized by source

Stem cell source	Stem cell type	References
Fetus	Various HSCs and organ-specific MSCs	[2, 35, 42, 50, 62, 63]
Umbilical cord blood	Umbilical cord blood HSCs (UCB HSC) Umbilical cord blood MSCs (UCB MSC) Endothelial progenitor cells (EPC) Multilineage stem cells (MLSC) Very small embryonic-like stem cells (VSEL)	[43, 50, 64–66]
Wharton's jelly	Umbilical cord MSCs (UCMSC) Type 1 Umbilical cord MSCs (UCMSC) Type 2	[4, 67, 68]
Amniotic fluid	Amniotic fluid HSCs (AF-HSC) Amniotic fluid MSCs (AF-MSC) Amniotic fluid stem cells (AFSC)	[4, 32, 42, 44, 50]
Amniotic membrane	Amniotic epithelial cells (AEC) Amniotic MSCs (AMSC) Amnion-derived stem cells (ADSC)	[4, 24, 36, 70]
Placenta	Chorionic MSCs (CMSC) Chorionic trophoblastic cells (CTC)	[24, 32, 49, 50, 70, 71]
Maternal blood	Pregnancy-associated progenitor cells (PAPC)	[72–74]

MSC mesenchymal stem (or stromal) cell, *HSC* hematopoietic stem cell

Fetal Tissue

Human fetal tissue can be isolated from the body of the fetus including its liver, bone marrow, kidney and mesonephric tissue, pancreas, spleen, and fetal blood [2, 24, 50]. Fetal tissue stem cells are heterogeneous in terms of their expressed cell surface markers and their differentiation potential, depending on their site of origin and their gestational age [50]. For example, fetal hepatic stem cells are less differentiated earlier versus later in gestation [62]. Also, the fetal liver is the primary source of HSCs early in gestation. As the fetus develops, the HSCs migrate to the bone marrow, likely through the fetal blood circulation. Thus, relative concentrations of HSCs in the fetal liver, blood, and bone marrow change with gestational age [50]. In addition, studies of MSCs in fetal tissues reveal that their overall numbers decline with gestational age [50].

Fetal stem cells from the fetus itself can be hard to obtain. Human fetal tissue can be obtained after an induced abortion, but this has raised ethical qualms regarding the rights and moral status of the fetus [42]. In addition, tissue from spontaneous abortion usually is limited in its use due to chromosomal abnormalities and other confounding factors such as infections or anoxia [32, 42]. Some authors have looked into xenotransplantation of fetal organs, taking advantage of fetal tissue's high concentration of stem cells and immunosuppressive nature, but without the ethical qualms of using human fetal organs [35]. Transplantation of early fetal pig organs such as the kidney and pancreas into rats has been found to add functional subunits to the host organs with varying requirements for immunosuppression [35, 63]. However, xenotransplantation into humans is fraught with its own unique technical and ethical challenges. Thus, focus has been primarily on other, less controversial sources of fetal stem cells including umbilical cord blood, Wharton's jelly, amniotic fluid, amniotic membranes, placenta, and maternal blood.

Umbilical Cord Blood

Fetal HSCs, during their migration from the liver to the bone marrow, can be found in umbilical cord blood, where they can be ethically and safely collected after birth [50]. Stem cells from UCB have, *in vitro*, differentiated into neural, cardiac, epithelial, hepatocytic, and dermal cell types [43]. Umbilical cord blood is a source of HSCs, MSCs, endothelial progenitor cells (EPC), multilineage stem cells (MLSC), and very small embryonic-like stem cells (VSEL). The UCB HSCs are used frequently in bone marrow repopulation therapies, while the MSCs have been noted to support the expansion of the HSCs [64]. Umbilical cord blood MLSCs can differentiate along all 3 germ lines, and maintain their phenotype after at least 80 population doublings [43]. Some lines of UCB EPCs have been maintained beyond 100 population doublings [65]. Very small embryonic-like stem cells are smaller, have more open chromatin and a relatively larger nucleus than MSCs, and are also found in adult bone marrow in addition to umbilical cord blood [66]. As described in previous sections, UCB is stored in tissue banks and used for many therapeutic purposes, largely those for which adult bone marrow is also used.

Wharton's Jelly

Wharton's jelly, first described by Thomas Wharton in 1656, is a proteoglycan rich connective tissue encasing the umbilical vessels of the umbilical cord [67]. Wharton's jelly contains umbilical cord mesenchymal stem cells (UCMSC). About 400,000 UCMSCs can be isolated per umbilical cord [68]. These cells can differentiate down mesenchymal lines of adipogenic, chondrogenic, and osteogenic cell types [4]. They also express markers of all three germ layers [54]. They can proliferate beyond 80 population doublings [4]. There are at least two UCMSC cell types, described as "type 1" and "type 2" by Karahuseyingolu and colleagues. Type 2 cells more easily differentiate into neuronal cell types than type 1 cells [68]. Like UCB, UCMSCs are being stored in tissue banks [68].

Amniotic Fluid

Obtained via amniocentesis without ethical objection, amniotic fluid is currently collected for diagnostic and therapeutic purposes. The composition of amniotic fluid in the first half of pregnancy is primarily derived from active transport of sodium and chloride across the amniotic membrane and fetal tissues with water passively following [32, 42]. In the second half of pregnancy, amniotic fluid is largely derived secondary to fetal micturition [32, 42]. A small amount of volume is also contributed by fetal respiratory and gastrointestinal tract secretions and excretions [32, 42]. Cells found in amniotic fluid are primarily differentiated epithelial cells from the urinary, respiratory, and gastrointestinal tracts, and from the epidermis [4, 32, 42]. In some pathologic states, other cells may be present in the amniotic fluid including neural cells in the presence of neural tube defects [69], and peritoneal cells in the presence of abdominal wall defects [42]. Both hematopoietic and mesenchymal stem cells have been isolated from amniotic fluid. The amniotic fluid MSCs (AF-MSC) express both mesodermal and ectodermal markers and can differentiate, at least, down both of these germ lines [4]. Amniotic fluid stem cells (AFSCs) have also been described, expressing the stem cell factor receptor, c-kit, as well as MSC-specific markers, and proliferating to more than 250 population doublings [44]. These AFSCs can differentiate into all 3 germ layers [44] and form embryoid bodies [50], and thus appear similar to embryonic stem cells. However, some authors believe that these cells are the same as the AF-MSCs described by other investigators [42].

Placenta and Amniotic Membrane

Stem cells of the placenta are obtainable via chorionic villus sampling which is regularly performed for diagnostic purposes without ethical objection. Amniotic membrane and placental stem cells may also be obtained immediately postpartum via procurement of the afterbirth. The fetal component of the placenta is the chorionic plate, composed of the amnion and chorion. Within the amnion, several cell types displaying various levels of "stemness" have been described, including amniotic epithelial cells (AEC), amniotic mesenchymal stem cells (AMSC), and amnion-derived stem cells. Amniotic epithelial cells express stem cell markers and are pluripotent *in vitro*, differentiating into neurons, astrocytes, glia, osteocytes, adipocytes, endothelial cells, cardiomyocytes, myocytes, hepatocytes, and pancreatic cells [24, 36], but they have demonstrated a particular tendency toward adipogenic [48] and neurogenic differentiation [4]. Amniotic mesenchymal stem cells can also be pluripotent *in vitro* [24], and also are more easily induced to adipogenic [70] and neurogenic lines [4]. Amnion-derived stem cells are less well described and require further study.

The chorion also is a source of stem cells including chorionic mesenchymal stem cells and chorionic trophoblastic cells. Chorionic MSCs (CMSC) can be isolated from chorionic villus sampling during first trimester diagnostic testing [50]. They have demonstrated similar pluripotency to AMSCs, however they have been less well studied, perhaps due to their relatively reduced proliferative capacity [24], although their mitotic rate is comparable [71]. The CMSCs have demonstrated a particular tendency toward chondrogenic, osteogenic, myogenic, and neurogenic cell lines [48, 70]. Chorionic trophoblastic cells have received little study thus far [49]. Like amniotic fluid stem cells, chorion-derived c-kit positive cells have been identified which appear to differentiate along all 3 germ layers, but like the AFSCs, further study is required to elucidate if this is a unique cell type [32].

Maternal Circulation

Fetal cells can be detected in the maternal circulation of mothers who have carried a fetus to term, even decades after the last pregnancy [72]. These pregnancy-associated progenitor cells (PAPC) are most easily identified when the fetus was male, and the Y chromosome can be used as a maker for cells of fetal origin [73, 74]. The presence of fetal cells among maternal cells is termed "fetal cell microchimerism" [74]. These PAPCs are heterogeneous, but some are thought to be fetal stem cells, expressing markers of pluripotency such as CD34 [74]. These cells differentiate and persist in the maternal tissues, including the bone marrow, thymus, heart, and circulating lymphocyte populations. They also are postulated to home to damaged maternal tissues such as the livers of mothers with cirrhosis [74]. Current hypotheses that these stem cells home to damaged maternal tissue are supported by early animal studies [74]. The potential that these PAPCs could affect healing on the mother is intriguing. However, hypotheses also exist that these PAPCs could cause autoimmune disease in the mother such as systemic lupus erythematosus or systemic sclerosis. Nevertheless, the assertion that these fetal stem cells could be a cause of maternal autoimmune disease has been challenged by the contrasting hypothesis that fetal stem cells are simply homing to tissue which was already diseased [73]. Due to low numbers of fetal stem cells (1–6 cells per mL of maternal venous blood [74]), maternal peripheral blood is not yet a practical source of these stem cells [42].

Fetal Stem Cells: Clinical Applications and Beyond

While clinical use of umbilical cord blood has been successfully used for over two decades to repopulate bone marrow after leukemia treatment or to treat hematological diseases [15, 16], fetal stem cells from various other sources are now being investigated for their therapeutic applications, with much of the progress occurring only within the last decade or so. Their low to non-existent immunoreactivity and absence of tumor formation make fetal stem cells ideal candidates for allo- and even xenotransplantation in regenerative therapies. Stem cells have also been observed to home to sites of injury where they engraft and participate in healing through their paracrine effects. Fetal stem cells are showing potential for treating diseases from nearly every organ system of the body (Table 1.2).

Table 1.2 Potential therapeutic applications of fetal stem cells

Organ system	Regenerated tissue	Disease model treated	Transplanted cell type	Host	Cell survival in host	References
Nervous system	Brain	Parkinson's Disease	hAECs	Rat	>14 Days	[75, 76]
			hUCMSCs	Rat	Unknown	[77]
		Hemorrhagic stroke	hUCB HSCs	Rat	>14 Days	[78]
		Ischemic stroke	hUCB HSCs	Mouse	Unknown	[87]
			Mouse AFSCs	Mouse	Unknown	[88]
	Spinal cord	Spinal cord injury	hUCB HSCs	Rat	3–5 Weeks	[57]
			hAECs	Monkey	>60 Days	[56]
	Brain and spinal cord	ALS	hUCB	Mouse	10–12 Weeks	[79]
		Krabbe's disease	hUCB	Human	Unknown	[80]
	Peripheral nerve	Mechanical ligation	hAF-MSCs	Rat	Unknown	[81]
		Crush injury	Rat AF-MSCs	Rat	10–30 Days	[82]
Cardiovascular system	Heart	Myocardial infarction	hAMSCs	Rat	>2 Months	[83]
			Porcine AMSCs	Pig	>30 Days	[84]
			hAFSC	Rat	>30 Days	[86]
			hUCB HSCs	Rat	>30 Days	[85]
	Peripheral vasculature	Buerger's disease	hUCB MSCs	Human/mouse	>30 Days	[89]
Respiratory system	Lung	Pulmonary fibrosis	hAMSCs, CMSCs, and AECs	Mouse	>14 Days	[91]
		Hyperoxic lung injury	hAFSCs	Mouse	>40 Days	[90]

(continued)

Table 1.2 (continued)

Organ system	Regenerated tissue	Disease model treated	Transplanted cell type	Host	Cell survival in host	References
Digestive system	Liver	None	hAECs	Mouse	>7 Days	[92]
		Cirrhosis	Fetal liver hMSCs	Human	Unknown	[93]
Urinary system	Kidney	ATN	hAFSCs	Mouse	>21 Days	[94]
Endocrine system	Pancreas	Diabetes	hUCB	Mouse	Unknown	[97]
			hAECs	Mouse	Unknown	[96]
			Fetal pancreatic hMSCs	Sheep	>27 Months	[95]
Skeletal system	Bone	Osteogenesis imperfecta	Fetal Blood hMSCs	Mouse	>12 Weeks	[98]
			Fetal Liver hMSCs	Human	>9 Months	[99]
Muscular system	Muscle	Chemical injury	hUCMSCs	Rat	>2 Weeks	[100]
		Duchenne muscular dystrophy	Fetal Blood hMSCs	Mouse	>19 Weeks	[101]
Integumentary system/eye	Eye	Progressive blindness	hUCMSCs	Rat	>70 Days	[104]

AEC amniotic epithelial cell, *AF-MSC* amniotic fluid mesenchymal stem cell, *AFSC* amniotic fluid stem cell, *ALS* amyotrophic lateral sclerosis, *AMSC* amniotic mesenchymal stem cell, *ATN* acute tubular necrosis, *CMSC* chorionic mesenchymal stem cell, *h* human, *HSC* hematopoietic stem cell, *MSC* mesenchymal stem cell, *UCB* umbilical cord blood, *UCMSC* umbilical cord mesenchymal stem cell

Nervous System

In the central nervous system, diseases such as Parkinson's disease, hemorrhagic stroke, spinal cord injury, amyotrophic lateral sclerosis (ALS), and Krabbe's disease are being investigated for potential fetal stem cell therapy. Human AECs have shown promise in treating Parkinson's disease through their ability to produce their own dopamine as well as their ability to secrete trophic factors which promote the survival of endogenous dopaminergic neurons. Transplantation of these cells has led to amelioration of Parkinsonian symptoms in mice [75, 76]. Human UCMSCs have shown similar promise [77]. In rat models of hemorrhagic stroke, transplantation of umbilical cord blood HSCs has led to a significant improvement in neurologic function [78]. Umbilical cord blood HSCs have also shown potential in treating spinal cord injury, the transplantation of which improved functional outcomes in rat models and promoted endogenous axon regeneration [57]. Similar results have been obtained in primate models of spinal cord injury using human AECs [56]. In ALS, the disease progression in mice has been slowed, and their lifespan increased, after transplantation of umbilical cord blood stem cells [79]. In all of the preceding studies, the stem cells were noted to survive *in vivo* at least a couple of weeks, providing evidence for their low immunologic profile. Human trials have been performed on infants with Krabbe's disease using umbilical cord blood transfusions. Transfusion of UCB increased the survival to 2 years of age of pre-symptomatic newborns from 40 to 100 %, and most of this treatment group developed age-appropriate cognitive function [80].

Aside from the central nervous system, peripheral nerve injuries may also be treated with fetal stem cells. Amniotic fluid MSCs transplanted into rat models of sciatic nerve injury caused improvement in target muscle function and decreased nerve conduction latency compared to controls [81]. This effect was shown to be mediated, at least in part, by secretion of neurotrophic factors [82].

Cardiovascular System

Fetal stem cells have also been studied in myocardial infarction (MI) models. After a myocardial infarction, the affected cardiac tissue becomes fibrotic and loses function. The regenerative capabilities of fetal stem cells have been applied to MI models with mixed results. Amniotic mesenchymal stem cells have differentiated into cardiomyocyte-like cells in rat models of MI *in vivo*, but did not beat on their own [83]. In another study, the same cell-type transplanted into a porcine model of MI failed to differentiate into cardiomyocytes, but did differentiate into endothelial and smooth muscle cell types [84]. Human amniotic fluid stem cells and UCB HSCs have successfully reduced infarct size and improved left ventricular function and neovascularization after transplantation into rat MI models [85, 86]. However, while the stem cells showed evidence of cardiomyocyte and endothelial differentiation, it did not appear that the bulk of the beneficial effect was from cardiomyocyte mass replacement by the stem cells but instead by cytokine production stimulating angiogenesis and the growth of nearby native tissue.

The neovascularization-promoting paracrine activities of stem cells have been exploited for other ischemic diseases such as ischemic stroke and Buerger's disease. Umbilical cord blood HSCs transplanted into mouse models of cerebrovascular infarct induced neovascularization of the ischemic zone and promoted host neurogenesis [87]. Improved neurologic function after ischemic stroke has been confirmed after transplantation of AFSCs in cerebrovascular ischemic/reperfusion injury mouse models [88]. For Buerger's disease, human trials of UCB MSC transplantation in affected patients led to resolution of rest pain, healing of necrotic skin lesions, and increased capillary number and size [89].

Respiratory System

Fetal stem cells have improved healing in mouse models of lung injury. Amniotic fluid stem cells have homed and engrafted into hyperoxia-damaged mouse lungs, where they differentiated into lung-specific cells [90]. Reduction in fibrosis formation in bleomycin-induced pulmonary fibrosis mouse models was achieved by administration of a mixture of AMSCs, CMSCs, and AECs, which engrafted into the lung tissue and aided in scar reduction [91]. These studies lend optimism to eventual treatment of progressive pulmonary diseases such as bronchopulmonary dysplasia, pulmonary fibrosis or COPD.

Digestive System

Cirrhosis has been investigated as a potential therapeutic target for fetal stem cells. Human AECs have been induced *in vitro* to hepatic cells. They expressed liver-specific cell markers including alpha-fetoprotein, and synthesized and excreted albumin [92]. This could make them potentially useful for aid in liver regeneration. Human translational research has also proved promising. Transplantation of fetal liver MSCs into human patients with cirrhosis improved their MELD scores as well as other clinical and biochemical parameters [93].

Genitourinary System

Acute tubular necrosis is a renal disease frequently encountered in hospitalized patients, occasionally requiring dialysis. Human AFSCs transplanted into a mouse model of rhabdomyolysis-induced acute tubular necrosis reduced the severity of renal injury in the acute phase [94]. This study demonstrated that administration of fetal stem cells can have a reno-protective effect.

Endocrine System

Given the profound strain on the medical system caused by the morbidity of diabetes, stem cell applications to treat this disease are being investigated. Human fetal pancreatic MSCs have been transplanted onto fetal sheep pancreases where they engrafted and began secreting insulin [95]. This yielded promising results for a potential way to treat diabetes with a continuous insulin source. However, given the ethical constraints surrounding the use of human fetal tissue, some have looked to xenographs. Porcine fetal pancreatic islets transplanted into rats and monkeys did not require immunosuppression to survive, making xenotransplantation into human recipients a therapeutic possibility [63]. Human extra-fetal sources of fetal stem cells have also been investigated for use in diabetes. Human AECs and UCB-derived stem cells have each been noted to secrete insulin and correct hyperglycemia after transplantation into the host diabetic mouse [96, 97].

Skeletal System

The administration of fetal stem cells can potentially treat progressive skeletal diseases such as osteogenesis imperfecta. Human MSCs isolated from fetal blood transplanted into prenatal mouse models of osteogenesis imperfecta reduced bone fracture rates and increased bone strength, thickness, and length compared to controls [98]. Human trials have been also been attempted. Human fetal liver MSCs transplanted into a human fetus with osteogenesis imperfecta were confirmed to engraft in the bone and differentiate along osteogenic lines [99].

Muscular System

Fetal stem cells can help repair injured or diseased muscle. Following chemical damage to rat muscle, transplantation of human UCMSCs led to engraftment and skeletal muscle differentiation [100]. These findings held promise for the development of a potential treatment for Duchenne muscular dystrophy. To trial this, MSCs from human fetal blood and bone marrow were each transplanted into dystrophic fetal mice, which similarly led to engraftment and myogenic differentiation, but unfortunately muscle recovery did not achieve a curative level [101].

Integumentary System/Eye

As discussed earlier in the chapter, amniotic membrane has been used as a temporary graft for burn wounds and ulcerations, taking advantage of the trophic factors and proangiogenic properties of the cells contained therein, as well as their immunomodulatory properties [102]. Human UCB-derived stem cells have been successfully induced to keratinocytes *in vitro* [103]. This is promising for potential therapeutic use in skin transplantation, especially in cases where limited native skin area is available for autografting [43].

Corneal injury therapy also has its origins in amniotic membrane grafting, which is still used in various forms for corneal ulcers and chemical burns as a basement membrane substitute to promote healing in patients [55]. Umbilical cord MSCs have also been investigated, the transplantation of which onto rat models of photoreceptor degeneration led to a reduced degree of degeneration [104].

Hematology/Oncology

Umbilical cord blood has been used for quite some time in bone marrow regeneration after leukemia treatment, and to treat hematologic diseases such as Fanconi's anemia [15, 50]. Fetal stem cells also have shown promise in actively fighting cancer. It has been observed that MSCs tend to home to tumors, making them potential vehicles for local administration of antineoplastic agent therapy [105]. Trials of adult bone marrow MSCs engineered to express a tumor apoptotic agent confirmed this hypothesis when a mouse model of metastatic lung cancer was completely cleared of metastatic tumor burden upon local administration of the specially engineered MSCs [105]. Umbilical cord MSCs engineered to express interferon-beta homed to the tumor in mouse models of breast cancer, and when given with 5-fluorouracil, the combination chemo/stem cell therapy reduced the size of the tumor beyond what was accomplished with single agent 5-FU chemotherapy alone [106]. This novel method of locally administering chemotherapy could potentially reduce systemic side effects and allow increased concentrations to be administered to the tumor.

Tissue Engineering

Aside from transplantation of suspensions of fetal stem cells into the host or grafting primordial organs for therapeutic application, others have investigated fetal tissue engineering. Tissue engineering, originally termed "chimeric neomorphogenesis" was first described in 1988 by Vacanti and colleagues [107]. In tissue engineering, a biologically active scaffolding is created which acts as a sort of extracellular matrix, and the scaffolding is seeded with cells which proliferate onto the structure.

The engineered tissue is then transplanted into the host to replace a loss of function. The low immunogenic profile of fetal stem cells makes them potentially less susceptible to immune rejection than standard organ transplantations. Focuses in the area of fetal tissue engineering have included the trachea, the diaphragmatic tendon, bone grafts, and heart valves.

Early studies of fetal cell use in tissue engineering were performed by Fauza and colleagues when, in 1998, they reported harvesting fetal lamb bladder or skin tissue, isolating the stem cells, seeding them on a matrix, and then autotransplating them upon birth. They performed this in bladder augmentation and skin transplantation scenarios with evidence of increased bladder function in lambs with extrophy, and quicker wound epithelialization in wounded lambs, respectively [108, 109]. Fauza and colleagues then turned their attention to the problem of congenital tracheal anomalies. Long-segment tracheal stenosis, atresia, and agenesis have all proven especially difficult to correct. In 2002, they attempted to transplant an engineered trachea lined with an expanded line of fetal auricular chondrocytes into an ovine model, with positive functional results [110]. In addition to obtaining stem cells from fetal tissue harvesting, researchers have also been able to successfully engineer cartilaginous tissue using amniotic MSCs [111], which may prove to be a safer method of achieving similar outcomes.

Large congenital diaphragmatic hernias (CDH) have also proven to be an especially difficult anomaly to correct since standard polytetrafluoroethylene (PTFE) repair of the defect is frequently complicated by hernia recurrence due to patient growth. In 2004, AF-MSCs were used to construct a diaphragmatic tendon for use in a CDH sheep model. This resulted in a reduced incidence of recurrent hernia [112].

Fetal tissue engineered heart valves have also been investigated as replacement heart valves. Mechanical heart valves require life-long anticoagulation therapy, bioprosthetic valves are prone to deterioration, and neither grow with the patient [113]. These disadvantages could theoretically be circumvented by living tissue grafts. To that end, Schmidt and colleagues have successfully engineered a heart valve using either UCMSCs or UCB-EPCs on a biodegradable scaffold [113].

Conclusion

Stem cell therapeutics is a rapidly progressing field of inquiry. From the earliest investigations in the first half of the twentieth century to the explosion of interest today, fetal stem cells have emerged as distinct from embryonic and adult stem cells. The unique intermediate properties of fetal stem cells as compared to embryonic and adult make them ideal candidates for many potential therapeutic applications. With the numerous promising discoveries in animal models and early human trials of fetal stem cells, the future is encouraging for the development of many fetal stem cell-derived therapies. However, much work remains in elucidating the mechanisms by which these stem cells exert their healing properties and in applying this to the design of novel treatments.

Abbreviations

AEC	Amniotic epithelial cell
AF-MSC	Amniotic fluid mesenchymal stem cell
AFSC	Amniotic fluid stem cell
ALS	Amyotrophic lateral sclerosis
AMSC	Amniotic mesenchymal stem cell
ATN	Acute tubular necrosis
CDH	Congenital diaphragmatic hernia
CMSC	Chorionic mesenchymal stromal cell
EPC	Endothelial progenitor cell
ESC	Embryonic stem cell
HLA	Human leukocyte antigen
HSC	Hematopoietic stem cell
iPC	Induced pluripotent cell
MI	Myocardial infarction
MLSC	Multilineage stem cell
MSC	Mesenchymal stem cell
PAPC	Pregnancy-associated progenitor cell
PTFE	Polytetrafluoroethylene
UCB	Umbilical cord blood
UCMSC	Umbilical cord mesenchymal stem cell
VSEL	Very small embryonic-like stem cell

References

1. Liechty KW, Adzick NS, Crombleholme TM. Diminished interleukin 6 (IL-6) production during scarless human fetal wound repair. Cytokine. 2000;12(6):671–6.
2. Guillot PV, Gotherstrom C, Chan J, Kurata H, Fisk NM. Human first-trimester fetal MSC express pluripotency markers and grow faster and have longer telomeres than adult MSC. Stem Cells. 2007;25(3):646–54.
3. Dominici M, Le Blanc K, Mueller I, Slaper-Cortenbach I, Marini F, Krause D, et al. Minimal criteria for defining multipotent mesenchymal stromal cells. The International Society for Cellular Therapy position statement. Cytotherapy. 2006;8(4):315–7.
4. Marcus AJ, Woodbury D. Fetal stem cells from extra-embryonic tissues: do not discard. J Cell Mol Med. 2008;12(3):730–42.
5. Davis JW. Skin transplantation with a review of 550 cases at the Johns Hopkins Hospital. Johns Hopkins Med J. 1910;15:307–96.
6. Stern W. The grafting of preserved amniotic membrane to burned and ulcerated skin. JAMA. 1913;13:973–4.
7. Sabella W. Use of fetal membranes in skin grafting. Med Rec NY. 1913;83:478–80.
8. de Rotth A. Plastic repair of conjunctival defects with fetal membranes. Arch Ophthalmol. 1940;23:522–5.
9. Hurst AF, Tanner WE, Osman AA. Addison's disease, with severe anaemia, treated by suprarenal grafting. Proc R Soc Med. 1922;15(Clin Sect):19–20.
10. Thomas ED, Lochte Jr HL, Lu WC, Ferrebee JW. Intravenous infusion of bone marrow in patients receiving radiation and chemotherapy. N Engl J Med. 1957;257(11):491–6.

11. Barr RD, Whang-Peng J, Perry S. Hemopoietic stem cells in human peripheral blood. Science. 1975;190(4211):284–5.
12. Ashton BA, Allen TD, Howlett CR, Eaglesom CC, Hattori A, Owen M. Formation of bone and cartilage by marrow stromal cells in diffusion chambers in vivo. Clin Orthop Relat Res. 1980;151:294–307.
13. Owen M, Friedenstein AJ. Stromal stem cells: marrow-derived osteogenic precursors. Ciba Found Symp. 1988;136:42–60.
14. Knudtzon S. In vitro growth of granulocytic colonies from circulating cells in human cord blood. Blood. 1974;43(3):357–61.
15. Broxmeyer HE, Gluckman E, Auerbach A, Douglas GW, Friedman H, Cooper S, et al. Human umbilical cord blood: a clinically useful source of transplantable hematopoietic stem/progenitor cells. Int J Cell Cloning. 1990;8 Suppl 1:76–89. discussion 89–91.
16. Rubinstein P, Taylor PE, Scaradavou A, Adamson JW, Migliaccio G, Emanuel D, et al. Unrelated placental blood for bone marrow reconstitution: organization of the placental blood program. Blood Cells. 1994;20(2–3):587–96. discussion 96–600.
17. Long GD, Laughlin M, Madan B, Kurtzberg J, Gasparetto C, Morris A, et al. Unrelated umbilical cord blood transplantation in adult patients. Biol Blood Marrow Transplant. 2003;9(12):772–80.
18. Pierce GB, Dixon Jr FJ. Testicular teratomas. I. Demonstration of teratogenesis by metamorphosis of multipotential cells. Cancer. 1959;12(3):573–83.
19. Stevens LC. Embryonic potency of embryoid bodies derived from a transplantable testicular teratoma of the mouse. Dev Biol. 1960;2:285–97.
20. Kleinsmith LJ, Pierce Jr GB. Multipotentiality of single embryonal carcinoma cells. Cancer Res. 1964;24:1544–51.
21. Martin GR. Isolation of a pluripotent cell line from early mouse embryos cultured in medium conditioned by teratocarcinoma stem cells. Proc Natl Acad Sci U S A. 1981;78(12):7634–8.
22. Thomson JA, Itskovitz-Eldor J, Shapiro SS, Waknitz MA, Swiergiel JJ, Marshall VS, et al. Embryonic stem cell lines derived from human blastocysts. Science. 1998;282(5391):1145–7.
23. Bradley A, Evans M, Kaufman MH, Robertson E. Formation of germ-line chimaeras from embryo-derived teratocarcinoma cell lines. Nature. 1984;309(5965):255–6.
24. Kmiecik G, Niklinska W, Kuc P, Pancewicz-Wojtkiewicz J, Fil D, Karwowska A, et al. Fetal membranes as a source of stem cells. Adv Med Sci. 2013;58(2):185–95.
25. Wilmut I, Schnieke AE, McWhir J, Kind AJ, Campbell KH. Viable offspring derived from fetal and adult mammalian cells. Nature. 1997;385(6619):810–3.
26. Gurdon JB, Elsdale TR, Fischberg M. Sexually mature individuals of Xenopus laevis from the transplantation of single somatic nuclei. Nature. 1958;182(4627):64–5.
27. Spinelli V, Guillot PV, De Coppi P. Induced pluripotent stem (iPS) cells from human fetal stem cells (hFSCs). Organogenesis. 2013;9(2):101–10.
28. Sgaramella V, Zinder ND. Dolly confirmation. Science. 1998;279(5351):635. 7–8.
29. Takahashi K, Yamanaka S. Induction of pluripotent stem cells from mouse embryonic and adult fibroblast cultures by defined factors. Cell. 2006;126(4):663–76.
30. Sumi T, Tsuneyoshi N, Nakatsuji N, Suemori H. Apoptosis and differentiation of human embryonic stem cells induced by sustained activation of c-Myc. Oncogene. 2007;26(38):5564–76.
31. Yu J, Vodyanik MA, Smuga-Otto K, Antosiewicz-Bourget J, Frane JL, Tian S, et al. Induced pluripotent stem cell lines derived from human somatic cells. Science. 2007;318(5858):1917–20.
32. Fauza D. Amniotic fluid and placental stem cells. Best Pract Res Clin Obstet Gynaecol. 2004;18(6):877–91.
33. Torricelli F, Brizzi L, Bernabei PA, Gheri G, Di Lollo S, Nutini L, et al. Identification of hematopoietic progenitor cells in human amniotic fluid before the 12th week of gestation. Ital J Anat Embryol. 1993;98(2):119–26.

34. Streubel B, Martucci-Ivessa G, Fleck T, Bittner RE. In vitro transformation of amniotic cells to muscle cells – background and outlook. Wien Med Wochenschr. 1996;146(9–10):216–7.
35. Hammerman MR. Classic and current opinion in embryonic organ transplantation. Curr Opin Organ Transplant. 2014;19(2):133–9.
36. Ilancheran S, Moodley Y, Manuelpillai U. Human fetal membranes: a source of stem cells for tissue regeneration and repair? Placenta. 2009;30(1):2–10.
37. Selman K, Kafatos FC. Transdifferentiation in the labial gland of silk moths: is DNA required for cellular metamorphosis? Cell Differ. 1974;3(2):81–94.
38. Quesenberry PJ, Colvin G, Dooner G, Dooner M, Aliotta JM, Johnson K. The stem cell continuum: cell cycle, injury, and phenotype lability. Ann N Y Acad Sci. 2007;1106:20–9.
39. Wang X, Willenbring H, Akkari Y, Torimaru Y, Foster M, Al-Dhalimy M, et al. Cell fusion is the principal source of bone-marrow-derived hepatocytes. Nature. 2003;422(6934):897–901.
40. Fan Y, Luo Y, Chen X, Li Q, Sun X. Generation of human beta-thalassemia induced pluripotent stem cells from amniotic fluid cells using a single excisable lentiviral stem cell cassette. J Reprod Dev. 2012;58(4):404–9.
41. Moschidou D, Mukherjee S, Blundell MP, Drews K, Jones GN, Abdulrazzak H, et al. Valproic acid confers functional pluripotency to human amniotic fluid stem cells in a transgene-free approach. Mol Ther. 2012;20(10):1953–67.
42. Turner CG, Fauza DO. Fetal tissue engineering. Clin Perinatol. 2009;36(2):473–88. xii.
43. van de Ven C, Collins D, Bradley MB, Morris E, Cairo MS. The potential of umbilical cord blood multipotent stem cells for nonhematopoietic tissue and cell regeneration. Exp Hematol. 2007;35(12):1753–65.
44. De Coppi P, Bartsch Jr G, Siddiqui MM, Xu T, Santos CC, Perin L, et al. Isolation of amniotic stem cell lines with potential for therapy. Nat Biotechnol. 2007;25(1):100–6.
45. Ilancheran S, Michalska A, Peh G, Wallace EM, Pera M, Manuelpillai U. Stem cells derived from human fetal membranes display multilineage differentiation potential. Biol Reprod. 2007;77(3):577–88.
46. Kaviani A, Perry TE, Dzakovic A, Jennings RW, Ziegler MM, Fauza DO. The amniotic fluid as a source of cells for fetal tissue engineering. J Pediatr Surg. 2001;36(11):1662–5.
47. Jones GN, Moschidou D, Puga-Iglesias TI, Kuleszewicz K, Vanleene M, Shefelbine SJ, et al. Ontological differences in first compared to third trimester human fetal placental chorionic stem cells. PLoS One. 2012;7(9):e43395.
48. Portmann-Lanz CB, Schoeberlein A, Huber A, Sager R, Malek A, Holzgreve W, et al. Placental mesenchymal stem cells as potential autologous graft for pre- and perinatal neuroregeneration. Am J Obstet Gynecol. 2006;194(3):664–73.
49. Parolini O, Alviano F, Bagnara GP, Bilic G, Buhring HJ, Evangelista M, et al. Concise review: isolation and characterization of cells from human term placenta: outcome of the first international Workshop on Placenta Derived Stem Cells. Stem Cells. 2008;26(2):300–11.
50. Abdulrazzak H, Moschidou D, Jones G, Guillot PV. Biological characteristics of stem cells from foetal, cord blood and extraembryonic tissues. J R Soc Interface. 2010;7 Suppl 6:S689–706.
51. Macian F, Im SH, Garcia-Cozar FJ, Rao A. T-cell anergy. Curr Opin Immunol. 2004;16(2):209–16.
52. Magatti M, De Munari S, Vertua E, Gibelli L, Wengler GS, Parolini O. Human amnion mesenchyme harbors cells with allogeneic T-cell suppression and stimulation capabilities. Stem Cells. 2008;26(1):182–92.
53. Li H, Niederkorn JY, Neelam S, Mayhew E, Word RA, McCulley JP, et al. Immunosuppressive factors secreted by human amniotic epithelial cells. Invest Ophthalmol Vis Sci. 2005;46(3):900–7.
54. Fong CY, Chak LL, Biswas A, Tan JH, Gauthaman K, Chan WK, et al. Human Wharton's jelly stem cells have unique transcriptome profiles compared to human embryonic stem cells and other mesenchymal stem cells. Stem Cell Rev. 2011;7(1):1–16.

55. Meller D, Pauklin M, Thomasen H, Westekemper H, Steuhl KP. Amniotic membrane transplantation in the human eye. Dtsch Arztebl Int. 2011;108(14):243–8.
56. Sankar V, Muthusamy R. Role of human amniotic epithelial cell transplantation in spinal cord injury repair research. Neuroscience. 2003;118(1):11–7.
57. Nishio Y, Koda M, Kamada T, Someya Y, Yoshinaga K, Okada S, et al. The use of hemopoietic stem cells derived from human umbilical cord blood to promote restoration of spinal cord tissue and recovery of hindlimb function in adult rats. J Neurosurg Spine. 2006;5(5):424–33.
58. Kim J, Kang HM, Kim H, Kim MR, Kwon HC, Gye MC, et al. Ex vivo characteristics of human amniotic membrane-derived stem cells. Cloning Stem Cells. 2007;9(4):581–94.
59. Harrison DE, Zhong RK, Jordan CT, Lemischka IR, Astle CM. Relative to adult marrow, fetal liver repopulates nearly five times more effectively long-term than short-term. Exp Hematol. 1997;25(4):293–7.
60. Taylor PA, McElmurry RT, Lees CJ, Harrison DE, Blazar BR. Allogenic fetal liver cells have a distinct competitive engraftment advantage over adult bone marrow cells when infused into fetal as compared with adult severe combined immunodeficient recipients. Blood. 2002;99(5):1870–2.
61. Cardoso AA, Li ML, Batard P, Hatzfeld A, Brown EL, Levesque JP, et al. Release from quiescence of CD34+ CD38- human umbilical cord blood cells reveals their potentiality to engraft adults. Proc Natl Acad Sci U S A. 1993;90(18):8707–11.
62. Nava S, Westgren M, Jaksch M, Tibell A, Broome U, Ericzon BG, et al. Characterization of cells in the developing human liver. Differentiation. 2005;73(5):249–60.
63. Hammerman MR. Xenotransplantation of embryonic pig kidney or pancreas to replace the function of mature organs. J Transplant. 2011;2011:501749.
64. in 't Anker PS, Noort WA, Kruisselbrink AB, Scherjon SA, Beekhuizen W, Willemze R, et al. Nonexpanded primary lung and bone marrow-derived mesenchymal cells promote the engraftment of umbilical cord blood-derived CD34(+) cells in NOD/SCID mice. Exp Hematol. 2003;31(10):881–9.
65. Ingram DA, Mead LE, Tanaka H, Meade V, Fenoglio A, Mortell K, et al. Identification of a novel hierarchy of endothelial progenitor cells using human peripheral and umbilical cord blood. Blood. 2004;104(9):2752–60.
66. Kucia M, Halasa M, Wysoczynski M, Baskiewicz-Masiuk M, Moldenhawer S, Zuba-Surma E, et al. Morphological and molecular characterization of novel population of CXCR4+ SSEA-4+ Oct-4+ very small embryonic-like cells purified from human cord blood: preliminary report. Leukemia. 2007;21(2):297–303.
67. Taghizadeh RR, Cetrulo KJ, Cetrulo CL. Wharton's Jelly stem cells: future clinical applications. Placenta. 2011;32 Suppl 4:S311–5.
68. Karahuseyinoglu S, Cinar O, Kilic E, Kara F, Akay GG, Demiralp DO, et al. Biology of stem cells in human umbilical cord stroma: in situ and in vitro surveys. Stem Cells. 2007;25(2):319–31.
69. Cremer M, Schachner M, Cremer T, Schmidt W, Voigtlander T. Demonstration of astrocytes in cultured amniotic fluid cells of three cases with neural-tube defect. Hum Genet. 1981;56(3):365–70.
70. Gucciardo L, Lories R, Ochsenbein-Kolble N, Done E, Zwijsen A, Deprest J. Fetal mesenchymal stem cells: isolation, properties and potential use in perinatology and regenerative medicine. BJOG. 2009;116(2):166–72.
71. Kaviani A, Perry TE, Barnes CM, Oh JT, Ziegler MM, Fishman SJ, et al. The placenta as a cell source in fetal tissue engineering. J Pediatr Surg. 2002;37(7):995–9. discussion -9.
72. Bianchi DW, Zickwolf GK, Weil GJ, Sylvester S, DeMaria MA. Male fetal progenitor cells persist in maternal blood for as long as 27 years postpartum. Proc Natl Acad Sci U S A. 1996;93(2):705–8.
73. Khosrotehrani K, Bianchi DW. Multi-lineage potential of fetal cells in maternal tissue: a legacy in reverse. J Cell Sci. 2005;118(Pt 8):1559–63.

74. Nguyen Huu S, Dubernard G, Aractingi S, Khosrotehrani K. Feto-maternal cell trafficking: a transfer of pregnancy associated progenitor cells. Stem Cell Rev. 2006;2(2):111–6.
75. Kakishita K, Elwan MA, Nakao N, Itakura T, Sakuragawa N. Human amniotic epithelial cells produce dopamine and survive after implantation into the striatum of a rat model of Parkinson's disease: a potential source of donor for transplantation therapy. Exp Neurol. 2000;165(1):27–34.
76. Kakishita K, Nakao N, Sakuragawa N, Itakura T. Implantation of human amniotic epithelial cells prevents the degeneration of nigral dopamine neurons in rats with 6-hydroxydopamine lesions. Brain Res. 2003;980(1):48–56.
77. Weiss ML, Medicetty S, Bledsoe AR, Rachakatla RS, Choi M, Merchav S, et al. Human umbilical cord matrix stem cells: preliminary characterization and effect of transplantation in a rodent model of Parkinson's disease. Stem Cells. 2006;24(3):781–92.
78. Nan Z, Grande A, Sanberg CD, Sanberg PR, Low WC. Infusion of human umbilical cord blood ameliorates neurologic deficits in rats with hemorrhagic brain injury. Ann N Y Acad Sci. 2005;1049:84–96.
79. Garbuzova-Davis S, Willing AE, Zigova T, Saporta S, Justen EB, Lane JC, et al. Intravenous administration of human umbilical cord blood cells in a mouse model of amyotrophic lateral sclerosis: distribution, migration, and differentiation. J Hematother Stem Cell Res. 2003;12(3):255–70.
80. Escolar ML, Poe MD, Provenzale JM, Richards KC, Allison J, Wood S, et al. Transplantation of umbilical-cord blood in babies with infantile Krabbe's disease. N Engl J Med. 2005;352(20):2069–81.
81. Pan HC, Yang DY, Chiu YT, Lai SZ, Wang YC, Chang MH, et al. Enhanced regeneration in injured sciatic nerve by human amniotic mesenchymal stem cell. J Clin Neurosci. 2006;13(5):570–5.
82. Pan HC, Cheng FC, Chen CJ, Lai SZ, Lee CW, Yang DY, et al. Post-injury regeneration in rat sciatic nerve facilitated by neurotrophic factors secreted by amniotic fluid mesenchymal stem cells. J Clin Neurosci. 2007;14(11):1089–98.
83. Zhao P, Ise H, Hongo M, Ota M, Konishi I, Nikaido T. Human amniotic mesenchymal cells have some characteristics of cardiomyocytes. Transplantation. 2005;79(5):528–35.
84. Sartore S, Lenzi M, Angelini A, Chiavegato A, Gasparotto L, De Coppi P, et al. Amniotic mesenchymal cells autotransplanted in a porcine model of cardiac ischemia do not differentiate to cardiogenic phenotypes. Eur J Cardiothorac Surg. 2005;28(5):677–84.
85. Hirata Y, Sata M, Motomura N, Takanashi M, Suematsu Y, Ono M, et al. Human umbilical cord blood cells improve cardiac function after myocardial infarction. Biochem Biophys Res Commun. 2005;327(2):609–14.
86. Yeh YC, Lee WY, Yu CL, Hwang SM, Chung MF, Hsu LW, et al. Cardiac repair with injectable cell sheet fragments of human amniotic fluid stem cells in an immune-suppressed rat model. Biomaterials. 2010;31(25):6444–53.
87. Taguchi A, Soma T, Tanaka H, Kanda T, Nishimura H, Yoshikawa H, et al. Administration of CD34+ cells after stroke enhances neurogenesis via angiogenesis in a mouse model. J Clin Invest. 2004;114(3):330–8.
88. Rehni AK, Singh N, Jaggi AS, Singh M. Amniotic fluid derived stem cells ameliorate focal cerebral ischaemia-reperfusion injury induced behavioural deficits in mice. Behav Brain Res. 2007;183(1):95–100.
89. Kim SW, Han H, Chae GT, Lee SH, Bo S, Yoon JH, et al. Successful stem cell therapy using umbilical cord blood-derived multipotent stem cells for Buerger's disease and ischemic limb disease animal model. Stem Cells. 2006;24(6):1620–6.
90. Carraro G, Perin L, Sedrakyan S, Giuliani S, Tiozzo C, Lee J, et al. Human amniotic fluid stem cells can integrate and differentiate into epithelial lung lineages. Stem Cells. 2008;26(11):2902–11.
91. Cargnoni A, Gibelli L, Tosini A, Signoroni PB, Nassuato C, Arienti D, et al. Transplantation of allogeneic and xenogeneic placenta-derived cells reduces bleomycin-induced lung fibrosis. Cell Transplant. 2009;18(4):405–22.

92. Sakuragawa N, Enosawa S, Ishii T, Thangavel R, Tashiro T, Okuyama T, et al. Human amniotic epithelial cells are promising transgene carriers for allogeneic cell transplantation into liver. J Hum Genet. 2000;45(3):171–6.
93. Khan AA, Shaik MV, Parveen N, Rajendraprasad A, Aleem MA, Habeeb MA, et al. Human fetal liver-derived stem cell transplantation as supportive modality in the management of end-stage decompensated liver cirrhosis. Cell Transplant. 2010;19(4):409–18.
94. Perin L, Sedrakyan S, Giuliani S, Da Sacco S, Carraro G, Shiri L, et al. Protective effect of human amniotic fluid stem cells in an immunodeficient mouse model of acute tubular necrosis. PLoS One. 2010;5(2):e9357.
95. Ersek A, Pixley JS, Goodrich AD, Porada CD, Almeida-Porada G, Thain DS, et al. Persistent circulating human insulin in sheep transplanted in utero with human mesenchymal stem cells. Exp Hematol. 2010;38(4):311–20.
96. Wei JP, Zhang TS, Kawa S, Aizawa T, Ota M, Akaike T, et al. Human amnion-isolated cells normalize blood glucose in streptozotocin-induced diabetic mice. Cell Transplant. 2003;12(5): 545–52.
97. Zhao Y, Wang H, Mazzone T. Identification of stem cells from human umbilical cord blood with embryonic and hematopoietic characteristics. Exp Cell Res. 2006;312(13):2454–64.
98. Guillot PV, Abass O, Bassett JH, Shefelbine SJ, Bou-Gharios G, Chan J, et al. Intrauterine transplantation of human fetal mesenchymal stem cells from first-trimester blood repairs bone and reduces fractures in osteogenesis imperfecta mice. Blood. 2008;111(3):1717–25.
99. Le Blanc K, Gotherstrom C, Ringden O, Hassan M, McMahon R, Horwitz E, et al. Fetal mesenchymal stem-cell engraftment in bone after in utero transplantation in a patient with severe osteogenesis imperfecta. Transplantation. 2005;79(11):1607–14.
100. Conconi MT, Burra P, Di Liddo R, Calore C, Turetta M, Bellini S, et al. CD105(+) cells from Wharton's jelly show in vitro and in vivo myogenic differentiative potential. Int J Mol Med. 2006;18(6):1089–96.
101. Chan J, Waddington SN, O'Donoghue K, Kurata H, Guillot PV, Gotherstrom C, et al. Widespread distribution and muscle differentiation of human fetal mesenchymal stem cells after intrauterine transplantation in dystrophic mdx mouse. Stem Cells. 2007;25(4):875–84.
102. Subrahmanyam M. Amniotic membrane as a cover for microskin grafts. Br J Plast Surg. 1995;48(7):477–8.
103. Kamolz LP, Kolbus A, Wick N, Mazal PR, Eisenbock B, Burjak S, et al. Cultured human epithelium: human umbilical cord blood stem cells differentiate into keratinocytes under in vitro conditions. Burns. 2006;32(1):16–9.
104. Lund RD, Wang S, Lu B, Girman S, Holmes T, Sauve Y, et al. Cells isolated from umbilical cord tissue rescue photoreceptors and visual functions in a rodent model of retinal disease. Stem Cells. 2007;25(3):602–11.
105. Loebinger MR, Eddaoudi A, Davies D, Janes SM. Mesenchymal stem cell delivery of TRAIL can eliminate metastatic cancer. Cancer Res. 2009;69(10):4134–42.
106. Rachakatla RS, Pyle MM, Ayuzawa R, Edwards SM, Marini FC, Weiss ML, et al. Combination treatment of human umbilical cord matrix stem cell-based interferon-beta gene therapy and 5-fluorouracil significantly reduces growth of metastatic human breast cancer in SCID mouse lungs. Cancer Invest. 2008;26(7):662–70.
107. Vacanti JP, Morse MA, Saltzman WM, Domb AJ, Perez-Atayde A, Langer R. Selective cell transplantation using bioabsorbable artificial polymers as matrices. J Pediatr Surg. 1988;23(1 Pt 2):3–9.
108. Fauza DO, Fishman SJ, Mehegan K, Atala A. Videofetoscopically assisted fetal tissue engineering: bladder augmentation. J Pediatr Surg. 1998;33(1):7–12.
109. Fauza DO, Fishman SJ, Mehegan K, Atala A. Videofetoscopically assisted fetal tissue engineering: skin replacement. J Pediatr Surg. 1998;33(2):357–61.
110. Fuchs JR, Terada S, Ochoa ER, Vacanti JP, Fauza DO. Fetal tissue engineering: in utero tracheal augmentation in an ovine model. J Pediatr Surg. 2002;37(7):1000–6. discussion -6.

111. Kunisaki SM, Fuchs JR, Steigman SA, Fauza DO. A comparative analysis of cartilage engineered from different perinatal mesenchymal progenitor cells. Tissue Eng. 2007;13(11): 2633–44.
112. Fuchs JR, Kaviani A, Oh JT, LaVan D, Udagawa T, Jennings RW, et al. Diaphragmatic reconstruction with autologous tendon engineered from mesenchymal amniocytes. J Pediatr Surg. 2004;39(6):834–8. discussion -8.
113. Schmidt D, Mol A, Odermatt B, Neuenschwander S, Breymann C, Gossi M, et al. Engineering of biologically active living heart valve leaflets using human umbilical cord-derived progenitor cells. Tissue Eng. 2006;12(11):3223–32.

Chapter 2
Feto-Maternal Cell Trafficking and Labor

S. Christopher Derderian, Cerine Jeanty, and Tippi C. MacKenzie

Introduction

Maternal-fetal cellular trafficking (MFCT) is a well described phenomenon during pregnancy in which maternal cells migrate into the fetus and fetal cells migrate into the mother [1–5]. The specific mechanisms leading to such trafficking and its lifelong consequences have fascinated scientists for decades and are still actively being investigated. For example, several groups have demonstrated an association between MFCT and both transplant tolerance and autoimmune disorders. Additionally, pregnancy complications have been shown to be associated with increased trafficking between the mother and fetus which are listed in Table 2.1. Innovative strategies to detect microchimerism have reinvigorated the interest in the field and will be outlined in this chapter. In this chapter, we will review implications of microchimerism, particularly as it relates to long-term consequences and pregnancy complications. Finally, we will explore the effects congenital abnormalities and fetal surgery have on maternal-fetal cellular trafficking.

Mechanisms of Cellular Trafficking

Maternal microchimerism (MMc) refers to the presence of maternal cells within the fetus. This has been demonstrated by the presence of cells of maternal origin within the liver, spleen, thymus, thyroid, and skin of neonates [6], indicating the placenta

S.C. Derderian, M.D. • C. Jeanty, M.D. • T.C. MacKenzie, M.D. (✉)
Department of Surgery and The Center for Maternal-Fetal Precision Medicine, University of California, San Francisco, 513 Parnassus Avenue, San Francisco, CA 94143-0570, USA
e-mail: Tippi.Mackenzie@ucsfmedctr.org

D.O. Fauza, M. Bani (eds.), *Fetal Stem Cells in Regenerative Medicine*, Stem Cell Biology and Regenerative Medicine,
DOI 10.1007/978-1-4939-3483-6_2

Table 2.1 Conditions associated with increased maternal fetal trafficking

Autoimmune processes
Diabetes mellitus-type I
Neonatal lupus congenital heart block
Multiple sclerosis
Hirschsprung's disease
Autoimmune thyroiditis
Primary biliary cirrhosis
Systemic lupus erythematous
Pregnancy complication
Preeclampsia
Intrauterine fetal growth restriction
Preterm labor
Iatrogenic
Open fetal intervention
Laparoscopic fetal intervention
Congenital anomalies
Aneuploidy
Congenital diaphragmatic hernia

is not a perfect barrier as previously imagined. As MMc has been found in various organs well into adulthood, cells of maternal origin must possess the capacity to self renew [5]. Some postulate that MMc results from multipotent mesenchymal stromal cell migration across the placenta, which is governed by vascular endothelial growth factor A (VEGF-A) [7], a potent stimulator of hematopoietic stem cell migration [8]. In fact, the concentration of VEGF-A is higher in the fetal circulation compared to the maternal, which likely promotes its transplacental migration [9]. Additionally, in a mouse model, we demonstrated high levels of MMc in circulation at mid-gestation which decreases over time, such that it is undetectable at birth [10]. In this model, inflammatory stimuli during pregnancy, such as fetal intervention, led to changes in the number and type of cells that traffic, including maternal T, which usually does not cross over at baseline [10]. These results suggest that alterations in trafficking are not a result of general leakiness at the maternal-fetal interface, which is further supported by experiments showing that chemokine gene silencing limits T cell trafficking [11].

Fetal microchimerism (FMc), on the other hand, refers to the presence of fetal cells within maternal tissues and blood and can also persist for decades after delivery [2]. Similar to MMc, fetal cells have been found in multiple organs including the liver, kidney, heart, and bone marrow [12, 13], though the exact mechanism by which fetal cells migrate into the maternal circulation remains elusive. Fetal cell-free DNA (fDNA) has also been observed within the maternal circulation, which is released from the placental trophoblast layer lining the maternal-fetal interphase. Apoptosis and cell necrosis at this interphase leads to the release of fDNA into the maternal circulation [14, 15], the implications of which are actively being explored.

Strategies to Detect Maternal-Fetal Cellular Trafficking

Fetal microchimerism was first observed in 1893 when fetal trophoblast cells were observed in a lung specimen from a woman who suffered from eclamspia [16]. Several decades later, in 1963, maternal cells were identified in a cord blood sample using fluorescently labeled maternal leukocytes [17]. Since then, our understanding of MFCT has improved in large part from advances in techniques to distinguish mixed populations of cells.

In recent years, investigators have applied the common technique of gene amplification with polymerase chain reaction (PCR) to identify microchimerism is the context of pregnancy. It has become a useful tool to detect and quantify fetal DNA within the maternal circulation [3, 5, 18, 19]. This method is in large part restricted to gender mismatches in which primers to loci on the Y chromosome are used to distinguish fetal from maternal DNA [18, 20]. Using PCR amplification, fetal DNA can be detected circulating within the maternal serum in 80 % of normal pregnancies [21] and has been isolated as early as 4 weeks postconception [22, 23].

An alternative strategy is to compare non-shared HLA-DR or Insertion-Deletion alleles between the fetus and mother. To compare these allelic differences between cell populations, paired maternal and cord blood is analyzed using quantitative real-time polymerase chain reaction (qRT-PCR). Non-shared alleles between the two cell populations are termed informative alleles as they provide a means to distinguish one set of genetic material from another. In trauma patients who were transfused multiple units of allogeneic blood, Lee and colleagues compared 12 HLA-DR and 12 Insertion-Deletion alleles [24]. From this study, they found that at least 1 informative allele could be determined in 99.5 % of patients. Applied to MFCT, this strategy has been used to quantify the number of fetal cells in the maternal circulation (or vice versa) [25]. While informative, this strategy requires examination of both maternal and fetal blood and is therefore usually only applicable after birth.

MFCT is a particularly critical field of investigation as it has the potential to improve noninvasive detection of fetal anomalies. Currently, clinicians rely on second trimester sonographic imaging to identify fetuses at risk for aneuploidy and congenital anomalies. Positive screening is followed by invasive procedures such as amniocentesis and chorionic villus sampling for diagnosis. Despite a diagnostic accuracy of 98–99 % [26], these procedures carry a risk to both the fetus and mother [27]. Prenatal diagnosis that does not disrupt the maternal-fetal interface may be accomplished by identifying and analyzing fetal DNA within the maternal circulation. Several European countries including the Netherlands, Sweden, and Denmark have already implemented this strategy to determine fetal gender and Rhesus D status [28].

In mice, it is possible to evaluate the number and types of cells that traffic using flow cytometry [10], but this is not yet possible in humans unless the HLA type is known,[29] and antibodies to such cell markers exist. Alternatively, cells may be sorted into groups (T cells, B cells, etc.) prior to PCR sequencing, a technique that has not yet been explored in pregnant woman. Currently, probing for fetal DNA is more feasible than isolating individual cells as the quantity of fetal DNA is much more than the number of fetal cells within the maternally circulating [15, 30]. Overcoming this barrier may help identify which cell populations are more influential during trafficking and whether particular populations are more prevalent in the setting of pregnancy complications.

Tolerogenic and Immunogenic Consequences of Microchimerism

Microchimerism can lead to a tolerogenic or immunogenic state. The presence of maternal cells in the fetus may play a role in fetal immune education and has been found to induce regulatory T cells to maternal antigen, which suppress the fetal immune response to the mother [29]. Tolerance to non-inherited maternal antigens has implications for transplantation tolerance later in life. For example, patients with biliary atresia, who have increased levels of MMc, have improved graft survival when they receive a maternal liver transplant compared to a paternal graft [31]. In acute leukemia, patient survival is increased and graft-versus-host disease is reduced when transplantation is with maternal stem cells [32, 33].

Conversely, microchimerism has been associated with autoimmune diseases in both mothers and children. Increased levels of MMc have been observed in childhood diseases, including diabetes mellitus-type I, neonatal lupus congenital heart block, multiple sclerosis [34], and Hirschsprung's disease [35]. Autoimmune diseases associated with FMc include systemic sclerosis in which fetal cells have been detected within both peripheral blood and skin lesions [36, 37], autoimmune thyroiditis, primary biliary cirrhosis, and systemic lupus erythematous [38]. It is important to note that a causal relationship has not been established and the association with microchimerism may indicate that microchimeric cells proliferate in response to the disease process.

Pregnancy Complications Associated with Cellular Trafficking

Several independent investigators have found an association between increased FMc and pregnancy complications, including preeclampsia, intrauterine growth restriction (IUGR), and preterm labor [19, 39–46]. These observations may reflect a maternal immune response to fetal antigens, or may simply be a marker of the increased inflammatory milieu in the host. Understanding the mechanisms that promote increased cellular trafficking may lead to therapies to offset the development of preterm labor and other pregnancy complications.

Preeclampsia

Preeclampsia is a significant cause of both fetal and maternal mortality during pregnancy [47] and is characterized by maternal hypertension and proteinuria after the 20th week of gestation [48]. The incidence ranges from 2 to 7 % in normal nulliparous females [49, 50], and increases to 18 % in those who have previously had preeclamsia [51]. Complications include placental abruption, renal failure, HELLP syndrome and even death.

The pathologic processes leadings to preeclampsia are thought to occur at the location of the placenta, as histological examination of placentas in preeclamptic patients frequently shows infarction and sclerotic arterioles with poor remodeling of the uterine spinal arteries [52]. In addition, hypoxic changes and oxidative stress at the feto-placental interface may lead to increased apoptosis and DNA released into the maternal circulation [53–57], particularly from the syncytiotrophoblast layer [58, 59].

Multiple groups have proposed an association between FMc and preeclampsia [60–66]. While some have found elevated level of maternally circulating fetal erythroblasts and placental syncytiotrophoblast microvesicles [67, 68], most studies have focused on increased levels of fDNA within the maternal circulation. Not only are levels elevated at delivery [64], but increased levels have been detected circulating within the maternal serum as early as the first trimester [65, 69]. Illanes and colleagues found that the quantity of maternal circulating fDNA measured between 11 and 14 weeks gestation directly correlated with the likelihood of developing preeclampsia [56], though other investigators have not found this association [70]. These conflicting observations warrant further investigation to not only standardize techniques but understand the process leading to fDNA release and how it may relate to the development of preeclampsia.

Maternal sampling for fDNA has been considered as a screening tool to predict preeclampsia. Preliminary results by Farina and colleagues found that increased levels of fDNA may be predictive in asymptomatic low risk patients during the second trimester [53]. They found that maternal serum levels of fDNA were 2.4-fold higher in mothers who developed preeclampsia compared to gestational age matched controls. As these are preliminary results, further studies are needed to determine the sensitivity of the assay as well as a cost analysis profile. If second trimester fDNA levels prove to be a useful screening tool, efforts may be focused towards monitoring patients at high-risk for developing preeclampsia or other complications associated with preeclampsia such as placental abruption, renal failure, and HELLP syndrome.

Intrauterine Fetal Growth Restriction

Intrauterine growth restriction is another complication of pregnancy effecting 3–7 % of births worldwide. It is defined by fetal weight below the 10th percentile for a given gestational age and may result in respiratory distress syndrome, intraventricular hemorrhage, necrotizing enterocolitis, and death. There are various underlying causes including both fetal (congenital abnormalities, chromosomal anomalies, and infection) and maternal (alcohol consumption, smoking, vascular disease, and malnutrition) origins. Like preeclampsia, IUGR may develop from abnormal placentation involving aberrant spiral artery development [57] with increased trophoblast cell apoptosis and necrosis as well as impaired oxygen and nutrient delivery to the fetus [71, 72].

While studies examining FMc in IUGR are limited and conflicting, some groups have found increased fetal erythroblasts and fDNA in maternal serum in cases of IUGR [39, 73], while others have not, despite using similar methods and patient populations [74, 75]. Conflicting results may be secondary to the various etiologies of IUGR. Perhaps maternal causes of IUGR, such as preeclampsia and vascular disease, result in abnormal placental development and trophoblast cell death, while fetal causes, such as aneuploidy and congenital abnormalities, do not significantly impact the placenta.

Preterm Labor

Spontaneous preterm labor occurs in approximately 12 % of births and is the converging end-product of various pathological processes [76]. Causes include intrauterine infections [77], placental vascular insufficiency [78, 79], uterine over-distention [80], and a shortened cervix [81, 82], resulting in the release of several cytokines and prostaglandins [83]. These inflammatory mediators promote the release of uterotonins which induce uterine contractions and proteases which result in cervical changes, culminating in preterm delivery [83].

Several groups have proposed an association between preterm labor and alterations in cellular trafficking [19, 42, 44]. For example, Leung and colleagues have implicated fDNA as a marker for preterm labor near the time of delivery [44]. The molecular pathway leading to labor in this population is unclear and further studies correlating cytokine and prostaglandin levels among patients with increased FMc may shed light into a more specific pathway. Although it is not clear whether these alterations are causally related to preterm birth, it has been suggested that increased fetal cell trafficking triggers the maternal immune response, which can induces labor [19].

Investigators have also directly explored the role of the maternal immune system in preterm labor. For example, Lee and colleagues demonstrated that women with circulating antibodies against fetal HLA class I or class II antigen, measured during the second trimester, were at increased risk for developing spontaneous preterm labor [84]. We recently found that MMc is also increased in mice undergoing preterm delivery as a result of LPS injection, with a particular increase in T cell trafficking if the fetuses are allogeneic to the mother [85]. Furthermore, we have seen that maternal T cells cause demise of allogeneic fetuses after fetal intervention, indicating the role of the maternal adaptive immune system in this pregnancy complication [86]. Taken together, preterm labor is a complex process that likely results from multiple mechanisms, including increases in the quantity of FMc and, possibly, an immune response between the mother and the fetus.

Fetal Surgery

Open fetal surgery was pioneered over 30 years ago and has since evolved with the advent of minimally invasive techniques. Fetal surgery has been shown to improve survival and long-term outcomes in disease processes such as twin-to-twin transfusion syndrome, myelomeningocele, and congenital diaphragmatic hernias [87–90]. However, fetal surgery often results in preterm delivery, which abrogates some of the benefits of the procedure. For example, a recent multi-center randomized control trial comparing the prenatal repair of myelomeningocele to standard postnatal repair, found that prenatal repair led to a reduced need for postnatal ventriculoperitoneal shunting as well as improved long term motor function and mental development [87] but frequently results in preterm delivery with a mean gestational age at delivery of 34.1 weeks compared to 37.3 in the standard postnatal control group.

Universal acceptance of fetal surgery for non-lethal congenital diseases has been hampered by the risk of pregnancy complications. These risks include preterm premature rupture of the membrane, placental abruption, uterine rupture, chorioamniotic separation, and preterm labor [87]. In fact, preterm delivery prior to 37 weeks gestation, even following minimally invasive procedures, exceeds 80 % [91, 92]. Since the latency period between the procedure and delivery typically ranges from 4 to 7 weeks [91, 92], it is possible that downstream events rather than the insult of the surgery itself leads to preterm labor. This observation led multiple groups to explore the effect of fetal intervention on MFCT [10, 25, 93, 94]. In a mouse model of fetal intervention, we reported that maternal cells traffic into the fetal circulation after fetal stem cell transplantation, with a particular increase in trafficking T cells in this context [10]. These cells have a functional consequence, in that they limit the stem cell engraftment into the fetus [10]. We have reported a similar findings in patients undergoing fetal surgery for the correction of myelomeningoceles: using PCR to genotype non-shared HLA-DR alleles between mother and fetus, we demonstrated increased trafficking of maternal cells within the fetal circulation following open fetal surgery for myelomeningocele repair [25]. These findings suggest that there is either increased trafficking of cells after fetal intervention or increased proliferation of trafficked cells in the inflammatory environment after fetal surgery. Interestingly, there was no increase in MMc if fetal intervention was performed at the time of birth, indicating that changes in microchimerism take some time to develop.

Increase in FMc during fetal surgery has been demonstrated in some studies, [94] but not others [25]. Following laser coagulation for twin-to-twin transfusion syndrome [94]. increased fDNA was found with longer operative times, increased number of vessels ablated and demise of 1 twin [94]. However, a study measuring circulating mRNA following fetal intervention did not demonstrate a difference between those who underwent fetal intervention and age matched controls [93]. The differences observed may reflect the challenge with detecting a very small pool of cells or genomic material within a large maternal blood volume. It is important to note that no study has proven a causal link between altered microchimerism and pregnancy complications. However, understanding the role of altered MFCT in the context of preterm labor and pregnancy complications may lead to treatments to abrogate such consequences.

Congenital Anomalies

Maternal-fetal cellular trafficking may also be influenced by aneuploidy and congenital anomalies. For example, levels of FMc are significantly higher in mothers carrying fetuses with trisomy 21 [95] and lower in those with trisomy 18, 13, or monosomy X [96]. In a study analyzing cord blood samples from infants with a congenital diaphragmatic hernia, we found an increased number of maternal cells in the fetal circulation at the time of birth which increased with disease severity [97]. These findings suggest that the presence of fetal anomalies may influence trafficking, possibly secondary to an inflammatory response from fetal distress.

Conclusion

In summary, there is striking evidence to suggest that pregnancy complications are associated with alterations in fetal microchimerism. The mechanisms leading to increased levels of trafficking remain a fascinating unanswered question in the field. Fetal and maternal inflammation and immune responses are likely critical players in this process and in the onset of pregnancy complications. New technologies will ideally unveil mechanistic pathways affected by MFCT and may provide targets for therapies to mitigate pregnancy complications. Beyond pregnancy, long-lived microchimerism may have additional consequences for tolerance and immunity in both the mother and her children.

References

1. Ariga H, Ohto H, Busch MP, Imamura S, Watson R, Reed W, et al. Kinetics of fetal cellular and cell-free DNA in the maternal circulation during and after pregnancy: implications for noninvasive prenatal diagnosis. Transfusion. 2001;41(12):1524–30. PubMed.
2. Bianchi DW, Zickwolf GK, Weil GJ, Sylvester S, DeMaria MA. Male fetal progenitor cells persist in maternal blood for as long as 27 years postpartum. Proc Natl Acad Sci U S A. 1996;93(2):705–8. PubMed Pubmed Central PMCID: 40117.
3. Gammill HS, Guthrie KA, Aydelotte TM, Adams Waldorf KM, Nelson JL. Effect of parity on fetal and maternal microchimerism: interaction of grafts within a host? Blood. 2010;116(15):2706–12. PubMed Pubmed Central PMCID: 2974583.
4. Hall JM, Lingenfelter P, Adams SL, Lasser D, Hansen JA, Bean MA. Detection of maternal cells in human umbilical cord blood using fluorescence in situ hybridization. Blood. 1995;86(7):2829–32. PubMed.

5. Maloney S, Smith A, Furst DE, Myerson D, Rupert K, Evans PC, et al. Microchimerism of maternal origin persists into adult life. J Clin Invest. 1999;104(1):41–7. PubMed Pubmed Central PMCID: 408407.
6. Srivatsa B, Srivatsa S, Johnson KL, Bianchi DW. Maternal cell microchimerism in newborn tissues. J Pediatr. 2003;142(1):31–5. PubMed.
7. Chen CP, Lee MY, Huang JP, Aplin JD, Wu YH, Hu CS, et al. Trafficking of multipotent mesenchymal stromal cells from maternal circulation through the placenta involves vascular endothelial growth factor receptor-1 and integrins. Stem Cells. 2008;26(2):550–61. PubMed.
8. Hattori K, Dias S, Heissig B, Hackett NR, Lyden D, Tateno M, et al. Vascular endothelial growth factor and angiopoietin-1 stimulate postnatal hematopoiesis by recruitment of vasculogenic and hematopoietic stem cells. J Exp Med. 2001;193(9):1005–14. PubMed Pubmed Central PMCID: 2193424.
9. Galazios G, Papazoglou D, Giagloglou K, Vassaras G, Koutlaki N, Maltezos E. Umbilical cord serum vascular endothelial growth factor (VEGF) levels in normal pregnancies and in pregnancies complicated by preterm delivery or pre-eclampsia. Int J Gynaecol Obstet. 2004;85(1):6–11. PubMed.
10. Nijagal A, Wegorzewska M, Jarvis E, Le T, Tang Q, MacKenzie TC. Maternal T cells limit engraftment after in utero hematopoietic cell transplantation in mice. J Clin Invest. 2011;121(2):582–92. PubMed Pubmed Central PMCID: 3026737.
11. Nancy P, Tagliani E, Tay CS, Asp P, Levy DE, Erlebacher A. Chemokine gene silencing in decidual stromal cells limits T cell access to the maternal-fetal interface. Science. 2012;336(6086):1317–21. PubMed Pubmed Central PMCID: 3727649.
12. O'Donoghue K, Chan J, de la Fuente J, Kennea N, Sandison A, Anderson JR, et al. Microchimerism in female bone marrow and bone decades after fetal mesenchymal stem-cell trafficking in pregnancy. Lancet. 2004;364(9429):179–82. PubMed.
13. Koopmans M, Kremer Hovinga IC, Baelde HJ, Fernandes RJ, de Heer E, Bruijn JA, et al. Chimerism in kidneys, livers and hearts of normal women: implications for transplantation studies. Am J Transplant. 2005;5(6):1495–502. PubMed.
14. Lo YM, Rainer TH, Chan LY, Hjelm NM, Cocks RA. Plasma DNA as a prognostic marker in trauma patients. Clin Chem. 2000;46(3):319–23. PubMed.
15. Lo YM, Tein MS, Lau TK, Haines CJ, Leung TN, Poon PM, et al. Quantitative analysis of fetal DNA in maternal plasma and serum: implications for noninvasive prenatal diagnosis. Am J Hum Genet. 1998;62(4):768–75. PubMed Pubmed Central PMCID: 1377040.
16. Lapaire O, Holzgreve W, Oosterwijk JC, Brinkhaus R, Bianchi DW. Georg Schmorl on trophoblasts in the maternal circulation. Placenta. 2007;28(1):1–5. PubMed.
17. Desai RG, Creger WP. Maternofetal passage of leukocytes and platelets in man. Blood. 1963;21:665–73. PubMed.
18. Al-Yatama MK, Mustafa AS, Ali S, Abraham S, Khan Z, Khaja N. Detection of Y chromosome-specific DNA in the plasma and urine of pregnant women using nested polymerase chain reaction. Prenat Diagn. 2001;21(5):399–402. PubMed.
19. Farina A, LeShane ES, Romero R, Gomez R, Chaiworapongsa T, Rizzo N, et al. High levels of fetal cell-free DNA in maternal serum: a risk factor for spontaneous preterm delivery. Am J Obstet Gynecol. 2005;193(2):421–5. PubMed.
20. Traeger-Synodinos J. Real-time PCR, for prenatal and preimplantation genetic diagnosis of monogenic diseases. Mol Aspects Med. 2006;27(2–3):176–91. PubMed.
21. Lo YM, Corbetta N, Chamberlain PF, Rai V, Sargent IL, Redman CW, et al. Presence of fetal DNA in maternal plasma and serum. Lancet. 1997;350(9076):485–7. PubMed.
22. Alberry M, Maddocks D, Jones M, Abdel Hadi M, Abdel-Fattah S, Avent N, et al. Free fetal DNA in maternal plasma in anembryonic pregnancies: confirmation that the origin is the trophoblast. Prenat Diagn. 2007;27(5):415–8. PubMed.
23. Gupta AK, Holzgreve W, Huppertz B, Malek A, Schneider H, Hahn S. Detection of fetal DNA and RNA in placenta-derived syncytiotrophoblast microparticles generated in vitro. Clin Chem. 2004;50(11):2187–90. PubMed.

24. Lee TH, Chafets DM, Reed W, Wen L, Yang Y, Chen J, et al. Enhanced ascertainment of microchimerism with real-time quantitative polymerase chain reaction amplification of insertion-deletion polymorphisms. Transfusion. 2006;46(11):1870–8. PubMed.
25. Saadai P, Lee TH, Bautista G, Gonzales KD, Nijagal A, Busch MP, et al. Alterations in maternal-fetal cellular trafficking after fetal surgery. J Pediatr Surg. 2012;47(6):1089–94. PubMed Pubmed Central PMCID: 3377979.
26. American College of Obstetricians and Gynecologists. ACOG Practice Bulletin No. 88, December 2007. Invasive prenatal testing for aneuploidy. Obstet Gynecol. 2007;110(6):1459–67. PubMed
27. Alfirevic Z, Sundberg K, Brigham S. Amniocentesis and chorionic villus sampling for prenatal diagnosis. Cochrane Database Syst Rev. 2003;3, CD003252. PubMed.
28. Scheffer PG, van der Schoot CE, Page-Christiaens GC, de Haas M. Noninvasive fetal blood group genotyping of rhesus D, c, E and of K in alloimmunised pregnant women: evaluation of a 7-year clinical experience. BJOG. 2011;118(11):1340–8. PubMed.
29. Mold JE, Michaelsson J, Burt TD, Muench MO, Beckerman KP, Busch MP, et al. Maternal alloantigens promote the development of tolerogenic fetal regulatory T cells in utero. Science. 2008;322(5907):1562–5. PubMed Pubmed Central PMCID: 2648820.
30. Krabchi K, Gros-Louis F, Yan J, Bronsard M, Masse J, Forest JC, et al. Quantification of all fetal nucleated cells in maternal blood between the 18th and 22nd weeks of pregnancy using molecular cytogenetic techniques. Clin Genet. 2001;60(2):145–50. PubMed.
31. Nijagal A, Fleck S, Hills NK, Feng S, Tang Q, Kang SM, et al. Decreased risk of graft failure with maternal liver transplantation in patients with biliary atresia. Am J Transplant. 2012;12(2):409–19. PubMed.
32. Stern M, Ruggeri L, Mancusi A, Bernardo ME, de Angelis C, Bucher C, et al. Survival after T cell-depleted haploidentical stem cell transplantation is improved using the mother as donor. Blood. 2008;112(7):2990–5. PubMed Pubmed Central PMCID: 2962448.
33. van Rood JJ, Loberiza Jr FR, Zhang MJ, Oudshoorn M, Claas F, Cairo MS, et al. Effect of tolerance to noninherited maternal antigens on the occurrence of graft-versus-host disease after bone marrow transplantation from a parent or an HLA-haploidentical sibling. Blood. 2002;99(5):1572–7. PubMed.
34. Nijagal A, MacKenzie TC. Clinical implications of maternal-fetal cellular trafficking. Semin Pediatr Surg. 2013;22(1):62–5. PubMed.
35. Kiefer AS, Lang TR, Hein MS, McNallan KT, Moir CR, Reed AM. Maternal microchimerism in Hirschsprung's disease. Am J Perinatol. 2012;29(2):71–8. PubMed.
36. Nelson JL, Furst DE, Maloney S, Gooley T, Evans PC, Smith A, et al. Microchimerism and HLA-compatible relationships of pregnancy in scleroderma. Lancet. 1998;351(9102):559–62. PubMed.
37. Scaletti C, Vultaggio A, Bonifacio S, Emmi L, Torricelli F, Maggi E, et al. Th2-oriented profile of male offspring T cells present in women with systemic sclerosis and reactive with maternal major histocompatibility complex antigens. Arthritis Rheum. 2002;46(2):445–50. PubMed.
38. Gammill HS, Nelson JL. Naturally acquired microchimerism. Int J Dev Biol. 2010;54(2–3):531–43. PubMed Pubmed Central PMCID: 2887685.
39. Al-Mufti R, Lees C, Albaiges G, Hambley H, Nicolaides KH. Fetal cells in maternal blood of pregnancies with severe fetal growth restriction. Hum Reprod. 2000;15(1):218–21. PubMed.
40. Gravett MG, Witkin SS, Haluska GJ, Edwards JL, Cook MJ, Novy MJ. An experimental model for intraamniotic infection and preterm labor in rhesus monkeys. Am J Obstet Gynecol. 1994;171(6):1660–7. PubMed.
41. Hamilton SA, Tower CL, Jones RL. Identification of chemokines associated with the recruitment of decidual leukocytes in human labour: potential novel targets for preterm labour. PLoS One. 2013;8(2):e56946. PubMed Pubmed Central PMCID: 3579936.
42. Jakobsen TR, Clausen FB, Rode L, Dziegiel MH, Tabor A. High levels of fetal DNA are associated with increased risk of spontaneous preterm delivery. Prenat Diagn. 2012;32(9):840–5. PubMed.

43. Lee J, Romero R, Xu Y, Kim JS, Topping V, Yoo W, et al. A signature of maternal anti-fetal rejection in spontaneous preterm birth: chronic chorioamnionitis, anti-human leukocyte antigen antibodies, and C4d. PLoS One. 2011;6(2):e16806. PubMed Pubmed Central PMCID: 3033909.
44. Leung TN, Zhang J, Lau TK, Hjelm NM, Lo YM. Maternal plasma fetal DNA as a marker for preterm labour. Lancet. 1998;352(9144):1904–5. PubMed.
45. Romero R, Avila C, Santhanam U, Sehgal PB. Amniotic fluid interleukin 6 in preterm labor. Association with infection. J Clin Invest. 1990;85(5):1392–400. PubMed Pubmed Central PMCID: 296584.
46. Romero R, Mazor M, Munoz H, Gomez R, Galasso M, Sherer DM. The preterm labor syndrome. Ann N Y Acad Sci. 1994;734:414–29. PubMed.
47. Roberts JM, Redman CW. Pre-eclampsia: more than pregnancy-induced hypertension. Lancet. 1993;341(8858):1447–51. PubMed.
48. Brown MA, Lindheimer MD, de Swiet M, Van Assche A, Moutquin JM. The classification and diagnosis of the hypertensive disorders of pregnancy: statement from the International Society for the Study of Hypertension in Pregnancy (ISSHP). Hypertens Pregnancy. 2001;20(1):IX–XIV. PubMed.
49. Hauth JC, Ewell MG, Levine RJ, Esterlitz JR, Sibai B, Curet LB, et al. Pregnancy outcomes in healthy nulliparas who developed hypertension. Calcium for Preeclampsia Prevention Study Group. Obstet Gynecol. 2000;95(1):24–8. PubMed.
50. Knuist M, Bonsel GJ, Zondervan HA, Treffers PE. Intensification of fetal and maternal surveillance in pregnant women with hypertensive disorders. Int J Gynaecol Obstet. 1998;61(2):127–33. PubMed.
51. Hnat MD, Sibai BM, Caritis S, Hauth J, Lindheimer MD, MacPherson C, et al. Perinatal outcome in women with recurrent preeclampsia compared with women who develop preeclampsia as nulliparas. Am J Obstet Gynecol. 2002;186(3):422–6. PubMed.
52. Lim KH, Zhou Y, Janatpour M, McMaster M, Bass K, Chun SH, et al. Human cytotrophoblast differentiation/invasion is abnormal in pre-eclampsia. Am J Pathol. 1997;151(6):1809–18. PubMed Pubmed Central PMCID: 1858365.
53. Farina A, Sekizawa A, Sugito Y, Iwasaki M, Jimbo M, Saito H, et al. Fetal DNA in maternal plasma as a screening variable for preeclampsia. A preliminary nonparametric analysis of detection rate in low-risk nonsymptomatic patients. Prenat Diagn. 2004;24(2):83–6. PubMed.
54. Hubel CA. Oxidative stress in the pathogenesis of preeclampsia. Proc Soc Exp Biol Med Soc Exp Biol Med. 1999;222(3):222–35. PubMed.
55. Redman CW, Sargent IL. Placental debris, oxidative stress and pre-eclampsia. Placenta. 2000;21(7):597–602. PubMed.
56. Roberts JM, Hubel CA. Oxidative stress in preeclampsia. Am J Obstet Gynecol. 2004;190(5):1177–8. PubMed.
57. Sheppard BL, Bonnar J. An ultrastructural study of utero-placental spiral arteries in hypertensive and normotensive pregnancy and fetal growth retardation. Br J Obstet Gynaecol. 1981;88(7):695–705. PubMed.
58. Hahn S, Huppertz B, Holzgreve W. Fetal cells and cell free fetal nucleic acids in maternal blood: new tools to study abnormal placentation? Placenta. 2005;26(7):515–26. PubMed.
59. Tjoa ML, Cindrova-Davies T, Spasic-Boskovic O, Bianchi DW, Burton GJ. Trophoblastic oxidative stress and the release of cell-free feto-placental DNA. Am J Pathol. 2006;169(2):400–4. PubMed Pubmed Central PMCID: 1698796.
60. Cotter AM, Martin CM, O'Leary JJ, Daly SF. Increased fetal DNA in the maternal circulation in early pregnancy is associated with an increased risk of preeclampsia. Am J Obstet Gynecol. 2004;191(2):515–20. PubMed.
61. Hahn S, Holzgreve W. Fetal cells and cell-free fetal DNA in maternal blood: new insights into pre-eclampsia. Hum Reprod Update. 2002;8(6):501–8. PubMed.
62. Leung TN, Zhang J, Lau TK, Chan LY, Lo YM. Increased maternal plasma fetal DNA concentrations in women who eventually develop preeclampsia. Clin Chem. 2001;47(1):137–9. PubMed.

63. Levine RJ, Qian C, Leshane ES, Yu KF, England LJ, Schisterman EF, et al. Two-stage elevation of cell-free fetal DNA in maternal sera before onset of preeclampsia. Am J Obstet Gynecol. 2004;190(3):707–13. PubMed.
64. Lo YM, Leung TN, Tein MS, Sargent IL, Zhang J, Lau TK, et al. Quantitative abnormalities of fetal DNA in maternal serum in preeclampsia. Clin Chem. 1999;45(2):184–8. PubMed.
65. Zhong XY, Holzgreve W, Hahn S. The levels of circulatory cell free fetal DNA in maternal plasma are elevated prior to the onset of preeclampsia. Hypertens Pregnancy. 2002;21(1):77–83. PubMed.
66. Zhong XY, Laivuori H, Livingston JC, Ylikorkala O, Sibai BM, Holzgreve W, et al. Elevation of both maternal and fetal extracellular circulating deoxyribonucleic acid concentrations in the plasma of pregnant women with preeclampsia. Am J Obstet Gynecol. 2001;184(3):414–9. PubMed.
67. Chua S, Wilkins T, Sargent I, Redman C. Trophoblast deportation in pre-eclamptic pregnancy. Br J Obstet Gynaecol. 1991;98(10):973–9. PubMed.
68. Holzgreve W, Ghezzi F, Di Naro E, Ganshirt D, Maymon E, Hahn S. Disturbed feto-maternal cell traffic in preeclampsia. Obstet Gynecol. 1998;91(5 Pt 1):669–72. PubMed.
69. Illanes S, Parra M, Serra R, Pino K, Figueroa-Diesel H, Romero C, et al. Increased free fetal DNA levels in early pregnancy plasma of women who subsequently develop preeclampsia and intrauterine growth restriction. Prenat Diagn. 2009;29(12):1118–22. PubMed.
70. Stein W, Muller S, Gutensohn K, Emons G, Legler T. Cell-free fetal DNA and adverse outcome in low risk pregnancies. Eur J Obstet Gynecol Reprod Biol. 2013;166(1):10–3. PubMed.
71. Huppertz B, Kadyrov M, Kingdom JC. Apoptosis and its role in the trophoblast. Am J Obstet Gynecol. 2006;195(1):29–39. PubMed.
72. Smith SC, Baker PN, Symonds EM. Increased placental apoptosis in intrauterine growth restriction. Am J Obstet Gynecol. 1997;177(6):1395–401. PubMed.
73. Smid M, Vassallo A, Lagona F, Valsecchi L, Maniscalco L, Danti L, et al. Quantitative analysis of fetal DNA in maternal plasma in pathological conditions associated with placental abnormalities. Ann N Y Acad Sci. 2001;945:132–7. PubMed.
74. Sekizawa A, Jimbo M, Saito H, Iwasaki M, Matsuoka R, Okai T, et al. Cell-free fetal DNA in the plasma of pregnant women with severe fetal growth restriction. Am J Obstet Gynecol. 2003;188(2):480–4. PubMed.
75. Crowley A, Martin C, Fitzpatrick P, Sheils O, O'Herlihy C, O'Leary JJ, et al. Free fetal DNA is not increased before 20 weeks in intrauterine growth restriction or pre-eclampsia. Prenat Diagn. 2007;27(2):174–9. PubMed.
76. Goldenberg RL, Culhane JF, Iams JD, Romero R. Epidemiology and causes of preterm birth. Lancet. 2008;371(9606):75–84. PubMed.
77. Gomez R, Ghezzi F, Romero R, Munoz H, Tolosa JE, Rojas I. Premature labor and intra-amniotic infection. Clinical aspects and role of the cytokines in diagnosis and pathophysiology. Clin Perinatol. 1995;22(2):281–342. PubMed.
78. Brar HS, Medearis AL, De Vore GR, Platt LD. Maternal and fetal blood flow velocity waveforms in patients with preterm labor: relationship to outcome. Am J Obstet Gynecol. 1989;161(6 Pt 1):1519–22. PubMed.
79. Strigini FA, Lencioni G, De Luca G, Lombardo M, Bianchi F, Genazzani AR. Uterine artery velocimetry and spontaneous preterm delivery. Obstet Gynecol. 1995;85(3):374–7. PubMed.
80. Joseph KS, Allen AC, Dodds L, Vincer MJ, Armson BA. Causes and consequences of recent increases in preterm birth among twins. Obstet Gynecol. 2001;98(1):57–64. PubMed.
81. Hassan SS, Romero R, Berry SM, Dang K, Blackwell SC, Treadwell MC, et al. Patients with an ultrasonographic cervical length < or =15 mm have nearly a 50 % risk of early spontaneous preterm delivery. Am J Obstet Gynecol. 2000;182(6):1458–67. PubMed.
82. Andersen HF, Nugent CE, Wanty SD, Hayashi RH. Prediction of risk for preterm delivery by ultrasonographic measurement of cervical length. Am J Obstet Gynecol. 1990;163(3):859–67. PubMed.
83. Lockwood CJ, Kuczynski E. Risk stratification and pathological mechanisms in preterm delivery. Paediatr Perinat Epidemiol. 2001;15 Suppl 2:78–89. PubMed.

84. Lee J, Romero R, Xu Y, Miranda J, Yoo W, Chaemsaithong P, et al. Detection of anti-HLA antibodies in maternal blood in the second trimester to identify patients at risk of antibody-mediated maternal anti-fetal rejection and spontaneous preterm delivery. Am J Reprod Immunol. 2013;70(2):162–75. PubMed.
85. Wegorzewska M, Le T, Tang Q, MacKenzie TC. Increased maternal microchimerism in the allogeneic fetus during LPS-induced preterm labor in mice. Chimerism. Chimerism 2015 Mar 16:1–6.
86. Wegorzewska M, Nijagal A, Wong CM, Le T, Lescano N, Tang Q, et al. Fetal intervention increases maternal T cell awareness of the foreign conceptus and can lead to immune-mediated fetal demise. J Immunol. 2014;192(4):1938–45. PubMed.
87. Adzick NS, Thom EA, Spong CY, Brock 3rd JW, Burrows PK, Johnson MP, et al. A randomized trial of prenatal versus postnatal repair of myelomeningocele. N Engl J Med. 2011;364(11):993–1004. PubMed Pubmed Central PMCID: 3770179.
88. Harrison MR, Keller RL, Hawgood SB, Kitterman JA, Sandberg PL, Farmer DL, et al. A randomized trial of fetal endoscopic tracheal occlusion for severe fetal congenital diaphragmatic hernia. N Engl J Med. 2003;349(20):1916–24. PubMed.
89. Rossi AC, D'Addario V. Laser therapy and serial amnioreduction as treatment for twin-twin transfusion syndrome: a metaanalysis and review of literature. Am J Obstet Gynecol. 2008;198(2):147–52. PubMed.
90. Ruano R, Yoshisaki CT, da Silva MM, Ceccon ME, Grasi MS, Tannuri U, et al. A randomized controlled trial of fetal endoscopic tracheal occlusion versus postnatal management of severe isolated congenital diaphragmatic hernia. Ultrasound Obstet Gynecol. 2012;39(1):20–7. PubMed.
91. Harrison MR, Sydorak RM, Farrell JA, Kitterman JA, Filly RA, Albanese CT. Fetoscopic temporary tracheal occlusion for congenital diaphragmatic hernia: prelude to a randomized, controlled trial. J Pediatr Surg. 2003;38(7):1012–20. PubMed.
92. Deprest J, Jani J, Gratacos E, Vandecruys H, Naulaers G, Delgado J, et al. Fetal intervention for congenital diaphragmatic hernia: the European experience. Semin Perinatol. 2005;29(2):94–103. PubMed.
93. Tjoa ML, Jani J, Lewi L, Peter I, Wataganara T, Johnson KL, et al. Circulating cell-free fetal messenger RNA levels after fetoscopic interventions of complicated pregnancies. Am J Obstet Gynecol. 2006;195(1):230–5. PubMed.
94. Wataganara T, Gratacos E, Jani J, Becker J, Lewi L, Sullivan LM, et al. Persistent elevation of cell-free fetal DNA levels in maternal plasma after selective laser coagulation of chorionic plate anastomoses in severe midgestational twin-twin transfusion syndrome. Am J Obstet Gynecol. 2005;192(2):604–9. PubMed.
95. Bianchi DW, Williams JM, Sullivan LM, Hanson FW, Klinger KW, Shuber AP. PCR quantitation of fetal cells in maternal blood in normal and aneuploid pregnancies. Am J Hum Genet. 1997;61(4):822–9. PubMed Pubmed Central PMCID: 1715976.
96. Rava RP, Srinivasan A, Sehnert AJ, Bianchi DW. Circulating fetal cell-free DNA fractions differ in autosomal aneuploidies and monosomy x. Clin Chem. 2014;60(1):243–50. PubMed.
97. Fleck S, Bautista G, Keating SM, Lee TH, Keller RL, Moon-Grady AJ, et al. Fetal production of growth factors and inflammatory mediators predicts pulmonary hypertension in congenital diaphragmatic hernia. Pediatr Res. 2013;74(3):290–8. PubMed.

Chapter 3
Paracrine Effects of Fetal Stem Cells

Mariusz Z. Ratajczak, Gabriela Schneider, and Janina Ratajczak

Introduction

Both prokaryotic and eukaryotic cells communicate and exchange information by employing various cell–cell contact mechanisms. In addition to a crosstalk mediated by adhesion molecules (e.g., integrins) and their corresponding ligands expressed on interacting cells, an important role in cell–cell communication play paracrine signals that involve secretion of soluble and non-soluble factors [1–6]. Accordingly, cells secrete several soluble factors including (1) peptide-based growth factors, cytokines, chemokines and enzymes that direct cell responses and modify surrounding microenvironment (e.g., metalloproteinases, enzymes processing extracellular ligands), (2) bioactive lipids (e.g., sphingosine-1-phosphate, creamide-1 phosphate, lysophosphatidic acid, eicosanoids) and (3) extracellular nucleotides (e.g., ATP, UTP) [1–6]. All these soluble paracrine factors play an important role in interaction between cells. In parallel, growing attention is recently focused on cell-to-cell communication that involves paracrine effects of cell-derived spherical membrane fragments called extracellular microvesicles (ExMVs), a mechanism that for many years has been largely ignored and overlooked [7–12].

Accordingly, both soluble factors as well as non-soluble ExMVs if released from the cells employed as cellular therapeutics in regenerative medicine seem to play an important role in improving the function of damaged organs [10, 13]. A growing

M.Z. Ratajczak, M.D., Ph.D., D.Sci. (✉) • G. Schneider, Ph.D., • J. Ratajczak, M.D., Ph.D., D.Sci.
Department of Medicine, James Graham Brown Cancer Center, University of Louisville,
500 S. Floyd Street, Louisville, KY 40202, USA
e-mail: mzrata01@louisville.edu; j0rata01@louisville.edu

D.O. Fauza, M. Bani (eds.), *Fetal Stem Cells in Regenerative Medicine*,
Stem Cell Biology and Regenerative Medicine,
DOI 10.1007/978-1-4939-3483-6_3

body of evidence indicate that soluble factors and ExMVs secreted from hematopoietic stem cells (HSCs), hematopoietic progenitor cells (HPCs), multipotent stroma cells (MSCs), or cardiac stem cells (CSCs) employed in various treatment strategies in regenerative medicine may (1) inhibit apoptosis of cells residing in the damaged tissues, (2) stimulate proliferation of cells that survived organ injury, and (3) stimulate vascularization of affected tissues [7–13].

More importantly, evidence accumulates that some of the beneficial therapeutic effects reported after application of intact cells (e.g., MSCs) could be achieved by using just ExMVs derived from these cells [13]. These pro-regenerative effects mediated by ExMVs are explained by the fact that these small, spherical membrane fragments (1) are enriched in bioactive lipids (e.g., sphingosine-1-phosphate), (2) may express anti-apoptoic and pro-stimulatory growth factors or cytokines (e.g., vascular endothelial growth factor [VEGF], stem cell factor [SCF], or stromal derived factor-1 [SDF-1]) on their surface, and (3) may deliver mRNA, regulatory miRNA, and proteins to the damaged tissues that improve overall cell function. Based on these observations, as mentioned above potential use of ExMVs, instead of whole cells, has become an exciting new concept in regenerative medicine [1, 13]. We will address this issue later on in our chapter in context of new possibilities of therapeutic application of embryonic stem cells (ESCs).

ESCs are pluripotent and as such cells may differentiate into cells belonging to three germ layers (meso-, ecto- and endoderm). In the last 25 years a significant effort has been involved to harness a potential of these cells in regenerative medicine. However, the major problem with clinical application of ESCs, is the risk that they will grow teratomas after injection into host [1]. Until this problem will be solved, ESCs remain merely an interesting object to study various mechanisms related to developmental biology and embryogenesis.

However, despite this limitation at the current point it would be possible to explore and harness paracrine effects of these cells in regenerative medicine and in this chapter we will discuss this intriguing possibility.

Regenerative Medicine Is Searching for Effective and Safe Pluripotent/Multipotent Stem Cells

The field of regenerative medicine is searching for a source of stem cells that can be safely and efficiently employed for regeneration of damaged organs (e.g., heart, liver, kidney, or neural tissue) [1]. In experimental animal models of organ damage (e.g., heart infarct, liver damage, ischemic kidney failure, or stroke) various types of stem cells isolated from adult tissues have been employed, including, as mentioned above, HSPCs, MSCs, and CSCs [1, 14–16]. Similar types of cells are employed in the clinic to treat patients to improve the function of damaged organs [1, 17–20].

Interestingly, while some beneficial effects have been reported following cell-based therapies, there is no solid evidence that the cells employed to regenerate damaged tissues truly give rise to organ-specific cell populations (e.g., new functional cardiomyocytes in heart, hepatocytes in liver, or tubular epithelium cells in kidney). Therefore, the concept that tissue committed stem cells, such as for example HSCs, are plastic and may trans-differentiate into cells from different germ layers (e.g., cardiomyocytes, neural cells, or hepatocytes) lacks solid experimental support [1, 14, 21–24]. Thus, this initially tempting concept of stem cell plasticity or stem cell trans-dedifferentiation has been challenged by several investigators and some of observed positive effects of stem cell therapy have been explained by other alternative mechanisms that will be shortly discussed below.

First, it is possible that some of the stem cell plasticity data could be explained simply by the phenomenon of cell fusion [14, 25]. Accordingly, the cells observed in damaged tissues that express markers of both the donor cells employed in treatment (e.g., HSPCs) and cells typical of the damaged organ (e.g., cardiomyocytes in heart damaged by infarct), could be heterokaryons, the result of the fusion of therapeutic cells with somatic host cells in the damaged organ. However, as it is today widely accepted cell fusion is an extremely rare phenomenon [14, 25].

Next, cells employed for therapy in regenerative medicine (e.g., mononuclear cells isolated from bone marrow, umbilical cord blood or mobilized peripheral blood) may, from the beginning, contain heterogeneous populations of stem cells. It is known that cells from the hematopoietic tissues for example are enriched in several types of non-hematopoietic stem/progenitor cells, including MSCs, endothelial progenitor cells (EPCs), as well as a population of pluripotent very small embryonic-like stem cells (VSELs) [1, 26].

Finally, what will be further discussed, some of the benefits observed in organ and tissue regeneration after infusion of therapeutic cells, could be explained by paracrine effects. It is well known that stem cells, for example, ESCs, HSPCs or MSCs, are a source of several trophic soluble and non-soluble (ExMVs) factors and that all these factors, if released from these cells, could inhibit apoptosis of damaged cells, promote tissue repair and vascularization [1, 27].

Stem Cells as a Source of Soluble Paracrine Factors

Several types of cells and in particular stem cells secrete several soluble factors and the repertoire of such anti-apoptotic, proliferation-stimulating, and pro-angiopoietic factors varies with the stem cell type to be employed for treatment. Many years ago, while studying stem cell-derived paracrine mechanisms, we demonstrated that purified normal human bone marrow (BM)- and mobilized peripheral blood (mPB)-derived $CD34^+$ HSPCs express mRNA for various growth factors, cytokines and chemokines. More importantly, we confirmed the expression of several of these factors in conditioned media using ELISA [1, 2]. Accordingly, we found mRNA transcripts for numerous growth factors (SCF, FLT3 ligand, FGF-2, VEGF, HGF, IGF-1, and TPO), cytokines (TNF-α, Fas-L, INF-α, IL-1, and IL-16), and chemokines (MIP-1α, MIP-1β, RANTES, MCP-2, MCP-3, MCP-4, IL-8, IP-10, MCD, and PF-4) and more importantly confirmed by ELISA the presence of VEGF, HGF,

FGF-2, SCF, FLT3 ligand, TPO, IL-16, IGF-1, TGF-β1, TGF- β2, RANTES, MIP-1α, MIP-1β, IL-8, and PF-4 proteins in media conditioned by these cells. Subsequently, in experimental settings *in vitro* we demonstrated that media conditioned by $CD34^+$ cells may inhibit apoptosis, stimulate proliferation, and chemoattract several types of cells, including endothelial cells [1, 2]. We have recently reported similar observations for human $CD133^+$ cells that are enriched for several types of stem cells including HSPCs, VSELs and endothelial progenitor cells (EPCs) [28]. To support this further, another group has shown that murine adipose tissue stem cells (ASCs)-derived conditioned media regenerate lung tissue microvascular injury, and had a similar therapeutic effect as intact ASCs [29].

Stem Cells as a Source of Non-soluble Paracrine Signals: A Role of Extracellular Microvesicles (ExMVs)

Both prokaryotic and eukaryotic cells communicate and exchange information by secreting ExMVs [6–13], a mechanism that for many years has been largely overlooked. Mounting evidence demonstrates also that ExMVs can even replace intact cells to improve the function of damaged organs in several tissue injury models [1, 9, 10].

ExMVs are shed from the cell surface of normal healthy or damaged cells during membrane blebbing and "hijack" both membrane components and the engulfed cytoplasmic contents [7–12]. Shedding of membrane-derived ExMVs is a physiological phenomenon that accompanies cell activation and growth. Interestingly, rapidly proliferating cells tend to secrete more ExMVs than slowly growing ones. This explains why for example ESCs are a rich source of these small circular membrane fragments. In parallel another source of ExMVs is the intracellular endosomal membrane compartment. These particular ExMVs, termed exosomes, are usually released from cells as secretory granules during the process of exocytosis [12, 30]. While ExMVs released from the surface membranes during membrane blebbing are relatively large (0.1–1 μm), exosomes are much smaller (30–100 nm) and appear more homogeneous in size. Overall, in conditioned media harvested from the cells, both types of ExMVs are always simultaneously present.

Ironically, for many years ExMVs have been largely overlooked, and regarded as apoptotic bodies or cell debris. Today, it is already acknowledged that ExMVs are secreted or shed by healthy and not dying cells, and are different than apoptotic bodies released from dying cells. ExMVs, as mentioned above, not only contain numerous proteins and lipids similar to those present in the membranes of the cells from which they originate, but since ExMV membranes engulf some cytoplasm during their generation bymembrane blebbing, they may also contain intracellular proteins, mRNA, and regulatory miRNA [1, 7–12]. In this transfer of mRNA or proteins, ExMVs behave as a naturally engineered "liposomes." Since cells under steady-state conditions tend to store mRNA and miRNA for later utilization under stress conditions, explains why they can release these molecules into the extracellular space "encapsulated" within MVs.

Overall, this ExMVs-mediated communication between cells developed very early in the course of eukaryocytic evolution, before soluble mediators-specific receptor signaling axes emerged.

Fetal Cells and Embryonic Stem Cells (ESCs) as a Rich Source of Soluble and Non-soluble Paracrine Factors: A Novel Tool to Expand *Ex Vivo* Adult Stem Cells

Fetal stem cells-derived ExMVs known as argosomes have been described to play an important role in embryogenesis as source of morphogens that are expressed on their surface (e.g., Hedgehog, Notch and Wingless) and are involved in tissue patterning and organ development [31]. These morphogens are released from producing fetal cells and distributed through adjacent tissue. However, some of them associate tightly with the cell membrane of argosomes and are dispersed over large distances through the developing tissues. Thus, the properties of argosomes are consistent with their being a vehicle for the spread of Wingless protein [31].

Since the maintenance of pluripotency and undifferentiated propagation of ESCs in *in vitro* cultures requires tight cell to cell contacts and effective intercellular signaling, we hypothesized that these cells secrete several paracrine signals to maintain their integrity and in particular we focused on ESCs derived ExMVs. Furthermore, it had been demonstrated that mature somatic cells co-cultured with intact ESCs or extracts from these cells undergo epigenetic changes [9, 13], however a mechanism involved in this phenomenon was not clearly explained when initially described.

We have hypothesized that these effects could be explained by a biological modification of the target cells via ESCs-derived ExMVs [27] and that ExMVs will express stem cell-specific molecules that may support self-renewal and expansion of adult cells. Intrigued by these observations, we investigated whether ESC-derived ExMVs could enter HPSCs as a kind of physiological "liposomes" and increase their pluripotency after delivering ESCs-derived mRNA. To address this hypothesis, we employed expansion of murine and human HPSCs as a model. We found that ExMVs isolated from murine ESCs (ES-D3) and human ESCs (CCTL14) in serum-free cultures significantly (1) enhanced survival and improved expansion of murine HSPCs, (2) upregulated the expression of early pluripotent markers (Oct-4, Nanog and Rex-1) and early hematopoietic stem cell (Scl, HoxB4 and GATA 2) markers in target cells. These effects were paralleled by ExMVs mediated phosphorylation of MAPK p42/44 and serine-threonine kinase AKT in expanded cells. The biological effects of ESCs-derived ExMVs were inhibited after heat inactivation or pretreatment with RNAse, indicating a major involvement of protein and mRNA components of ESCs-derived ExMVs in the observed phenomena [27].

Of note, in these experiments we reported for the first time a mechanism of horizontal transfer of mRNA between cells because ExMVs transferred mRNA after transfer to target cells has been translated into the corresponding proteins [27]. We also found that both murine ES-D3 cell-derived ExMVs and human CCTL14 cell-derived ExMVs are highly enriched in Wnt-3 protein and express mRNA for several early pluripotent transcription factors at much higher level as compared to ESCs from which they originated [1, 27]. This selective increase in mRNA content in ESCs-derived ExMVs compared to parental ES cells confirms the presence of a mechanism that enriches ExMVs in mRNA molecules before their shedding from parental cells. Based on this conditioned media harvested from *in vitro* cultured ESCs enriched in ExMVs as well as several soluble factors could be employed as a new tool to expand adult stem cells for application in regenerative medicine. Studies to identify other biologically active components of ESCs-derived ExMV in addition to mRNAs coding several stem cell-specific transcription factors and Wnt-3 protein are in progress.

Towards Development of Engineered ESCs-Derived ExMVs: A Novel Tool to Regenerate Damaged Tissues

Based on the fact that ExMVs have similar beneficial effects in regenerative medicine therapy as the intact cells from which they are derived [10], it would be possible to engineer and modify ExMVs to employ them more efficiently for tissue organ regeneration *in vivo*. Several possibilities for how to make this approach more efficient are shown in Fig. 3.1. For example, ExMVs could be isolated for potential application in regenerative medicine from a large-scale *ex vivo* expansion of cells (e.g., ESC) cultured in appropriate generators. These ExMV-generating ESCs could be genetically modified in order to produce custom-engineered ExMVs more suitable for therapy. For this purpose as ExMVs producing cells lines could be employed non-differentiated ESC or already ESC-derived differentiated cell lines.

First, as depicted in Fig. 3.1a, it should be possible to expand ESC that lack genes encoding histocompatibility antigens. This approach would minimize the possibility of cross-immunization with donor HLA antigens. Second, ExMVs producer ESC lines (e.g., ESCs) could be transduced with genes that overexpress on their surface (1) peptides that protect target cells in damaged organs from apoptosis and stimulate proliferation of residual remaining cell population (e.g., SCF or Notch ligands) or (2) factors that effectively induce angiogenesis (e.g., VEGF, FGF-2, or SDF-1) (Fig. 3.1b). Third, ExMVs producer ESC lines could be enriched for mRNA and regulatory miRNA species that, after delivery to the damaged tissues, promote regeneration (Fig. 3.1c). We speculate that ExMVs derived from ESC cultured in hypoxic conditions would be enriched in mRNAs and miRNAs that promote angiogenesis. Finally, we envision that ExMVs producer ESCs lines could be enriched for molecules that facilitate their tropism to the damaged organ and subsequently promote retention of ExMVs in the damaged tissues (Fig. 3.1d). Taking advantage of epigenetic memory in ESCs-differentiated cells, one can also envision that, for example, ExMVs from ESC differentiated into epidermal cell line would preferentially affect regeneration of damaged skin (e.g., after burns), ExMVs isolated from

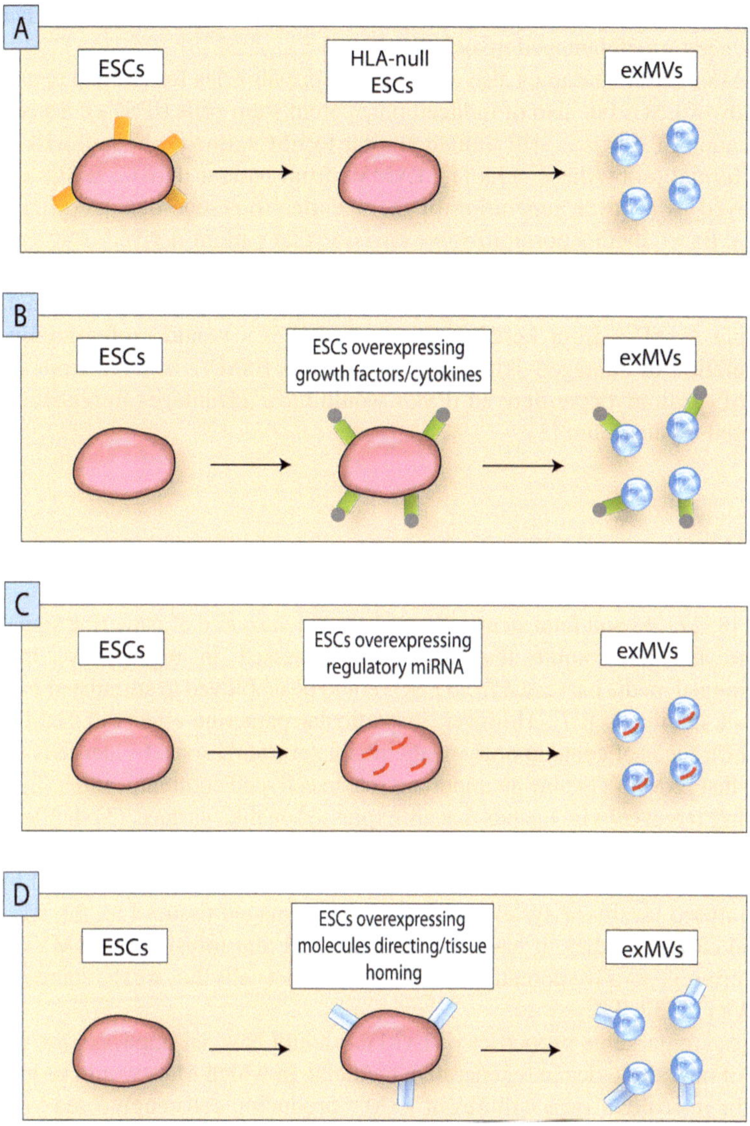

Fig. 3.1 Different approaches to generating from ESCs more efficient pro-regenerative ExMVs *in vitro*. ExMVs could be harvested from *in vitro* cultures of ExMV-producing ESCs lines. Such cell lines may be modified to obtain ExMVs that (1) do not express HLA antigens (panel **a**), (2) are enriched for growth factors, cytokines, and chemokines that promote regeneration of damaged organs (panel **b**), (3) are enriched in mRNA and regulatory miRNA facilitating regeneration of damaged tissues and/or promoting angiogenesis (panel **c**), and (4) express molecules that direct them to, and subsequently be retained in, damaged tissues (panel **d**)

the supernatants of cardiomyocyte-differentiated cell lines would have advantages in regeneration of damaged myocardium.

ExMVs-based therapies also open up new possibilities for clinical applications not only of ESCs but also of induced pluripotent stem cells (iPSCs). Since *in vivo* application of iPSCs is still limited similar to ESCs, due to the high risk of teratoma formation by these cells [1], ExMVs from patient-derived iPSCs could be employed as a novel generation of therapeutics to rescue damaged organs and tissues. Based on this possibility, we envision that patient-derived iPSCs could be employed as ExMV-producing cells. Moreover, similarly as for ESCs epigenetic memory of cells employed for generation of iPSCs, one can also envision that, for example, ExMVs from keratinocyte-derived iPSCs would preferentially affect regeneration of damaged skin (e.g., after burns), ExMVs isolated from supernatants of cardiomyocyte-derived iPSCs would have advantages in regeneration of damaged myocardium [1].

Conclusions

ESCs [8, 27, 28] and fetal stem cells [31] are a rich source of paracrine signals both soluble and non-soluble that could be harnessed in regenerative medicine. Conditioned media harvested from ESCs could be employed to stimulated expansion of adult stem cells [27]. However, in particular paracrine effects of ExMVs have become a focus of contemporary medicine for several reasons. Evidence is accumulating that ExMVs (1) play an important role in cell–cell communication, (2) directly stimulate target cells by ligands that are expressed on their surface, (3) shuttle mRNA, regulatory miRNA, proteins, and organelles between cells [7–12].

Strong evidence is accumulating that ExMVs are abundantly secreted by stem cells infused locally or systemically to rescue damaged tissues [13, 27, 32, 33]. In several elegant studies, it has been demonstrated that infusion of ExMVs has the same pro-regenerative potential as infusion of intact cells that are the source of these ExMVs [13, 32, 33].

These remarkable properties of ExMVs should have an impact in the development of new strategies in regenerative medicine in which MVs would be harvested from large-scale *in vitro* cultures of ExMV-producing cells engineered to overexpress appropriate growth factors, cytokines, surface molecules, mRNA, and miRNA that inhibit apoptosis of target cells and promote neovascularization of damaged tissues. Such custom-engineered "super ExMVs" could become a new class of cell-derived therapeutics in regenerative medicine and pave a new way for clinical application of ESCs and ESCs-derived cell lines as ExMVs producing cells in regenerative medicine.

Acknowledgments This work was supported by NIH grants 2R01 DK074720 and R01HL112788, the Stella and Henry Endowment, and Maestro grant 2011/02/A/NZ4/00035 to MZR.

References

1. Ratajczak MZ, Kucia M, Jadczyk T, Greco NJ, Wojakowski W, Tendera M, et al. Pivotal role of paracrine effects in stem cell therapies in regenerative medicine: can we translate stem cell-secreted paracrine factors and microvesicles into better therapeutic strategies? Leukemia. 2012;6:1166–73.
2. Majka M, Janowska-Wieczorek A, Ratajczak J, Ehrenman K, Pietrzkowski Z, Kowalska MA, et al. Numerous growth factors, cytokines, and chemokines are secreted by human CD34(+) cells, myeloblasts, erythroblasts, and megakaryoblasts and regulate normal hematopoiesis in an autocrine/paracrine manner. Blood. 2001;97:3075–85.
3. Janowska-Wieczorek A, Majka M, Ratajczak J, Ratajczak MZ. Autocrine/paracrine mechanisms in human hematopoiesis. Stem Cells. 2001;19:99–107.
4. Sahoo S, Klychko E, Thorne T, Misener S, Schults KM, Millay M, et al. Exosomesfrom human CD34+ stem cells mediate their proangiopoietic paracrine activity. Circ Res. 2011;109:724–8.
5. Lataillade JJ, Clay D, Bourin P, Hérodin F, Dupuy C, Jasmin C, et al. Stromal cell-derived factor 1 regulates primitive hematopoiesis by suppressing apoptosis and by promoting G(0)/G(1) transition in CD34(+) cells: evidence for an autocrine/paracrine mechanism. Blood. 2002;99:1117–29.
6. Ratajczak J, Wysoczynski M, Hayek F, Janowska-Wieczorek A, Ratajczak MZ. Membrane-derived microvesicles (MV): important and underappreciated mediators of cell to cell communication. Leukemia. 2006;20:1487–95.
7. George JN, Thoi LL, McManus L, Reimann TA. Isolation of human platelet membrane microparticles from plasma and serum. Blood. 1982;60:834–9.
8. Yuan A, Faber EL, Rapoport AL, Tejada D, Deniskin R, Akhmedov NB, et al. Transfer of miRNA by embryonic stem cell microvesicles. PLoS One. 2009;4:e4722.
9. Quesenberry PJ, Dooner MS, Aliotta JM. Stem cell plasticity revisited: the continuum marrow model and phenotypic changes mediated by microvesicles. Exp Hematol. 2010;38:581–92.
10. Camussi G, Deregibus MC, Tetta C. Paracrine/endocrine mechanism of stem cells on kidney repair: role of microvesicle-mediated transfer of genetic information. Curr Opin Nephrol Hypertens. 2010;19:7–12.
11. Beaudoin AR, Grondin G. Shedding of vesicular material from the cell surface of eukaryotic cells: different cellular phenomena. Biochim Biophys Acta. 1991;1071:203–19.
12. Fevrier B, Raposo G. Exosomes: endosomal-derived vesicles shipping extracellular messages. Curr Opin Cell Biol. 2004;16:415–21.
13. Gatti S, Bruno S, Deregibus MC, Sordi A, Cantaluppi V, Tetta C, et al. Microvesicles derived from human adult mesenchymal stem cells protect against ischemia-reperfusion-induced acute and chronic kidney injury. Nephrol Dialysis Transplant. 2011;26:1474–83.
14. Friedman RS, Krause DS. Regeneration and repair: new findings in stem cell research and aging. Ann N Y Acad Sci. 2009;1172:88–94.
15. Joyce N, Annett G, Wirthlin L, Olson S, Bauer G, Nolta JA. Mesenchymal stem cells for the treatment of neurodegenerative disease. Regen Med. 2010;5:933–46.
16. Ratajczak MZ, Zuba-Surma EK, Wysoczynski M, Wan W, Ratajczak J, Wojakowski W, et al. Hunt for pluripotent stem cell – regenerative medicine search for almighty cell. J Autoimmun. 2008;30:151–62.
17. Tang XL, Rokosh DG, Guo Y, Bolli R. Cardiac progenitor cells and bone marrow-derived very small embryonic-like stem cells for cardiac repair after myocardial infarction. Circ J. 2010;74:390–404.
18. Tendera M, Wojakowski W, Ruzyłło W, Chojnowska L, Kepka C, Tracz W, et al. Intracoronary infusion of bone marrow-derived selected CD34+CXCR4+ cells and non-selected mononuclear cells in patients with acute STEMI and reduced left ventricular ejection fraction: results of randomized, multicentre Myocardial Regeneration by Intracoronary Infusion of

Selected Population of Stem Cells in Acute Myocardial Infarction (REGENT) Trial. Eur Heart J. 2009;30:1313–21.
19. Corti S, Locatelli F, Donadoni C, Strazzer S, Salani S, Del Bo R, et al. Neuroectodermal and microglial differentiation of bone marrow cells in the mouse spinal cord and sensory ganglia. J Neurosci Res. 2002;70:721–33.
20. Petersen BE, Bowen WC, Patrene KD, Mars WM, Sullivan AK, Murase N, et al. Bone marrow as a potential source of hepatic oval cells. Science. 1999;284:1168–70.
21. Wagers AJ, Sherwood RI, Christensen JL, Weissman IL. Little evidence for developmental plasticity of adult hematopoietic stem cells. Science. 2002;297:2256–9.
22. Murry CE, Soonpaa MH, Reinecke H, Nakajima H, Nakajima HO, Rubart M, et al. Haematopoietic stem cells do not transdifferentiate into cardiac myocytes in myocardial infarcts. Nature. 2004;428:664–8.
23. Castro RF, Jackson KA, Goodell MA, Robertson CS, Liu H, Shine HD. Failure of bone marrow cells to transdifferentiate into neural cells in vivo. Science. 2002;297:1299.
24. Kucia M, Ratajczak J, Ratajczak MZ. Are bone marrow stem cells plastic or heterogenous – that is the question. Exp Hematol. 2005;33:613–23.
25. O'Malley K, Scott EW. Stem cell fusion confusion. Exp Hematol. 2004;32:131–4.
26. Kucia M, Reca R, Campbell FR, Zuba-Surma E, Majka M, Ratajczak J, et al. A population of very small embryonic-like (VSEL) CXCR4(+)SSEA-1(+)Oct-4+ stem cells identified in adult bone marrow. Leukemia. 2006;20:857–69.
27. Ratajczak J, Miekus K, Kucia M, Zhang J, Reca R, Dvorak P, et al. Embryonic stem cell-derived microvesicles reprogram hematopoietic progenitors: evidence for horizontal transfer of mRNA and protein delivery. Leukemia. 2006;20:847–56.
28. Ratajczak J, Kucia M, Mierzejewska K, Marlicz W, Pietrzkowski Z, Wojakowski W, et al. Paracrine proangiopoietic effects of human umbilical cord blood-derived purified CD133+ cells-implications for stem cell therapies in regenerative medicine. Stem Cells Dev. 2013;22:422–30.
29. Myers TJ, Granero-Molto F, Longobardi L, Li T, Yan Y, Spagnoli A. Mesenchymal stem cells at the intersection of cell and gene therapy. Expert Opin Biol Ther. 2010;10:1663–79.
30. Greenwalt TJ. The how and why of exocytic vesicles. Transfusion. 2006;46:143–52.
31. Greco V, Hannus M, Eaton S. Argosomes: a potential vehicle for the spread of morphogens through epithelia. Cell. 2001;106:633–45.
32. Del Tatto M, Ng T, Aliotta JM, et al. Marrow cell genetic phenotype change induced by human lung cancer cells. Exp Hematol. 2011;39:1072–80.
33. Herrera MB, Fonsato V, Gatti S, et al. Human liver stem cell-derived microvesicles accelerate hepatic regeneration in hepatectomized rats. J Cell Mol Med. 2010;14(6B):1605–18.

Chapter 4
Immunogenicity and Immunomodulation of Fetal Stem Cells

Stephen E. Sherman and David A. Hess

Introduction

In the 1970s, Alexander Friedenstein and colleagues were the first to identify clonogenic, plastic adherent cells from murine bone marrow termed colony forming units of fibroblasts (CFU-F). These cells differentiated into multiple mesodermal cell types including osteocytes, chondrocytes and adipocytes, but did not contribute to hematopoietic tissue [1, 2]. Soon after, similar cells from adult human bone marrow were characterized (reviewed in [3]) and were shown to engraft multiple human tissues (mesenchymal and nonmesenchymal) after systemic infusion. Multiple tissues within the human body, including both fetal and adult sources, have been shown to contain mesodermal precursors capable of differentiating into cartilage, bone, adipose, and muscle tissue [4–6]. In subsequent studies these cells were given many names including marrow stromal stem cells or most accurately multipotent stromal cells (MSC) based on their diverse differentiative potential. In 1991, Arnold Caplan coined the term mesenchymal stem cells to describe these cells, highlighting their putative developmental origin and considerable therapeutic potential [7]. These different nomenclatures have caused confusion and controversy in the field because not all bones are derived from embryonic mesenchymal tissue, and the self-renewal capacity of MSC remains highly disputed [8, 9]. In this chapter, MSC is used

S.E. Sherman, B.Sc.
Department of Physiology and Pharmacology, Western University,
1151 Richmond Street, London, ON, Canada, N6A 5B7

D.A. Hess, Ph.D. (✉)
Krembil Centre for Stem Cell Biology, Robarts Research Institute, London, ON, Canada

Department of Physiology and Pharmacology, University of Western Ontario,
1151 Richmond Street, London, ON, Canada, N6A 3K7
e-mail: dhess@robarts.ca

D.O. Fauza, M. Bani (eds.), *Fetal Stem Cells in Regenerative Medicine*,
Stem Cell Biology and Regenerative Medicine,
DOI 10.1007/978-1-4939-3483-6_4

interchangeably to represent both multipotent stromal cell and mesenchymal stem cell nomenclatures.

In addition to the supportive role of fibroblasts in human tissues, MSC are thought to play an instructive role in a paracrine fashion to aid in the repair and regeneration of organs, and in the modulation of the immune system at sites of tissue injury [10–13]. MSC have been shown to exert their effects via both soluble factors and direct cell–cell contact, demonstrating their ability to ameliorate autoimmune disease conditions through anti-inflammatory mediators and to support regenerative processes [14–16]. However, MSC from various adult sources, including human bone marrow, have been shown to possess considerable variability in the degree of immunosuppression and in the secretion of regenerative factors depending on the health status of the individual from which the cells are obtained [17]. Because of this tissue-specific heterogeneity, preclinical research has focused on the regenerative prowess and immunomodulatory capacity of MSC isolated from tissues of early ontogeny that are untouched by chronic disease pathologies [18]. Therefore, the goal of this chapter is to review our current understanding of the molecular interactions between MSC and cells of the innate and the adaptive immune system. Furthermore, we focus on highlighting the unique immunogenicity and immunomodulation by MSC derived from umbilical cord/placental, and prenatal sources.

MSC Isolation and Characterization

The isolation of MSC has been successfully achieved from adult human tissues such as bone marrow, adipose, kidney, liver and more recently from umbilical cord, placental and Wharton's jelly samples [5, 19–21]. Currently, the most commonly studied MSC are from adult human bone marrow, while isolation from widely accessible human umbilical cord and placental samples have more been recently reported [22, 23]. The conventional means of MSC isolation is through plastic adherence of heterogeneous mononuclear cells derived from homogenates of the aforementioned sources. Subsequent culturing of this adherent fraction results in the propagation of a non-hematopoietic (CD45-negative) population that takes on a fibroblast-like morphology [22, 24]. These cells are highly proliferative in serum containing cultures and express the cell surface markers CD73, CD90, and CD105, without significant expression of the hematopoietic markers CD45, CD34, CD14, CD19, and HLA-DR. This cell surface expression profile is internationally recognized in defining the purity of MSC during culture [25]. Also, outlined in these minimal criteria for MSC is the ability to efficiently generate mature osteocytes, chondrocytes and adipocytes under differentiative culture conditions.

Despite establishment of these unifying standards, MSC-like populations of variable purity demonstrate significant heterogeneity in cell surface marker expression when isolated from different tissue sources. Unfortunately, little progress has been made in identifying uniquely-expressed cell surface markers *in situ* as a way to prospectively purify MSC from human sources. Unlike hematopoietic precursors, the MSC field has yet to develop an accepted means of isolating MSC from tissues using cell-surface markers or enzymatic functions. Recently, the isolation of a perivascular associated MSC subset (also known as pericytes) has been demonstrated using melanoma cell adhesion molecule or CD146 expressed *in situ* and to varying degrees during *ex vivo* culture [26]. In culture these perivascular MSC, which are depleted of CD34-expressing endothelial cells, meet the minimal criteria to define MSC by cell surface marker expression and differentiative capacity. Notably, CD146+ MSC can be isolated from both adult and placental tissues at varying frequencies. Likewise, the nerve growth factor receptor (CD271) may represent another cell-surface marker expressed on MSC with active immunomodulatory properties [27]. Whether or not these markers can be used interchangeably to purify MSC from umbilical cord of placental tissues remains a topic of debate. Perhaps the most consistent method to purify both adult and perinatal-derived MSC is through the conservation of known progenitor cell functions. One such function is aldehyde dehydrogenase activity, a cytosolic enzyme highly expressed in precursors from multiple lineages that protects essential, or long-lived cells against oxidative environmental insults. High ALDH activity (ALDHhi) has been demonstrated as a conserved function in primitive cells from hematopoietic, mesenchymal, endothelial, and neural progenitor lineages [28]. Indeed, human adult BM ALDHhi cells form CFU-F at a frequency of 1 colony in approximately 1500 cells. By using ALDH function, either before or after MSC expansion, we can prospective isolate progenitor cells higher in the MSC hierarchy, thereby reducing variability between samples and unwanted heterogeneity as cells are expanded ex vivo.

There also exist inconsistencies in characterizing MSC after culture as associated adhesion molecules are not lineage specific, and different markers are variably expressed under the different culture conditions (serum containing and serum-free) employed by the field. For example, CD73 and CD105 are also expressed on plastic adherent endothelial precursors, and MSC from umbilical cord blood (UCB) demonstrate inherently low expression of CD90 [29]. These inherent discrepancies characterizing MSC is thought, in part, to be due to differences in the source tissue from which the MSC were derived [30]. Thus, different MSC clones may be predisposed to a specific lineage within a putative developmental hierarchy akin to the highly characterized hematopoietic lineage. Although these cells are able to show multipotent differentiation potential into bone, cartilage and fat [31, 32], a more extensive look into the immunomodulatory function and regenerative capabilities of MSC relative to tissue source would be important for the development of a well-defined functional hierarchy within highly heterogeneous MSC cultures.

MSC Variability Among Sources and Stages of Ontogeny

Mesenchymal stem cells, over the course of development, can be visualized in a hierarchical structure [11, 13]. Starting from fertilization, the blastocyst contains pluripotent cells capable of forming any tissue in the human body. As these cells begin to divide and specialize, they lose both self-renewal and differentiative capacity, becoming a more mature and committed progenitor cell that expand to form various tissues of the human body and remain throughout adulthood to replenish lost cells during tissue repair and growth. Therefore, the hierarchical structure of stem cells becomes more evident as the lineage differentiates towards maturity [33]. Therefore, isolation of MSC from later stages in ontogeny is likely to yield heterogeneous populations throughout the hierarchy that may account for greater variability in functional studies. As such, when isolating MSC from earlier stages of ontogeny, or from a more purified progenitor population, we begin to see their true functional potentials.

Another factor that contributes to adult MSC functional variability is mutational senescence. As MSC divide over the lifespan of the organism, gene expression patterns are changed and telomere function can impact cellular senescence [34]. For example, there are considerable differences in the proliferative potential, growth patterns, telomere length, and lifespan of MSC derived from older versus younger patients. There is also a significant increase in the onset of proliferative senescence *in vitro* when comparing MSC obtained from older versus younger individuals [35]. Lastly, it is becoming evident that the functional capacities of MSC derived from adult bone marrow are very patient specific. For example, the level of secreted anti-inflammatory cytokines and the level of T-cell inhibition vary extensively between bone marrow samples [17]. These differences could arise from epigenetic changes as a result of imprinting or environmental factors including lifestyle and chronic disease comorbidities [36]. Finally, underlying disease is another factor that may affect the functional properties of secretory progenitor cell types [37]. For these reasons, obtaining MSC earlier in ontogeny is expected to improve homogeneity and subsequent functional potency of these cells in the clinical setting.

Minimal Criteria to Define MSC; Bone Marrow versus Fetal versus Cord Blood Derived MSC

The isolation of MSC from adult, fetal or perinatal sources is accomplished through very similar procedures. To obtain a single cell suspension from fetal tissues such as the fetal liver, the tissues are homogenized and strained through a filter [38]. Umbilical cord blood-derived MSC has been isolated from the umbilical vein using similar protocols as BM MSC [29] but require the addition of collagenase prior to release adherent cells *in situ* [19]. Once the cells are in single cell suspension, mononuclear cells are separated through density gradient centrifugation and plated for selection via plastic adherence. While fetal tissues contain a higher frequency of MSC relative to adult sources [38], the major disadvantage to fetal stem cells becoming a viable source of MSC is that there remains ethical controversy regarding the accrual of cells from pre-natal sources. These ethical barriers delay the use of fetal stem cells for widespread therapeutic purposes. However, umbilical cord and placental-derived MSC are ethically obtained at birth from normally discarded material, and can be propagated efficiently for therapeutic application.

Unlike fetal-derived MSC, umbilical cord or placental-derived MSC demonstrate similar cell-surface marker expression and functional capacities when compared to adult BM-derived MSC. Firstly, common MSC markers such as CD105, CD73, and CD29 are present on both bone marrow and cord-derived MSC [39], and both sources differentiate into the three aforementioned mesenchymal lineages [29, 40]. Importantly, both *in vitro* and *in vivo* characterization demonstrate the ability of MSC from bone marrow and umbilical cord to modulate immune activity [39, 41]. Studies comparing marrow, adipose, and umbilical cord-derived MSC have shown that immune regulating cytokine secretion is similar between these sources [42], suggesting that these functional criteria may represent the best way to compare and contrast MSC from different sources and from varying stages of ontogeny.

An additional source of MSC with immunomodulatory potential is from Wharton's jelly. Wharton's jelly MSC demonstrate similar expression patterns for several immunogenic markers as BM-derived MSC [21]. Although the knowledge base regarding MSC immunomodulatory function is heavily biased towards adult BM-derived MSC, new studies are emerging that demonstrate the immunomodulatory capabilities of MSC are conserved in tissues of earlier ontogeny, indicating their immense potential in therapeutic applications. However, proof-of-concept studies still need to be conducted to highlight the differences and similarities of adult BM versus umbilical cord or placental sources. Nonetheless, full-term umbilical cord, placental and Wharton's jelly represent attractive sources of MSC for widespread clinical use due to the lack of immunogenicity elicited after transplantation and a high degree of immunomodulatory effects observed in pre-clinical studies.

Early MSC Transplantation Trials to Modulate Immune Function

The first clinical trial aimed at supporting hematopoiesis used autologous MSC during myeloablative therapies for breast cancer, and demonstrated the ability to safely transplant MSC free from side effects or adverse reactions [43]. As MSC became recognized for their immunomodulatory properties, MSC became ideal candidates for treating the hematopoietic transplantation-induced complication graft-versus-host disease (GVHD). Clinical studies have demonstrated that infusion of allogeneic or autologous MSC increased survival rates in steroid-resistant GVHD patients without MSC-related toxicity or ectopic tissue formation [44–46]. Other trials have also been conducted using MSC to treat Crohn's disease-related fistulas, resulting in improved fistula recovery post-surgery and increased quality of life for the patients [47, 48].

Type I diabetes is another autoimmune disease where MSC therapy currently holds promise. MSC have been shown to increase beta cell mass in the injured pancreas of STZ-treated mice, allowing for partial restoration of blood glucose levels [49, 50]. Alongside endogenous regeneration, MSC may help to inhibit the autoimmune response towards beta cells, making it a very attractive option as a cellular therapy for type I diabetes [51]. With a focus on MSC, companies such as Osiris Therapeutics are currently performing larger-scale clinical trials to combat serious autoimmune and inflammatory diseases including possible cell-based implants for diabetics (http://www.osiris.com/clinical.php).

The Immune System Simplified

Co-ordinated immunity can be divided into two inter-related systems: the innate and adaptive immune system [52]. The innate immune system is the body's primitive defense mechanism that responds immediately to pathogens and generally results in inflammation to destroy the pathogen. In general, the innate immune system does not involve 'memory' to pathogens and can be found in many primitive organisms. The innate immune system includes physical barriers (i.e., skin, mucous etc.) and phagocytic responder cells such as natural killer (NK) cells, macrophages and granulocytes. These cell types engulf particles and/or infected cells, and secrete cytokines generating inflammation [53]. In contrast, the adaptive immune system is mainly comprised of T- and B-lymphocytes and associated antigen presenting cells such as dendritic cells (DC) that prevent the propagation of the pathogen and form immunological memory to antigens associated with specific pathogens. Antigens are unique protein sequences that are foreign to the body that are generally recognized by antibodies or presented on cells via the major histocompatibility complexes (MHC). When an antigen is detected in the human body, both the innate and adaptive immune system work together to eliminate the pathogen, infected cells, and prevent the re-entry of the pathogen upon subsequent exposures [54, 55].

Innate Immune Cells

The cells of the innate immune system rely on signals in the microenvironment such as chemokines and cytokines to home to in areas of infection. NK cells mediate the lysis of cells that do not express 'self' antigens presented on MHC class I cell surface molecules, typically expressed on all nucleated somatic cell [56, 57]. In contrast, adaptive T-lymphocytes depend upon the presence of MHC class II molecules (expressed primarily on immune cells) on an antigen presenting cell (APC) cell surface. Thus, NK cells can act without inflammatory signals and is thought to be one of the fastest immune-cell responders. NK cells also have the ability to recognize antibodies bound to foreign antigens, and to elaborate various chemokines and cytokines [58], which aid in the homing and activation of other innate immune cells alongside cells of the adaptive immune system.

Macrophages are cells that can scavenge whole cells, debris, and pathogens through phagocytosis and degrade these products with cytolytic enzymes. Macrophages must first be activated with pro-inflammatory cytokines in order to become phagocytic. After phagocytosis, macrophages process the antigens and present them on the cell surface via the MHC class II molecules. Thus, macrophages are also considered to be APC that activate adaptive immunity. DC are also considered to be a bridge between adaptive and the innate immune systems. Similarly, DC function to present epitopes (the part of an antigen that is recognized by the adaptive immune system) through MHC class II complexes. DC are also known as professional APC as there is a much greater concentration of MHC class II on the extended processes of DC for enhanced antigen presentation. In addition, DC possess the ability to detect minute concentrations of antigens in the microenvironment. Both DC and macrophages arise via the differentiation of monocytes, another important cell type of the innate immune system. In response to chemokines, monocytes traverse through tissues (like macrophages) and respond to inflammatory cytokines produced at the site of infection to enhance the both innate and adaptive immune responses [57].

The Adaptive Immune Cells

The adaptive immune system is comprised of 2 main cell types: T- and B-lymphocytes. These adaptive immune cells are constantly sampling cells for recognition of "self" antigens. Foreign antigens, when detected, elicit a response from the adaptive immune system to eliminate the non-self or pathogen epitope expression and to subsequently form "memory" of the specific antigen upon subsequent exposure [59]. B-lymphocytes are a branch of the adaptive immune system responsible for antibody-mediated or humoral immunity. B-lymphocytes can bind directly to antigens and require interaction with T cells in order to differentiate into plasma cells. Once differentiated, plasma cells mass-produce antibodies specific to antigenic epitopes, allowing the innate immune system to detect and destroy the antigen. A small subset of plasma cells persist in the body for a long period of time as "memory B cells," and enhance the response upon re-exposure to a specific pathogen [60].

T-lymphocytes are arguably the backbone of the adaptive immune system because they serve to aid in the humoral immune response as well as mediate cytotoxic cell-mediated immunity. T cells become activated through antigens presented primarily within MHC class II complexes on APC. Depending on the MHC complex involved (class I or II), T cell subsets will respond to the stimuli through different pathways [61]. When T-lymphocytes interact with antigens presented on MHC class I complexes, CD8+ cytotoxic T cell formation is favored and respond to lyse the infected cell [62]. When T cells become activated by antigens presented on MHC class II complexes on APC, they can also differentiate into CD4+ helper T cells. Helper T cells are responsible for aiding in the activation of various immune cell types via secretion of soluble factors and through direct cell contact. Thus, the recruitment and differentiation of innate immune cells and B cells is greatly dependent on the intricate balance between CD4+ and CD8+ T cells in the microenvironment [57]. Lastly, naïve T cells are also able to differentiate into regulatory T lymphocytes (Treg) [63]. Regulatory T cells are thought to be primarily CD4+/CD25+, and function by secreting anti-inflammatory cytokines and factors which promote macrophage differentiation into an M2 phenotype. Anti-inflammatory molecules secreted by M2 macrophages in turn help the immune system to taper down an inflammatory response. Treg are also capable of generating immune tolerance to certain antigens [64].

MSC Immunogenicity

MSC are known to elicit very low immunogenicity, meaning they typically evade surveillance by the immune system after transplantation. This low immunogenicity is regulated by low endogenous expression of MHC class II complexes in their resting state, a characteristic unique to MSC that evade the immune system [62, 65]. However, there are experimental conditions whereby MSC can be targeted for deletion by the immune system. IL-2-activated NK cells are able to target and lyse MSC, however, when pre-treated with IFN-γ, MSC may also evade pre-activated NK cells via anti-inflammatory cytokine secretion discussed in further detail below [66]. Thus, in the allogeneic transplantation setting, there exists a finite balance between NK cell mediated targeting of MSC versus the suppression of NK cell cytolytic activity towards MSC and neighboring cells. Surprisingly, after pre-treatment with IFN-γ which typically upregulates MHC class II expression, MSC still do not elicit an immune response from mixed leukocyte reactions (MLR) or purified T cells [31, 67]. This likely due to a combination of two reasons:

1. During exposure to IFNγ, MHC class II expression is temporarily decreased on MSC [68], enabling MSC to avoid initial detection by T-lymphocytes.
2. Even when recognized by the immune system MSC evade activated immune cells by the secretion of factors that inhibit T cell activation and cytolytic activities [31, 67, 69].

Although counter-intuitive, MSC secrete several chemokines in response to inflammation that recruit cells of both the innate and adaptive immune system, yet possess low immunogenicity [70]. Despite luring in cells of the immune system, MSC remain undetected presumably through modification of MHC complex expression and by the secretion of anti-inflammatory factors. Of note, MSC polarize cells of the innate immune system, such as DC and macrophages, to a regulatory or anti-inflammatory state (Fig. 4.1) [14, 71]. Macrophages are of special interest as MSC have been reported to promote M2 differentiation, a phenotype associated with non-phagocytic and anti-inflammatory macrophage functions [17, 72, 73]. In addition,

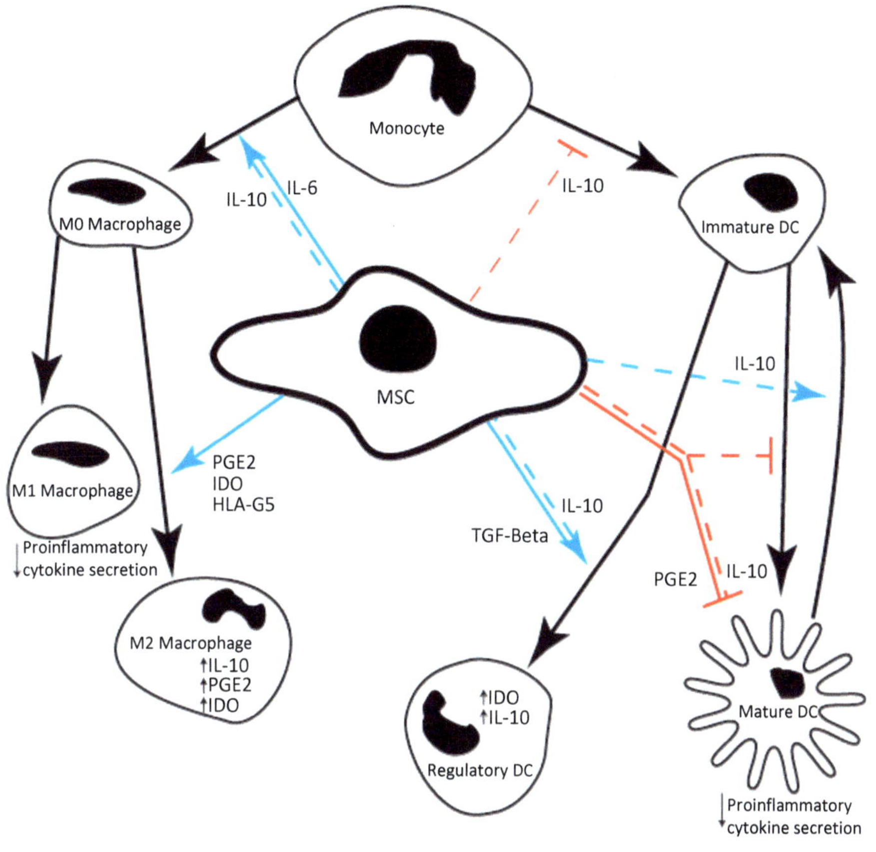

Fig. 4.1 A schematic of the influence of MSC on the monocyte lineage and its downstream progeny. *Black arrows* represent the cellular differentiation/polarization of immune effector cells to an immunocompetent phenotype. *Blue arrows* represent the facilitation of immune cell maturation while *red blunted lines* represent the inhibition of differentiation. *Solid lines* represent direct effects from factors secreted by MSC. In contrast, *dotted lines* represent indirect effects by cytokines also present in the microenvironment. MSC secrete IDO, PGE2 and TGF-beta that act to increase IL-10 production by M2 macrophages. Collectively, these effectors act to dampen the inflammatory activities of T-lymphocytes, B-lymphocytes and NK-cells in the microenvironment

MSC may similarly promote the presence of a regulatory subset of DC. This remains controversial as a definitive population regulatory DC has yet to be identified by cell surface markers, however, evidence is emerging that DC can be skewed towards an anti-inflammatory cytokine profile [41, 74, 75]. A “regulatory DC” theoretically promote the production of Treg under the appropriate conditions [74, 76]. Overall, when DC are exposed to MSC, there is a marked reduction in antigen presentation efficiency and subsequent stimulation of immune cells such that the potential detection of MSC is reduced [77].

Although generally considered lowly immunogenic, MSC have also been reported to stimulate the immune system under specific conditions [78]. MSC have been shown to activate purified CD4+ T lymphocytes when incubated with low concentrations of IFN-γ and foreign antigens *in vitro* [68]. Conversely, IFN-γ pre-treated MSC without the presence of exogenous antigens inhibits T lymphocyte proliferation [39, 79, 80]. Currently, the low immunogenicity of MSC holds clinical promise because MSC may be considered an "off-the-shelf" cellular product that seemingly does not require HLA phenotyping, as allogenic MSC show long term engraftment in the bone marrow of baboons without immune rejection [81].

Factors Implicated in MSC Immunomodulation

For a concise general description of the main factors involved in MSC-mediated immunomodulation, refer to Table 4.1. For a conceptual schematic on the interaction between the major immunomodulatory factors during MSC exposure to inflammatory effectors, please refer to Fig. 4.2.

Transforming Growth Factor-β (TGF-β)/Hepatocyte Growth Factor (HGF)

The anti-inflammatory cytokine TGF-β is constitutively produced by MSC and CD14+ monocytes [82]. MSC secretion of TGF-β was increased after cell-contact only with T-lymphocytes; however, the factors responsible for this response have not been identified [83]. TGF-β incubated with CD4+ T helper cells blocks cytotoxic activity [84]. When MSC are directly in contact with T cells, TGF-β supplementation further inhibits T-lymphocyte proliferation [83]. TGF-β has also been found to play a role in the formation of Treg as the addition of neutralizing antibodies towards TGF-β resulted in a significant decrease in Treg marker expression [85]. Likewise, TGF-β aids in the polarization of DC resulting in an increased amount of IL-10 secreted (Fig. 4.1) [86]. However, TGF-β, like other anti-inflammatory factors secreted from MSC, may not act in isolation as the addition of neutralizing antibodies only leads to partial restoration of T cell activation [87, 88].

Like TGF-β, HGF is also constitutively produced by MSC and levels have been observed to increase upon co-culture with T cells [42]. As proof of concept, both TGF-β and HGF co-administration to activated T cells resulted in inhibition of T cell proliferation [87]. Neutralizing antibodies towards HGF partially reversed the inhibition of MLR containing T cells, irradiated allogenic peripheral blood leukocytes (PBL), and MSC [87]. However, similar experiments performed using adipose-derived MSC, the upregulation of TGF-β and HGF was not observed, highlighting important secretory differences between MSC isolated from different adult sources [89].

Table 4.1 Cytokines involved in immune modulation mediated by MSC

Factor	Activators	Secreted by	Function	Cells affected
TGF-beta	Direct contact with T cells	MSC	Inhibits activated T cells	T cells
		Monocytes	Blocks NK cell cytotoxicity	NK cells
		Treg lymphocytes	Aids in regulatory DC formation	DCs
			Aids in Treg formation	
HGF	Direct contact with T cells	MSC	Inhibits activated T cells	T cells
		Monocytes		
PGE2	Co-culture with NK cells	MSC	Inhibits NK cell proliferation in concert with TGF-beta and IDO	NK cells
	PGE2	M2 macrophages	Enhances Th1 & M2 macrophage polarization	B cells
	IFN-gamma/ TNF-alpha		Linked to IDO expression in immature DCs	Macrophages
			Apoptosis of activated/ mature T cells	T cells
			Inhibits mature B-cell proliferation	
IDO	IFN-gamma/ TNF-alpha	M2 macrophages	Inhibits T cell proliferation	Monocytes
	Stabilized by IL-10	Regulatory DCs	inhibits NK cell proliferation	Macrophages
		MSC	Polarization to regulatory immune cells	T cells
				NK cells
IL-10	PGE2	M2 macrophages	Decreases proinflammatory cytokine secretion	Monocytes
	TGF-beta	Regulatory DCs	Decreases cell surface expression of co-stimulatory molecules	Macrophages
	IDO	Tregs	Decreases MHC class II membrane localization	DCs
	HLA-G5	Monocytes	Aids in regulatory immune cell polarization	
HLA-G5	IL-10	MSC	Decreases proinflammatory cytokine secretion	T cells
		M2 macrophages	Aids in regulatory immune cell production	NK cells
			Prevents NK cell-mediated cytolysis	
			Prevents CD8+ CTL proliferation	

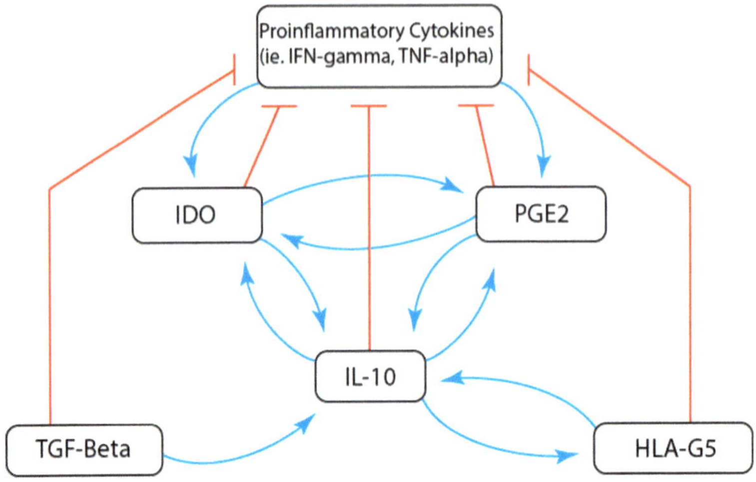

Fig. 4.2 Interplay between pro- and anti-inflammatory cytokines involved in the modulation of the immune response by MSC. *Blue arrows* represent stimuli that result in the up-regulation of a given factor in the microenvironment. *Red blunted lines* represent inhibition of pro-inflammatory cytokine production by resident immune cells. The immunomodulatory signaling cascade is initiated via pro-inflammatory cytokine secretion by infiltrating immune cells in the niche

Indoleamine 2,3-Deoxygenase (IDO)

IDO is a catabolic enzyme that converts the essential amino acid tryptophan to kynurenine. The depletion of tryptophan and/or the addition of kynurenine has been shown to inhibit the proliferation of lymphocytes and pathogens [90, 91]. The lack of tryptophan and/or the presence of kynurenine evoked by IDO has been demonstrated to induce both T lymphocyte apoptosis and inhibition of immune cell proliferation [91, 92]. IDO is secreted by MSC and APC such as macrophages and immature DC in response to IFN-γ. In contrast to TGF-β and HGF, IDO is not constitutively produced in MSC [93, 94]. In MSC and DC, IDO expression is also maintained/stabilized by the presence of IL-10 in the microenvironment [74, 95]. Many studies show that IDO inhibits T cell proliferation in both mice and humans [93, 96]. When the IDO inhibitor, 1-methyl-tryptophan, is added to MLR containing MSC, T lymphocyte proliferation was partially restored [17]. IDO is also an important inhibitor of fetal rejection, as the addition of 1-methyl-tryptophan to a surrogate mouse mothering allogenic offspring resulted in enhanced immune rejection of the fetuses [94].

IDO also enables the polarization of M2 macrophages and increases the number of Treg in culture, further implicating its important role in MSC-mediated immunomodulation [17, 76]. Furthermore, IDO has also been found to act with PGE2 to inhibit NK cell proliferation [97]. However, it is important to emphasize that MSC-mediated IDO secretion is not absolutely required for immune suppression. Human

MSC deficient in the IFN-γ receptor (and therefore unable to secrete IDO) were still able to inhibit PBMC proliferation via alternative mechanisms [93]. Although IDO acting on its own enhances immune cell apoptosis, MSC secretion of IDO does not enhance apoptosis but primarily inhibits T cell proliferation via tryptophan depletion [92].

Interleukin 10 (IL-10)/Human Leukocyte Antigen G5 (HLA-G5)

IL-10 a key anti-inflammatory cytokine mainly produced by M2 macrophages. Although there have been controversial reports describing the secretion of IL-10 directly by MSC [30, 77, 83], most studies demonstrate the lack of IL-10 secretion from MSC even under the influence of various pro-inflammatory cytokines such as IFN-γ and TNF-α [17, 31, 41, 67, 82]. However, the presence of IL-10 was markedly increased by the addition of MSC to MLR, suggesting that MSC induce IL-10 production by macrophages. Indeed, one group showed that IL-10$^{-/-}$ MSC were still able to increase the levels of IL-10 in MLR [98]. Only after the MLR have been depleted of macrophages was there a notable decrease in the levels of IL-10.

IL-10 is generally secreted by immune-regulatory cell types including M2 macrophages, regulatory DC, and Treg (Fig. 4.3) [86, 99, 100]. The effect of IL-10 functions mainly on monocytes and subsequently affects other cell types downstream [72]. In general, IL-10 reduces the expression of co-stimulatory molecules on the cell surface of monocytes, macrophages, and DC preventing T-cell activation [101, 102]. In addition, IL-10 decreases MHC class II localization on myeloid cell types as a result of aberrant recycling and recruitment to the cell membrane. In contrast, IL-10 does not affect MHC class I expression on myeloid cells [102]. Due to the lack of MHC class II expression and reduced co-stimulatory molecules on the cell surface of APC, local T lymphocytes demonstrate reduced activation efficiency. Finally, IL-10 aids in the conversion of DC and macrophages into IL-10 secreting cell types in an autoregulatory fashion (Fig. 4.3) [75, 103]. Thus, IL-10 added to MLR inhibits proliferation while IL-10 supplementation does not affect the proliferation of immune cells stimulated using the mitogen, phytohaemagluttanin (PHA) [104].

Another function stimulated by IL-10 is the release of HLA-G5 from MSC. MSC express HLA-G5 on the cell surface in very low quantities and IL-10 increases HLA-G5 secretion [105, 106]. IL-10 secretion by immune cells is also increased by the addition of HLA-G5, suggesting these effectors operate in a positive feedback loop to decrease the immune response [107]. HLA-G5 has been observed to prevent NK cell-mediated cytolysis and decrease IFN-γ secretion [105, 108]. The addition of neutralizing antibodies toward HLA-G5 to MLR including MSC and PBL resulted in the decrease of Treg formation [105]. Also, HLA-G5 has been shown to act on cytotoxic CD8+ T cells to prevent their proliferation [109].

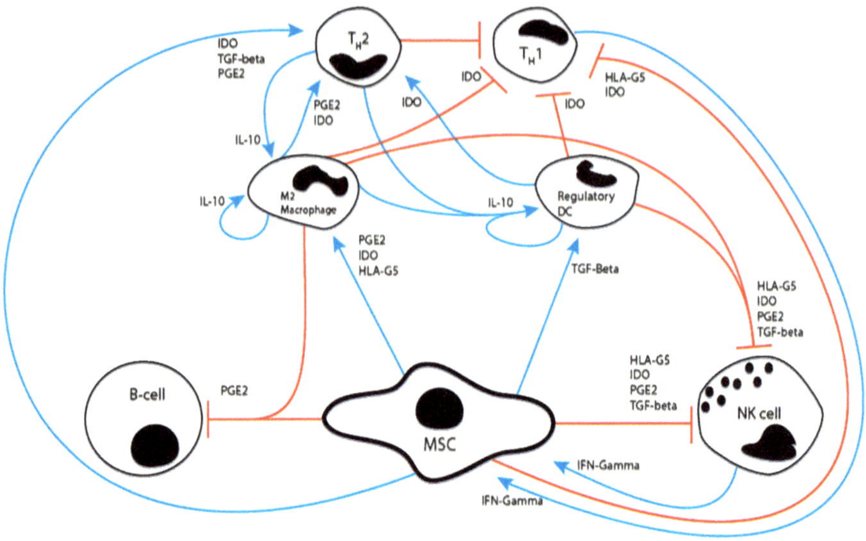

Fig. 4.3 A schematic of the immunomodulatory signaling networks impacted MSC. *Blue arrows* represent the facilitation of regulatory immune cell function while *red blunted lines* represent the inhibition of immune cell proliferation and pro-inflammatory cytokine secretion. IFNγ acts on MSC to increase the production of IDO, PGE2 and TGF-β that act to increase IL-10 production by M2 macrophages and Th2 lymphocytes. Collectively, these effectors act to dampen the inflammatory activities of activated T-lymphocytes, B-lymphocytes and NK-cells. IL-10 and TGF-beta secretion may also promote the formation of regulatory dendritic cells to further reduce antigen presentation in the microenvironment

Prostaglandin E2 (PGE2)

Prostaglandins are lipid molecules that demonstrate pleiotropic effects throughout the body. Prostaglandin production begins with the conversion of arachidonic acid via cyclooxygenase 1/2 (COX-1 and COX-2). Downstream prostaglandin synthases generate specific prostaglandins [110]. Immunologists have demonstrated the addition of indomethacin, a COX inhibitor, does not inhibit the proliferation of CD3/CD28-stimulated PBMC [67] or the proliferation of NK cells [97]. However, blocking prostaglandin E2 (PGE2) concurrently with IDO inhibition restored NK cell proliferation indicating the importance of the combination of these effectors [97]. Similarly, the combination of TGF-β and PGE2 inhibited NK cell proliferation [14]. While PGE2 is constitutively produced by MSC [67], levels of PGE2 are increased upon co-culture of MSC with NK cells, PBMC, and/or exposure to IFN-γ [14, 41, 42, 67].

Similar to TGF-β, PGE2 promotes Treg formation as the inhibition of PGE2 decreases Treg production *in vitro* [85]. CD4+ T cells treated with PGE2 up-regulate FOXP3 expression, an accurate determinant of the Treg function [31, 111]. PGE2 signaling also moderates immune responses through reduced IL-2 secretion and by causing apoptosis in activated T-cells. Furthermore, PGE2 inhibits the proliferation of immature B-cells while having little effect on mature B-cells [110]. *In vivo* models have demonstrated that PGE2 inhibition in mice rendered MSC ineffective at protecting against autoimmune encephalomyelitis. In this model, PGE2 was also shown to be linked to an increase in IDO expression in DC [112].

In macrophages, PGE2 decreases the production of pro-inflammatory cytokines (TNF-α, IL-1β, IL-8, IL-12) and up-regulates IL-10 production [113]. This supports the concept that PGE2 polarizes macrophages to an M2 regulatory phenotype [98]. Even though PGE2 does not directly inhibit MLR or PBMC stimulated with CD3/CD28, PGE2 is able to inhibit lymphocyte proliferation when stimulated by PHA [67, 104]. Thus, through the formation of regulatory immune cells and the cooperative actions of other anti-inflammatory molecules, PGE2 potently moderates the inflammatory response.

Other Factors

Although the aforementioned molecules comprise the major effectors contributing towards immunomodulation by MSC, some of the other less studied factors include: vascular endothelial growth factor (VEGF), Lymphocyte function-associated antigen 3 (LFA-3), and B7 homolog 1 (B7-H1). Furthermore, there are likely other MSC-secreted factors yet to be discovered that may function in an immunomodulatory fashion. Briefly, VEGF, LFA-3, and B7-H1 are all factors secreted by MSC [31, 41, 114]. VEGF is a pro-angiogenic molecule secreted by MSC and other cell types, including cancers in response to pro-inflammatory cytokines [41, 115, 116]. In the case of cancer, VEGF has been reported to inhibit immune responses, making this an attractive molecule to study immune regulatory properties in MSC [117]. Recombinant LFA-3 has been used to inhibit T cell proliferation while promoting and enabling Treg formation [118]. Finally, upon activation with IFN-γ, MSC have been shown to up-regulate B7-H1 expression, a co-stimulatory molecule proposed to play a role in immunosuppression [114, 119]. Many of these factors represent active areas of investigation. As investigating the immunomodulatory effects of MSC progresses, precautions should be taken to ensure that further controversy does not arise due to differences in experimental methods.

The Big Picture: MSC Immunomodulation and Immunogenicity

MSC can be obtained from various different tissues of the body and from different points in development [5, 19–21]. MSC from both adult and fetal sources lack immunogenicity and have potent immunomodulatory effects. Importantly, MSC isolated from different tissues demonstrate variable effectiveness in regulating the immune response. Similarly, there are differences in the immuno modulatory functions observed between individual BM-MSC samples [17]. Sample variability may be reduced by using cells from early ontogeny, such as UCB-derived MSC, as these cells function similarly to BM-MSC, and have not been exposed to environmental stressors that facilitate mutational senescence. Nonetheless, MSC derived from different sources appear to modulate the immune system after exposure to IFN-γ and other pro-inflammatory cytokines [14, 120]. In response to inflammatory stimuli, MSC from perinatal or adult sources secrete anti-inflammatory cytokines, acting as a switch to dampen the immune response through a positive-feedback loop (Figs. 4.2 and 4.3).

MSC from both perinatal and adult sources constitutively secrete TGF-β, HGF, PGE2, and possibly other factors that influence the immune response [42, 67, 82]. However, PGE2, and IDO are known to increase after MSC exposure to pro-inflammatory cytokines and work in concert with secreted factors such as TGF-β to moderate inflammation in immune effector cells [14, 41, 85]. As a demonstrated, conditioned media from MSC co-cultured with PBMC directly inhibits PBMC proliferation [67]. Similarly, transwell experiments also demonstrate the immunomodulatory effects of MSC-secreted factors, activated by exposure to inflammatory signals [87, 121]. When MSC are directly co-cultured with immune cells, there is a more profound inhibitory response which indicates both soluble factors and cell–cell contact is required to achieve optimal inhibition of the immune response [67, 71, 87]. The efficacy by which MSC in co-culture modulate immune responses show their promise as potential cellular therapies for moderating autoimmune and inflammatory diseases in the clinic.

Chemokines secreted by MSC attract immune cells to MSC, so that their effects can be maximized through close proximity [70, 71]. Collectively, IDO and PGE2 are factors that taper the immune response via activities on multiple immune cell types (Fig. 4.3) [70, 71]. IDO secretion by MSC is up-regulated in areas of inflammation by IFN-γ, causing the liberation of tryptophan catabolites and inhibition of T-cell proliferation [91, 92]. IDO has also been observed to polarize T cells into Treg, macrophages into M2 macrophages, and works in conjunction with PGE2 to modulate NK cell proliferation [17, 76, 97]. PGE2 is also up-regulated upon MSC stimulation with pro-inflammatory cytokines and works in conjunction with TGF-β to inhibit NK cell proliferation and to induce Treg formation [14, 85]. PGE2 will also increase IDO expression in regulatory DC [112] and will induce macrophages polarization into the M2 phenotype, which further secrete PGE2 [98, 113]. As a result of both IDO and PGE2 secretion, IL-10 is subsequently secreted from Treg, M2 macrophages, and regulatory DC [86, 99, 100]. The inhibitory effect of IL-10 is mediated through the down-regulation of co-stimulatory molecules and MHC class II complexes on APC, retarding their ability to stimulate T cells [101, 102, 122]. IL-10 not only aids in the conversion of more regulatory cell types, but it also acts back on MSC to increase HLA-G5 secretion [105, 106]. HLA-G5 acts to prevent NK cell cytolysis of neighbouring cells and to prevent further secretion of IFN-γ [105, 108]. HLA-G5 also acts to prevent CD8+ cytotoxic T cell proliferation and to promote Treg formation [109]. MSC are capable of promoting the differentiation of monocytes to macrophages and inhibiting the differentiation of monocytes to DC (Fig. 4.1) [17, 39]. Mature DC become polarized to a more immature state when cultured in the presence of MSC [77]. MSC have also been shown to regulate B-cell activation through soluble factors as demonstrated in transwell experiments [15]. Collectively, the direct effects of anti-inflammatory cytokines alongside regulatory immune cell formation and associated decreased pro-inflammatory cytokine secretion minimize the immune response through a positive-feedback mechanism, a powerful means to inhibit the local immune responses (Figs. 4.2 and 4.3).

BM Versus UCB Versus Fetal MSC Immunomodulation

There have been few studies to put the immunomodulatory and immunogenic properties of fetal MSC in perspective with adult or umbilical cord-derived MSC [123]. The differentiation potential of UCB-derived MSC has been demonstrated to be similar to adult sources [32, 39]. Like adult MSC, both UCB and fetal MSC express MHC class I but not MHC class II molecules [67, 68, 105]. Both adult and fetal MSC do not express co-stimulatory molecules and fail to induce proliferation of allogenic PBL when pre-exposed to IFN-γ. In addition, pre-exposure of IFN-γ to both adult and fetal MSC results in similar immunosuppressive effects [79].

Both BM-MSC and UCB-MSC are able to polarize macrophages to a M2, anti-inflammatory phenotype. This became evident as depletion of the CD14+ population in MLR with allogenic UCB-MSC resulted in increased lymphocyte proliferation [124]. UCB-MSC have also been shown to reduce DC function and inhibit PBMC proliferation through both cell-contact and soluble factors [39]. Wharton's jelly-derived MSC also inhibit T-cell proliferation by similar mechanisms as BM-MSC [42]. However, the immunological properties of Wharton's jelly MSC needs to be investigated further to identify more similarities and differences relative to BM-MSC before using this as a reliable source of MSC.

As of yet, fetal-derived MSC are not recognized as a reliable source of MSC for cellular therapies. To make fetal MSC more relevant, the secretome of fetal MSC relative to UCB-MSC or BM-MSC under various conditions needs to be compared. Furthermore, it would be appropriate to investigate the ability of fetal-derived MSC to be able to polarize immune cell types into regulatory cell types to ascertain whether these influences are cell-contact dependent. Because of the ethical concerns surrounding the derivation of fetal-derived MSC, UCB-MSC offers an 'off the shelf' alternative. However, more research needs to go into integrating the immunomodulatory effects of MSC from early ontogeny in pre-clinical transplantation models of autoimmune diseases.

Summary and Future Perspectives

Adult BM, umbilical cord and placental derived MSC share potent paracrine mechanisms for the formation of a localized immunosuppressive niche. Multiple MSC-secreted effectors, such as TGF-β, PGE2, IDO and HLA-G5, have been well documented to modulate the functions of mature effector cell types resulting in generalized reductions in NK cell and T cell-mediated proliferation and cytolytic activities. Importantly, these MSC-secreted effectors also shift the balance towards immunoregulatory phenotypes, by promotion of M2 macrophage, Th2 lymphocyte and regulatory DC production. Collectively, these regulatory immune cell types actively secrete IL-10 which further dampen inflammatory immune cell function. Thus, transplantation of immunomodulatory MSC, from term umbilical cord or placental samples, represent a potent strategy for the development of immunomodulatory cell-based therapies.

Although considerable progress has been made to characterize immunomodulatory effectors secreted by MSC *in vitro*, additional research is required to realize the potential of MSC immunomodulation in clinical trials. First, prospective isolation using unique cell surface markers or conserved stem cell functions is needed to identify MSC subtypes with the most potent immunomodulatory effects and to understand the role of these cells in a putative MSC hierarchy. Second, genomic and proteomic studies to determine global cytokine secretory patterns by MSC isolated from adult versus perinatal sources would identify the optimal source of MSC for immunomodulatory therapies. Third, additional *in vivo* transplantation studies are warranted to determine proof-of-concept and to determine whether MSC implantation can dampen local autoimmune reactivity during autoimmunity. Collectively, accrual of this knowledge will pave the way for the development of rational clinical trials to "tip the balance" in favour of tissue regeneration/repair versus immune-mediated destruction during autoimmunity.

References

1. Friedenstein AJ, et al. Precursors for fibroblasts in different populations of hematopoietic cells as detected by the in vitro colony assay method. Exp Hematol. 1974;2(2):83–92.
2. Friedenstein AJ, Gorskaja JF, Kulagina NN. Fibroblast precursors in normal and irradiated mouse hematopoietic organs. Exp Hematol. 1976;4(5):267–74.
3. Prockop DJ. Marrow stromal cells as steam cells for nonhematopoietic tissues. Science. 1997;276(5309):71–4.
4. Pittenger MF. Multilineage potential of adult human mesenchymal stem cells. Science. 1999;284(5411):143–7.
5. Jiang Y, et al. Pluripotency of mesenchymal stem cells derived from adult marrow. Nature. 2002;418(6893):41–9.
6. Phadnis SM, et al. Human bone marrow-derived mesenchymal cells differentiate and mature into endocrine pancreatic lineage in vivo. Cytotherapy. 2011;13(3):279–93.
7. Caplan AI. Mesenchymal stem cells. J Orthop Res. 1991;9(5):641–50.
8. Bianco P, et al. "Mesenchymal" stem cells in human bone marrow (skeletal stem cells): a critical discussion of their nature, identity, and significance in incurable skeletal disease. Hum Gene Ther. 2010;21(9):1057–66.
9. da Silva Meirelles L, Caplan AI, Nardi NB. In search of the in vivo identity of mesenchymal stem cells. Stem Cells. 2008;26(9):2287–99.
10. Maitra B, et al. Human mesenchymal stem cells support unrelated donor hematopoietic stem cells and suppress T-cell activation. Bone Marrow Transplant. 2004;33(6):597–604.

11. Salem HK, Thiemermann C. Mesenchymal stromal cells: current understanding and clinical status. Stem Cells. 2010;28(3):585–96.
12. Si Y, et al. Infusion of mesenchymal stem cells ameliorates hyperglycemia in type 2 diabetic rats: identification of a novel role in improving insulin sensitivity. Diabetes. 2012;61(6):1616–25.
13. Caplan AI, Correa D. The MSC: an injury drugstore. Cell Stem Cell. 2011;9(1):11–5.
14. Nauta AJ, Fibbe WE. Immunomodulatory properties of mesenchymal stromal cells. Blood. 2007;110(10):3499–506.
15. Corcione A, et al. Human mesenchymal stem cells modulate B-cell functions. Blood. 2006;107(1):367–72.
16. Gopurappilly R, Bhat V, Bhonde R. Pancreatic tissue resident mesenchymal stromal cell (MSC)-like cells as a source of in vitro islet neogenesis. J Cell Biochem. 2013;114(10):2240–7.
17. Francois M, et al. Human MSC suppression correlates with cytokine induction of indoleamine 2,3-dioxygenase and bystander M2 macrophage differentiation. Mol Ther. 2012;20(1):187–95.
18. Fong CY, et al. Human Wharton's jelly stem cells have unique transcriptome profiles compared to human embryonic stem cells and other mesenchymal stem cells. Stem Cell Rev. 2011;7(1):1–16.
19. Covas DT, et al. Isolation and culture of umbilical vein mesenchymal stem cells. Braz J Med Biol Res. 2003;36:1179–83.
20. Zuk PA, et al. Human adipose tissue is a source of multipotent stem cells. Mol Biol Cell. 2002;13(12):4279–95.
21. Anzalone R, et al. New emerging potentials for human Wharton's jelly mesenchymal stem cells: immunological features and hepatocyte-like differentiative capacity. Stem Cells Dev. 2010;19(4):423–38.
22. Mareschi K, et al. Isolation of human mesenchymal stem cells: bone marrow versus umbilical cord blood. Haematologica. 2001;86(10):1099–100.
23. Wexler SA, et al. Adult bone marrow is a rich source of human mesenchymal 'stem' cells but umbilical cord and mobilized adult blood are not. Br J Haematol. 2003;121(2):368–74.
24. Bieback K, et al. Critical parameters for the isolation of mesenchymal stem cells from umbilical cord blood. Stem Cells. 2004;22(4):625–34.
25. Dominici M, et al. Minimal criteria for defining multipotent mesenchymal stromal cells. The International Society for Cellular Therapy position statement. Cytotherapy. 2006;8(4):315–7.
26. Crisan M, et al. A perivascular origin for mesenchymal stem cells in multiple human organs. Cell Stem Cell. 2008;3(3):301–13.
27. Kuci Z, et al. Clonal analysis of multipotent stromal cells derived from CD271+ bone marrow mononuclear cells: functional heterogeneity and different mechanisms of allosuppression. Haematologica. 2013;98(10):1609–16.
28. Capoccia BJ, et al. Revascularization of ischemic limbs after transplantation of human bone marrow cells with high aldehyde dehydrogenase activity. Blood. 2009;113(21):5340–51.
29. Lee OK, et al. Isolation of multipotent mesenchymal stem cells from umbilical cord blood. Blood. 2004;103(5):1669–75.
30. Ryan JM, et al. Interferon-gamma does not break, but promotes the immunosuppressive capacity of adult human mesenchymal stem cells. Clin Exp Immunol. 2007;149(2):353–63.
31. Batten P, et al. Human mesenchymal stem cells induce T cell anergy and downregulate T cell allo-responses via the TH2 pathway: relevance to tissue engineering human heart valves. Tissue Eng. 2006;12(8):2263–73.
32. Erices A, Conget P, Minguell JJ. Mesenchymal progenitor cells in human umbilical cord blood. Br J Haematol. 2000;109(1):235–42.
33. Rao MS, Mattson MP. Stem cells and aging: expanding the possibilities. Mech Ageing Dev. 2001;122(7):713–34.

34. Mitchell KE, et al. Matrix cells from Wharton's jelly form neurons and glia. Stem Cells. 2003;21(1):50–60.
35. Stenderup K, et al. Aging is associated with decreased maximal life span and accelerated senescence of bone marrow stromal cells. Bone. 2003;33(6):919–26.
36. Jaenisch R, Bird A. Epigenetic regulation of gene expression: how the genome integrates intrinsic and environmental signals. Nat Genet. 2003;33(Suppl):245–54.
37. Fadini GP, et al. Diabetes impairs progenitor cell mobilisation after hindlimb ischaemia-reperfusion injury in rats. Diabetologia. 2006;49(12):3075–84.
38. Campagnoli C. Identification of mesenchymal stem/progenitor cells in human first-trimester fetal blood, liver, and bone marrow. Blood. 2001;98(8):2396–402.
39. Wang M, et al. The immunomodulatory activity of human umbilical cord blood-derived mesenchymal stem cells in vitro. Immunology. 2009;126(2):220–32.
40. Muraglia A, Cancedda R, Quarto R. Clonal mesenchymal progenitors from human bone marrow differentiate in vitro according to a hierarchical model. J Cell Sci. 2000;113(Pt 7):1161–6.
41. Aggarwal S, Pittenger MF. Human mesenchymal stem cells modulate allogeneic immune cell responses. Blood. 2005;105(4):1815–22.
42. Najar M, et al. Mesenchymal stromal cells use PGE2 to modulate activation and proliferation of lymphocyte subsets: combined comparison of adipose tissue, Wharton's Jelly and bone marrow sources. Cell Immunol. 2010;264(2):171–9.
43. Koc ON, et al. Rapid hematopoietic recovery after coinfusion of autologous-blood stem cells and culture-expanded marrow mesenchymal stem cells in advanced breast cancer patients receiving high-dose chemotherapy. J Clin Oncol. 2000;18(2):307–16.
44. Lazarus HM, et al. Cotransplantation of HLA-identical sibling culture-expanded mesenchymal stem cells and hematopoietic stem cells in hematologic malignancy patients. Biol Blood Marrow Transplant. 2005;11(5):389–98.
45. Le Blanc K, et al. Mesenchymal stem cells for treatment of steroid-resistant, severe, acute graft-versus-host disease: a phase II study. Lancet. 2008;371(9624):1579–86.
46. Ringden O, et al. Mesenchymal stem cells for treatment of therapy-resistant graft-versus-host disease. Transplantation. 2006;81(10):1390–7.
47. Duijvestein M, et al. Autologous bone marrow-derived mesenchymal stromal cell treatment for refractory luminal Crohn's disease: results of a phase I study. Gut. 2010;59(12):1662–9.
48. Ciccocioppo R, et al. Autologous bone marrow-derived mesenchymal stromal cells in the treatment of fistulising Crohn's disease. Gut. 2011;60(6):788–98.
49. Lee RH, et al. Multipotent stromal cells from human marrow home to and promote repair of pancreatic islets and renal glomeruli in diabetic NOD/scid mice. Proc Natl Acad Sci. 2006;103(46):17438–43.
50. Bell GI, et al. Transplanted human bone marrow progenitor subtypes stimulate endogenous islet regeneration and revascularization. Stem Cells Dev. 2012;21(1):97–109.
51. Abdi R, et al. Immunomodulation by mesenchymal stem cells: a potential therapeutic strategy for type 1 diabetes. Diabetes. 2008;57(7):1759–67.
52. Fearon DT, Locksley RM. The instructive role of innate immunity in the acquired immune response. Science. 1996;272(5258):50–3.
53. Hoffmann JA. Phylogenetic perspectives in innate immunity. Science. 1999;284(5418):1313–8.
54. Janeway Jr CA, Medzhitov R. Innate immune recognition. Annu Rev Immunol. 2002;20:197–216.
55. Banchereau J, Steinman RM. Dendritic cells and the control of immunity. Nature. 1998;392(6673):245–52.
56. Mills CD. M1 and M2 macrophages: oracles of health and disease. Crit Rev Immunol. 2012;32(6):463–88.
57. Banchereau J, et al. Immunobiology of dendritic cells. Annu Rev Immunol. 2000;18:767–811.
58. Vivier E, et al. Functions of natural killer cells. Nat Immunol. 2008;9(5):503–10.

59. Gray D, et al. B-T lymphocyte interactions in the generation and survival of memory cells. Immunol Rev. 1996;150:45–61.
60. Slifka MK, et al. Humoral immunity due to long-lived plasma cells. Immunity. 1998;8(3):363–72.
61. McMenamin C, Holt PG. The natural immune response to inhaled soluble protein antigens involves major histocompatibility complex (MHC) class I-restricted CD8+ T cell-mediated but MHC class II-restricted CD4+ T cell-dependent immune deviation resulting in selective suppression of immunoglobulin E production. J Exp Med. 1993;178(3):889–99.
62. Swain SL. T cell subsets and the recognition of MHC class. Immunol Rev. 1983;74(1):129–42.
63. Shevach EM. CD4+ CD25+ suppressor T cells: more questions than answers. Nat Rev Immunol. 2002;2(6):389–400.
64. Perez VL, et al. Induction of peripheral T cell tolerance in vivo requires CTLA-4 engagement. Immunity. 1997;6(4):411–7.
65. Papamichail M, et al. Natural killer lymphocytes: biology, development, and function. Cancer Immunol Immunother. 2004;53(3):176–86.
66. Spaggiari GM, et al. Mesenchymal stem cell-natural killer cell interactions: evidence that activated NK cells are capable of killing MSCs, whereas MSCs can inhibit IL-2-induced NK-cell proliferation. Blood. 2006;107(4):1484–90.
67. Tse WT, et al. Suppression of allogeneic T-cell proliferation by human marrow stromal cells: implications in transplantation. Transplantation. 2003;75(3):389–97.
68. Chan JL, et al. Antigen-presenting property of mesenchymal stem cells occurs during a narrow window at low levels of interferon-gamma. Blood. 2006;107(12):4817–24.
69. Le Blanc K, et al. Mesenchymal stem cells inhibit and stimulate mixed lymphocyte cultures and mitogenic responses independently of the major histocompatibility complex. Scand J Immunol. 2003;57(1):11–20.
70. Li W, et al. Mesenchymal stem cells: a double-edged sword in regulating immune responses. Cell Death Differ. 2012;19(9):1505–13.
71. Bernardo ME, Fibbe WE. Mesenchymal stromal cells: sensors and switchers of inflammation. Cell Stem Cell. 2013;13(4):392–402.
72. Melief SM, et al. Multipotent stromal cells induce human regulatory T cells through a novel pathway involving skewing of monocytes toward anti-inflammatory macrophages. Stem Cells. 2013;31(9):1980–91.
73. Eggenhofer E, Hoogduijn MJ. Mesenchymal stem cell-educated macrophages. Transplant Res. 2012;1(1):12.
74. Munn DH, et al. Potential regulatory function of human dendritic cells expressing indoleamine 2,3-dioxygenase. Science. 2002;297(5588):1867–70.
75. Wakkach A, et al. Characterization of dendritic cells that induce tolerance and T regulatory 1 cell differentiation in vivo. Immunity. 2003;18(5):605–17.
76. Chung DJ, et al. Indoleamine 2,3-dioxygenase-expressing mature human monocyte-derived dendritic cells expand potent autologous regulatory T cells. Blood. 2009;114(3):555–63.
77. Jiang XX, et al. Human mesenchymal stem cells inhibit differentiation and function of monocyte-derived dendritic cells. Blood. 2005;105(10):4120–6.
78. Stagg J, et al. Interferon-gamma-stimulated marrow stromal cells: a new type of nonhematopoietic antigen-presenting cell. Blood. 2006;107(6):2570–7.
79. Le Blanc K. Immunomodulatory effects of fetal and adult mesenchymal stem cells. Cytotherapy. 2003;5(6):485–9.
80. Meisel R, et al. Human bone marrow stromal cells inhibit allogeneic T-cell responses by indoleamine 2,3-dioxygenase-mediated tryptophan degradation. Blood. 2004;103(12):4619–21.
81. Devine SM, et al. Mesenchymal stem cells are capable of homing to the bone marrow of non-human primates following systemic infusion. Exp Hematol. 2001;29(2):244–55.
82. Beyth S, et al. Human mesenchymal stem cells alter antigen-presenting cell maturation and induce T-cell unresponsiveness. Blood. 2005;105(5):2214–9.

83. Nasef A, et al. Identification of IL-10 and TGF-beta transcripts involved in the inhibition of T-lymphocyte proliferation during cell contact with human mesenchymal stem cells. Gene Expr. 2007;13(4–5):217–26.
84. Yamagiwa S, et al. A role for TGF-beta in the generation and expansion of CD4+CD25+ regulatory T cells from human peripheral blood. J Immunol. 2001;166(12):7282–9.
85. English K, et al. Cell contact, prostaglandin E(2) and transforming growth factor beta 1 play non-redundant roles in human mesenchymal stem cell induction of CD4+CD25(High) forkhead box P3+ regulatory T cells. Clin Exp Immunol. 2009;156(1):149–60.
86. Zhang M, et al. Splenic stroma drives mature dendritic cells to differentiate into regulatory dendritic cells. Nat Immunol. 2004;5(11):1124–33.
87. Di Nicola M. Human bone marrow stromal cells suppress T-lymphocyte proliferation induced by cellular or nonspecific mitogenic stimuli. Blood. 2002;99(10):3838–43.
88. Groh ME, et al. Human mesenchymal stem cells require monocyte-mediated activation to suppress alloreactive T cells. Exp Hematol. 2005;33(8):928–34.
89. Cui L, et al. Expanded adipose-derived stem cells suppress mixed lymphocyte reaction by secretion of prostaglandin E2. Tissue Eng. 2007;13(6):1185–95.
90. Taylor MW, Feng GS. Relationship between interferon-gamma, indoleamine 2,3-dioxygenase, and tryptophan catabolism. FASEB J. 1991;5(11):2516–22.
91. Terness P, et al. Inhibition of allogeneic T cell proliferation by indoleamine 2,3-dioxygenase-expressing dendritic cells: mediation of suppression by tryptophan metabolites. J Exp Med. 2002;196(4):447–57.
92. Mellor AL, Munn DH. IDO expression by dendritic cells: tolerance and tryptophan catabolism. Nat Rev Immunol. 2004;4(10):762–74.
93. Gieseke F, et al. Human multipotent mesenchymal stromal cells inhibit proliferation of PBMCs independently of IFNgammaR1 signaling and IDO expression. Blood. 2007;110(6):2197–200.
94. Munn DH. Prevention of allogeneic fetal rejection by tryptophan catabolism. Science. 1998;281(5380):1191–3.
95. Finger EB, Bluestone JA. When ligand becomes receptor – tolerance via B7 signaling on DCs. Nat Immunol. 2002;3(11):1056–7.
96. Yang SH, et al. Soluble mediators from mesenchymal stem cells suppress T cell proliferation by inducing IL-10. Exp Mol Med. 2009;41(5):315–24.
97. Spaggiari GM, et al. Mesenchymal stem cells inhibit natural killer-cell proliferation, cytotoxicity, and cytokine production: role of indoleamine 2,3-dioxygenase and prostaglandin E2. Blood. 2008;111(3):1327–33.
98. Nemeth K, et al. Bone marrow stromal cells attenuate sepsis via prostaglandin E(2)-dependent reprogramming of host macrophages to increase their interleukin-10 production. Nat Med. 2009;15(1):42–9.
99. Melief SM, et al. Multipotent stromal cells skew monocytes towards an anti-inflammatory interleukin-10-producing phenotype by production of interleukin-6. Haematologica. 2013;98(6):888–95.
100. Fontenot JD, et al. Regulatory T cell lineage specification by the forkhead transcription factor foxp3. Immunity. 2005;22(3):329–41.
101. Allavena P, et al. IL-10 prevents the differentiation of monocytes to dendritic cells but promotes their maturation to macrophages. Eur J Immunol. 1998;28(1):359–69.
102. Koppelman B, et al. Interleukin-10 down-regulates MHC class II alphabeta peptide complexes at the plasma membrane of monocytes by affecting arrival and recycling. Immunity. 1997;7(6):861–71.
103. Mantovani A, et al. The chemokine system in diverse forms of macrophage activation and polarization. Trends Immunol. 2004;25(12):677–86.
104. Rasmusson I, et al. Mesenchymal stem cells inhibit lymphocyte proliferation by mitogens and alloantigens by different mechanisms. Exp Cell Res. 2005;305(1):33–41.
105. Selmani Z, et al. Human leukocyte antigen-G5 secretion by human mesenchymal stem cells is required to suppress T lymphocyte and natural killer function and to induce CD4+CD25highFOXP3+ regulatory T cells. Stem Cells. 2008;26(1):212–22.

106. Moreau P, et al. IL-10 selectively induces HLA-G expression in human trophoblasts and monocytes. Int Immunol. 1999;11(5):803–11.
107. Kanai T, et al. Soluble HLA-G influences the release of cytokines from allogeneic peripheral blood mononuclear cells in culture. Mol Hum Reprod. 2001;7(2):195–200.
108. Rouas-Freiss N, et al. The alpha1 domain of HLA-G1 and HLA-G2 inhibits cytotoxicity induced by natural killer cells: is HLA-G the public ligand for natural killer cell inhibitory receptors? Proc Natl Acad Sci U S A. 1997;94(10):5249–54.
109. Kapasi K, et al. HLA-G has a concentration-dependent effect on the generation of an allo-CTL response. Immunology. 2000;101(2):191–200.
110. Harris SG, et al. Prostaglandins as modulators of immunity. Trends Immunol. 2002;23(3):144–50.
111. Baratelli F, et al. Prostaglandin E2 induces FOXP3 gene expression and T regulatory cell function in human CD4+ T cells. J Immunol. 2005;175(3):1483–90.
112. Matysiak M, et al. Immunoregulatory function of bone marrow mesenchymal stem cells in EAE depends on their differentiation state and secretion of PGE2. J Neuroimmunol. 2011;233(1–2):106–11.
113. Hinz B, Brune K, Pahl A. Prostaglandin E(2) upregulates cyclooxygenase-2 expression in lipopolysaccharide-stimulated RAW 264.7 macrophages. Biochem Biophys Res Commun. 2000;272(3):744–8.
114. Sheng H, et al. A critical role of IFNgamma in priming MSC-mediated suppression of T cell proliferation through up-regulation of B7-H1. Cell Res. 2008;18(8):846–57.
115. Zhang ZG, et al. VEGF enhances angiogenesis and promotes blood-brain barrier leakage in the ischemic brain. J Clin Invest. 2000;106(7):829–38.
116. Carmeliet P. VEGF as a key mediator of angiogenesis in cancer. Oncology. 2005;69 Suppl 3:4–10.
117. Gabrilovich DI, et al. Production of vascular endothelial growth factor by human tumors inhibits the functional maturation of dendritic cells. Nat Med. 1996;2(10):1096–103.
118. Yamashita K, et al. A dimeric form of soluble recombinant sheep LFA-3(CD58) inhibits human T-cell proliferation by generating regulatory T cells. Immunopharmacology. 1997;37(2–3):209–20.
119. Augello A, et al. Bone marrow mesenchymal progenitor cells inhibit lymphocyte proliferation by activation of the programmed death 1 pathway. Eur J Immunol. 2005;35(5):1482–90.
120. Shi Y, et al. How mesenchymal stem cells interact with tissue immune responses. Trends Immunol. 2012;33(3):136–43.
121. Rasmusson I, et al. Mesenchymal stem cells inhibit the formation of cytotoxic T lymphocytes, but not activated cytotoxic T lymphocytes or natural killer cells. Transplantation. 2003;76(8):1208–13.
122. Redpath S, et al. Murine cytomegalovirus infection down-regulates MHC class II expression on macrophages by induction of IL-10. J Immunol. 1999;162(11):6701–7.
123. Gotherstrom C, et al. Immunomodulatory effects of human foetal liver-derived mesenchymal stem cells. Bone Marrow Transplant. 2003;32(3):265–72.
124. Cutler AJ, et al. Umbilical cord-derived mesenchymal stromal cells modulate monocyte function to suppress T cell proliferation. J Immunol. 2010;185(11):6617–23.

Chapter 5
Embryonic Stem Cells and Fetal Development Models

Monika Nowak-Imialek and Heiner Niemann

Introduction

In the laboratory mouse, four different types of pluripotent stem cells have been successfully established, including embryonic stem cells (ES cells) from the inner cell mass (ICM) of blastocysts, epiblast stem cells (EpiSCs) from the developing epiblast, embryonic germ cells from primordial germ cells (EGCs) and embryonic carcinoma cells (ECCs) from teratocarcinomas [38]. The development of the mammalian embryo begins with the formation of the totipotent zygote from which all embryonic and placental tissues are derived. Following cleavage division and formation of the blastocyst, with the two compartments, the trophectoderm and the inner cell mass (ICM), cells of the ICM gradually lose totipotency and give rise to the three embryonic lineages, but are not able to form the placenta, and are called pluripotent. Pluripotency is lost after implantation in the uterus when the epiblast epithelializes and is being prepared for gastrulation and organogenesis [8]. The first population of stem cells that can be identified in the developing fetus at the onset of gastrulation are the primordial germ cells (PGCs) located in the proximal part of the epiblast [67]. Thereafter PGCs migrate into the genital ridges to form ovaries or testis, initiate cell division, meiosis and differentiation (oogenesis or spermatogenesis) and ultimately form female or males gametes. PGCs from genital ridges and spermatogonial stem cells from testis can be converted into pluripotent germ cells under suitable *in vitro* conditions. The presence of female germ line stem cells (FGSCs), also called oogonial stem cells (OSCs), in the adult mammalian ovary which are able to forming new oocytes and follicles after birth is controversially discussed in the literature (see [27]).

M. Nowak-Imialek, D.V.M. • H. Niemann, D.V.M., Ph.D. (✉)
Department of Biotechnology, Institute of Farm Animal Genetics (FLI),
Mariensee, Neustadt 31535, Germany
e-mail: monika.nowak@fli.bund.de; heiner.niemann@fli.bund.de

D.O. Fauza, M. Bani (eds.), *Fetal Stem Cells in Regenerative Medicine*,
Stem Cell Biology and Regenerative Medicine,
DOI 10.1007/978-1-4939-3483-6_5

Pluripotent stem cells are a unique type of cells because they remain undifferentiated indefinitely under appropriate *in vitro* culture conditions and can be induced to form virtually all cell types of the mammalian organism. These properties render pluripotent cell lines a valuable tool for research in developmental biology and make them attractive for application in therapies for many currently incurable illnesses and in regenerative medicine [93]. True so-called bona fide pluripotent ES cells, which can colonize the germ line and produce functional germ cells, have only been described in the laboratory mouse [23, 61] and rat [14, 59]. Well-characterized pluripotent cells isolated from early embryos have been established from monkey and human embryos [106, 107]. However, germ line contribution of these cell lines could not be proven, mainly due to ethical and legal restrictions.

Prior to clinical application of pluripotent cells or tissue, their survival and functional integration, their long-term genetic stability and the absence of tumorigenic potential must be assessed in suitable pre-clinical animal models. The domestic pig is considered as an excellent experimental model in pre-clinical trials of cell therapy because of its genetic, morphological and physiological similarity with humans [10, 52, 71]. Until now, no true germ line competent pluripotent stem cell lines have been reported in farm animals. In most cases, pluripotent stem cells derived from farm animals have failed to maintain or reach the pluripotent state and were therefore called "pluripotent stem-like cells" or "putative stem cells" [76]. It is not yet clear, whether this is due to deficiencies of current *in vitro* culture conditions, which do not support proliferation of farm animals derived stem cells and maintenance of pluripotency, or due to the lack of knowledge about factors regulating stem cells in other species than human and rodents, or the lack of appropriate markers which can be used for identification of pluripotent stem cells in pigs [33].

The pathways that regulate cell renewal and pluripotency in stem cells have been studied in detail in mouse and human [94]. A better understanding of cell signalling events in porcine pluripotent cells may help to improve *in vitro* culture conditions and allow for the establishment and prolonged culture of bona fide pluripotent stem cells from pigs [34]. Pluripotency is maintained by a dense network of different transcription factors and is influenced by specific signalling pathways [70]. Considerable differences in cell signalling exist between mouse and human ES cells. In the mouse, pluripotency is mainly regulated by Janus kinases (JAK)/signal transducers and activators of transcription (STAT) signalling and WNT and bone morphogenetic protein (BMT) signalling. In contrast, human transcription regulation of pluripotency critically depends on fibroblast growth factor (FGF) and transforming growth factor (TGF)-b/ACTIVIN/NODAL signalling [34, 70]. In the pig, pluripotent pathways are not well defined, but it has been shown that FGF and ACTIVIN/NODAL signalling rather than LIF/JAK/STAT3 are important in regulating pluripotency in the porcine inner cell mass or epiblast [2, 35]. Despite numerous attempts, stable proliferating porcine ESC lines with all pluripotent characteristics, including germ line competence have not yet been established. *In vitro* culture protocols optimized for mouse and human ESCs did not support proliferation or maintenance of pluripotency of cultured porcine cells [11, 76]. Reliable markers for identification of pluripotent stem cells in pig are not yet available. We have recently

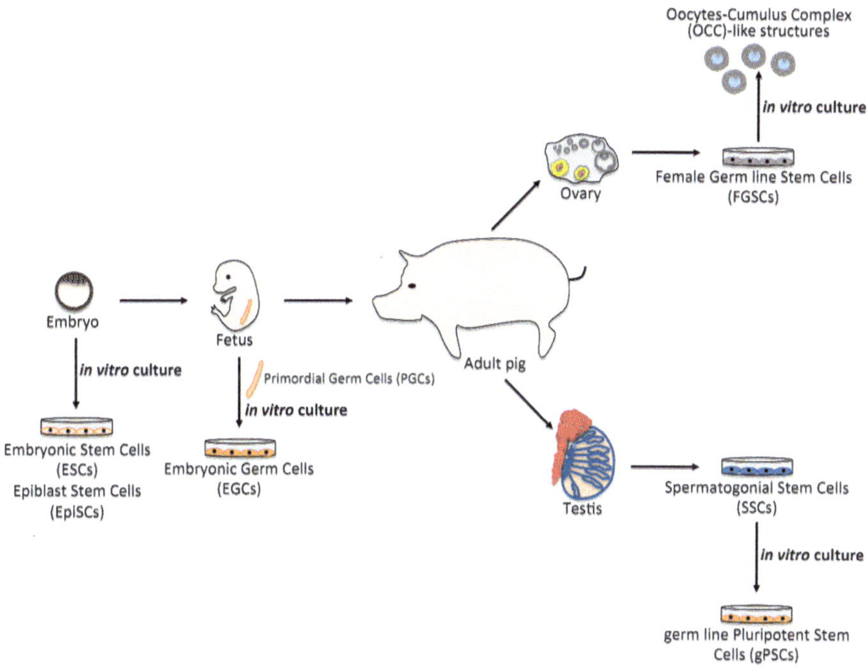

Fig. 5.1 Potential sources for pluripotent stem cells in the pig. Porcine pluripotent stem cells can be established from the inner cell mass of blastocysts (ESCs) or developing epiblast (EpiSCs). During fetal development pluripotent embryonic germ cells (EGCs) can be derived *in vitro* from primordial germ cells (PGCs) of genital ridges. In addition several groups have derived pluripotent stem cell-like cells from gonads, including germ line pluripotent stem cells (gPSCs) from *in vitro* cultured spermatogonial stem cells (SSCs) from testes or female germ line stem cells from ovary (FGSCs)

established an Oct4-EGFP transgenic pig model to facilitate identification and monitoring of porcine pluripotent cells *in vivo* and *in vitro*. This chapter focuses on the recent progress in the derivation of porcine pluripotent cells from embryonic and fetal tissue (Fig. 5.1) and summarizes the production and validation of Oct4-EGFP transgenic pigs that were used as model for study of porcine pluripotent cells (Fig. 5.2).

Porcine Embryonic and Epiblast Stem Cells

More than 20 years ago, first attempts to isolate porcine embryonic stem cells from *in vivo* embryos [24, 72, 73, 91], *in vitro* produced blastocysts or embryos after parthenogenetic activation (see [9, 11, 12, 33, 46, 105, 110]) have been reported. It has been shown that porcine ES-like cells grow either as flatted, polygonal epithelial-like [16, 66, 89], or as fibroblast or trophoblast-like cells [101]. Multiple cell lines

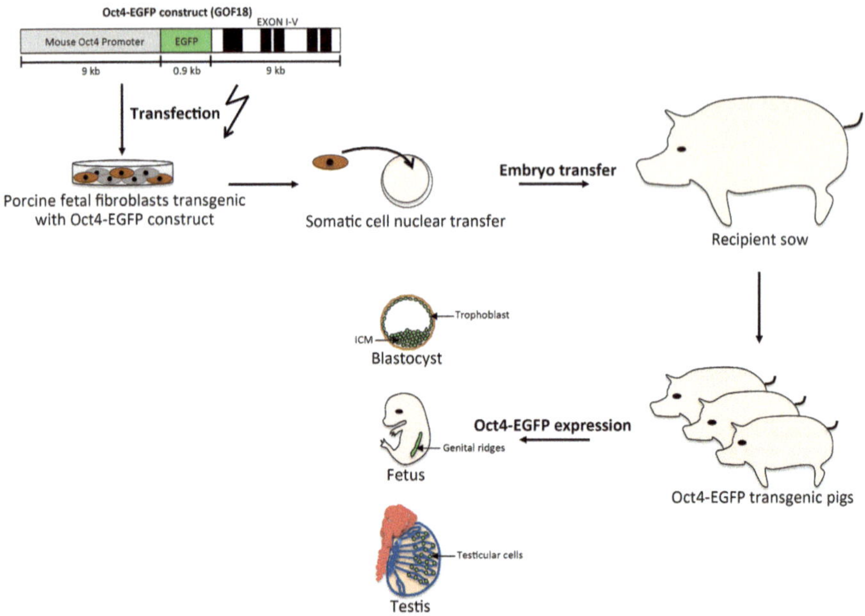

Fig. 5.2 The Oct4-EGFP transgenic pig model for monitoring of OCT4 expressing cells in pig. The murine Oct4-EGFP (GOF-18/EGFP, OG2) construct containing the genomic clone of the entire Oct4 promoter (9 kb) fused to the EGFP cDNA and approximately 9 kb of the Oct4 exon/intron region has been used for the generation of Oct4-EGFP transgenic pigs. Stable transfected porcine fetal fibroblasts with the OG2 construct were used as donor cells for somatic cell nuclear transfer. Reconstructed embryos were transferred to recipients sows which delivered Oct4-EGFP transgenic pigs. Expression of the EGFP reporter was confined to the inner cell mass (ICM) and trophectoderm of blastocyst, in genital ridges isolated from OG2 fetuses and in testicular cells from testes of adult OG2 boars

have been isolated and were maintained over extended periods of time *in vitro* [90], formed embryoid bodies and expressed alkaline phosphatase (AP), the first basic marker for identification of stem cells [2, 16, 58, 103]. However, isolated stem cell-like lines were poorly characterized mainly at early passages based on morphology and expression of a small number of pluripotent stem cell markers, including OCT4, NANOG, SOX2 or SSEA-1 [2, 6, 109, 112, 113, 117]. ES-like colonies could be established from ICMs isolated from somatic cell nuclear transfer (SCNT) derived blastocysts [47, 48]. These cells showed typical ES-like cells morphology with compact shape and distinct borders. Most of colonies grow for 2–5 passages and then differentiated or degenerated; two cell lines could be established. These lines expressed OCT4 and surface marker proteins (SSEA-1, SSEA-4, TRA-1–60 and TRA-1–81), formed embryoid bodies *in vitro*, but failed to generate teratomas.

Unfortunately, most of the established porcine cell lines had only a limited capacity to differentiate *in vitro* into derivatives of the three germ layers [2, 6, 112, 113]. The established putative porcine stem cell lines did not meet essential criteria of pluripotency, including teratoma and chimera formation with germ line contribution. Only few groups have characterized their cell lines by *in vivo* differentiation assay. In one case, injection of ES-like cells into nude mice resulted in teratomas [39]. Three studies demonstrated production of chimeric pigs after injection of putative porcine ES cells into blastocysts [16, 112, 115]. Contribution to the three germ layers was minor only and could be determined by sensitive PCR analysis or by coat spotting, whereas germ line chimerism could not be demonstrated.

Another source for isolation of pluripotent stem cells from embryos is the epiblast. There is one report on the isolation of cell lines from porcine epiblast (EpiSCs) [2]. These cell lines expressed the core pluripotency factors OCT4, NANOG, SOX2, and NODAL. Similar to mouse EpiSC and human ESC, cultured porcine epiblast-like cells critically depend on ACTIVIN/NODAL signalling for self-renewal [70]. But teratomas and chimera formation were not reported.

Similar methods as those employed in the mouse and human have been used in most studies on the derivation and *in vitro* culture of porcine pluripotent stem cell lines from porcine embryos. However, these were not compatible with extended proliferation or true pluripotency. One likely explanation could be differences in preimplantation development and formation of the placenta which is non-invasive eptheliochorial in the pig versus invasive hemochorial in human and mouse [36, 49]. The porcine inner cell mass (ICM) is formed on day 4–5 after fertilization in the blastocyst. The formation of porcine epiblast starts later and takes longer than in mouse and humans, the epiblast appears after 6–7 days and is visible until day 12 post fertilization of porcine development [35]. Porcine implantation starts on day 14 (d14) of development compared to d5 and d8 in mice and humans, respectively [33]. Even 24 years after the first report on the derivation of porcine ES-like cells, porcine ES cells which meet all criteria for pluripotency, including chimera formation with germ line contribution have not been reported.

Porcine Embryonic Germ Cells (pEGCs)

Embryonic germ cells derived from cultured PGCs could be an alternative source of pluripotent cells since pluripotent ES cells are not available in pigs. PGCs are unipotent progenitors which ultimately form sperm or eggs, and can first be identified during gastrulation [62]. They form clusters of 50–100 cells at the base of the allantois [65] and subsequently migrate into the forming genital ridges of early fetuses [3, 40] where they proliferate and start to differentiate according to embryonic sex. PGCs isolated from genital ridges are not pluripotent, and do not contribute to chimeras following blastocyst injection [100]. However, when mouse PGCs were cultured *in vitro* in the presence of growth factors and cytokines, including LIF (Leukemia Inhibitor Factor), SCF (Stem Cell Factor) and bFGF (basic Fibroblast Growth Factor) they were epigenetically reprogrammed to pluripotent cells and called embryonic germ cells (EGCs) [63]. These cells share several important characteristics with ESCs, such as self-renewal, morphology, high activity of alkaline phosphatase, expression of pluripotency genes and surface antigens and the potency to form teratomas and to contribute to the germ line in chimeras after injection into host blastocysts [50, 63, 100].

More recently, mouse EGCs could be established with high efficiency using the two-inhibitor (2i) culture system (inhibitors for autocrine mitogen-activated protein kinase (MAPK) signalling and glycogen synthase kinase) and supplementation with LIF. These EGC lines contribute extensively to healthy chimeric mice, including the germ line [57]. It has been shown that stage of embryonic development is a critical factor for successful derivation of murine EGC lines. Mouse EGC lines have been successfully established from d8.5–13.5 embryos [22, 63, 98].

In the pig, EGCs have been derived from fetuses between days 20 and 28 of pregnancy [17, 86, 92, 97]. Derivation of EGCs from later stages (d30) has been unsuccessful [17]. On day 30 porcine EGCs colonize the genital ridges and inhibit cell division. Establishment of new DNA methylation marks in PGCs at later stages of development may have affected successful derivation of EGCs [17, 87]. The first attempts towards successful *in vitro* culture and characterization of porcine EGCs was performed in medium without any growth factors and the resulting cell lines resembled mouse EGCs, expressed alkaline phosphatase and differentiated *in vitro* into various cell type [97]. After injection of these PGCs into blastocysts one chimeric piglet was born, verified by the presence of skin spots, but germ line contribution could not be detected. This prompted a series of experiments with various combinations of growth factors and cytokines such as LIF, SCF, and bFGF and different feeder cells for identification of the best *in vitro* culture condition for porcine EGCs [21, 55, 69, 85, 108]. Porcine EGCs proliferated over 54 passages when the medium was supplemented with knockout serum replacement (KSR), LIF, SCF and bFGF [85]. Most porcine EGC lines expressed pluripotent stem cell markers, including AP, SSEA-1, SSEA-3, SSEA-4, TRA-1-81 and OCT4 [17, 20, 69, 85]. Some of these lines formed embryoid body-like structures and differentiated into three germ layers *in vitro* [17, 86, 108]. Few groups tested pluripotency of porcine EGCs in chimera experiments [92, 96, 97], but no germ line contribution has been detected. In one case, injected EGCs were found with very low efficiency (14 %) in fetal gonadal tissue by sensitive PCR and blotting techniques [69].

The potential of EGCs for the production of transgenic pigs has been demonstrated [1, 92, 95]. Porcine EGCs were stably transfected with an EGFP construct [1, 92, 95] and integrated into the inner cell mass of host blastocysts after injection [95] or gave rise to blastocysts expressing GFP after use as donor cells in SCNT [1]. Transgenic EGC-like cells expressing GFP contributed to the germ layers in chimeric fetuses and stillborn piglets. Moreover, somatic, but not germ line chimerism, has been reported from blastocyst injection using porcine EGCs [92]. Published evidence indicates that porcine EGCs with true pluripotent characteristics, which are equivalent to mouse ESCs, have not yet been established.

Germ Line Stem Cells from Adult Pigs

The genital ridge is the somatic precursor of both the ovaries and the testes. In the genital ridges PGCs proliferate and start differentiation according to the sex of the embryos. PGCs have the potential to develop either as meiotic oocytes surrounded by a single layer of somatic cells, all together forming the primordial follicle, or as pro-spermatogonia enclosed from precursors of Sertoli and peritubular myoid cells (pre-seminiferous tubules) [64]. In the adult male testis, a population of spermatogonial stem cells (SSCs) possesses the ability of self-renewal and to produce progenitor cells, which differentiate into mature spermatozoa [13]. SSCs isolated from fetal and adult mouse testis can be propagated *in vitro* for prolonged period of time [45]. Similarly to PGCs, unipotent mouse SSCs have been converted under specific *in vitro* culture conditions into pluripotent cells called germ line derived pluripotent stem (gPS) cells [51].

It has long been postulated that ovaries do not have the capability to replenish the fixed pool of oocytes during adult life, which in turn results in progressive loss of follicles [81]. This long-held concept was challenged and a population of rare female germ line or oogonial stem cells (FGSCs) with germ cell morphology, self-renewal capacity similar to the SSCs in the testes and development to mature oocytes under *in vitro* culture was identified [42, 116]. However, the presence of FGSCs is still controversially discussed (see [27]).

The recent progress in SSCs and FGSCs isolation, characterization, *in vitro* culture and manipulation in rodents prompted scientists to optimize these procedures in farm animals. Isolation and long-term culture of germ line stem cells from testis and/or ovaries of farm animals would offer a new source of stem cells, which could be used as substitute for the lacking germ line competent pluripotent ES cells. The recent progress in isolation, characterization, and *in vitro* culture of porcine SSCs and FGSCs is provided below.

Porcine Testis-Derived Germ Stem Cells

Information about *in vitro* culture conditions that support maintenance and proliferation of porcine gonocytes or spermatogonial stem cells (SSCs) is scarce. SSCs arise from gonocytes in the juvenile testis, which originate from PGCs from genital ridges appearing during fetal development. Gonocytes represent a very rare population within the testicular cells and comprise ~1 % of the neonatal testicular cells in the newborn rat [111]. During the first week after birth gonocytes resume proliferation and migrate to the basement membrane of the seminiferous tubules and differentiate into SSCs (reviewed by [68]). The timing of the transition of gonocytes into SSCs is difficult to assess and starts approximately at the age of 2 months in pigs [30]. Gonocytes isolated from mouse testis proliferated over 5 months in the *in vitro* culture in the presence of cytokines and growth factors, including LIF, GDNF (Glial cell line-Derived Neurotrophic Factor), EGF (Epidermal Growth Factor), and bFGF on mitomycin C inactivated mouse embryonic feeder cells [45]. Porcine gonocytes have been successfully isolated from neonatal testis and could be maintained *in vitro* without growth factors for a maximum of 1 week [29, 30]. Primary porcine gonocyte cultures expressed pluripotency markers such as SSEA-1, NANOG, OCT4 and specific germ cell markers such as DBA (Dolichos Biflorus Agglutinin), KIT (know as c-KIT) and ZBTB16 (Zinc finger and BTB domain-containing protein 16). Interestingly, freshly isolated gonocytes either did not or only weakly expressed pluripotency determining transcription factors [29]. Gonocytes from primary cultures formed teratomas containing tissue from the three germ layers after subcutaneous injection in nude mice. In another study, *in vitro* culture of porcine gonocytes in culture medium supplemented with EGF and bFGF resulted in two cell lines resembling mouse SSCs which could be maintained for nine passages [54]. Porcine gonocytes could be successfully cultured *in vitro* using culture medium supplemented with 1 % of fetal bovine serum (FBS), EGF and bFGF [125].

Recent studies in mouse suggested that pluripotency persists in spermatogonial stem cells, which constitutes the founder cell population for spermatogenesis in the adult testis [43]. Protocols for efficient isolation and prolonged *in vitro* culture of SSCs have been described in rodents [43, 44, 119]. Likewise, SSCs are an extremely rare population in the testis and only 0.02–0.03 % of the total mouse testicular cells have stem cells activity [104]. Interestingly, mouse SSCs from both juvenile and adult testis can be converted *in vitro* into pluripotent stem cells called germ line pluripotent stem cells (gPS cells) [51]. Pluripotency of mouse gPS cells was confirmed by both, *in vitro* and *in vivo* differentiation, germ line contribution in chimeras and germ line transmission to the next generation [51].

Little is known about the *in vitro* culture conditions that support long term maintenance and proliferation of porcine SSCs. Survival of porcine SSCs *in vitro* was enhanced in medium supplemented with SCF and GM-CSF (Granulocyte Macrophage-Colony Stimulating Factor), but stem cell characteristics of porcine SSCs were not reported [19]. Another hurdle towards establishing SSCs cultures is the lack of informative porcine markers. PGP 9.5 (Protein Gene Product 9.5) and DBA seem to be reliable markers and were consistently expressed in porcine gonocytes/SSCs [30, 60]. However, *in vitro* culture conditions for porcine gonocytes or SSCs are largely elusive; medium supplemented with growth factors successfully used for rodents did not support porcine SSCs proliferation. To convert porcine SSCs into pluripotent germ line derived stem cells, it is necessary to maintain porcine gonocytes/SSCs for long-term in *in vitro* culture.

Porcine Ovary-Derived Germ Stem Cells

The dogma that the ovaries contain at birth a fixed number of mitotically active oocytes and there are no female germ line stem cells was published more then 60 years ago [129] and had not been challenged for many decades. Only recently, the availability of female germ line (or oogonial) stem cells (FGSCs) in the adult ovary has emerged as a controversial issue in the field of reproductive science (reviewed by [18]). Experimental evidence for the presence of putative FGSCs in adult ovaries was shown in mouse [42, 80, 116, 128], human [4, 116], rats [126] and monkey [118]. Human oogonial stem cells was expanded *in vitro* for several months and spontaneously formed oocytes *in vitro*, which had the capacity to become fertilized [116]. Freshly isolated mouse and human FGSCs expressed high levels of genes specific for early germ cells (Blimp1, Dppa3, Iftitm3), and the catalytic subunit of telomerase (Tert) [116].

However, the results are not yet conclusive, because the existence of extra-ovarian FGSCs in the adult mouse originating from peripheral blood cells or from ovarian surface epithelium (OSE) cannot be ruled out [41, 56, 82, 114]. This warrants further study into the origin and features of the putative FGSCs. Putative porcine FGSCs were located in the Theca layers and were positive for markers specific for pluripotent and germ line cells (OCT4, SSEA-4, SSEA-3, c-KIT, c-MYC, KLF4, SOX2, NANOS2, CD49f and VASA) [5]. Moreover, they maintained similar characteristics as mouse FGSCs and ESCs over a 4-months *in vitro* culture period. Porcine FGSCs differentiated *in vitro* into many different cell types, including adipocyte-like cells, cardiac or neuronal cells, but no oocytes were observed. Putative porcine PGCs-like cells isolated from ovarian surface epithelium of adult pig ovaries were generated *in vitro* from Vasa-positive VSEL stem cells. Porcine PGCs-like cells expressed germ and stem cell markers like Fragilis, THY-1, SSEA-4 and c-KIT after 1 week of *in vitro* culture. Surprisingly, after 4 weeks of culture these cells started also to express ESCs markers, like NANOG, SOX2, REX1, c-MYC and KLF4, suggesting dedifferentiation of the cells [15]. Most likely, the establishment and maintenance of FGSCs cultures require co-culture with somatic cells from the ovaries. Porcine PGC-like cells could be maintained *in vitro* for 6 months without loss of proliferation potential. About 0.1 % of porcine PGC-like cells cultured *in vitro* under differentiation conditions formed oocyte-cumulus complex (OCC)-like structures [15]. However, the origin, regulation of proliferation and *in vitro* differentiation of porcine FGSCs remains to be unequivocally demonstrated.

Transgenic Pig Models for Tracing of Stem Cells during Embryonic and Fetal Development

Information on the characteristics, maintenance and self-renewal of porcine pluripotent stem cells is limited which hampers the ability to develop pig models for pre-clinical testing of novel cell therapies. Knowledge on stem cell self-renewal and development is largely based on studies performed in rodents. However, mouse models frequently do not mimic the human situation well enough. The use of fluorescent proteins driven by stem or germ cell specific promoters for the production of transgenic animals could provide an important tool for labelling mammalian stem cells in culture or *in vivo* in whole organisms [31, 32]. The main advantage of fluorescent reporters is the non-invasive live imaging for localization and monitoring stem cells and their isolation using fluorescence activated cell sorter (FACS) techniques without additional staining. Transgenic mice expressing fluorescent protein driven by stem cell specific promoters have been widely used to monitoring stem or germ cell behaviour ([28, 77, 78, 83, 120, 122]; reviewed by [37, 102]). The Stella-GFP transgene with fluorescence restricted exclusively to the germ line has been useful to visualizing mouse PGCs *in vivo* and at derivation of PGCs from ES cells [83]. Neurogenin 3-GFP transgenic mice have been used in the study of germ cells in adult mouse testes [122]. The GFP positive cells were localized near the basement membrane and represented undifferentiated spermatogonial stem cells. Transgenic mice expressing GFP driven by the Oct4 promoter (Oct4-EGFP mice) have emerged as extremely valuable model for *in vivo* imaging of either pluripotent stem cells in preimplantation embryos, or germ cells during migration and colonization of genital ridges or cells with stem cell properties in ovaries and testes [25, 79, 120, 123, 124].

The Oct4 gene is a member of the POU (Pit-Oct-Unc) family of transcription factors that is crucial for transcription regulation during preimplantation development and is involved in controlling self-renewal and maintenance of pluripotency [7]. In the mouse, the Oct4-EGFP transgene mimics exactly the expression profile of the endogenous Oct4 gene [123]. Recently, Oct4-EGFP rabbits have also been produced and characterized [121]. The Oct4-EGFP transgenic pigs are the only available large animal model for monitoring pluripotency [74]. Oct4-EGFP transgenic pigs were successfully produced by using the murine 18 kb genomic sequence of the mouse Oct4 gene fused to the enhanced green fluorescent (EGFP) cDNA (GOF-18/EGFP) [74]. These Oct4-EGFP transgenic pigs are equivalent to the long established Oct4-EGFP mouse model. Expression of the EGFP reporter was found in ICM and trophectoderm in blastocysts produced by somatic cell nuclear transfer with Oct4-EGFP transgenic porcine fibroblasts or collected 5 days after mating of a wild-type sow to an Oct4-EGFP transgenic boar (Fig. 5.3). During fetal development expression of the Oct4-EGFP transgene was restricted to germ cells isolated from genital ridges of 25 days old porcine fetuses (Fig. 5.3; [74]). In adult pigs the transgene was expressed in testicular cells, but not in mature spermatozoa. This is consistent with findings in male Oct4-EGFP transgenic mice, where Oct4 expression is found in post-proliferative pro-spermatogonia after birth and after onset of spermatogenesis is maintained in undifferentiated A spermatogonial stem cells (SSCs), before it is down-regulated during germ cell differentiation [84].

Purification of EGFP positive cells is necessary for identifying and characterizing OCT4 expressing cells in porcine testis. Pig testes are composed of the germ cells, incl. spermatogonial stem cells, the meiotic spermatocytes, spermatids and the somatic cell compartment with the Sertoli and Leydig cells, both playing a major role for functional spermatogenesis. Information on porcine SSCs is scarce; their morphology, specific markers and *in vitro* culture conditions are largely unknown.

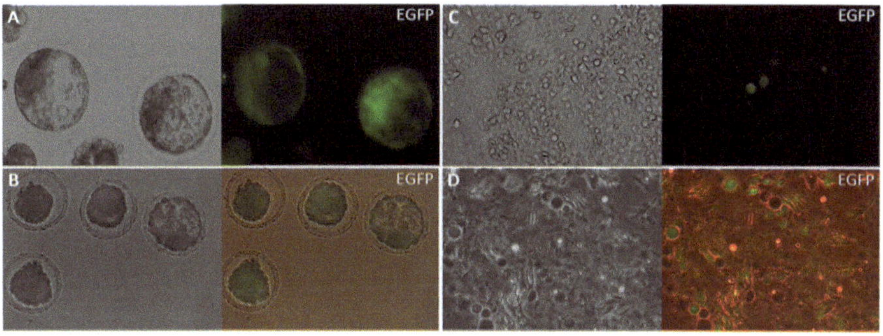

Fig. 5.3 Reactivation of the Oct4-EGFP (OG2) transgene in pig. EGFP expression has been found in porcine cloned blastocysts *in vitro* from day 5 or 6 (**a**) and in *in vivo* blastocysts day 6 (**b**). Thereafter, in genital ridges isolated from Oct4-EGFP transgenic fetuses at day 25 (**c**) and in testicular cells from adult boars (**d**). Bright-field image (*left*) and corresponding EGFP fluorescence image (*right*)

The Oct4-EGFP transgenic pig model has proven useful for identification, visualization and isolation of EGFP expressing cells in our laboratory (Nowak-Imialek, unpublished data). To define the origin of EGFP expressing cells, we isolated these cells from adult transgenic testis using fluorescence-activated cell sorting (FACS)-based techniques. Gene expression analysis of isolated EGFP positive cells demonstrated the presence of genes specific for undifferentiated (OCT4, UTF1, FGFR3, PGP 9.5, THY-1, SALL4 and GFRα1) but also for differentiated (BOLL and PRM2) germ cells. Markers specific for Sertoli cells (VIMENTIN) and Leydig cells (LHCGR) were not observed. To verify the localization of EGFP positive cells in seminiferous tubules, we performed immunohistochemical detection of EGFP in adult pig testis. Unlike the Oct4-EGFP reporter mouse model, GFP protein was not found in spermatogonia attached to the basement membrane of seminiferous tubules, but instead were found in differentiated germ cells, including spermatocytes and spermatids. These results show that Oct4-EGFP expression in testis differs between mouse and porcine Oct4-EGFP transgenic models (Nowak-Imialek, unpublished data).

A promising application of the Oct4-EGFP transgenic pigs or cells thereof will be in reprogramming studies, where the EGFP expression can readily identify pluripotent cells. Somatic cells from the Oct4-EGFP transgenic pigs were subjected to different reprogramming protocols to test their usefulness for monitoring the epigenetic reprogramming process [53, 74]. The usefulness of the transgene for monitoring reprogramming was first demonstrated by fusion of porcine Oct4-EGFP fibroblasts with pluripotent mouse ES cells [74]. The resulting inter-species hybrids formed aggregated colonies typical for murine ES cells, showed a high proliferation rate and reactivated the EGFP fluorescence after 3 days. However, mouse-pig hybrids were unstable and lost EGFP fluorescence during *in vitro* culture. Probably incompatibilities between mouse and porcine genome after cell fusion caused a loss of porcine chromosomes [75].

Porcine Oct4-EGFP fibroblasts were also reprogrammed employing viral transduction or by using the non-viral Sleeping Beauty transposon system to deliver the reprogramming factors Oct4, Sox2, Klf4, and cMyc [53, 74]. Successful reprogramming to the pluripotent state was indicated by changes in cell morphology and reactivation of the Oct4-EGFP reporter. The transposon-reprogrammed induced pluripotent stem (iPS) cells showed long-term proliferation *in vitro* over 40 passages, expressed embryonic stem cells related transcription factors, including OCT4, NANOG, SOX2, REX1, ESRRB, DPPA5, and UTF1 and surface markers of pluripotency, including SSEA-1 and TRA-1-60. *In vitro* differentiation resulted in derivatives of the three germ layers and after injection of putative iPS cells under the skin of immunodeficient mice teratomas were observed [53]. However, the variable morphology of porcine iPS cells at later passages and the low number of Oct4-EGFP positive cells indicated that maintenance of pluripotency in reprogrammed porcine cells was not consistently achieved under our *in vitro* culture conditions.

These data show that the availability of Oct4-EGFP transgenic pigs provides a useful monitoring system for studying factors which are critical for the maintenance of pluripotency and should thus facilitate establishing a culture system that is compatible with long-term proliferation of porcine pluripotent cells.

Another germ cell specific transgenic reporter is the Stimulated Retinoic Acid 8 (STRA8), which is expressed in mouse gonocytes and premeiotic spermatocytes [99, 127]. Stra8-EYFP pigs showed a stronger signal in premeiotic cells and early spermatocytes than in other cell types. These transgenic pigs may also be useful for germ cell transplantation and studies of *in vitro* spermatogenesis [99]. However, Stra8 marks also a differentiated cell population in mouse testes, which is in contrast to the Oct4 gene that is exclusively expressed in spermatogonial stem cells [26].

Conclusions and Perspectives

Germ line competent pluripotent stem cell lines were successfully established from laboratory mouse and rat. Although the laboratory mouse is an excellent model for basic research, its short life span, small size and high inbreeding limits application for novel cellular therapies of regenerative medicine. The domesticated pig is an attractive large animal model for pre-clinical testing of safety and efficiency of stem cell-based therapies [53]. However, porcine pluripotent stem cell-like characteristics are only partially similar to those of true pluripotent stem cells and they could not be maintained for extended periods of time *in vitro*. Recently, significant advances have been made in the reprogramming of porcine somatic cells into induced pluripotent stem cells (iPSCs). However, so far porcine iPSCs and ESCs failed to meet the full set of criteria for pluripotency [76, 88]. A major challenge is to determine optimal *in vitro* culture conditions, which are different from those commonly used for pluripotent cells from rodents and human. The limited information on key signalling pathways and growth factors involved in regulation of self-renewal and pluripotency in porcine stem cells prevents identification of cell culture conditions supporting long-term proliferation of these cells. The Oct4-EGFP transgenic pigs that were generated in our laboratory facilitate the derivation of germ line competent pluripotent stem cells from domesticated pigs. These pigs provide a unique tool for analysing the origin and properties of OCT4 expressing cells *in vivo*, but also for establishing effective *in vitro* culture systems for pluripotent cells.

References

1. Ahn KS, Won JY, Heo SY, Kang JH, Yang HS, Shim H. Transgenesis and nuclear transfer using porcine embryonic germ cells. Cloning Stem Cells. 2007;9:461–8.
2. Alberio R, Croxall N, Allegrucci C. Pig epiblast stem cells depend on activin/nodal signaling for pluripotency and self-renewal. Stem Cells Dev. 2010;6:484–95.
3. Anderson R, Copeland TK, Schöler H, Heasman J, Wylie C. The onset of germ cell migration in the mouse embryo. Mech Dev. 2000;91:61–8.
4. Anderson RA, McLaughlin M, Woods DC, Tilly JL, Telfer EE. Evaluation of oogonial stem cells and neo-oogenesis in girls and women with Turner syndrome: a pilot study. Hum Reprod. 2013;28(Suppl):52.
5. Bai Y, Yu M, Hu Y, Qiu P, Liu W, Zheng W, Peng S, Hua J. Location and characterization of female germline stem cells (FGSCs) in juvenile porcine ovary. Cell Prolif. 2013;46:516–28.
6. Blomberg LA, Schreier LL, Talbot NC. Expression analysis of pluripotency factors in the undifferentiated porcine inner cell mass and epiblast during in vitro culture. Mol Reprod Dev. 2008;75:450–63.
7. Boiani M, Schöler HR. Regulatory networks in embryo-derived pluripotent stem cells. Nat Rev Mol Cell Biol. 2005;6:872–84.
8. Boroviak T, Nichols J. The birth of embryonic pluripotency. Philos Trans R Soc Lond B Biol Sci. 2014;369(1657):pii:20130541.
9. Brevini TA, Antonini S, Cillo F, Crestan M, Gandolfi F. Porcine embryonic stem cells: facts, challenges and hopes. Theriogenology. 2007;68 Suppl 1:S206–13.
10. Brevini TA, Antonini S, Pennarossa G, Gandolfi F. Recent progress in embryonic stem cell research and its application in domestic species. Reprod Domest Anim. 2008;43 Suppl 2:193–9.
11. Brevini TA, Pennarossa G, Gandolfi F. No shortcuts to pig embryonic stem cells. Theriogenology. 2010;74:544–50.
12. Brevini TA, Tosetti V, Crestan M, Antonini S, Gandolfi F. Derivation and characterization of pluripotent cell lines from pig embryos of different origins. Theriogenology. 2007;67:54–63.
13. Brinster RL, Zimmermann JW. Spermatogenesis following male germ-cell transplantation. Proc Natl Acad Sci U S A. 1994;91:11298–302.
14. Buehr M, Meek S, Blair K, Yang J, Ure J, Silva J, McLay R, Hall J, Ying QL, Smith A. Capture of authentic embryonic stem cells from rat blastocysts. Cell. 2008;135:1287–98.
15. Bui HT, Van Thuan N, Kwon DN, Choi YJ, Kang MH, Han JW, Kim T, Kim JH. Identification and characterization of putative stem cells in the adult pig ovary. Development. 2014;141:2235–44.
16. Chen LR, Shiue YL, Bertolini L, Medrano JF, BonDurant RH, Anderson GB. Establishment of pluripotent cell lines from porcine preimplantation embryos. Theriogenology. 1999;52:195–212.
17. Cong Y, Ma J, Sun R, Wang J, Xue B, Wang J, Xie B, Wang J, Hu K, Liu Z. Derivation of putative porcine embryonic germ cells and analysis of their multi-lineage differentiation potential. J Genet Genomics. 2013;40:453–64.
18. De Felici M, Barrios F. Seeking the origin of female germline stem cells in the mammalian ovary. Reproduction. 2013;146:R125–30.

19. Dirami G, Ravindranath N, Pursel V, Dym M. Effects of stem cell factor and granulocyte macrophage-colony stimulating factor on survival of porcine type A spermatogonia cultured in KSOM. Biol Reprod. 1999;61:225–30.
20. Dong X, Tsung H, Mu Y, Liu L, Chen H, Zhang L, Wang H, Feng S. Generation of chimeric piglets by injection of embryonic germ cells from inbred Wuzhishan miniature pigs into blastocysts. Xenotransplantation. 2013;21:140–8.
21. Durcova-Hills G, Prelle K, Muller S, Stojkovic M, Motlik J, Wolf E, Brem G. Primary culture of porcine PGCs requires LIF and porcine membrane-bound stem cell factor. Zygote. 1998;6:271–5.
22. Durcova-Hills G, Tokunaga T, Kurosaka S, Yamaguchi M, Takahashi S, Imai H. Immunomagnetic isolation of primordial germ cells and the establishment of embryonic germ cell lines in the mouse. Cloning. 1999;1:217–24.
23. Evans MJ, Kaufman MH. Establishment in culture of pluripotential cells from mouse embryos. Nature. 1981;292:154–6.
24. Evans MJ, Notarianni E, Laurie S, Moor RM. Derivation and preliminary characterization of pluripotent cell lines from porcine and bovine blastocysts. Theriogenology. 1990;33:125–8.
25. Garcia T, Hofmann MC. Isolation of undifferentiated and early differentiating type A spermatogonia from Pou5f1-GFP reporter mice. Methods Mol Biol. 2012;825:31–44.
26. Geijsen N, Hochedlinger K. gPS navigates germ cells to pluripotency. Cell Stem Cell. 2009;5:3–4.
27. Gheorghisan-Galateanu AA, Hinescu ME, Enciu AM. Ovarian adult stem cells: hope or pitfall? J Ovarian Res. 2014;7:71.
28. Giuili G, Tomljenovic A, Labrecque N, Oulad-Abdelghani M, Rassoulzadegan M, Cuzin F. Murine spermatogonial stem cells: targeted transgene expression and purification in an active state. EMBO Rep. 2002;3:753–9.
29. Goel S, Fujihara M, Tsuchiya K, Takagi Y, Minami N, Yamada M, Imai H. Multipotential ability of primitive germ cells from neonatal pig testis cultured in vitro. Reprod Fertil Dev. 2009;21:696–708.
30. Goel S, Sugimoto M, Minami N, Yamada M, Kume S, Imai H. Identification, isolation, and in vitro culture of porcine gonocytes. Biol Reprod. 2007;77:127–37.
31. Habermann FA, Wuensch A, Sinowatz F, Wolf E. Reporter genes for embryogenesis research in livestock species. Theriogenology. 2007;68 Suppl 1:S116–24.
32. Hadjantonakis AK, Gertsenstein M, Ikawa M, Okabe M, Nagy A. Generating green fluorescent mice by germline transmission of green fluorescent ES cells. Mech Dev. 1998;76:79–90.
33. Hall V. Porcine embryonic stem cells: a possible source for cell replacement therapy. Stem Cell Rev. 2008;4:275–82.
34. Hall VJ. Early development of the porcine embryo: the importance of cell signalling in development of pluripotent cell lines. Reprod Fertil Dev. 2012;25:94–102.
35. Hall VJ, Christensen J, Gao Y, Schmidt MH, Hyttel P. Porcine pluripotency cell signaling develops from the inner cell mass to the epiblast during early development. Dev Dyn. 2009;238:2014–24.
36. Hall VJ, Jacobsen JV, Rasmussen MA, Hyttel P. Ultrastructural and molecular distinctions between the porcine inner cell mass and epiblast reveal unique pluripotent cell states. Dev Dyn. 2010;239:2911–20.
37. Hammond SS, Martin A. Tools for the genetic analysis of germ cells. Genesis. 2009;47:617–27.
38. Han DW, Tapia N, Joo JY, Greber B, Araúzo-Bravo MJ, Bernemann C, Ko K, Wu G, Stehling M, Do JT, Schöler HR. Epiblast stem cell subpopulations represent mouse embryos of distinct pregastrulation stages. Cell. 2010;143:617–27.
39. Hochereau-de Reviers MT, Perreau C. In vitro culture of embryonic disc cells from porcine blastocysts. Reprod Nutr Dev. 1993;33:475–83.
40. Irie N, Tang WW, Surani AM. Germ cell specification and pluripotency in mammals: a perspective from early embryogenesis. Reprod Med Biol. 2014;13:203–15.

41. Johnson J, Bagley J, Skaznik-Wikiel M, Lee HJ, Adams GB, Niikura Y, Tschudy KS, Tilly JC, Cortes ML, Forkert R, Spitzer T, Iacomini J, Scadden DT, Tilly JL. Oocyte generation in adult mammalian ovaries by putative germ cells in bone marrow and peripheral blood. Cell. 2005;122:303–15.
42. Johnson J, Canning J, Kaneko T, Pru JK, Tilly JL. Germline stem cells and follicular renewal in the postnatal mammalian ovary. Nature. 2004;428:145–50. Erratum in: Nature. 2004;430:1062.
43. Kanatsu-Shinohara M, Lee J, Inoue K, Ogonuki N, Miki H, Toyokuni S, Ikawa M, Nakamura T, Ogura A, Shinohara T. Pluripotency of a single spermatogonial stem cell in mice. Biol Reprod. 2008;78:681–7.
44. Kanatsu-Shinohara M, Muneto T, Lee J, Takenaka M, Chuma S, Nakatsuji N, Horiuchi T, Shinohara T. Long-term culture of male germline stem cells from hamster testes. Biol Reprod. 2008;78:611–7.
45. Kanatsu-Shinohara M, Ogonuki N, Inoue K, Miki H, Ogura A, Toyokuni S, Shinohara T. Long-term proliferation in culture and germline transmission of mouse male germline stem cells. Biol Reprod. 2003;69:612–6.
46. Keefer CL, Pant D, Blomberg L, Talbot NC. Challenges and prospects for the establishment of embryonic stem cell lines of domesticated ungulates. Anim Reprod Sci. 2007;98:147–68.
47. Kim HS, Son HY, Kim S, Lee GS, Park CH, Kang SK, Lee BC, Hwang WS, Lee CK. Isolation and initial culture of porcine inner cell masses derived from in vitro-produced blastocysts. Zygote. 2007;15:55–63.
48. Kim S, Kim JH, Lee E, Jeong YW, Hossein MS, Park SM, Park SW, Lee JY, Jeong YI, Kim HS, Kim YW, Hyun SH, Hwang WS. Establishment and characterization of embryonic stem-like cells from porcine somatic cell nuclear transfer blastocysts. Zygote. 2010;18:93–101.
49. King GJ, Atkinson BA, Robertson HA. Implantation and early placentation in domestic ungulates. J Reprod Fertil Suppl. 1981;31:17–30.
50. Knowles BB, Aden DP, Solter D. Monoclonal antibody detecting a stage-specific embryonic antigen (SSEA-1) on preimplantation mouse embryos and teratocarcinoma cells. Curr Top Microbiol Immunol. 1978;81:51–3.
51. Ko K, Tapia N, Wu G, Kim JB, Bravo MJ, Sasse P, Glaser T, Ruau D, Han DW, Greber B, Hausdörfer K, Sebastiano V, Stehling M, Fleischmann BK, Brüstle O, Zenke M, Schöler HR. Induction of pluripotency in adult unipotent germline stem cells. Cell Stem Cell. 2009;5:87–96.
52. Kues WA, Niemann H. The contribution of farm animals to human health. Trends Biotechnol. 2004;22:286–94.
53. Kues WA, Herrmann D, Barg-Kues B, Haridoss S, Nowak-Imialek M, Buchholz T, Streeck M, Grebe A, Grabundzija I, Merkert S, Martin U, Hall VJ, Rasmussen MA, Ivics Z, Hyttel P, Niemann H. Derivation and characterization of sleeping beauty transposon-mediated porcine induced pluripotent stem cells. Stem Cells Dev. 2013;22:124–35.
54. Kuijk EW, Colenbrander B, Roelen BA. The effects of growth factors on in vitro-cultured porcine testicular cells. Reproduction. 2009;138:721–31.
55. Lee CK, Piedrahita JA. Effects of growth factors and feeder cells on porcine primordial germ cells in vitro. Cloning. 2000;2:197–205.
56. Lee H-J, Selesniemi K, Niikura Y, Niikura T, Klein R, Dombkowski DM, Tilly JL. Bone marrow transplantation generates immature oocytes and rescues long-term fertility in a preclinical mouse model of chemotherapy-induced premature ovarian failure. J Clin Oncol. 2007;25:3198–204.
57. Leitch HG, Blair K, Mansfield W, Ayetey H, Humphreys P, Nichols J, Surani MA, Smith A. Embryonic germ cells from mice and rats exhibit properties consistent with a generic pluripotent ground state. Development. 2010;137:2279–87.
58. Li M, Zhang D, Hou Y, Jiao L, Zheng X, Wang WH. Isolation and culture of embryonic stem cells from porcine blastocysts. Mol Reprod Dev. 2003;65:429–34.
59. Li P, Tong C, Mehrian-Shai R, Jia L, Wu N, Yan Y, Maxson RE, Schulze EN, Song H, Hsieh CL, Pera MF, Ying QL. Germline competent embryonic stem cells derived from rat blastocysts. Cell. 2008;135:1299–310.

60. Luo J, Megee S, Rathi R, Dobrinski I. Protein gene product 9.5 is a spermatogonia-specific marker in the pig testis: application to enrichment and culture of porcine spermatogonia. Mol Reprod Dev. 2006;73:1531–40.
61. Martin G. Isolation of a pluripotent cell line from early mouse embryos cultured in medium conditioned by teratocarcinoma stem cells. Proc Natl Acad Sci U S A. 1981;78:7634–8.
62. Matsui Y. Developmental fates of the mouse germ cell line. Int J Dev Biol. 1998;42:1037–42.
63. Matsui Y, Zsebo K, Hogan BL. Derivation of pluripotential embryonic stem cells from murine primordial germ cells in culture. Cell. 1992;70:841–7.
64. McClellan K, Gosden R, Taketo T. Continuous loss of oocytes throughout meiotic prophase in the normal mouse ovary. Dev Biol. 2003;258:334–48.
65. McLaren A. Germ and somatic cell lineages in the developing gonad. Mol Cell Endocrinol. 2000;163:3–9.
66. Miyoshi K, Taguchi Y, Sendai Y, Hoshi H, Sato E. Establishment of a porcine cell line from in vitro-produced blastocysts and transfer of the cells into enucleated oocytes. Biol Reprod. 2000;62:1640–6.
67. Montiel E, Guillomot M, Rojas M, Bustos-Obregón E, Flechon J. Primordial germ cell characterization by immunohistochemistry of vasa-homologue protein in preimplantational rabbit embryos. Int J Dev Biol. 2001;45(S1):141–2.
68. Moore H, Udayashankar R, Aflatoonian B. Stem cells for reproductive medicine. Mol Cell Endocrinol. 2008;288:104–10.
69. Mueller S, Prelle K, Rieger N, Petznek H, Lassnig C, Luksch U, Aigner B, Baetscher M, Wolf E, Mueller M, Brem G. Chimeric pigs following blastocyst injection of transgenic porcine primordial germ cells. Mol Reprod Dev. 1999;54:244–54.
70. Ng HH, Surani MA. The transcriptional and signalling networks of pluripotency. Nat Cell Biol. 2011;13:490–6.
71. Niemann H. Pigs as model systems for biomedical research. In: Rodriguez-Martinez H, Soede NM, Flowers WL, editors. Control of pig reproduction IX. Leicestershire: Context Products Ltd.; 2013. p. 267–86.
72. Notarianni E, Galli C, Laurie S, Moor RM, Evans MJ. Derivation of pluripotent, embryonic cell lines from the pig and sheep. J Reprod Fertil Suppl. 1991;43:255–60.
73. Notarianni E, Laurie S, Moor RM, Evans MJ. Maintenance and differentiation in culture of pluripotential embryonic cell lines from pig blastocysts. J Reprod Fertil Suppl. 1990;41:51–6.
74. Nowak-Imialek M, Kues WA, Petersen B, Lucas-Hahn A, Herrmann D, Haridoss S, Oropeza M, Lemme E, Schöler HR, Carnwath JW, Niemann H. Oct4-enhanced green fluorescent protein transgenic pigs: a new large animal model for reprogramming studies. Stem Cells Dev. 2011;20:1563–75.
75. Nowak-Imialek M, Kues WA, Rudolph C, Schlegelberger B, Taylor U, Carnwath JW, Niemann H. Preferential loss of porcine chromosomes in reprogrammed interspecies cell hybrids. Cell Reprogram. 2010;12:55–65.
76. Nowak-Imialek M, Niemann H. Pluripotent cells in farm animals: state of the art and future perspectives. Reprod Fertil Dev. 2012;25:103–28.
77. Ohinata Y, Payer B, O'Carroll D, Ancelin K, Ono Y, Sano M, Barton SC, Obukhanych T, Nussenzweig M, Tarakhovsky A, Saitou M, Surani MA. Blimp1 is a critical determinant of the germ cell lineage in mice. Nature. 2005;436:207–13.
78. Ohinata Y, Sano M, Shigeta M, Yamanaka K, Saitou M. A comprehensive, non-invasive visualization of primordial germ cell development in mice by the Prdm1-mVenus and Dppa3-ECFP double transgenic reporter. Reproduction. 2008;136:503–14.
79. Ohmura M, Yoshida S, Ide Y, Nagamatsu G, Suda T, Ohbo K. Spatial analysis of germ stem cell development in Oct-4/EGFP transgenic mice. Arch Histol Cytol. 2004;67:285–96.
80. Pacchiarotti J, Maki C, Ramos T, Marh J, Howerton K, Wong J, Pham J, Anorve S, Chow YC, Izadyar F. Differentiation potential of germ line stem cells derived from the postnatal mouse ovary. Differentiation. 2010;79:159–70.

81. Pan Y. A new tool to generate transgenic rats using female germline stem cells from postnatal ovaries. Mol Hum Reprod. 2014;20:283–5.
82. Parte S, Bhartiya D, Telang J, Daithankar V, Salvi V, Zaveri K, Hinduja I. Detection, characterization, and spontaneous differentiation in vitro of very small embryonic-like putative stem cells in adult mammalian ovary. Stem Cells Dev. 2011;20:1451–64.
83. Payer B, Chuva de Sousa Lopes SM, Barton SC, Lee C, Saitou M, Surani MA. Generation of stella-GFP transgenic mice: a novel tool to study germ cell development. Genesis. 2006;44:75–83.
84. Pesce M, Wang X, Wolgemuth DJ, Schöler H. Differential expression of the Oct-4 transcription factor during mouse germ cell differentiation. Mech Dev. 1998;71:89–98.
85. Petkov SG, Anderson GB. Culture of porcine embryonic germ cells in serum-supplemented and serum-free conditions: the effects of serum and growth factors on primary and long-term culture. Cloning Stem Cells. 2008;10:263–76.
86. Petkov SG, Marks H, Klein T, Garcia RS, Gao Y, Stunnenberg H, Hyttel P. In vitro culture and characterization of putative porcine embryonic germ cells derived from domestic breeds and Yucatan mini pig embryos at days 20–24 of gestation. Stem Cell Res. 2011;6:226–37.
87. Petkov SG, Reh WA, Anderson GB. Methylation changes in porcine primordial germ cells. Mol Reprod Dev. 2009;76:22–30. Erratum in: Mol Reprod Dev. 2009:76:522.
88. Petkov SG, Hyttel P, Niemann H. The small molecule inhibitors PD0325091 and CHIR99021 reduce expression of pluripotency-related genes in putative porcine induced pluripotent stem cells. Cell Reprogram. 2014;16:235–40.
89. Piedrahita JA, Anderson GB, Bondurant RH. Influence of feeder layer type on the efficiency of isolation of porcine embryo-derived cell lines. Theriogenology. 1990;34:865–77.
90. Piedrahita JA, Anderson GB, Bondurant RH. On the isolation of embryonic stem cells: comparative behavior of murine, porcine and ovine embryos. Theriogenology. 1990;34:879–901.
91. Piedrahita JA, Anderson GB, Martin GR, Bondurant RH, Pashen RL. Isolation of embryonic stem cell-like colonies from porcine embryos. Theriogenology. 1988;29:286.
92. Piedrahita JA, Moore K, Oetama B, Lee CK, Scales N, Ramsoondar J, Bazer FW, Ott T. Generation of transgenic porcine chimeras using primordial germ cell-derived colonies. Biol Reprod. 1998;58:1321–9.
93. Prelle K, Zink N, Wolf E. Pluripotent stem cells—model of embryonic development, tool for gene targeting, and basis of cell therapy. Anat Histol Embryol. 2002;3:169–86.
94. Rao M. Conserved and divergent paths that regulate self-renewal in mouse and human embryonic stem cells. Dev Biol. 2004;275:269–86.
95. Rui R, Qiu Y, Hu Y, Fan B. Establishment of porcine transgenic embryonic germ cell lines expressing enhanced green fluorescent protein. Theriogenology. 2006;65:713–20.
96. Rui R, Shim H, Moyer AL, Anderson DL, Penedo CT, Rowe JD, BonDurant RH, Anderson GB. Attempts to enhance production of porcine chimeras from embryonic germ cells and preimplantation embryos. Theriogenology. 2004;61:1225–35.
97. Shim H, Gutiérrez-Adán A, Chen LR, BonDurant RH, Behboodi E, Anderson GB. Isolation of pluripotent stem cells from cultured porcine primordial germ cells. Biol Reprod. 1997;57:1089–95.
98. Shim SW, Han DW, Yang JH, Lee BY, Kim SB, Shim H, Lee HT. Derivation of embryonic germ cells from post migratory primordial germ cells, and methylation analysis of their imprinted genes by bisulfite genomic sequencing. Mol Cells. 2008;25:358–67.
99. Sommer JR, Jackson LR, Simpson SG, Collins EB, Piedrahita JA, Petters RM. Transgenic Stra8-EYFP pigs: a model for developing male germ cell technologies. Transgenic Res. 2012;21:383–92.
100. Stewart CL, Gadi I, Bhatt H. Stem cells from primordial germ cells can reenter the germ line. Dev Biol. 1994;161:626–8.
101. Strojek RM, Reed MA, Hoover JL, Wagner TE. A method for cultivating morphologically undifferentiated embryonic stem cells from porcine blastocysts. Theriogenology. 1990;33:901–13.

102. Suzuki H, Dann CT, Rajkovic A. Generation of a germ cell-specific mouse transgenic CHERRY reporter, Sohlh1-mCherryFlag. Genesis. 2013;51:50–8.
103. Talbot NC, Rexroad Jr CE, Pursel VG, Powell AM, Nel ND. Culturing the epiblast cells of the pig blastocyst. In Vitro Cell Dev Biol Anim. 1993;29A:543–54.
104. Tegelenbosch RA, de Rooij DG. A quantitative study of spermatogonial multiplication and stem cell renewal in the C3H/101F1 hybrid mouse. Mutat Res. 1993;290:193–200.
105. Telugu BP, Ezashi T, Roberts RM. The promise of stem cell research in pigs and other ungulate species. Stem Cell Rev. 2010;6:31–41.
106. Thomson JA, Itskovitz-Eldor J, Shapiro SS, Waknitz MA, Swiergiel JJ, Marshall VS, Jones JM. Embryonic stem cell lines derived from human blastocysts. Science. 1998;282:1145–7. Erratum in: Science. 282:1827.
107. Thomson JA, Kalishman J, Golos TG, Durning M, Harris CP, Becker RA, Hearn JP. Isolation of a primate embryonic stem cell line. Proc Natl Acad Sci U S A. 1995;92:7844–8.
108. Tsung HC, Du ZW, Rui R, Li XL, Bao LP, Wu J, Bao SM, Yao Z. The culture and establishment of embryonic germ (EG) cell lines from Chinese mini swine. Cell Res. 2003;13:195–202.
109. Vackova I, Madrova J. Porcine embryonic stem-like cells, an animal model for human stem cell therapy. Hum Reprod. 2006;21 Suppl 1:163.
110. Vackova I, Ungrova A, Lopes F. Putative embryonic stem cell lines from pig embryos. J Reprod Dev. 2007;53:1137–49.
111. van Dissel-Emiliani FM, de Rooij DG, Meistrich ML. Isolation of rat gonocytes by velocity sedimentation at unit gravity. J Reprod Fertil. 1989;86:759–66.
112. Vassiliev I, Vassilieva S, Beebe LF, Harrison SJ, McIlfatrick SM, Nottle MB. In vitro and in vivo characterization of putative porcine embryonic stem cells. Cell Reprogram. 2010;12:223–30.
113. Vassiliev I, Vassilieva S, Beebe LF, McIlfatrick SM, Harrison SJ, Nottle MB. Development of culture conditions for the isolation of pluripotent porcine embryonal outgrowths from in vitro produced and in vivo derived embryos. J Reprod Dev. 2010;56:546–51.
114. Virant-Klun I, Zech N, Rozman P, Vogler A, Cvjeticanin B, Klemenc P, Malicev E, Meden-Vrtovec H. Putative stem cells with an embryonic character isolated from the ovarian surface epithelium of women with no naturally present follicles and oocytes. Differentiation. 2008;76:843–56.
115. Wheeler MB. Development and validation of swine embryonic stem cells: a review. Reprod Fertil Dev. 1994;6:563–8.
116. White YA, Woods DC, Takai Y, Ishihara O, Seki H, Tilly JL. Oocyte formation by mitotically active germ cells purified from ovaries of reproductive-age women. Nat Med. 2012;18:413–21.
117. Wianny F, Perreau C, Hochereau de Reviers MT. Proliferation and differentiation of porcine inner cell mass and epiblast in vitro. Biol Reprod. 1997;57:756–64.
118. Wolff EF, Libfraind LL, Weitzel P, Woods DC, Feng Y, Tilly JL, DeCherney AH, Tisdale JF. Oogonial stem cells generate mature oocytes in an autologous Rhesus macaque transplantation model. Reprod Sci. 2014;21(Suppl):119A.
119. Wu Z, Falciatori I, Molyneux LA, Richardson TE, Chapman KM, Hamra FK. Spermatogonial culture medium: an effective and efficient nutrient mixture for culturing rat spermatogonial stem cells. Biol Reprod. 2009;81:77–86.
120. Yeom YI, Fuhrmann G, Ovitt CE, Brehm A, Ohbo K, Gross M, Hübner K, Schöler HR. Germline regulatory element of Oct-4 specific for the totipotent cycle of embryonal cells. Development. 1996;122:881–94.
121. Yin M, Fang Z, Jiang W, Xing F, Jiang M, Kong P, Li Y, Zhou X, Tang L, Li S, Chen X. The Oct4 promoter-EGFP transgenic rabbit: a new model for monitoring the pluripotency of rabbit stem cells. Int J Dev Biol. 2013;57:845–52.
122. Yoshida S, Takakura A, Ohbo K, Abe K, Wakabayashi J, Yamamoto M, Suda T, Nabeshima Y. Neurogenin3 delineates the earliest stages of spermatogenesis in the mouse testis. Dev Biol. 2004;269:447–58.

123. Yoshimizu T, Sugiyama N, De Felice M, Yeom YI, Ohbo K, Masuko K, Obinata M, Abe K, Schöler HR, Matsui Y. Germline-specific expression of the Oct-4/green fluorescent protein (GFP) transgene in mice. Dev Growth Differ. 1999;41:675–84.
124. Youn H, Kim SH, Choi KA, Kim S. Characterization of Oct4-GFP spermatogonial stem cell line and its application in the reprogramming studies. J Cell Biochem. 2013;114:920–8.
125. Zheng Y, Tian X, Zhang Y, Qin J, An J, Zeng W. In vitro propagation of male germline stem cells from piglets. J Assist Reprod Genet. 2013;30:945–52.
126. Zhou L, Wang L, Kang JX, Xie W, Li X, Wu C, Xu B, Wu J. Production of fat-1 transgenic rats using a post-natal female germline stem cell line. Mol Hum Reprod. 2014;20:271–81.
127. Zhou Q, Nie R, Li Y, Friel P, Mitchell D, Hess RA, Small C, Griswold MD. Expression of stimulated by retinoic acid gene 8 (Stra8) in spermatogenic cells induced by retinoic acid: an in vivo study in vitamin A-sufficient postnatal murine testes. Biol Reprod. 2008;79:35–42.
128. Zou K, Yuan Z, Yang Z, Luo H, Sun K, Zhou L, Xiang J, Shi L, Yu Q, Zhang Y, Hou R, Wu J. Production of offspring from a germline stem cell line derived from neonatal ovaries. Nat Cell Biol. 2009;11:631–6.
129. Zuckerman S. The number of oocytes in the mature ovary. Recent Prog Horm Res. 1951;6:63–108.

Chapter 6
Fetal Cell Reprogramming and Transformation

Jeong Mook Lim and Ji Yeon Ahn

Introduction

The Importance and Necessity of Fetal Cell Reprogramming

Most organisms consist of tissues that originate from either ectodermal (epiblastic) or endodermal (hypoblastic) germ cell layers. These two layers subsequently interact with each other to form the mesoblast, a middle germinal layer of undifferentiated cells in the embryo that develops into the mesoderm. Ectodermal, endodermal and mesodermal cells then begin to differentiate into various cell types where the micro-environment to which these trigeminal cells are exposed, plays a pivotal role in determining cell differentiation pathways. Changes in epigenetic status occur in progenitor cells and subsequently lead to cell transformation and differentiation. Embryonic cells have the plasticity that allows for cell reprogramming and differentiation, regulated by epigenetic changes. In addition, they neither begin to yield their own cell lineage nor transform into the cells of different lineages without environmental stimulation. Therefore, specific signals and/or suitable microenvironments serve as

J.M. Lim, D.V.M., Ph.D. (✉)
Departments of Agricultural Biotechnology, Seoul National University, Seoul 151-921, South Korea

Cancer Research Institute, Seoul National University Hospital, Seoul 110-742, South Korea

Research Institutes of Agriculture and Life Sciences, Seoul National University, Seoul 151-921, South Korea
e-mail: limjm@snu.ac.kr

J.Y. Ahn, Ph.D.
Departments of Agricultural Biotechnology, Seoul National University, Seoul 151-921, South Korea
e-mail: ahnji@snu.ac.kr

D.O. Fauza, M. Bani (eds.), *Fetal Stem Cells in Regenerative Medicine*, Stem Cell Biology and Regenerative Medicine,
DOI 10.1007/978-1-4939-3483-6_6

prerequisite factors for cell reprogramming and differentiation. Understanding the microenvironment-driving specific signal transduction pathways may provide the tools needed for manipulating the epigenetic reprogramming and morphological transformation *in vitro* or *in vivo*. However, the signals also interact with each other in various ways or make clusters for ending action, emphasizing the complexity of reprogramming mechanisms.

It is generally accepted that fully differentiated cells lose their plasticity under normal physiological state. However, non-physiologic stimulants provoke the transformation of terminally differentiated cells, which yield abnormal or dysfunctional cells. Recent data further demonstrate such environmental stress can make terminally differentiated cells to be reprogrammed, which can lead to normal regeneration. In some cases, un-harmonized or incomplete reprogramming may induce genetic alterations that could lead to a pathological status. Factors that induce epigenetic reprogramming and phenotypic transformation are important for regulating cell differentiation and organogenesis.

Methods to Induce Cell Reprogramming

Originating from the same genetic makeup, different cell types express specific genes responsible for different functions. In addition, the somatic cells of various phenotypes have the capacity to de-differentiate into the stem cell lines of the same genetic and cellular background. While several methods have been used to induce cellular reprogramming, these methods can be classified into two categories: direct genetic modification and *de novo* reprogramming by environmental regulation.

Direct Genetic Modification

Ectopic Gene Delivery

In 2006, Yamanaka's team established colony-forming cells from skin fibroblasts by introducing the genes *Oct4*, *Sox2*, *Klf4* and *c-Myc,* using retroviral vectors. These cells were termed induced pluripotent stem cells (iPSCs) due to their shared similarities with embryonic stem cells in terms of cellular characteristics and vigorous proliferation profile [1]. Viral vectors can be readily used for ectopic gene delivery to generate iPSC lines. For instance, Moloney murine leukemia virus (MMLV) vector can be used for reprogramming fibroblasts. The MMLV retroviral vector provides the space required for viral packaging signal, transcription and processing factors. However, the target cells become exposed to a carcinogenic environment, which may lead to oncogenic activity. Also, if the viral transgenes are not eliminated after the completion of delivery, they can randomly reactivate and induce heterogeneous diversity [2, 3].

Several studies have shown an increase in the efficacy of the established iPSC methods and a concomitant reduction of the risk of genetic disturbances. For example, lentiviral vector systems originating from HIV have been developed for this purpose. Unlike MMLV vector, these vectors can convey transgenes to both dividing and non-dividing cells, resulting in higher reprogramming efficiency. As this lentiviral vector cannot be silenced in target cells after gene delivery, it may act as a stronger carcinogen than retroviral vectors. These issues have led to the development of more efficient tools for gene delivery, as listed below [4].
Integrin-defective viral vector. This vector system is a powerful tool for generating integrated-free, cell transformation. Two integrin-defective viral delivery systems, adenoviral and Sendai viral vectors, have been developed for the purpose of generating iPS cells. This vector system does not yield any risk of having transgenes in the generated reprogrammed cells. Since these vectors are defective, lower genomic integration with high efficiency is one of the advantages of this method. However, this technique needs repeated infection for certain cell types and it shows slow kinetics of reprogramming [5, 6].

Piggyback (PB) transposon. This technique enhances stable integration of non-viral constructs. The components of PB transposon are transposon-containing, donor plasmid, which also contains co-transfected helper plasmid expressing transposase for the transfer of genes. It is usually active in pluripotent stem cells of several species, including mouse and human, and it has a potential of higher genomic integration efficiency than the random integration obtained from plasmids. A significant advantage of Piggyback system is its ability to erase transgenes even though genomic integration is required in the protocol. Transgenes can be deleted precisely by transposase without modifying the sequence of the integrated site. Integration-free, cell reprogramming can be achieved by the Piggyback system [7].

Transient episomal delivery. This method is based on direct delivery of non-replicating [8, 9] or replicated episomal vector [10], which can be used in order to avoid weakening of integration-defective virus. The acquisition of pluripotency in terminally differentiated, somatic cells was difficult by the overexpression of the four genes reported originally and further efforts were made to overexpress an additional 2–3 genes by episomal delivery. Nevertheless, the efficiency was not improved due to a number of reasons particularly, an intrinsic factor of the vector itself that incurs a rapid silencing of transgenes due to quick and extensive methylation of the vector sequence [11].

Transfection of plasmids/mini-circle DNAs. This method is accomplished by the non-viral integration of plasmids/mini-circle DNAs that include the required genes to reprogram target cells [12, 13]. This method allows for the manipulation of much larger DNA fragments so that it is possible to clone the cDNA enclosing up to five reprogramming genes. The technique is quite simple, but it requires multiple rounds of transfections, which is labor-intensive. Low efficiency in genomic integration leads to lower risk of transgene reactivation, compared with viral vector systems. Inevitably, utilizing this system in reprogramming cells to iPSCs leads to lower efficiency.

RNA Delivery

There are two types of RNA delivery: mRNA [14] and miRNA [15]. Direct delivery of synthetic mRNA comes with a big advantage for cell reprogramming where plasmid or viral vectors can be avoided completely. In this system, synthetic RNAs encode reprogramming genes, making it possible to accomplish cell reprogramming. Utilization of this method of delivery requires modified RNAs that are transcribed *in vitro* for leaving from the endogenous system of target cells and by this deletion, much higher efficiency than any of non-integrative techniques can be achieved. However, the risk of oncogene activation is high with the utilization of this system, because high dosages of genes are needed for direct mRNA delivery. The technique of using miRNA has been exploited recently, but the effectiveness of this protocol is still controversial.

Protein-Based Reprogramming

This method was suggested following the first report on the success of somatic cell reprogramming by cell extracts [16–18], which encouraged researchers to look for key proteins that can be used in cell reprogramming. Several groups successfully generated iPSCs by repeated exposure of fibroblasts to the recombinant proteins [17, 19]. Defined stoichiometry and the optimal concentration of transcription factors can lead to reprogramming target cells by using designed proteins. This method also has an advantage in avoiding genetic manipulation of target cells by utilizing non-integrated materials. However, the protein-based method has been shown to be inefficient due to the requirement of large amount of purified recombinant proteins and less reproducibility. Recent reports show that iPS cell derivation can be easily induced by the use of cell penetrating protein (CPP) [20–22].

Exposure to Small Molecules

This simple and efficient method does not require genomic integration. Instead, it utilizes small molecules or chemicals [23]. Furthermore, the small molecules or chemicals employed in this method interfere specifically with the regular function of the intermediates in signal and metabolic pathways and accelerate reprogramming procedures. It can reduce the risk of any mutative side-effects. However, to utilize this technique for cell reprogramming, sufficient information about the exact pathways involved in reprogramming from differentiated cells to pluripotent cells is needed.

De Novo Reprogramming by Environmental Regulation

There are several techniques to induce somatic cell reprogramming into iPSCs without direct genetic manipulation through using cellular niche and acellular microenvironments.

Somatic Cell Nuclear Transfer (SCNT)

SCNT involves transferring the nucleus of a somatic cell into enucleated oocytes for reprogramming. The transferred nucleus is immediately reprogrammed under the exposure of mature ooplasm and assumes the role of the nucleus to the fertilized oocyte. However, the ooplasmic factors that induce such reprogramming of the somatic nucleus remain elusive [24]. Reconstructed embryos following SCNT then begin to develop by the reprogramming into more advanced stage of development than the zygote stage. Derivation of stem cells from this reconstructed oocyte make it possible to establish immune-specific ESCs for the reprogrammed cells [25–27]. Such immune-specific ESC line is more clinical than the ESCs derived from natural fertilization, because it acquires immune-specificity without direct genetic manipulation. However, SCNT opens the risk of individual cloning of humans, raising significant ethical issues. A significant decrease in developmental competence of reconstructed oocytes with increasing genetic abnormally is another problem with using this method [28, 29].

Somatic-Stem Cell Fusion

The fusion of somatic cells and stem cells is used as a technique to reprogram the fused somatic cells into pluripotent cells [30]. Embryonic carcinoma cells (ECCs), embryonic germ cells (EGCs) and ESCs are all good sources for fusion-based reprogramming technology. However, it is not clear if the pluripotency acquired by the fused cells is clinically applicable due to its genetic and cytoplasmic instability (either tetraploidy or bi-nuclear cytoplasm).

Environmental Cue (Environmental Stress)

Wakayama and his colleague recently reported that pluripotent stem cells can be generated by inducing transformation in $CD45^+$ hematopoietic cell under environmental stress [31]. Although these results were not reproducible by other labs, it is possible that environmental factors can induce cell reprogramming and transformation. The transformation of normal cells into cancer cells after long-term exposure of specific carcinogens is an example and various inducers for cell transformation have been elucidated. However, it is not clear whether such induction can be used for acquiring cell plasticity like fetal cells because the reprogramming procedure can increase the potential risk of genetic damage resulting in unexpected mutations.

Cell-to-Cell Interaction

This is a reprogramming technique that utilizes direct cell-to-cell contact without genetic modification. A Korean group reported that the co-culture of fibroblasts and ovarian stromal cells can induce cell reprogramming leading to the generation of

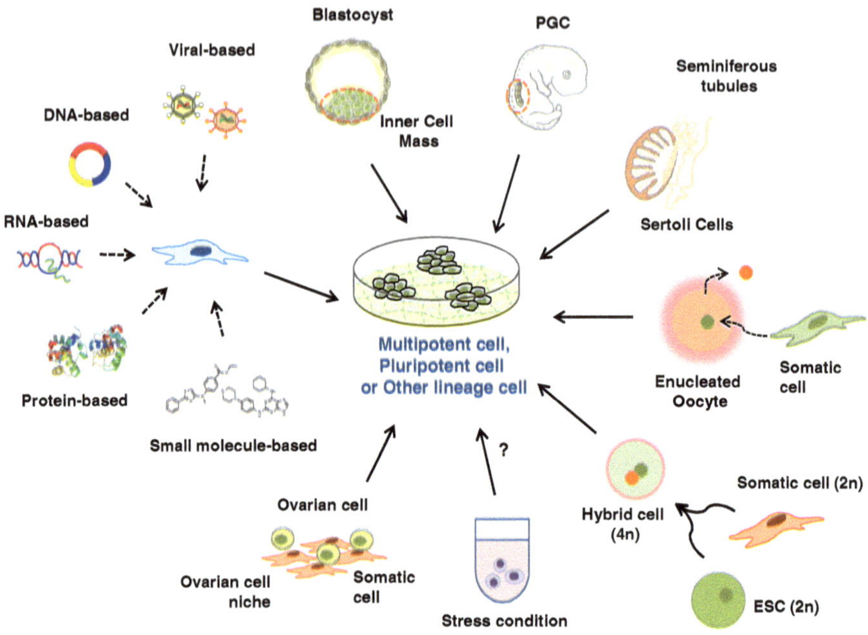

Fig. 6.1 Suggested methods for reprogramming somatic cells. Somatic cell reprogramming can be induced by different environments or artificial treatments such as the manipulation/culture of inner cells mass cells of blastocyst, primordial germ cell, Sertoli cell in seminiferous tubule, somatic cell or embryonic cell nuclear transfer, the regulation of cell niche or overexpression of reprogramming transcription factors

colony-forming cells, which show similar characteristics to ESCs [32]. However, this is a technique in its preliminary stages due to the observed aberrant genotype and cellular phenotype. Another issue with this technique is the interaction of a mixed cell population, which significantly reduces reproducibility and confers a difficulty in establishing a standard operation protocol (Fig. 6.1).

Reprogramming Methods for Fetal Cells

Most cells retrieved at the periods of early embryogenesis serve as the progenitors of terminally differentiated somatic cells, consisting of various tissues and under a specific cellular or acellular environment, progenitor cells begin to differentiate into the cells of lower lineages. It is widely accepted that undifferentiated and differentiating fetal cells have a better cell plasticity than terminally-differentiated adult cells due to the early embryonic environment [33–35]. Furthermore, the superior plasticity of fetal somatic cells over adult somatic cells of the same type has been recently confirmed [36]. For example, fetal hepatocytes can be reprogrammed into iPSCs at a frequency 50-fold higher than adult hepatocytes. Adult hepatocytes have been reprogrammed into iPSCs with six genes, whereas fetal hepatocytes only need three (OCT4, SOX2, NANOG) or four genes (OCT4, SOX2, NANOG and LIN28 or OCT4, SOX2, KLF4 and C-MYC) for reprogramming. Therefore, it can be suggested that the proposed technologies for cell reprogramming can be employed for fetal somatic cells with better results than those obtained with adult somatic cells. Probably, methods employed for fetal cells reprogramming are different from those employed for adult cells reprogramming and care should be taken in selecting the optimal reprogramming protocol according to the stage of development and the lineage of the differentiated cells Also, cell type and lineage influence cell plasticity. For example, the mesenchymal cells derived from mesodermal cells can differentiate into various cell types in response to specific environments, which may have better cell plasticity than other embryonic cells that are committed to undergo a single lineage of differentiation. Use of mesenchymal cells classified as embryonic connective tissue cells is a good choice for securing embryo-derived, plastic cells.

Phenotypical Reprogramming by Cell Transformation

Fetal tissues mainly include progenitor cells before completion of organogenesis. In adult tissues, there are undifferentiated cells with the potential to differentiate under specific conditions. Examples include putative germ cells in reproductive tissue and immature oocytes. Understanding the normal differentiation process of these cells during organogenesis may help define the conditions required to differentiate these cells *in vitro*. Also, understanding physiological or pathological mesenchymal-to-epithelial transition (MET) or epithelial-to-mesenchymal transition (EMT) may provide information about the mechanisms involved in organogenesis. Elucidation on EMT/MET in embryonic and adult tissues contributes to developing the technology to control cellular reprogramming of not only stromal and epithelial cells, but also iPS cells and even ESCs.

Germ Cell and Germ Cell-Derived Stem Cells

Primordial Germ Cell

The germ cells in multicellular eukaryotes serve as progenitors of the male or female gametes that can differentiate into all somatic cells. They undergo both meiosis and mitosis, but sometimes they become inactivated. Primordial germ cell (PGC) is derived from the epiblast of primitive ectoderm. PGC migrates into the endodermal yolk sac wall and finally homes into the genital ridge derived from mesoderm to give rise to the male and female gonad. PGC multiplies by mitotic divisions during migratory periods, but further incorporates anatomically with the genital ridge after sex differentiation. Male PGC then becomes differentiated into the spermatogonia following Y-chromosome activation, while the female differentiates into the oogonia. Thereafter, those cells continue to mitotically proliferate during prenatal periods [37–40]. The isolation and selection of PGCs from mixed cell population of reproductive tissue is critical. Cell surface markers such as SSEAs along with sorting instruments such as FACS and MACS are required for such isolations. PGC specification is also dependent on the expression of BMP4 and BMP8b from the extraembryonic ectoderm [41–43], and other PGC-specific factors are currently under investigation [44–47].

Embryonic Germ Cell (EGC)

PGC is the founder cells of the germ cell lineage called germline. Both embryonic cell and germline cell of early stage maintain full developmental potency and directly relate to somatic differentiation [48–52]. The germ cell-derived PGC is also the progenitor of teratocarcinoma, and they can establish embryonic germ cell (EGC) with pluripotent activity *in vitro*. PGC acquires similar properties to that of ESCs under a specific environment, which links PGC to EGC transformation *in vitro*. However, the underlying mechanism is still unknown. Acquiring self-renewal activity without spontaneous differentiation is a prerequisite factor for the establishment of EGC *in vitro*. Therefore, the PGC/EGC culture medium is specially designed for stem cell self-renewal and several factors to induce self-renewal signals are supplemented. Conventional culture systems have been employed for the stem cell establishment and feeder cells have been used for supporting the proliferation activity of PGCs/EGCs [53, 54].

In vivo tests are essential for cell pluripotency and contribution of putative pluripotent cell to germline chimera development in animal models is considered as a critical parameter. Cell aggregation [55], blastocyst injection [56–58] and tetraploid complementation [59] are the main protocols for the validation of germline chimerism. ESC and possibly EGCs maintain this capacity *in vitro* [54, 60, 61]. Thus, despite their origins, EGC may be indistinguishable from ESC at the molecular level [49, 62–64]. A number of studies confirm that ESC can differentiate into PGC [65], while it is uncertain whether the PGCs can differentiate into all somatic cell types. Unlike EGC, it is not clear that the PGC contributes to germline chimaera production following injection into the early embryo, because there may be a different state of pluripotency between PGC and pre-implantation epiblast. PGC may only have a unipotent or a bipotent activity, which only gives rise to male and female germ cells. However, PGC-to-EGC reprogramming leads to acquired pluripotency [64]. As a matter of fact, PGC expresses pluripotent genes and they subsequently have a capacity to be reprogrammed into pluripotent stem cells [66].

Spermatogonial Stem Cell (SSC)

Male germ cell sequentially differentiates into gonocyte and spermatogonium. The spermatogonium matures into spermatozoon via spermatogenesis, through mitotic and meiotic cell divisions. Transient cells appearing during spermatogenesis consist of male germline lineages. Spermatogenesis lasts over one's life in the seminiferous tubules of the testes, beginning at puberty. In fact, this process consists of spermatocytogenesis and spermiogenesis, which undergoes both genetic and phenotypic transformations including nuclear condensation, ejection of the cytoplasm and formation of the acrosome and flagellum [67]. The development of spermatogonia is classified into multiple stages and the cells of several stages are considered as the SSCs. In spermatogonia, their incomplete cytokinesis at the specific stages results in forming the syncytium that connects mother and daughter cells and these cytoplasmic bridges help supply cellular substances to haploid, daughter cells. The Sertoli cell and the secretory substances in the seminiferous tubules are the modulator of spermatogenesis. GDNF stimulates the self-renewal of SSCs [68], while BMP4 shows an antagonistic function compared with GDNF [69]. GDNF activation upregulates the expression of several transcription factor-encoding genes such as *Bcl6b*, *Etv5*, and *Lhx1* [70], while the non-GDNF-stimulated *Plzf* and *Taf4b* regulate the function of SSCs [70]. FGF9 inhibits meiosis by regulating fetal gonocyte and prenatal spermatogonium [71].

The spermatogenesis is the combined process of SSC self-renewal and differentiation. Similar to other adult stem cells, SSCs are rare, representing less than 0.03 % of the total cell population [72] and it is extremely difficult to define cell characteristics. The only way to identify SSCs is to monitor their biological capacity to produce or to maintain spermatogenesis. The role of SSC as pluripotent cells has only been reported in mice and chicken [73]. In chicken, SSCs isolated *in vitro* can induce germline transmission by transplantation into developing embryo [74, 75]. The gonocyte, as SSC progenitor, is a cell subcategorized into mitotic (M)-prospermatogonium, T1-prospermatogonium or T2-propsermatogonium [76], which colonizes the basement membrane of the seminiferous tubules. Gonocytes resume proliferation immediately after birth [77]. T2-prospermatogonium is a cell either initiating spermatogenesis or maintaining SSC population [76, 78, 79]. There are lots of spermatogonia subtypes mainly consisting of type A, intermediate and type B spermatogonia [80–82]. Type A spermatogonium of early stage is undifferentiated and alternative SSC self-renewal may be explained by either the A0-to-A1 cell transformation [83–85] or the Adark-to-Apale transformation [86, 87] according to species. The mitotic A0 spermatogonium observed rarely in A0 spermatogonia is considered as the 'reserve stem cells.' This cell type does not contribute to inducing spermatogenesis at normal condition, but only it is activated when spermatogenesis is disrupted. In contrast, the A1–A4 spermatogonia are considered as 'active stem cells' and A4 spermatogonium can determine its cell fate to self-renew or to differentiate by dividing into A1 spermatogonium or intermediate spermatogonium, respectively. Such a cellular process is a unique model for cellular reprogramming and phenotypic changes, which show cell plasticity occurred in both fetal and adult tissues.

Reprogramming of Oogonium-Derived, Immature Oocyte Growing in Ovarian Follicle

The activation of Y-chromosome dependent genes results in sex difference, and the ovaries of female gonad generate lots of germ cell-derived cells before birth. In contrast to male, the generation of gonadal germ cells in female is immediately ceased by birth, germ cell-dependant generation of primordial follicles is not possible after birth. Thus, total number of oogonia (the progenitors of mature oocytes) in a female is fixed at birth. Thus only a limited number of oogonia (a few hundreds in human) differentiate into developmentally competent, mature oocytes. The rest of female germ cells arrest development during the meiotic process of oogenesis.

It is controversial whether female germline stem cell is present in the ovaries or not. Recently, some researchers have claimed that adult germline stem cell being turned over from primordial follicle can continuously supply the ovarian follicle pool, which is identified as a mitotically active cell positive for germ cell-specific markers. The establishment of ovarian (oogonial) stem cells has been reported with reference to developmentally competent oocyte [88–90]. Primordial follicles are highly stable; therefore, neither germ-line stem cell activity nor transformation activity can be detected even under specific conditions. Some researchers denied the presence of ovarian stem cells *in vivo*, but cell culture might induce the dedifferentiation of ovarian cells following cellular reprogramming [91, 92]. Alternatively, there are putative plastic cells in the other part of ovarian tissue. In our previous data, the presence of Oct4/Nanog-positive cells in ovarian stromal tissue is detected (Gong et al., 2010). ESC-like, colony-forming cells were subsequently derived from the culture of stromal cell populations in culture, but they may lack the ability to induce germline transmission. Despite the uncertainty on the presence of germline stem cells, the previous reports demonstrate the presence of 'reprogrammable' cells that are positive for stemness-specific markers. If germline stem cells are present in adult ovaries, there are two applications as the progenitor of various functional cells for regenerating damaged tissues and as the source of developmentally competent oocytes for reproductive purposes. Formation of oocyte-like structure and ovarian follicles by the culture and tissue transplantation may support the presence of germline stem cells, but more evidence is required for leading clinical applications.

Preantral Follicle Culture

To increase the feasibility of deriving developmentally competent oocytes female germline stem cells, manipulating ovarian follicles is absolutely required as a supporting technology. In fact, successful recruiting of developmentally arrested immature oocytes for oocyte maturation is the best choice for securing abundant numbers of developmentally competent oocytes. Yielding of developmentally competent oocytes by culturing primary follicles has much better feasibility than culturing of secondary follicles, because of their availability. Since the follicle growth is regulated by various autocrine, paracrine and endocrine factors, different culture protocol for each stage of follicle development may be required for optimizing folliculogenesis *in vitro* [93–96]. Studies of ovarian follicle culture conducted in various species such as hamsters [97], sheep [98], cattle [99], humans [100], and mice [101] have demonstrated limitations in the follicle-culture system. Eppig and his colleagues produced live offspring in mice after IVF of mature oocytes derived from *in vitro* secondary follicle culture [102, 103]. They established a simplified culture system for early

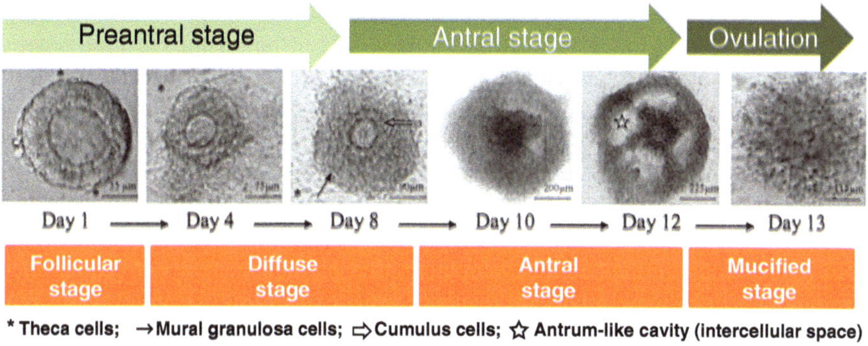

Fig. 6.2 Development and differentiation of an ovarian follicle cultured for 12 days. Early preantral follicle at the follicular stage is retrieved and after being cultured for 4 days, the follicle develops to late preantral follicle (named as a diffuse stage). During the second half of the culture, mural and cumulus cells of the follicle differentiate, which subsequently leads to the formation of antral-like cavities (named as antral stage). The follicle differentiates into a pre-ovulatory follicle and 18 hours after the hormonal stimulation on day 12, oocyte in the follicle of mucified stage extrudes beyond the follicular cell mass, which becomes free-floating in the culture droplet. (*parts of the figure are from Adriaens et al., 2004*) [108]

prenatal follicles using a drop-culture system [104]. The development of primary into secondary follicles using neonatal ovarian cultures has been reported by Sadeu et al. [105]. The derivation of developmentally competent oocytes by culturing primary follicles was reported by Lenie et al. [95]. Immune-specific stem cells can be derived from parthenogenetic activation of oocytes derived from the culture of primary and secondary follicles [106, 107]. A number of model studies can be possible by utilizing a chemically defined, culture medium [94, 107, 108] (Fig. 6.2).

Parthenogenesis for Reprogramming of Mature Oocyte

Immune specificity of living cells to their donors or immune tolerance of patients to biomaterials are essential factors for clinical application of regenerative technology. Since the beginning of stem cell research, the establishment of patient-specific ESCs has been considered as one of the most important factors to advance their use in tissue regeneration. SCNT has been considered to establish the patient-specific ESCs, but several factors, including ethical related issues, have hindered the feasibility of SCNT-based, ESC-related technology. As one of the alternatives for deriving patient-specific cells, oocyte parthenogenesis can be considered and combining with the follicle culture technology described in previous section, its feasibility can be extended. The technique of parthenogenesis is a method for reprogramming of oogonium-derived, mature oocyte to zygote without normal fertilization by activation of degeneration-fated chromatid following meiosis. Autologous stem cells were established by the parthenogenetic activation of ovulated oocytes in human [109], primates [110] and other animal species [111]. The first establishment of the autologous ESCs derived from combined methods of parthenogenesis and ovarian follicle culture was made in mice [112, 113]. It can be postulated that genetic background is an important factor for the ESC establishment and parental inheritance influences its efficiency. In mouse, changes in paternal inheritance without altering maternal heredity significantly affect stemness-related gene expression [114]. In addition, microenvironment during *in vitro*-folliculogenesis and post-parthenogenetic development is also important [107]. Exposure to antioxidants during early oogenesis not only reduces the production of intracytoplasmic ROS, but also improves preimplantation development after parthenogenetic activation. Such promoting effect may relate to regulate gene expression. Of relevance, Wnt signaling-related genes are among major effecter of antioxidant treatment.

Next question is whether the ESCs derived from parthenogenesis is compatible to the ESCs derive from naturally fertilized embryos. No phenotypic difference between normally fertilized and parthenogenetic ESCs was detected. The parthenogenetic ESCs have a homozygous genome with minimal crossover-associated heterozygosity [93], so they can be employed for patient-specific cell and tissue therapy. Unfortunately, the molecular signature of the parthenogenetic ESCs has been poorly investigated to date. Under certain genetic background and microenvironment, however, the alteration of gene expression induced by parthenogenesis is similar to or quantitatively less than that induced by strain difference. No significant difference in stem cell characteristics, including self-renewal and differentiation, was detected in parthenogenetic ESCs when compared to normally fertilized ESCs [106]. Nevertheless, to obtain conclusive results on the clinical feasibility of parthenogenetic ESCs, large scale experiments are required. Analytical systems can be useful for monitoring clinical feasibility of various autologous stem cells such as induced pluripotent stem cells, as well as parthenogenetic ESCs [109, 115–117].

EMT/MET and Fetal Somatic Cell Reprogramming

General Aspect of EMT/MET

All somatic tissues are derived from the three germinal layers of epiblast-derived ectoderm, hypoblast-derived endoderm and mesoblast-derived mesoderm. Germ layers differentiate into the specific cell types following a lineage that is independent of other layers. Under specific microenvironment, mesoderm-derived mesenchymal cells transform into epithelial cells, presenting a different phenotype within ectodermal lineage. Conversely, ectoderm-derived epithelial cells transform into mesenchymal cells. Both mesenchymal-to-epithelial cell transition (MET) and epithelial-to-mesenchymal cell transition (EMT) occur reversibly during embryogenesis. Functionally, epithelial cells have a more stationary state than mesenchymal cells, by aligning tight junction-mediated, planar array with cell polarization. The epithelial cell expresses cell–cell adhesion markers such as E-cadherin [118]. In contrast, mesenchymal cell is motile and plastic, and it shows multipolar or spindle-shaped without complete cell-to-cell contact. Mesenchymal cell has an ability to invade through extracellular matrix and, vimentin, fibronectin, N-cadherin, Twist and Snail have been employed as markers of mesenchymal cells [119]. The finding of EMT and MET by either phenotype observation or marker expression confirms cell plasticity and transformation into other cell lineages, which is directly related to cell reprogramming.

Besides normal development, MET occurs in oncogenesis, cancer metastasis and even somatic cell reprogramming into iPS cells. In embryogenesis, MET is usually seen in nephrogenesis [120], somitogenesis [121], cardiogenesis [122] and hepatogenesis [123]. For the conduction of MET, epithelium-associated genes are upregulated and the mesenchyme-associated genes are downregulated. Upon the reprogramming into iPS cells, fibroblasts must undergo MET to successfully begin the initiation phase of reprogramming. Epithelial-associated genes such as E-cadherin/Cdh1, Cldns-3, -4, -7 and -11, and Ocln, and epithelial cell adhesion molecule such as Epcam and Crb3 are all upregulated before turning-on Nanog. Mesenchymal-associated genes such as Snail, Slug, Zeb-1 and -2, and N-cadherin are also downregulated following the iPS cell establishment. Exogenous TGF-β1 is an inhibitory factor of MET and it subsequently blocks iPS cell reprogramming [124].

EMT is a biological process used by epithelial cells to increase cellular plasticity, accompanied with losing cell adhesion. Thus, the expression of E-cadherin and its related gene is decreased during this process. There are several transcription factors for inducing EMT. Snail and Slug are repressors of E-cadherin and their expression induces EMT. The S, T, G and B transcription factors are also known to induce EMT. The process begins when the epithelial cell of a high stage carcinoma undergoes the mutation into mesenchymal cell. This mesenchymal cell then enters the blood stream through capillaries, which travels throughout the body (i.e., metastasis). Ectopic expression of Klf4 in iPS cell reprogramming may be specifically responsible for inducing E-cadherin expression [125], which to some degree, is similar to cancer cell transformation (Fig. 6.3).

Fetal Fibroblast as a Cell Transformation Inducer

Fibroblasts are common cell types in connective tissue, providing tissue integrity via structural maintenance and tissue metabolism support. Fibroblasts are mesoderm-and-mesenchymal in origin and they synthesize all extracellular components consisting of ground substances and fibers such as collagen, glycoproteins, glycosaminoglycans and reticular and elastic fibers, and various cytokines. (http://ghr.nlm.nih.gov/glossary=fibroblast). Each fibroblast has a branched cytoplasm surrounding an elliptical, speckled nucleus having two or more nucleoli. Activity of fibroblasts can be evaluated by the number of rough endoplasmic reticulum (rER) and rER number becomes large when a fibroblast is activated. Fibroblasts often align each other in parallel clusters when they cover a large space. Unlike epithelial cells, fibroblasts neither form flat monolayers nor get restricted by a polarizing attachment to the basal lamina. Fibroblasts show migration activity, while epithelial cell do not. In addition, fibroblasts play an important role in tissue healing process;

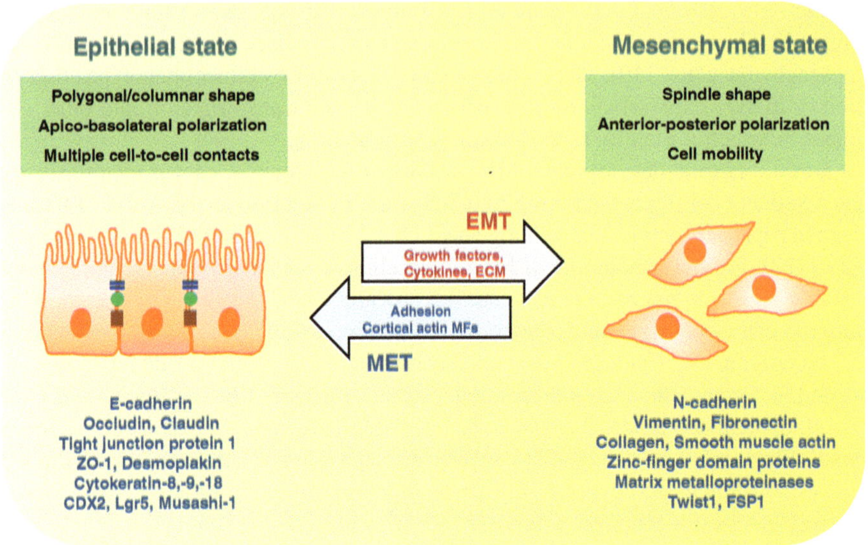

Fig. 6.3 The morphological and phenotypic changes during epithelial-mesenchymal transition (EMT) and the reverse process mesenchymal-epithelial transition (MET). EMT and MET entails profound morphological and phenotypic changes to a cell. These processes are regulated by effectors such as growth factors, cytokines, extracellular matrix (ECM), adhesion, cortical actin microfilaments (MFs) and most commonly used epithelial and mesenchymal markers

they are morphologically heterogeneous and their phonotype varies according to their location and activity. Fibroblasts often function as a basal lamina component under certain situations.

Based on their nature, embryonic fibroblasts are often used as feeder cells for nurturing various cells, including stem cells. However, fibroblasts retrieved from various tissues show both genetic and phenotypic plasticity in mice [126]. Fibroblasts have plasticity and positional memory and maintain original phenotype of their previous location at least for few generations [127]. Fibroblasts induce MET, while under specific conditions, fibroblast-transformed epithelial cells undergo EMT. Tissue damage stimulates the mitosis of fibroblasts, leading to activate cellular plasticity. Thus, it seems that fibroblasts readily adopt cell reprogramming either by induction or environmental stimulation.

Fetal Stem Cell (FSC)

Considering the function of stem cells during organogenesis, it is believed that they can be retrieved from fetal tissues at various stages of development. Furthermore, the observation of fetal brain cell transplantation into Parkinson's disease patients [128, 129] have led to the conclusion that stem cells in the brain tissues of aborted fetuses have regeneration capacity. However, the function of the putative brain stem cells cannot be normalized. There is a large scale migration of stem cells into different organs to complete formation of living organism during ontogeny. Thus, FSCs could be isolated at all stages of development and they are considered as the best cell suitable for cell reprogramming due to their intermediate properties between embryonic and adult stem cells. The characteristics of FSCs are defined as more primitive, higher proliferation rate, greater plasticity and more energetic telomerase activity compared to their adult counterparts [130, 131]. It has recently been reported that reprogramming kinetics of somatic cells differ based on cell types [33–35].

Hematopoietic Stem Cells (HSCs)

HSCs participate in the maintenance of hematopoiesis by generating all hematopoietic lineages throughout the lifespan [132]. Fetal HSCs exist in a number of fetal organs such as blood, liver, bone marrow, and umbilical cord. Early formation of hematopoietic organs occurs at the dorsal embryonic aorta in the region of aorta-gonad-mesonephros during embryogenesis, while they rapidly migrate into embryonic liver and finally home into bone marrow. The spatiotemporal properties of the cells are important for their migration activity and of changing in main primary site of hematopoiesis during fetal development. Fetal blood-derived HSCs retrieved from the first trimester are more primitive and have stronger potential to differentiate than the circulating HSCs retrieved from adults and even later stages of development in ontogeny [133–136]. They proliferate more rapidly than the HSCs retrieved from cord blood and adult bone marrow, and they have the ability to produce all hematopoietic lineages [137]. The number of HSCs circulating in fetal blood increases from the first trimester to the second trimester (known for completion of cell migration) and to initiate hematopoiesis in the bone marrow of fetuses [138]. During the second trimester, the population of HSC-containing CD34+ cells in the blood is about 4 %, which is similar to the population at the first trimester. In other fetal organs, there are 16.5 % of total nucleated cells in the bone marrow, 6 % in the liver, 5 % in the spleen and 1.1 % in the thymus [139]. After the completion of hematopoietic function in the bone marrow at the third trimester, relative population of CD34+ cells in the blood gradually decreases.

Morphologically, both fetal and adult HSCs resemble small lymphocytes, which have non-adherent round cells with a low cytoplasm-to-nucleus ratio. Their immunophenotypical characteristics are determined by co-expression of specific surface markers, which concomitantly lack several lineage commitment markers. There are many differences between the human and mouse hematopoietic cell markers for the commonly accepted type of HSCs. Surface antigen makers of HSCs generally expressed in human are $CD34^{+}$, $CD59^{+}$, Thy1/$CD90^{+}$, $CD38^{low/-}$, C-kit/$CD117^{+}$, lin^{-}. On the other hand, HSCs in mouse express $CD34^{low/-}$, SCA-1^{+}, Thy$1.1^{+/low}$, $CD38^{+}$, C-kit^{+}, lin^{-}. In addition, such combination of marker specificity does not cover the stem cells of all species and further development is necessary for the identification of novel stem cells markers in each species. Relevant efforts have recently led to the development of new markers such as SLAM (signaling lymphocyte activation molecule) family and rhodamine 123 [140, 141]. Hoechst 33342 is used to identify side population for detection of HSCs. Isolated fetal HSCs can be manipulated as a single unit and can be reprogrammed into other cell type without extended culture *in vitro*. $CD133^+$ immature mononuclear cells of umbilical cord blood cells express lower levels of *Oct4*, *Sox2*, *Nanog* and *Cripto*, and $CD133^+/CD34^+$, and immature mononuclear cells yield high rate of reprogramming, using non-integrating plasmids [142]. Banking of cord blood cells is available without decreased efficiency of reprogramming after freezing and thawing (0.027–0.05 %) [143].

Fetal liver is a reliable source of HSCs. The HSCs retrieved from the fetal liver have a higher cloning efficiency and generate more progenitors than the HSCs retrieved from adult bone marrow [144–146]. The HSCs retrieved from fetal liver at the first trimester express both hematopoietic and pancreatic markers. In the second trimester, the translocation of HSCs to other hematopoietic organs occurs, and the cells expressing specific hepatic markers appear [147]. Umbilical cord blood (UCB) is also a suitable source for fetal HSCs and the use of these tissues has ethical advantages due to their non-invasive usage and being discarded as bio-waste. Approximately 1 % of the mononuclear cord blood cells express the CD34 antigen and frequency of more primitive cells that expressed CD34 antigen, while lacking of CD38 antigen, is greater than that of adult bone marrow or cytokine-mobilized peripheral blood. HSCs derived from cord blood are more primitive and have a greater proliferation capacity due to their longer telomere length [148]. Besides, these cells can express neuronal proteins and can differentiate into neuron-like cells or glial cells [149].

Mesenchymal Stem Cells (MSCs)

Similar to adult MSC, fetal MSCs are included in the stromal cells of mesoderm-derived mesenchymal cells, which implies their differentiation ability into mesoderm-lineaged cells. In adults, MSCs were first isolated from the bone marrow, and subsequently found in other connective tissues such as adipose tissue, dental pulp, muscle, and liver and brain. Fetal MSCs were first identified in the liver, blood and bone marrow cells of the first trimester fetuses [150]. The isolated MSCs are fibroblast-like cells that originated from multipotent common mesenchymal precursor cells, and are considered as the supporting cells for hematopoiesis. These cells can be collected from a variety of fetal tissues such as bone marrow, liver, lung, kidney, thymus, dermis, pancreas and spleen as well as from the extra-embryonic tissues such as placenta, cord blood, amniotic fluid and Wharton's jelly of the umbilical cord. The quantity of MSCs in a fetus becomes different according to the stage of fetal development, which may results from the migration of hematopoietic cells. Approximately 0.4 % of nucleated cells in fetal blood are MSCs at the seventh week of gestation, after which it sharply decreases [150, 151].

Generally, fetal MSCs have higher proliferation and differentiation potential than adult MSCs. These cells have high growth kinetics (fetal 30–35 h vs. adult 80–100 h) [152] and a greater ability to differentiate into mesoderm cell lineage such as bones, muscles, and to trans-differentiate into neuronal cells such as oligodendrocytes, compared with adult MSCs [153, 154]. General characteristics of fetal MSCs is defined as

(1) the fibroblast-like cells that have plastic adherence and spindle-like morphology, (2) the cells showing extensive self-renewal capacity *in vitro* (3) the cells that have the capacity to differentiate into multiple cell lineages and (4) the cells expressing specific sets of surface markers such as CD29 (β1-integrin), CD73 (SH3 and SH4), CD105 (SH2), CD44 (HCAM1), CD90 (thy-1) of early bone marrow progenitor cell marker and extracellular matrix proteins of vimentin, laminin and fibronectin without the expression of hematopoietic cell (CD14, CD34, CD45) and endothelial cell (von Willebrand factor) markers. Recent reports on the expression of several pluripotency markers in some subpopulation of fetal MSC have provoked great interests in their stemness properties, compared with those of adult MSCs. Fetal MSCs express baseline level of Oct4, Nanog, Rex-1, SSEA-3, SSEA-4, Tra-1-60 and Tra-1-81 [152, 155]. There are different immunophenotypical characteristics of fetal MSCs according to retrieval sites and developmental stages. Forty-four percentage of total cell population are $CD34^+/CD45^-$ cells in fetal lung in early second trimester, while only 4.8–12.6 % exists in the bone marrow, spleen and liver [156]. Mesonephric MSCs do not express hematopoietic markers such as CD45 and CD34, but are highly positive for vimentin, laminin and type I collagen of mesenchymal cell markers [157]. There are several differences in fetal MSCs retrieved from different sites and at different stages. Lower potential to trigger osteogenesis is detected in the MSCs retrieved from the liver at the first and second trimesters than the MSCs retrieved from the blood at the first trimester and the spleen, lung and bone marrow at the second trimester [156]. On the other hand, the MSCs retrieved from the pancreas at the second trimester can differentiate into mesoderm-derived lineage [158] and successful transplantation after differentiation of the MSCs into pancreatic β-cells has been reported in sheep [159]. Metanephric MSCs can induce osteogenesis and myogenesis *in vitro* and further detect hemopoiesis and hepatocytogenesis after *in vivo*-transplantation [157].

Extra-embryonic tissues such as placenta and amniotic fluid are extra source of MSCs, and their usage can also avoid ethical disputes around the use of fetal tissues. Comparatively, larger number of MSCs exists in amniotic membrane and fluid and these extra-embryonic MSCs are positive for Oct4, Nanog, Sox-2 and SSEA-4 pluripotent stem cell markers [131, 148, 160]. Putatively pluripotent, amnion-derived MSCs expressing Sox2, Klf4 and c-Myc do not form teratoma *in vivo*, providing them with an excellent therapeutic feasibility in terms of cell safety and pluripotency [131, 161].

Neural Stem Cells (NSCs)

NSCs have the capacity to differentiate into neurons, astrocytes and oligodendrocytes. Compared with adult NSCs, fetal NSCs can be cultured for extended durations with strong self-renewal activity. Human fetal NSC can be isolated from the brains of aborted fetuses, using $CD133^+$, $CD34^-$ and $CD45^-$ and they can be subsequently cultured in form of neurospheres, while sustaining their cellular plasticity [162, 163]. NSCs express abundant levels of pluripotent associated genes, and this property yields an advantage in triggering cellular reprogramming. There have been several reports on superiority of NSCs for cell reprogramming [164, 165].

Genetic and Cellular Aspect After Cell Reprogramming or Transformation

Studying genetics and cellular plasticity of fetal cells contributes to a better understanding of reprogramming. At the same time, better models can be developed for studying fetal cell dynamics, as an important step towards clinical trials. Considering the important role of microenvironment in the induction of cell plasticity, either co-culture system or 3D culture may become a powerful model system for cell reprogramming and transformation. The mesoderm-derived cells may also be better choices for selecting the effector cell of reprogramming than single endodermal or ectodermal cell lineage. Recent reports on monitoring cell plasticity [32, 166], have employed a co-culture system of ovarian (stromal) cells as well as embryonic fibroblasts. The ovary consists of endoderm-based germ cell and its derived cells and mesoderm-based stromal cells. Fibroblasts have both genetic and cellular plasticity under various conditions; therefore, both ovarian stromal cells and fibroblasts can be employed as either an effector or a supporting cell type for reprogramming. In their initial experiments, ovarian cell served as the effector of cell transformation, and fetal fibroblasts were employed as support for ovarian cell transformation. Excessive number of stromal cells, maturing oocytes and ovarian follicles were eliminated before culture, which avoided rolling-up of cultured cells due to mass cell number, and only limited number of stromal cells were co-cultured with fetal fibroblasts. As a result, autologous ESC-like cells were derived from the co-culture system and the origin of the established colony-forming cells appeared to be the ovary. Based on these results, follow up experiments were designed by using embryonic fibroblasts as the effector of the transformation and ovarian cells as the helper cell for reprogramming.

Cellular Aspects of Cell Transformation

In another important study, ESC-like, colony-forming cells were derived from fibroblasts and subjected to short-tandom repeat microsatellite analysis. All lines established maintained greater than 20 subpassages, Factorial analysis clearly demonstrates that a significant model effect on the aggregation of fibroblasts was detected in co-cultured cell types, whereas the origin, strain, gender of fibroblasts and the strain of co-cultured cells with fibroblasts did not influence the aggregation. The potential relevance of stress-related or apoptosis-related environmental factors was also raised. The colony-forming fibroblasts were positive for ESC-specific markers and expressed ESC-specific genes with telomerase activity, while immunostaining of fibroblasts with mesenchymal cell-specific (CD44), stem cell-specific (Oct-4 and Nanog), germ cell-specific (Vasa and Fragilis), follicular cell-specific (AMH) and hematopoietic cell-specific (CD45) markers was not positive. Transmission electron microscopy showed that there was prominently different morphology between the fibroblasts before and after the colony formation. The colony-forming cells had similar ultrastructure with reference to ESCs: large ratio of nucleus to cytoplasmic volume, less developed membrane microvilli and cytoplasmic micro-organelles, and multiple nucleoli with distinct nuclear membrane in the nucleus. Approximately 30–50 % imprinted gene *Igf2* in the established lines was methylated. Formation of embryoid bodies *in vitro* and teratomas in immune-deficient mice were detected. The colony-forming cells also induced tissue-specific cell differentiation and showed ability similar ESC-like to differentiate into follicle structures *in vitro*. Somatic chimeras were detected after the transfer of the blastocysts derived from the aggregated embryos with fibroblast-derived ESC-like cells. However, progeny test did not yield germline chimera. Unexpectedly, karyotype of the established cells was all tetraploidy with XX sex chromosome. No tetraploid cells were found in their progenitor fibroblasts. Deletions and translocations were detected at multiple chromosomal sites in the ESC-like cells, while the deleted chromatids appeared sporadic and no typical chromosome abnormality patterns were detected. Change or breakdown of cell cycle checkpoint was detected in the established cells, which showed less phosphorylation of p53 level in the fibroblast-derived, ESC-like cells than E14 or R1 ESCs or the non-transformed fibroblasts.

These results support the notion of environment-driven cell plasticity in fibroblasts. From a different viewpoint, it is possible that the fibroblast-derived, ESC-like cells did not arise due to cell transformation, but they resulted from rapid proliferation of terminally-differentiated cells mixed with fibroblasts. The breakdown of cell cycle checkpoint in the established cells after fibroblast reprogramming was found, which may allow the aneuploid cells to survive. Probably, altering the cell cycle before transformation under a specific microenvironment may be an important step in the establishment of fibroblast reprogramming. These results further suggest a new strategy for establishing patient-specific pluripotent cells of desired genotype from human fetal somatic tissues.

Genetic Aspects and Gene Expression After Cell Transformation

SNP genotyping data demonstrated that CFFs had both homologous and heterologous recombination of genomic SNPs, and complete heterozygous SNPs were detected in EGCs, ESCs, and MFFs of the B6D2F1 strain [32]. Parthenogenetic ESC lines showed both homozygous and heterozygous chromosome recombination. As expected, only homozygotic SNP loci were detected in the fibroblasts of maternal or paternal origin. Pyrosequencing analysis of the methylation status showed the difference among fibroblast-derived, ESC-like cells, iPSCs, parthenogenetic ESCs, ESCs, EGCs and MEFs. Global gene expression analysis using cDNA microarrays shows that fibroblast-derived colonies are similar to ESCs and iPSCs, but not fibroblasts. Apparent difference in gene expression is also detected among the colonies, iPSCs and MFFs, and iPSCs. Interestingly, fibroblast-transformed colonies are closer to ESCs than to iPSCs, suggesting that the use of ectopic transcription factors to generate iPSCs may influence transcriptional regulation. On the other hand, as both fibroblast-derived colonies and iPSCs are induced by cell reprogramming, the pattern of gene expression is further different from ESC derived from normally fertilized embryo. Results of gene ontology analysis demonstrate that both factor- and environment-based reprogramming use a similar molecular signaling pathway of cellular immortalization. Although conclusive statements cannot be made using these results, at least, it can confirm either the genetic and cellular plasticity of fibroblasts or mixed cell populations. It is still possible that other fetal cells mixed in fibroblast populations such as primordial germ cell and mesenchymal cell of fibroblast progenitors involve the reprogramming and cell transformation observed in this study [32] (Fig. 6.4).

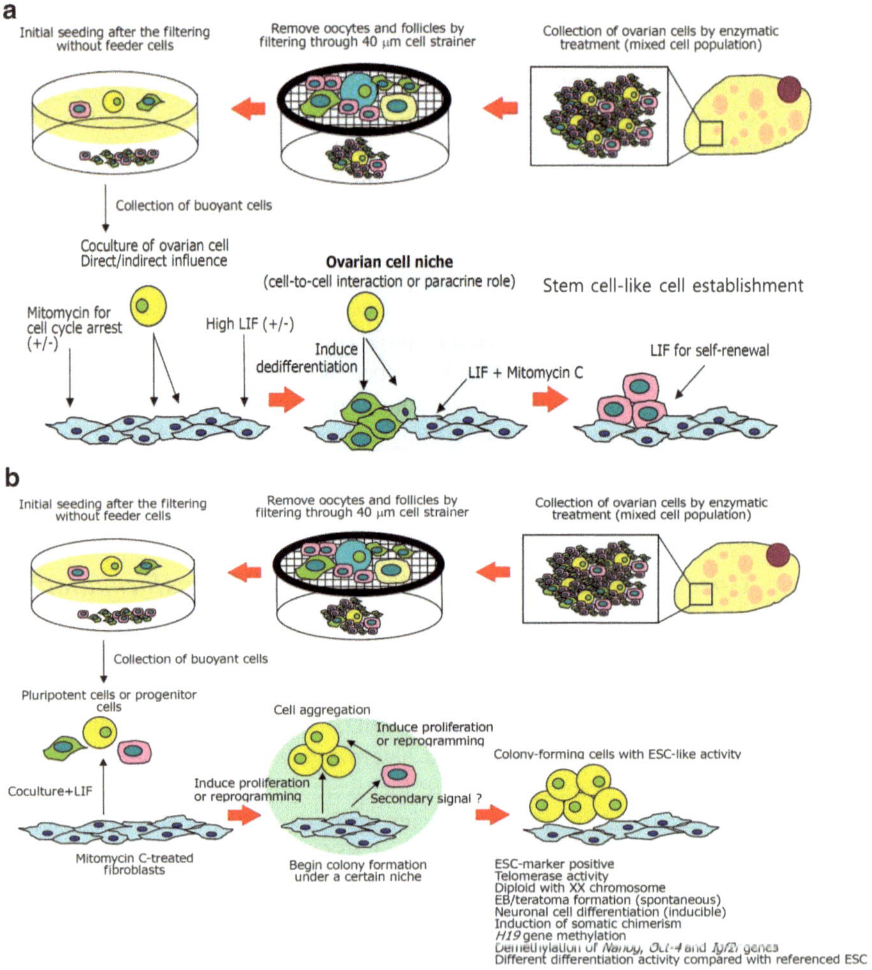

Fig. 6.4 Putative methods to establish non-embryonic, pluripotent cells by cell-to-cell interaction. (**a**) Establishment of fibroblast-derived, colony-forming cells having stem cell-like morphology and activity by potential interaction with the cells of other type under an ovarian cell niche. (**b**) Establishment of ovarian cell-derived, colony-forming cell by co-culture of fibroblasts

Potential Application and Conclusion

Stem cells are an important resource for developing future medical technology. By using the most updated technology for stem cell manipulation, however, it is unlikely that stem cell technology reaches to the level of therapeutic biomaterials. In other words, most scientists still inquire whether stem cells can be applied to both clinics and pharmaceutical industries. This skepticism is based on the fact that each individual has the genetic diversity of unlimited range, and anatomical similarity and phenotypic homogeneity at the cell, tissue and organ levels do not guarantee the homogeneity of cellular properties and activity under a specific microenvironment.

The solution to regulate genetic and cellular diversity of stem cells is a prerequisite for optimizing stem cell technology to the clinical level. The use of prenatal tissues including embryonic tissue, germ cells and primordial cells, and the progenitor of terminally differentiated cells provides great advantages for improving stem cell engineering and stem cell-based therapy. Discovery of 'reprogrammable' prenatal cells in adult system and designing of its manipulation technology for cell reprogramming significantly expand the source of clinically feasible stem cells. The use of prenatal cells for regulating various activities of functional cells in adult tissues further provide a key for regulating stemness of multipotent and pluripotent cells and even contributes to elucidating tumorigenesis and cancer therapy via controlling of undifferentiated cancer stem cells. Research on prenatal cell manipulation greatly contributes to increasing the efficiency of iPS cells or cell transformation technologies.

Clinical Applications

Establishment of standard operation protocols (SOPs) is a key factor for developing stem cell-derived products. Due to genetic diversity leading to phenotype change and cellular functions, however, it is difficult to establish universal SOPs for stem cell manipulation. On the other hand, the induction of a pluripotent stem cell state by manipulating terminally differentiated cells exposes target cells to various environmental stresses. In some cases, features such as tumor cell-like activity and genetic aneuploidy appear after cell manipulation. Prenatal cells can provide an opportunity for patient-specific stem cell-based therapy as well as access to more stem cell sources. For instance, large scale studies with prenatal cells may facilitate establishing proper SOPs for patient-specific stem cells. The elucidation on genetic and cellular response of prenatal stem cells to microenvironment not only completes the development of optimal protocols, but also contributes to securing the safe use of stem cells in patients.

It seems that the use of immune-compatible stem cells is an inevitable choice for improving the clinical applications of stem cells under the current technology. Autologous tissue is the exclusive resource for the immune-compatible stem cells.

The technology for modulating cell fate via cellular reprogramming and transformation further expands the use of prenatal stem cells and their differentiated progenies. Alteration of cell phenotype to derive clinical grade cells should be induced without incurring any abnormalities. Prenatal cells have a better cellular plasticity than neonatal or adult cells [167] and are excellent sources of cellular reprogramming and transformation. Relevant technology of prenatal cell transformation greatly contributes to retrieving clinical grade stem cell, which retains immune-compatibility.

Animal Prenatal Tissue as an Alternative for Therapeutic Purposes

Although a strict regulation on animal experimentation is established, accessibility to animal tissues is better than human samples at least from the ethical viewpoint. Prenatal tissues such as mouse embryonic fibroblasts have widely been employed for human laboratory work and for genetic manipulation with less ethical concern than human prenatal tissue. To date, a wide range of studies involving animal fetal tissues have been undertaken. However, the genetic relevance of animal models to human diseases needs to be well-defined. For example, chicken is considered as the same order of phylogenic tree as human and due to their technical accessibility, chicken eggs have long been used as a classical contributor for producing pharmaceutical biomaterials and a development model. Recent advances further expanded the importance of chicken eggs for the studying stem cell trafficking and differentiation. Pig has been also considered as an animal model with high physiological similarities to humans compared to other species (with the exception of primates). The size of pig genome is similar to that of human [168, 169] and physical co-localization of genetic loci on the same chromosome within species is much larger (more than three times) between humans and pigs than between humans and mice [170]. Nevertheless, mice and rats remain as the main species establishing genetically inbred lines. Primates serve as the ultimate preclinical animal models in pharmaceutical and biomaterial industries.

Conclusion

The use of prenatal cells for reprogramming and transformation not only contributes to increasing the clinical feasibility of patient-specific stem cell engineering, but also expands stem cell sources. While a wide range of studies is conducted using prenatal stem cells in a patient-specific manner, careful approaches are necessary to establish proper protocols for cellular reprogramming applicable to patients. The current parameters used to verify the normal status of stem cells are not sufficient with respect to their various functions. Innovative parameters to monitoring stem cell need to be established to better facilitate the use of prenatal stem cells.

Abbreviations

BMP	Bone morphogenetic protein
cDNA	Complementary DNA
CFFs	Colony forming fibroblasts
CPP	Cell penetrating protein
DNA	Deoxyribonucleic acid
ECCs	Embryonic carcinoma cells
EGCs	Embryonic germ cells
EMT	Epithelial-to-mesenchymal transition
ESCs	Embryonic stem cells
FACS	Fluorescent activated cell sorter
FGF	Fibroblast growth factor
FSCs	Fetal stem cells
GDNF	Glial-derived neurotrophic factor
HIV	Human immunodeficiency virus
HSCs	Hematopoietic stem cells
iPSCs	Induce pluripotent stem cells
MACS	Magnetic activated cell sorter
MET	Mesenchymal-to-epithelial transition
miRNA	microRNA
MMLV	Moloney murine leukemia virus
mRNA	Messenger RNA
MSCs	Mesenchymal stem cells
NSCs	Neural stem cells
PB	Piggyback
PGCs	Primordial germ cells
RNA	Ribonucleic acid
ROS	Reactive oxygen species
SCNT	Somatic cell nuclear transfer
SLAM	Signaling lymphocyte activation molecule
SNP	Single-nucleotide polymorphism
SOP	Standard operation protocol
SSCs	Spermatogonial stem cells
TGF	Transforming growth factor
UCB	Umbilical cord blood

References

1. Takahashi K, Yamanaka S. Induction of pluripotent stem cells from mouse embryonic and adult fibroblast cultures by defined factors. Cell. 2006;126(4):663–76.
2. Wernig M, et al. In vitro reprogramming of fibroblasts into a pluripotent ES-cell-like state. Nature. 2007;448(7151):318–24.

3. Brambrink T, et al. Sequential expression of pluripotency markers during direct reprogramming of mouse somatic cells. Cell Stem Cell. 2008;2(2):151–9.
4. Sommer CA, et al. Induced pluripotent stem cell generation using a single lentiviral stem cell cassette. Stem Cells. 2009;27(3):543–9.
5. Fusaki N, et al. Efficient induction of transgene-free human pluripotent stem cells using a vector based on Sendai virus, an RNA virus that does not integrate into the host genome. Proc Jpn Acad Ser B Phys Biol Sci. 2009;85(8):348–62.
6. Stadtfeld M, et al. Induced pluripotent stem cells generated without viral integration. Science. 2008;322(5903):945–9.
7. Wilson MH, Coates CJ, George Jr AL. PiggyBac transposon-mediated gene transfer in human cells. Mol Ther. 2007;15(1):139–45.
8. Okita K, et al. Generation of mouse induced pluripotent stem cells without viral vectors. Science. 2008;322(5903):949–53.
9. Gonzalez F, et al. Generation of mouse-induced pluripotent stem cells by transient expression of a single nonviral polycistronic vector. Proc Natl Acad Sci U S A. 2009;106(22):8918–22.
10. Yu J, et al. Human induced pluripotent stem cells free of vector and transgene sequences. Science. 2009;324(5928):797–801.
11. Hu K. All roads lead to induced pluripotent stem cells: the technologies of iPSC generation. Stem Cells Dev. 2014;23(12):1285–300.
12. Si-Tayeb K, et al. Generation of human induced pluripotent stem cells by simple transient transfection of plasmid DNA encoding reprogramming factors. BMC Dev Biol. 2010;10:81.
13. Narsinh KH, et al. Generation of adult human induced pluripotent stem cells using nonviral minicircle DNA vectors. Nat Protoc. 2011;6(1):78–88.
14. Warren L, et al. Highly efficient reprogramming to pluripotency and directed differentiation of human cells with synthetic modified mRNA. Cell Stem Cell. 2010;7(5):618–30.
15. Anokye-Danso F, et al. Highly efficient miRNA-mediated reprogramming of mouse and human somatic cells to pluripotency. Cell Stem Cell. 2011;8(4):376–88.
16. Freberg CT, et al. Epigenetic reprogramming of OCT4 and NANOG regulatory regions by embryonal carcinoma cell extract. Mol Biol Cell. 2007;18(5):1543–53.
17. Kim D, et al. Generation of human induced pluripotent stem cells by direct delivery of reprogramming proteins. Cell Stem Cell. 2009;4(6):472–6.
18. Hansis C, et al. Nuclear reprogramming of human somatic cells by xenopus egg extract requires BRG1. Curr Biol. 2004;14(16):1475–80.
19. Ding X, et al. The polycomb protein Ezh2 impacts on induced pluripotent stem cell generation. Stem Cells Dev. 2014;23(9):931–40.
20. Kabouridis PS. Biological applications of protein transduction technology. Trends Biotechnol. 2003;21(11):498–503.
21. Sandgren S, Cheng F, Belting M. Nuclear targeting of macromolecular polyanions by an HIV-Tat derived peptide. Role for cell-surface proteoglycans. J Biol Chem. 2002;277(41): 38877–83.
22. Liang G, He J, Zhang Y. Kdm2b promotes induced pluripotent stem cell generation by facilitating gene activation early in reprogramming. Nat Cell Biol. 2012;14(5):457–66.
23. Hou P, et al. Pluripotent stem cells induced from mouse somatic cells by small-molecule compounds. Science. 2013;341(6146):651–4.
24. Kono T. Nuclear transfer and reprogramming. Rev Reprod. 1997;2(2):74–80.
25. Wakayama T, et al. Differentiation of embryonic stem cell lines generated from adult somatic cells by nuclear transfer. Science. 2001;292(5517):740–3.
26. Byrne JA, et al. Producing primate embryonic stem cells by somatic cell nuclear transfer. Nature. 2007;450(7169):497–502.
27. Wakayama T, et al. Full-term development of mice from enucleated oocytes injected with cumulus cell nuclei. Nature. 1998;394(6691):369–74.
28. Couldrey C, Lee RS. DNA methylation patterns in tissues from mid-gestation bovine foetuses produced by somatic cell nuclear transfer show subtle abnormalities in nuclear reprogramming. BMC Dev Biol. 2010;10:27.

29. Ross PJ, Cibelli JB. Bovine somatic cell nuclear transfer. Methods Mol Biol. 2010;636: 155–77.
30. Cowan CA, et al. Nuclear reprogramming of somatic cells after fusion with human embryonic stem cells. Science. 2005;309(5739):1369–73.
31. Obokata H, et al. Stimulus-triggered fate conversion of somatic cells into pluripotency. Nature. 2014;505(7485):641–7.
32. Lee ST, et al. Transformation of somatic cells into stem cell-like cells under a stromal niche. Faseb J. 2013;27(7):2644–56.
33. Tan KY, et al. Efficient generation of iPS cells from skeletal muscle stem cells. PLoS One. 2011;6(10):e26406.
34. Kleger A, et al. Increased reprogramming capacity of mouse liver progenitor cells, compared with differentiated liver cells, requires the BAF complex. Gastroenterology. 2012;142(4):907–17.
35. Eminli S, et al. Differentiation stage determines potential of hematopoietic cells for reprogramming into induced pluripotent stem cells. Nat Genet. 2009;41(9):968–76.
36. Hansel MC, et al. Increased reprogramming of human fetal hepatocytes compared with adult hepatocytes in feeder-free conditions. Cell Transplant. 2014;23(1):27–38.
37. Alberts B, Johnson A, Lewis J, Raff M, Roberts K, Walter P. Molecular biology of the cell. New York: Garland Science; 2002.
38. Twyman R. Developmental biology. Oxford: Bios Scientific Publishers; 2001.
39. Cinalli RM, Rangan P, Lehmann R. Germ cells are forever. Cell. 2008;132(4):559–62.
40. Kunwar PS, Lehmann R. Developmental biology: germ-cell attraction. Nature. 2003;421(6920):226–7.
41. Ginsburg M, Snow MH, McLaren A. Primordial germ cells in the mouse embryo during gastrulation. Development. 1990;110(2):521–8.
42. Lawson KA, et al. Bmp4 is required for the generation of primordial germ cells in the mouse embryo. Genes Dev. 1999;13(4):424–36.
43. Ying Y, Qi X, Zhao GQ. Induction of primordial germ cells from murine epiblasts by synergistic action of BMP4 and BMP8B signaling pathways. Proc Natl Acad Sci U S A. 2001;98(14):7858–62.
44. Mei J, et al. C1q-like factor, a target of miR-430, regulates primordial germ cell development in early embryos of Carassius auratus. Int J Biol Sci. 2013;10(1):15–24.
45. Magnusdottir E, et al. A tripartite transcription factor network regulates primordial germ cell specification in mice. Nat Cell Biol. 2013;15(8):905–15.
46. Okamura D, et al. REST and its downstream molecule Mek5 regulate survival of primordial germ cells. Dev Biol. 2012;372(2):190–202.
47. Childs AJ, et al. LIN28 is selectively expressed by primordial and pre-meiotic germ cells in the human fetal ovary. Stem Cells Dev. 2012;21(13):2343–9.
48. Leitch HG, Smith A. The mammalian germline as a pluripotency cycle. Development. 2013;140(12):2495–501.
49. Leitch HG, Tang WW, Surani MA. Primordial germ-cell development and epigenetic reprogramming in mammals. Curr Top Dev Biol. 2013;104:149–87.
50. Monk M. A stem-line model for cellular and chromosomal differentiation in early mouse development. Differentiation. 1981;19(2):71–6.
51. Monk M, et al. Preimplantation sexing and diagnosis of hypoxanthine phosphoribosyl transferase deficiency in mice by biochemical microassay. Am J Med Genet. 1990;35(2):201–5.
52. Pesce M, et al. Differential expression of the Oct-4 transcription factor during mouse germ cell differentiation. Mech Dev. 1998;71(1–2):89–98.
53. Shim H, Anderson GB. In vitro survival and proliferation of porcine primordial germ cells. Theriogenology. 1998;49(3):521–8.
54. Resnick JL, et al. Long-term proliferation of mouse primordial germ cells in culture. Nature. 1992;359(6395):550–1.
55. Nagy A, Nagy K, Gertsenstein M. Production of mouse chimeras by aggregating pluripotent stem cells with embryos. Methods Enzymol. 2010;476:123–49.

56. Gardner RL. Clonal analysis of early mammalian development. Philos Trans R Soc Lond B Biol Sci. 1985;312(1153):163–78.
57. Gardner RL. Mouse chimeras obtained by the injection of cells into the blastocyst. Nature. 1968;220(5167):596–7.
58. Gardner RL, Rossant J. Investigation of the fate of 4-5 day post-coitum mouse inner cell mass cells by blastocyst injection. J Embryol Exp Morphol. 1979;52:141–52.
59. Wang ZQ, et al. Generation of completely embryonic stem cell-derived mutant mice using tetraploid blastocyst injection. Mech Dev. 1997;62(2):137–45.
60. Nichols J, Smith A. Naive and primed pluripotent states. Cell Stem Cell. 2009;4(6):487–92.
61. Matsui Y, Zsebo K, Hogan BL. Derivation of pluripotential embryonic stem cells from murine primordial germ cells in culture. Cell. 1992;70(5):841–7.
62. Leitch HG, et al. Embryonic germ cells from mice and rats exhibit properties consistent with a generic pluripotent ground state. Development. 2010;137(14):2279–87.
63. Sharova LV, et al. Global gene expression profiling reveals similarities and differences among mouse pluripotent stem cells of different origins and strains. Dev Biol. 2007;307(2):446–59.
64. Durcova-Hills G, Surani A. Reprogramming primordial germ cells (PGC) to embryonic germ (EG) cells. Curr Protoc Stem Cell Biol. 2008;Chapter 1:Unit1A3.
65. Turnpenny L, et al. Evaluating human embryonic germ cells: concord and conflict as pluripotent stem cells. Stem Cells. 2006;24(2):212–20.
66. Leitch HG, et al. Rebuilding pluripotency from primordial germ cells. Stem Cell Rep. 2013;1(1):66–78.
67. Johnson L. Efficiency of spermatogenesis. Microsc Res Tech. 1995;32(5):385–422.
68. Ryu BY, et al. Conservation of spermatogonial stem cell self-renewal signaling between mouse and rat. Proc Natl Acad Sci U S A. 2005;102(40):14302–7.
69. Itman C, et al. All in the family: TGF-beta family action in testis development. Reproduction. 2006;132(2):233–46.
70. Oatley JM, Brinster RL. Regulation of spermatogonial stem cell self-renewal in mammals. Annu Rev Cell Dev Biol. 2008;24:263–86.
71. Barrios F, et al. Opposing effects of retinoic acid and FGF9 on Nanos2 expression and meiotic entry of mouse germ cells. J Cell Sci. 2010;123(Pt 6):871–80.
72. Tegelenbosch RA, de Rooij DG. A quantitative study of spermatogonial multiplication and stem cell renewal in the C3H/101 F1 hybrid mouse. Mutat Res. 1993;290(2):193–200.
73. Sato T, et al. In vitro sperm production from mouse spermatogonial stem cell lines using an organ culture method. Nat Protoc. 2013;8(11):2098–104.
74. Jung JG, et al. Development of novel markers for the characterization of chicken primordial germ cells. Stem Cells. 2005;23(5):689–98.
75. Jung JG, et al. Identification, culture, and characterization of germline stem cell-like cells in chicken testes. Biol Reprod. 2007;76(1):173–82.
76. Urven LE, et al. Differential gene expression in fetal mouse germ cells. Biol Reprod. 1993;48(3):564–74.
77. Clermont Y, Perey B. Quantitative study of the cell population of the seminiferous tubules in immature rats. Am J Anat. 1957;100(2):241–67.
78. Kluin PM, de Rooij DG. A comparison between the morphology and cell kinetics of gonocytes and adult type undifferentiated spermatogonia in the mouse. Int J Androl. 1981;4(4):475–93.
79. Yoshida S, et al. The first round of mouse spermatogenesis is a distinctive program that lacks the self-renewing spermatogonia stage. Development. 2006;133(8):1495–505.
80. Roosen-Runge EC, Giesel Jr LO. Quantitative studies on spermatogenesis in the albino rat. Am J Anat. 1950;87(1):1–30.
81. Clermont Y, Leblond CP. Renewal of spermatogonia in the rat. Am J Anat. 1953;93(3):475–501.
82. Monesi V. Autoradiographic study of DNA synthesis and the cell cycle in spermatogonia and spermatocytes of mouse testis using tritiated thymidine. J Cell Biol. 1962;14:1–18.

83. Clermont Y, Bustos-Obregon E. Re-examination of spermatogonial renewal in the rat by means of seminiferous tubules mounted "in toto". Am J Anat. 1968;122(2):237–47.
84. Dym M, Clermont Y. Role of spermatogonia in the repair of the seminiferous epithelium following x-irradiation of the rat testis. Am J Anat. 1970;128(3):265–82.
85. Clermont Y, Hermo L. Spermatogonial stem cells in the albino rat. Am J Anat. 1975;142(2):159–75.
86. Clermont Y, Leblond CP. Differentiation and renewal of spermatogonia in the monkey, Macacus rhesus. Am J Anat. 1959;104:237–73.
87. Clermont Y, Antar M. Duration of the cycle of the seminiferous epithelium and the spermatogonial renewal in the monkey Macaca arctoides. Am J Anat. 1973;136(2):153–65.
88. Tilly JL, Johnson J. Recent arguments against germ cell renewal in the adult human ovary: is an absence of marker gene expression really acceptable evidence of an absence of oogenesis? Cell Cycle. 2007;6(8):879–83.
89. Tilly JL, Telfer EE. Purification of germline stem cells from adult mammalian ovaries: a step closer towards control of the female biological clock? Mol Hum Reprod. 2009;15(7):393–8.
90. Zheng W, Zhang H, Liu K. The two classes of primordial follicles in the mouse ovary: their development, physiological functions, and implications for future research. Mol Hum Reprod. 2014;20(4):286–92.
91. Johnson J, et al. Germline stem cells and follicular renewal in the postnatal mammalian ovary. Nature. 2004;428(6979):145–50.
92. Johnson J, et al. Oocyte generation in adult mammalian ovaries by putative germ cells in bone marrow and peripheral blood. Cell. 2005;122(2):303–15.
93. Kim IW, et al. Derivation of developmentally competent oocytes by the culture of preantral follicles retrieved from adult ovaries: maturation, blastocyst formation, and embryonic stem cell transformation. Fertil Steril. 2009;92(5):1716–24.
94. Choi JH, et al. Generation of viable embryos and embryonic stem cell-like cells from cultured primary follicles in mice. Biol Reprod. 2011;85(4):744–54.
95. Lenie S, et al. A reproducible two-step culture system for isolated primary mouse ovarian follicles as single functional units. Biol Reprod. 2004;71(5):1730–8.
96. Cortvrindt R, Smitz J, Van Steirteghem AC. Assessment of the need for follicle stimulating hormone in early preantral mouse follicle culture in vitro. Hum Reprod. 1997;12(4):759–68.
97. Makris A, Ryan KJ. Progesterone, androstenedione, testosterone, estrone, and estradiol synthesis in hamster ovarian follicle cells. Endocrinology. 1975;96(3):694–701.
98. Cecconi S, et al. In vitro development of sheep preantral follicles. Biol Reprod. 1999;60(3):594–601.
99. Echternkamp SE, et al. Ovarian follicular development in cattle selected for twin ovulations and births. J Anim Sci. 2004;82(2):459–71.
100. Gougeon A. Dynamics of follicular growth in the human: a model from preliminary results. Hum Reprod. 1986;1(2):81–7.
101. Liu J, et al. Maturation of mouse primordial follicles by combination of grafting and in vitro culture. Biol Reprod. 2000;62(5):1218–23.
102. Cortvrindt R, Smitz J, Van Steirteghem AC. In-vitro maturation, fertilization and embryo development of immature oocytes from early preantral follicles from prepuberal mice in a simplified culture system. Hum Reprod. 1996;11(12):2656–66.
103. Eppig JJ, Schroeder AC. Capacity of mouse oocytes from preantral follicles to undergo embryogenesis and development to live young after growth, maturation, and fertilization in vitro. Biol Reprod. 1989;41(2):268–76.
104. Choi JK, Agarwal P, He X. In vitro culture of early secondary preantral follicles in hanging drop of ovarian cell-conditioned medium to obtain MII oocytes from outbred deer mice. Tissue Eng Part A. 2013;19(23–24):2626–37.
105. Sadeu JC, Adriaenssens T, Smitz J. Expression of growth differentiation factor 9, bone morphogenetic protein 15, and anti-Mullerian hormone in cultured mouse primary follicles. Reproduction. 2008;136(2):195–203.

106. Lee ST, et al. Establishment of autologous embryonic stem cells derived from preantral follicle culture and oocyte parthenogenesis. Fertil Steril. 2008;90(5):1910–20.
107. Gong SP, et al. Improved establishment of autologous stem cells derived from preantral follicle culture and oocyte parthenogenesis. Stem Cells Dev. 2008;17(4):695–712.
108. Adriaens I, Cortvrindt R, Smitz J. Differential FSH exposure in preantral follicle culture has marked effects on folliculogenesis and oocyte developmental competence. Hum Reprod. 2004;19(2):398–408.
109. Revazova ES, et al. Patient-specific stem cell lines derived from human parthenogenetic blastocysts. Cloning Stem Cells. 2007;9(3):432–49.
110. Cibelli JB, et al. Parthenogenetic stem cells in nonhuman primates. Science. 2002;295(5556):819.
111. Hirabayashi M, et al. Derivation of embryonic stem cell lines from parthenogenetically developing rat blastocysts. Stem Cells Dev. 2014;23(2):107–14.
112. Kaufman MH, Barton SC, Surani MA. Normal postimplantation development of mouse parthenogenetic embryos to the forelimb bud stage. Nature. 1977;265(5589):53–5.
113. Kaufman MH, et al. Establishment of pluripotential cell lines from haploid mouse embryos. J Embryol Exp Morphol. 1983;73:249–61.
114. Gong SP, et al. Change in gene expression of mouse embryonic stem cells derived from parthenogenetic activation. Hum Reprod. 2009;24(4):805–14.
115. Kawahara M, et al. High-frequency generation of viable mice from engineered bi-maternal embryos. Nat Biotechnol. 2007;25(9):1045–50.
116. Mai Q, et al. Derivation of human embryonic stem cell lines from parthenogenetic blastocysts. Cell Res. 2007;17(12):1008–19.
117. Ju JY, et al. Establishment of stem cell lines from nuclear transferred and parthenogenetically activated mouse oocytes for therapeutic cloning. Fertil Steril. 2008;89(5 Suppl):1314–23.
118. Navarro P, et al. A role for the E-cadherin cell-cell adhesion molecule during tumor progression of mouse epidermal carcinogenesis. J Cell Biol. 1991;115(2):517–33.
119. Onder TT, et al. Loss of E-cadherin promotes metastasis via multiple downstream transcriptional pathways. Cancer Res. 2008;68(10):3645–54.
120. Davies JA. Mesenchyme to epithelium transition during development of the mammalian kidney tubule. Acta Anat (Basel). 1996;156(3):187–201.
121. Nakaya Y, et al. Mesenchymal-epithelial transition during somitic segmentation is regulated by differential roles of Cdc42 and Rac1. Dev Cell. 2004;7(3):425–38.
122. Nakajima Y, et al. Mechanisms involved in valvuloseptal endocardial cushion formation in early cardiogenesis: roles of transforming growth factor (TGF)-beta and bone morphogenetic protein (BMP). Anat Rec. 2000;258(2):119–27.
123. Li B, et al. Evidence for mesenchymal-epithelial transition associated with mouse hepatic stem cell differentiation. PLoS One. 2011;6(2):e17092.
124. Samavarchi-Tehrani P, et al. Functional genomics reveals a BMP-driven mesenchymal-to-epithelial transition in the initiation of somatic cell reprogramming. Cell Stem Cell. 2010;7(1):64–77.
125. Yori JL, et al. Kruppel-like factor 4 inhibits epithelial-to-mesenchymal transition through regulation of E-cadherin gene expression. J Biol Chem. 2010;285(22):16854–63.
126. Lim JM, Gong SP. Somatic cell transformation into stem cell-like cells induced by different microenvironments. Organogenesis. 2013;9(4):245–8.
127. Chang HY, et al. Diversity, topographic differentiation, and positional memory in human fibroblasts. Proc Natl Acad Sci U S A. 2002;99(20):12877–82.
128. Lindvall O. Neural transplantation: a hope for patients with Parkinson's disease. Neuroreport. 1997;8(14):3–10.
129. Kordower JH, et al. Neuropathological evidence of graft survival and striatal reinnervation after the transplantation of fetal mesencephalic tissue in a patient with Parkinson's disease. N Engl J Med. 1995;332(17):1118–24.
130. Abdulrazzak H, et al. Biological characteristics of stem cells from foetal, cord blood and extraembryonic tissues. J R Soc Interface. 2010;7 Suppl 6:S689–706.

131. De Coppi P, et al. Isolation of amniotic stem cell lines with potential for therapy. Nat Biotechnol. 2007;25(1):100–6.
132. Weissman IL, Shizuru JA. The origins of the identification and isolation of hematopoietic stem cells, and their capability to induce donor-specific transplantation tolerance and treat autoimmune diseases. Blood. 2008;112(9):3543–53.
133. Linch DC, et al. Studies of circulating hemopoietic progenitor cells in human fetal blood. Blood. 1982;59(5):976–9.
134. Zauli G, et al. In vitro growth of human fetal CD34+ cells in the presence of various combinations of recombinant cytokines under serum-free culture conditions. Br J Haematol. 1994;86(3):461–7.
135. Tocci A, et al. CD34+ cells from first-trimester fetal blood are enriched in primitive hemopoietic progenitors. Am J Obstet Gynecol. 2003;188(4):1002–10.
136. Shields LE, Andrews RG. Gestational age changes in circulating CD34+ hematopoietic stem/progenitor cells in fetal cord blood. Am J Obstet Gynecol. 1998;178(5):931–7.
137. Campagnoli C, et al. Circulating hematopoietic progenitor cells in first trimester fetal blood. Blood. 2000;95(6):1967–72.
138. Clapp DW, et al. Molecular evidence that in situ-transduced fetal liver hematopoietic stem/progenitor cells give rise to medullary hematopoiesis in adult rats. Blood. 1995;86(6):2113–22.
139. Lim FT, Kanhai HH, Falkenburg JH. Characterization of the human CD34+ hematopoietic progenitor cell compartment during the second trimester of pregnancy. Haematologica. 2005;90(2):173–9.
140. Oguro H, Ding L, Morrison SJ. SLAM family markers resolve functionally distinct subpopulations of hematopoietic stem cells and multipotent progenitors. Cell Stem Cell. 2013;13(1):102–16.
141. Wagner-Souza K, et al. Rhodamine 123 efflux in human subpopulations of hematopoietic stem cells: comparison between bone marrow, umbilical cord blood and mobilized peripheral blood CD34+ cells. Int J Mol Med. 2008;22(2):237–42.
142. Chou BK, et al. Efficient human iPS cell derivation by a non-integrating plasmid from blood cells with unique epigenetic and gene expression signatures. Cell Res. 2011;21(3):518–29.
143. Broxmeyer HE, et al. Hematopoietic stem/progenitor cells, generation of induced pluripotent stem cells, and isolation of endothelial progenitors from 21- to 23.5-year cryopreserved cord blood. Blood. 2011;117(18):4773–7.
144. Rollini P, et al. Long-term expansion of transplantable human fetal liver hematopoietic stem cells. Blood. 2004;103(3):1166–70.
145. Gallacher L, et al. Identification of novel circulating human embryonic blood stem cells. Blood. 2000;96(5):1740–7.
146. Holyoake TL, Nicolini FE, Eaves CJ. Functional differences between transplantable human hematopoietic stem cells from fetal liver, cord blood, and adult marrow. Exp Hematol. 1999;27(9):1418–27.
147. Nava S, et al. Characterization of cells in the developing human liver. Differentiation. 2005;73(5):249–60.
148. Delaney C, Ratajczak MZ, Laughlin MJ. Strategies to enhance umbilical cord blood stem cell engraftment in adult patients. Expert Rev Hematol. 2010;3(3):273–83.
149. McGuckin CP, et al. Umbilical cord blood stem cells can expand hematopoietic and neuroglial progenitors in vitro. Exp Cell Res. 2004;295(2):350–9.
150. Campagnoli C, et al. Identification of mesenchymal stem/progenitor cells in human first-trimester fetal blood, liver, and bone marrow. Blood. 2001;98(8):2396–402.
151. Yu M, et al. Mid-trimester fetal blood-derived adherent cells share characteristics similar to mesenchymal stem cells but full-term umbilical cord blood does not. Br J Haematol. 2004;124(5):666–75.
152. Guillot PV, et al. Human first-trimester fetal MSC express pluripotency markers and grow faster and have longer telomeres than adult MSC. Stem Cells. 2007;25(3):646–54.

153. Chan J, et al. Galectin-1 induces skeletal muscle differentiation in human fetal mesenchymal stem cells and increases muscle regeneration. Stem Cells. 2006;24(8):1879–91.
154. Kennea NL, et al. Differentiation of human fetal mesenchymal stem cells into cells with an oligodendrocyte phenotype. Cell Cycle. 2009;8(7):1069–79.
155. Zhang ZY, et al. Superior osteogenic capacity for bone tissue engineering of fetal compared with perinatal and adult mesenchymal stem cells. Stem Cells. 2009;27(1):126–37.
156. in't Anker PS, et al. Mesenchymal stem cells in human second-trimester bone marrow, liver, lung, and spleen exhibit a similar immunophenotype but a heterogeneous multilineage differentiation potential. Haematologica. 2003;88(8):845–52.
157. Almeida-Porada G, et al. Differentiative potential of human metanephric mesenchymal cells. Exp Hematol. 2002;30(12):1454–62.
158. Hu Y, et al. Isolation and identification of mesenchymal stem cells from human fetal pancreas. J Lab Clin Med. 2003;141(5):342–9.
159. Ersek A, et al. Persistent circulating human insulin in sheep transplanted in utero with human mesenchymal stem cells. Exp Hematol. 2010;38(4):311–20.
160. Yen BL, et al. Placenta-derived multipotent cells differentiate into neuronal and glial cells in vitro. Tissue Eng Part A. 2008;14(1):9–17.
161. Anchan RM, et al. Amniocytes can serve a dual function as a source of iPS cells and feeder layers. Hum Mol Genet. 2011;20(5):962–74.
162. Kennea NL, Mehmet H. Neural stem cells. J Pathol. 2002;197(4):536–50.
163. Uchida N, et al. Direct isolation of human central nervous system stem cells. Proc Natl Acad Sci U S A. 2000;97(26):14720–5.
164. Kim JB, et al. Direct reprogramming of human neural stem cells by OCT4. Nature. 2009;461(7264):649–53.
165. Kim JB, et al. Oct4-induced pluripotency in adult neural stem cells. Cell. 2009;136(3): 411–9.
166. Gong SP, et al. Embryonic stem cell-like cells established by culture of adult ovarian cells in mice. Fertil Steril. 2010;93(8):2594–601. 2601 e1–9.
167. Pappa KI, Anagnou NP. Novel sources of fetal stem cells: where do they fit on the developmental continuum? Regen Med. 2009;4(3):423–33.
168. Johansson M, Ellegren H, Andersson L. Comparative mapping reveals extensive linkage conservation—but with gene order rearrangements—between the pig and the human genomes. Genomics. 1995;25(3):682–90.
169. Goureau A, et al. Human and porcine correspondence of chromosome segments using bidirectional chromosome painting. Genomics. 1996;36(2):252–62.
170. Rettenberger G, et al. Visualization of the conservation of synteny between humans and pigs by heterologous chromosomal painting. Genomics. 1995;26(2):372–8.

Part II
Amniotic Fluid and Membrane

Chapter 7
Historical Perspectives

Joseph A. Brazzo III

Introduction

"What's past is prologue", were the infamous words of Antonio in William Shakespeare's *The Tempest*, Act 2, Scene 1. In context, the present state is not dictated by fate but is the product of events that occurred in the past; hence such past events set the stage for the present and thus parallel that of prologue [1]. It is historical perspective that serves as a template of prologue in hindsight. More specifically, historical perspective is more than a mere collection of historical facts but details the process of change and evolution of knowledge and inquiry as it relates to current knowledge and research of a specific field or topic [2]. Also, current states of knowledge and research, themselves, serve as prologue; and, thus, heavily dictate future perspectives [1, 2].

Stem cells from the amniotic membrane (AM) and amniotic fluid (AF) are among the most promising in biomedical research, standing at the forefront of numerous developments currently in an intense state of flux [3]. Such cells represent populations that are unique and versatile in properties and potentials, rendering them distinctively valuable to various aspects of regenerative medicine, tissue engineering, cellular and gene therapies, as discussed in detail throughout this textbook [3, 4]. There is debate as to when and where the very first identifications of AM and AF cells took place. It is widely accepted, however, that in-depth characterizations of such cells began a little over 70 years ago [5, 6]. While diagnostic applications of AF cells are several decades old, potential or documented therapeutic uses of these cells are relatively recent. The historical perspective offered in this chapter is not meant to be an exhaustive and all-inclusive account of all developments involving

J.A. Brazzo III, M.S. (✉)
Department of Surgery, Boston Children's Hospital,
300 Longwood Ave, Enders 720.2, Boston, MA 02115, USA
e-mail: joseph.brazzo@childrens.harvard.edu

D.O. Fauza, M. Bani (eds.), *Fetal Stem Cells in Regenerative Medicine*,
Stem Cell Biology and Regenerative Medicine,
DOI 10.1007/978-1-4939-3483-6_7

these cells. Rather, it is focused on an overview of select events that mark transition points in the research and acquired knowledge of such cells, which may not be already addressed in other chapters.

Ancient Perspectives

The fluid that bathes the developing fetus within the womb of the mother intrigued and mystified many in antiquity. An immersion of fascination with the AF during this period engendered mysticism, folklore, superstitions, and even religious doctrine, which would expand both time and cultures. Even today these beliefs remain ever so strongly with the cultures and religions from which they arose. Also, scientific inquiry of AF during this period, though scarce, would occur. Below is a very brief description of some of these beliefs and inquiries on the AF in ancient and pre-modern times—by no means is it a comprehensive review.

According to religious doctrine of most denominations, water is a universally nurturing and motherly entity that gives form to all of earthly existence. Similarly, tribal cultures of indigenous populations that continue to exist today place water at their spiritual epicenter. Likewise, within the doctrines of formal religion and tribal cultures, AF is viewed as a sacred and holy entity, and its role vital for the creation of life. Interestingly, such views were not too far off given today's current understanding of AF in both developmental biology and regenerative medicine.

Amniotic fluid is referenced most notably in the religious scriptures the Upanishads, ancient Buddhist text, and the Christian Bible. Of these texts, it's the Upanishads that are the oldest, dating back to 1700 BCE. Referencing the AF is the Chāndoya Upanishad, translated, "you are that", which seeks to explain the origin of the universe. In Chāndoya Upanishad we are told that the universe develops from an egg that split in two entities upon hatching, each of which representing individual membranes. Within these two membranes is a fluid, that of a nourishing Mother Ocean and the fluid that bathes the fetus, the amniotic fluid. Similarly, in Buddhism, Guan Yin, a Bodhisattva of compassion, is typically depicted holding a willow branch in one hand and a flask that contains and pours the "Dew of Immortality" in the other. This 'Dew' or fluid within the flask of which Guanyin holds is a nutritive broth or milk of elixir properties thought to be the same fluid within the mother's womb, the AF. In Christianity many believe that birth is actually a re-birth of a past sinful soul and that the AF cleanses the soul of all sin, a requirement to enter the Kingdom or gates of Heaven. Other sectors of Christianity believe that a simple baptism with holy water is enough to save the soul without re-birth after death.

Tribal cultures that reference AF include the Yorùbá and Akan cultures of the West African coast whose origins date back to seventh century BCE and twelfth century AD, respectively. Both cultures include an emphasis on a vast array of rituals, prayers, stories, and spiritual beliefs. In Yorùbá culture, the Goddess Yemoja, the mother of waters, gives rise to everything water, including the AF. And, it's her influence on forces through the AF that drive the creation and sustainability of a new life. Across the West African lands to the east in Ghana and Cote d'Ivoire are the people of the Akan culture. The Akan people believe that all celestial bodies and universal entities are governed by divine spirit-forces, *Abosom*. According to Akan people, the energy of a nurturing mother is equivalent to the energy of the moon, *Bosom*. Equivalently, AF is that of earth's oceans, which is continuously under the influence of the invisible forces of the moon. Thus, it's the moon's huge effect and influence on the rising and falling of the ocean's tide; as the AF is a conductor of the mother's mood, affecting the tides of the AF, and thus having a huge influence not only the physiology of the fetus but also the spirits of the child. Also, the salt of the oceans is that of God's tears, meant to cleanse and purify the souls of the people. Similarly, it is the salt in AF that holds a new life in a state of purity.

Little was known scientifically about AF before the twentieth century. This can be seen in the scant reference to AF in scientific literature during that time. The first reference to AF from a purely scientific standpoint can be contributed to Aristotle of Ancient Greece, who initially described AF stained with meconium as associated with fetal death. The term meconium meaning "opium-like" is believed to derive from the color and appearance associated with meconium stained AF. Aristotle described such an appearance of being black with a thick consistency very similar to processed opium. This is one of the only mentions of AF in known ancient Greek scientific texts, with no other significant reference until the late Middle Ages. In the sixteenth century, Realdo Columbus, an Italian anatomist and surgeon, correctly concluded that the fetus was protected by AF, but incorrectly postulated that the AF was solely the product of fetal sweat. And in the seventeenth century, the English physician William Harvey believed that the developing fetus received its nourishment from the AF by swallowing it, and by diffusion through pores in the skin of the fetus. Today, we know that the fetus does indeed swallow AF, but not as a source of nourishment. Lastly, Nicolas van Hoboken, a Dutch anatomist and physician, through the study of cow fetuses, also postulated that the nourishment of the fetus was through the AF. He also incorrectly postulated that small pores in the chorionic membrane, which allowed passage of molecules from the uterus or maternal blood circulation, were the derivation of such nourishing property of the AF. Other curious insights from antiquity and/or various cultures have been reported, though not in any organized or consistent fashion. Much like most aspects of fetal development, it was not until less than 100 years ago that the unique properties of the AF and AM started to be unveiled.

Amniotic Membrane

A Word on Applications Prior to Cell Characterization

Scientific investigation on the AM became prominent at the turn of the twentieth century [7]. This increase in scientific inquiry pushed aside the mysticism, folklore, and superstitions that entrenched the state of knowledge on the AM prior [7, 8]. The use of the AM as a skin graft in the 1910s marked the first documented scientific investigation of the membrane [7, 9]. Thereafter, research on the AM occurred intermittently in the first half of the twentieth century, and was confined specifically to clinical applications [7, 9–19]. Reference to AM cells in these early studies is loose or lacking. Indeed, it is unclear when AM cells were first recognized. Traditionally, identification of AM cells is attributed to studies which occurred in the 1950s with the advent of electron microscopy [5].

In 1910 Dr. John Davis of the Johns Hopkins Hospital transplanted and grafted "pieces of lining of the amniotic sac" [7, 9]. The results were unpromising; however, Davis defined the technical limitations of his day and suggested the use of AM in skin transplantation contingent on advancements in surgical and procurement techniques [9]. Following, in 1913, Stern and Sabella applied AM tissue to both skin burns and ulcers in independent studies, although they often collaborated, sharing data and strategies [7, 10, 11]. Unfortunately, following these initial studies on the AM, scientific investigation on the AM would fizzle and not reemerge for some twenty years until the utilization of AM in the creation of an artificial vagina [12].

In 1934, Brindeau, an Italian physician, used the AM as an epidermal lining for the creation of an artificial vagina in a patient with müllerian agenesis [12]. Five years later, Bruger et al. would use the AM in a more extensive construction of an artificial vagina [13]. During the transition from the 1930s to the 1940s, AM-derived tissue was used to prevent meningo-cerebral adhesions after lacerations to the head, including gunshot wounds, depressed fracture of the skull, and craniotomy procedures [14]. These studies suggested a reduction in meningo-cerebral adhesions and recommended the use of AM for adhesion prevention in other surgical procedures [14].

At approximately the same time, the advent of AM utilization in the field of ophthalmology can be attributed to Rotth [15]. He described the replacement of necrotic conjunctiva of a single eye with fetal membranes including AM and chorionic membrane (CM) obtained from caesarian sections. He believed the AM would be easily converted and well tolerated once implanted into the eye because it was "most similar" to the conjunctiva membrane. Although his initial results were not exactly promising, he pressed on with further investigation, claiming that the implanted embryonic tissue, specifically the AM, were histologically similar to the epithelium of the conjunctiva, even if histological analyses were not described in that report [15]. A little later, in 1946, Sorsby and Symons replaced necrotic conjunctiva induced by caustic agents with human AM [16]. That clinical experience was putatively encouraging; however, further experiments using rabbits and lime as a caustic agent treated with AM were unsuccessful [16].

In 1950, Troensegaard-Hansen reported his experience with grafted amnion onto chronic leg ulcers of individuals suffering from peripheral vascular disease, with healing of the leg ulcer and a reduction in intermittent claudication [17]. Additional reports from the same author or group later in the 1950s continued to depict favorable results in peripheral vascular disease [18, 19].

Soon thereafter, interest on AM in clinical applications would once again fade, only to reemerge in the early 1970s with a large focus on wound healing [20].

At this point in time, however, research on AM cell characterization and properties would be more refined. It is unclear why it would take scientists several decades to initiate microscopic and biochemical characterization of AM cells, despite the availability of clcctron microscopy in the early 1950s [5].

Cell Characterization

Max Knoll and Ernst Ruska invented the transmission electron microscope (TEM) in 1931 [21]. That invention made it possible for the first time the visualization and localization of cell and organelle structures [22]. The first TEM micrograph of a biological sample was published in 1934, yet it wasn't until 1951 that Albert Claude published the first TEM of an intact cell [22]. The initial description of the morphology of cells of the AM by TEM can be attributed to Bautzmann and Schröder in 1955. This initial study characterized the structure of amniotic membrane epithelial (AME) cells and Houfbauer cells found in the AM mesenchymal stroma [5]. As this study was reported in German, the extent of characterization of such cells is not easily available. It is unclear whether this report referenced previous experiments characterizing AM cells not present in the literature.

In 1960 Bourne first characterized the AM into distinct layers using electron microscopy [23]. Such layers included the AME, basement membrane, fibroblast layer, compact layer, and the spongy layer [23]. In Bourne's study, cells of the AM exist in two of these five layers, namely the AME and fibroblast layers. As mentioned above, Bautzmann and Schröder using TEM in 1955 first characterized AME cells; however, AME cells were characterized first in reference to their respective layer in the AM only 5 years later, by Bourne in 1960. The ontogeny and degree of homogeneity of AME cells remained unknown at that time [24].

Interestingly, prior to the specific cell characterization of AM cells in the 1950s, and following the many early investigations of AM in clinical applications, virologists heavily investigated the role of the AM in response to infectious and virulent pathogens [25, 26]. In the late 1940s and early 1950s this breed of scientists was eager to find suitable media to cultivate various virulent virus and bacterial strains. In 1955, Zitcer et al. identified and introduced trypsinized human AM as a superior cell source for cultivation of infectious virus and bacterial strains. For the next decade the human AM would be used to research the cytopathological effects of various infectious microbial strains. These studies microscopically examined AM cells to observe and record cytological changes. However, they loosely reference AM cell characteristics. Instead, they sought to observe specifically cellular pathological states upon infection. Ultimately, these early virology studies did not characterize AM cells or their properties and thus are not considered under that historical context [25, 26].

Phenotype Pattern(s)

Although there is a myriad of studies that have brought to light the structural, biochemical, and functional characteristics of AME cells, the debate as to the cellular uniformity of the AM persists to this day [24]. Do AME cells constitute a homogeneous or heterogeneous population, after all?

In 1965, Thomas reported two morphologically distinct AME cell types based on ultrastructural differences in organelle presence and cellular cytoplasm darkness [27]. He concluded the presence of two distinct A cell types: light and dark cells. Thus, it was therefore suggested that light and dark cell types had different roles as amniotic epithelial cells. This study was the first to support a heterogeneous AME cell population [24, 27]. Lister et al. would refute such a claim in 1968, and would be the study to initiate the perpetual debate as to AME cell uniformity [24, 28]. More specifically, Lister et al. described the ultrastructure of a single, homogeneous amniotic epithelial layer scarce of apical microvilli, complex intercellular microvilli with desmosomes, large nuclei, abundant glycogen, and infrequent mitochondria, endoplasmic reticulum, and Golgi lamellae [28]. Further studies, including from Armstrong et al., Wynn and French in 1968 and McCoshen et al. in 1981would support a heterogeneous cell population of amniotic epithelia [24]; however, studies refuting such a heterogeneous population included those from Sinha et al. 1971, Hempel et al. 1972, King et al. 1980 and Sonek et al. in 1991 [24]. Interestingly, in 1972 it was suggested by Hoyes et al. that studies with results supporting a heterogeneous population of AME cells based on cell morphology alone were the product of inappropriate fixations, along with other methodological and technical differences [29]. In 2003, Iwasaki et al. wished to put to rest this debate by using enzyme-histochemistry, tracer permeability analysis, and freeze-substitution fixation. He concluded that AME cells were homogeneous [24]. This study, however, did not end the controversy. Indeed, despite the advent of much more sophisticated methods of (stem or not) cell characterization, a final answer on this point still remains elusive [30–32].

Stem Cells

Prior to 2003, very little was known about stem cells and/or progenitor cells of the AM [30, 31, 34]. A study in 1996 showing human AME antigen positivity to neuro- and glial-specific antibodies was the first to hint at AME multilineage potential, specifically for neuro and glial lineages [33]. There is some confusion in the historical precedents of AM stem cell isolation, due to the fact that studies supposedly focused on placental cells often did not properly distinguish the derivation of the mesenchymal cell population of the placenta, and thus could have inadvertently included stem cell populations from the AM [30, 34]. Even when the amnion is removed prior to placental cell culture, remnants of AM mesenchyme can remain [34, 35]. In 2004, In'T Anker et al. isolated and better characterized mesenchymal stem cells from term AM [30]. A plethora of studies on mesenchymal and epithelial AM stem cells followed [31, 32].

Translational Developments

The therapeutic potential of AM cells, particularly of AM stem cells, is currently thought to be vast. It is covered in detail in other sessions of this book, yet the following is a list of some of these developments, exclusive of the use of the AM as a whole or graft:

- Bone constructs *in vivo* [36]
- Tendon-like structures [36]
- Pancreatic beta-cells [36]
- Hepatocyte-related functions *in vivo* [36]
- Cardiomyocyte-related functions *in vivo* [36, 37, 40]
- Chondrocyte differentiation *in vivo* and *in vitro* [36]
- Pancreatic beta-islet cells differentiation *in vivo* [36, 37]
- Neuro progenitor cells, neurons, and glial cells [36, 37]
- Auditory system cells [36]
- Type II pneumocytes lung epithelium differentiation [31]
- Cholangiocytes [31] (Biliary tract epithelia)
- *In vitro* differentiation: adipogenesis, osteogenesis, chondrogenesis [38]
- Inhibit proliferation of T cell and monocyte differentiation [39]
- Antitumor therapeutic [40]

Amniotic Fluid

Advent of Amniocentesis

The study of the AF at the turn of the twentieth century can be mostly attributed to the advent of amniocentesis [6, 41]. German physicians Prochownick, Von Schatz and Lambi performed the first documented amniocentesis in the 1877 to relieve pressure on a fetus suffering from diagnosed hydramnios [6, 41]. Following, Prochownick analyzed AF obtained from amniocentesis for sodium chloride, solid substances, and urea content [41]. In 1891, Schroeder studied the content of solid substances, ash, and albumin in AF samples obtained from amniocentesis [41]. There is no mention of any cell content or analysis in these studies [41].

Prenatal Screening: The Identification of AF Cells

In 1956, Fritz Fuchs and Polv Riis showed that fetal sex could be determined from AF cells through Barr body identification via amniocentesis [6]. Details of their methods for isolation and characterization of AF cells to identify fetal sex remain unknown. That study is considered the first to document the presence of cells in the AF, given the lack of comparable reports in the literature prior. Soon after that study, cells of the AF were further analyzed and initially classified by Sachs et al. [42] via smeared histology of sampled AF, with no cell culture [42]. Three types of epithelial cells were suggested: basal cells with vesicular nuclei and green-staining cytoplasm; precornified and cornified cells with vesicular to pyknotic; and keratinized cells with advanced nuclear degeneration; along with a large proportion of anucleated cells [42].

Widespread prenatal screening through amniocentesis begun in the late 1960s, mostly geared towards research in reproductive health and "early confirmation of pregnancy" [43]. At that time, culturing of AF cells was performed almost exclusively for genetic karyotyping [44]. In 1966, Steele and Breg described more systematic genetic screening/karyotyping of cultured AF cells [45]. Their cultures revealed two distinct morphological classes of cells: epithelial-like and fibroblast-like cells [45]. Not long thereafter, in 1969, Nelson and Emery in Edinburgh, Scotland, proposed modifications which optimized AF cell growth [44]. That was when adherence of AF cells to glass surfaces was first documented. They also proposed that AF cell growth was most viable in samples obtained at less than 20 weeks of gestation [44].

Over the course of the following two decades, culture techniques would drastically improve, increasing both the viability and growth of cultured AF cells [46–49]. This set the stage for further identification and understanding of different populations of AF cells [45–47, 49].

Cell Culture Modifications and Cell Classifications

Following the initial classification of attaching, colony-forming cells by Steele and Breg in 1966, numerous other forms of classification were proposed, with most broadly classifying cultured AF cells into three groups: attaching, colony-forming amniotic fluid cells; attaching non-colony forming, non-proliferative amniotic fluid cells; and non-attaching amniotic fluid cells [50–54]. In 1974, Hoehn et al. conducted further investigation on the AF cell culture types as defined by Steele and Breg [51].They identified three main classes of AF cells based on cellular morphology: fibroblast-like cells (F-type cells), epitheloid cells (E cells), and amniotic fluid specific cells (AF-specific cells) [51]. F-type cells were spindle-shaped cells, exhibiting high growth potential. E cells exhibited intimate cell-to cell contact, were resistant to trypsin detachment, and showed poor growth potential. AF-specific cells were pleomorphic, and exhibited intermediate growth potentials compared to F-type cells and E cells. Simultaneously, in 1974 Sutherland et al. cultured AF of second trimester pregnancies [52]. An inverted microscope was used to characterize cells based on morphology. Five cell types were readily identifiable and characterized. The first cell type identified was a macrophage. This cell type could be readily identified in primary culture but could not be sub-cultured. And, overtime, macrophage cells became degenerate and overgrown by other cell types. Also, three epithelial-like cells were identified: Epitheloid Type I, II, and III cells [52]. Epitheloid Type I cells could be sub-cultured but did not grow as colonies after passage. Epitheloid Type II cells were morphologically large in size and multinucleated with a fibrillar appearance, and were present in all passages. Epitheloid Type III cells morphologically resembled Epitheloid Type I cells, yet could be sub-cultured. The last cell type identified was the fibroblast-like cell [52]. This cell type was indistinguishable morphologically from embryonic or lung fibroblasts, and readily sub-cultured. Interestingly, cultures with cell colonies with a more distinct fibroblastic appearance lived longer. However, fibroblast culture colonies with a large presence of epitheloid type cells saw a larger decline in life or expectancy [52].

In 1981, Virtanen et al. characterized cultured AF cells with antibodies against intermediate filaments in indirect immunofluorescence microscopy [53]. Wishing to further refine AF cell classification beyond simplistic cell morphologies as was done in the previous studies [45, 51, 52], Virtanen et al. used antibodies against vimentin, keratin, desmin, and GFA [53]. Using this technique, the authors characterized five subtypes of epithelial cell types (E cells) and a fibroblastic cell type (fibroblastoid cell). E1 cells were the most common cell type in culture, pleomorphic in shape and size, fibroblast-like in morphology with bright fibrillar keratin-positive cytoplasms, exhibiting no cell-to-cell interaction, and were vimentin positive. The E1 cell subtype has been said to be the same cell as the AF specific cells previously classified by Sutherland et al. [51, 53]. E2 cells exhibited fibrillar organization of keratin and vimentin, lacked cell-to-cell organization or interaction, were large and flat in morphology, exhibited low growth and proliferation, and were present in all cell cultures, however, low in number. E3 cells were indistinguishable from E1 cells in size and morphology, but exhibited keratin organization and cell-to-cell interaction, and diffuse vimentin-positive fibrils in the cytoplasm. E4 cells were classified by their rapid growth in culture, exhibited keratin organization and cell-to-cell interaction, and were present in very small amounts in culture. E5 cells were morphologically large, multinucleated cells dispersed among the other cell types exhibiting bright keratin fibers, and stained for vimentin antibodies. Fibroblastoid cells were identified as vimentin-positive, keratin-negative cells. Cultures that showed Fibroblastoid cells as the dominant cell type in culture were very limited in number. Morphologically, E1, E3, and F-type cells were indistinguishable in culture, and were differentiated by keratin and vimentin expression and cell-to-cell interaction [53].

In 1982 a study described and compared naturally found, uncultured second trimester AF cells [54]. This study would be one of only a few that evaluated uncultured AF cells since the initial evaluation two decades prior [42, 54]. It revealed the presence of goblet cells, urothelial cells, histiocytes, macrophages, placental amniotic cells, and umbilical cord cells. Unfortunately, the vast majority of cells in natural AF are washed off during the cell culture process, thus explaining why information on all AF cells, i.e. including those uncultured, is lacking even today [58]. It is believed that a large part of the washed off, non-adhering cells in culture are exfoliated squamous cells, including identified goblet cells, histiocytes, and macrophages [54].

Cells in Abnormal Pregnancies

With the increasing use of AF cells in prenatal diagnostic screening, the urge to identify congenital and genetic pathologies unidentifiable through karyotyping and cellular morphology brought about the identification of properties of AF cells which had been previously unknown [55, 56]. The initial investigation of AF cell characterization in abnormal pregnancies can be attributed to Sutherland, Brock, and Scrimgeour in 1973, when the authors investigated the cellular content in AF of diagnosed anencephaly and found large numbers of macrophages therein [55]. In 1975, the same authors also identified a large amount of macrophages in AF samples in the setting of spina bifida [57]. In 1980, rapidly adhering cells from AF of a pregnancy with diagnosed ancephaly were positive for glial-specific antibodies, and thus heavily suggestive of a neural origin of such cells [58]. Numerous future studies would support and validate the presence of neural and glial cells in AF in the presence of Neural Tube Defects (NTDs), with this type of scrutiny eventually becoming a common option for the prenatal diagnosis of NTDs [59]. The realization that the cellular profile of the AF could be of diagnostic value n NTDs led to the search for other peculiar AF cell types in the setting of different congenital anomalies. Just as one additional example, congenital abdominal wall defects have been shown to contain peritoneal cells which are not present in normal pregnancies [56].

Stem Cells

The presence of progenitor cells in AF has long been supported by a plethora of studies that show subpopulations of cells that exhibit multilineage differentiation, self-renewal, and cytological markers expressed by embryonic and/or adult stem cells [3, 60–63, 70]. Interestingly, several studies prior would incrementally buttress the existence of progenitor stem cells in AF [64–67]. However, at the time of these prior studies (1990s), progenitor and stem cell research was a fledgling science, and thus the identifying characteristics of such cells were very loose [68]. The presence of progenitor cells in first trimester amniotic fluid was first proposed in 1993 [64]. Torricelli et al. cultured amniotic fluid samples from the 7th to 13th weeks of gestation obtained from voluntary abortions. Cells were identified in culture of amniotic fluid samples of gestational weeks equal to or greater than 11 weeks. The morphology of identified cells in the 11th week of gestation included small nucleated, round cells, most consistent with hematopoietic progenitor cells. The identification of progenitor cells in amniotic fluid by this study was the first ever to signify a possible presence of stem cells therein [64]. Over the course of the following decade or so, such inklings would accumulate, providing increasingly more evidence for the existence of AF progenitor and stem cells [65–67]. In 1996, Streubel et al. exposed AF culture cells to a supernatant line of rhamdomysarcoma cells [65]. Over the course of 6 weeks, hybrid AF cultured cells expressed muscle proteins, including dystrophin. This was the first documentation of multilineage differentiation potential of AF cells, in that case for myogenesis [65]. In 1999, a study conducted by Mosquera et al. investigated cellular telomere and telomerase activity of human AF cells [66]. Telomere length was shown to decrease; however, there was an increase in telomerase activity, providing further evidence for a presence of progenitor cells and stem cells in AF. Various studies since the early to mid-2000s better described different populations of progenitor and stem cells, an endeavor that continues to expand to this day.

The first study to propose a therapeutic use for AF progenitor cells was published in 2001 by Kaviani et al., in which a mesenchymal population of ovine AF cells was shown to be amenable to the fabrication of engineered constructs [67]. That study was soon followed by a similar one, only involving human cells [69]. Soon thereafter, the first application *in vivo* of a construct engineered from AF mesenchymal stem cells was reported by the same group [70], and then followed by numerous others, from that and other groups.

The provenance of AF stem cell identification is rarely defined by a single initial study. Rather, it typically denotes a blurry past with multiple contributors. The identification of AF stem cells is no exception, as it can be related to a collection of studies that have identified different characteristics of AF stem cells over the course of several years, including the initial studies discussed earlier [65–67]. To many, the defining moment of AF stem cell identification has been the discovery of the transcription factor (TF) Oct-4 positivity in a subpopulation of AF cells in 2003 by Prusa et al. [60, 62, 71, 72]. Recently, however, the validity of Oct-4 as a marker of pluripotency in AF stem cells and thus that as a true, and total marker of "stemness" has been brought to light, challenging previous observations [73]. Also recently, certain unique stem cells not normally present in the AF have been identified in the setting of disease, raising additional interesting prospects for translational developments [74].

Despite all the almost feverish activity surrounding AF and AM stem cells, it was not until only a few years ago that a biological role for at least some of these cells was uncovered, when ubiquitous AF mesenchymal stem cells were shown to play a germane role in the fetal wound healing process [75]. It is surprising that, until that study, not much seemed to have been attempted in the way of understanding why stem cells would be present in the AF or in the AM, in that one would be hard pressed to assume that their presence there was merely an epiphenomenon.

Perhaps expectedly for such a young field, the nomenclature surrounding AF stem cells is currently confusing. The terms amniotic stem cells (AFSC), amniotic fluid mesenchymal stem cells (AFMSC), amniotic mesenchymal stem cells (AMSC), and mesenchymal amniocytes are often used interchangeably. Some suggest that AF cells expressing the stem cell factor receptor CD117 or c-kit should be considered different from AFMSCs and thus be called AFSCs [71]. On the other hand, as discussed in more detail in other chapters, the amniotic fluid harbors more than one population of stem cells, rendering the term AFSC imprecise. In addition, AFSCs have a mesenchymal profile on flow cytometry and so-called AFMSCs can also express CD117. Regardless of such understandable nomenclature immaturity, the perspectives involving AF and AM stem cells are unquestionably promising and only beginning to be explored, as discussed in various other chapters of this textbook. The following years should add stimulating developments to this already engaging history.

References

1. Shakespeare W, David HH. The tempest. New Haven, CT: Yale UP; 1955.
2. Lawrence BS. Historical perspective: using the past to study the present. Acad Manag Rev. 1984;9(2):307–12.
3. Roubelakis MG, Trohatou O, Anagnou NP. Amniotic fluid and amniotic membrane stem cells: marker discovery. Stem Cells Int. 2012;2012:1–9.
4. Fauza DO. Amniotic fluid and placental stem cells. Best Pract Res. 2004;18(6):877–91.
5. Bautzmann H, Schroder R. Comparative studies on structure and function of the amnion; new findings on the human amnion including its fee connective tissue or so-called Hofbauer cells. Zeitsch Anat Entwicklung. 1955;119(1):7–22.
6. Woo J. A short history of amniocentesis, fetoscopy and chorionic villus sampling. www.ob-ultrasound.net/amniocentesis.html
7. Trelford JD, Trelford-Sauder M. The amnion in surgery, past and present. Am J Obstet Gyenecol. 1979;134(7):833–45.
8. Longo LD, Reynolds LP. Some historical aspects of understanding placental development, structure, and function. Int J Dev Biol. 2010;54(2-3):237–55.
9. Davis JW. Skin transplantation with a review of 550 cases at the Johns Hopkins Hospital. Johns Hopkins Med J. 1910;15:307–96.
10. Sabella W. Use of fetal membranes in skin grafting. Med Rec N Y. 1913;83:478–80.
11. Stern W. The grafting of preserved amniotic membrane to burned and ulcerated skin surfaces substituting skin grafts. J Am Med Assoc. 1913;13:973–4.
12. Brindeau A. Création d'un vagin artificiel à l'aide des membranes ovulaires d'un oeuf à terme. Gynecol Obstet. 1934;29:385.
13. Burger K. Weitere Erfahrungen über die kunstliche Scheidenbildung mit Eihäuten. Zentralbl Gynaekol. 1947;69:1153.
14. Chao Y, Humphreys S, Penfield W. A new method of preventing adhesions: the use of amnioplastin after craniotomy. Br Med J. 1940;1:517–9.
15. De Roth A. Plastic repair of conjunctival defects with fetal membranes. Arch Ophthalmol. 1940;23:522–5.
16. Sorsby A, Symons HM. Amniotic membrane grafts in caustic burns of the eye. Br J Ophthalmol. 1946;30:337–45.
17. Troensegaard-Hansen E. Amniotic grafts in chronic skin ulceration. Lancet. 1950;1:859–60.

18. Shaw S, Troensegaard-Hansen E. Chronic ulceration of the leg treated with amnion and chloramphenicol in propylene glycol. Lancet. 1952;2:225–6.
19. Troensegaard-Hansen E. Amnion implantation in peripheral vascular disease. Br Med J. 1956;2:262–8.
20. Trelford JD, Hanson FW, Anderson DG, Mendel V. Wound healing and the amniotic membrane. J Med. 1975;6(5-6):383–8.
21. Palucka, T. Overview of electron microscopy. History of recent science and technology. The Dibner Institute for the History of Science and Technology, 10 Dec 2002. Web. http://authors.library.caltech.edu/5456/1/hrst.mit.edu/hrs/materials/public/ElectronMicroscope/EM_HistOverview.htm. Accessed 12 Mar 2014.
22. Masters B (2009) History of the electron microscope in cell biology. Wiley Online Library. 10.1002/9780470015902.a0021539/abstract.
23. Bourne GL. The microscopic anatomy of the human amnion and chorion. Am J Obstet Gynecol. 1960;79:1070–3.
24. Iwasaki R, Matsubara S, Takizawa T, Takayama T, Yashiro T, Suzuki M. Human amniotic epithelial cells are morphologically homogeneous: enzymehistochemical, tracer, and freeze-substitution fixation study. Eur J Histochem. 2003;47(3):223–32.
25. Bernkopf H, Rosin A. Cytopathologic changes in tissue cultures of human amnionic cells infected with poliomyelitis, Coxsakie, and ECHO viruses. Am J Pathol. 1957;33(6):1215–27.
26. Duncan IBR, Bell EJ. Human-amnion-tissue culture in the routine virus laboratory. Br Med J. 1961;2:863–6.
27. Thomas CE. The ultrastructure of the human amnion epithelium. J Ultrastruct Res. 1965;13(1):65–83.
28. Lister UM. Ultrastructure of the human amnion, chorion and fetal skin. J Obstet Gynaecol Br Commonw. 1968;75(3):327–41.
29. Hoyes AD. Fine structure of human amniotic epithelium following short-term preservation in vitro. J Anat. 1972;111(Pt. 1):43–54.
30. Anker PS, Scherjon SA, Kleijburg-van der Keur C, de Groot-Swings GM, Claas FH, Fibbe WE, Kanhai HH. Isolation of mesenchymal stem cells of fetal or maternal origin from human placenta. Stem Cells. 2004;22(7):1338–45.
31. Miki T. Amnion-derived stem cells: in quest of clinical applications. Stem Cell Res Ther. 2011;2(3):25.
32. Rennie K, Gruslin A, Hengstschläger M, Pei D, Cai J, Nikaido T, Bani-Yaghoub M. Applications of amniotic membrane and fluid in stem cell biology and regenerative medicine. Stem Cells Int. 2012;2012:721538.
33. Sakuragawa N, Thangavel R, Mizuguchi M, Hirasawa M, Kamo I. Expression of markers for both neuronal and glial cells in human amniotic epithelial cells. Neurosci Lett. 1996;209(1):9–12.
34. Zhang X, Nakaoka T, Nishishita T, Watanabe N, Igura K, Shinomiya K, Takahashi TA, Yamashita N. Efficient adeno-associated virus-mediated gene expression in human placenta-derived mesenchymal cells. Microbiol Immunol. 2003;47(1):109–16.
35. Kohnen G, Kertschanska S, Demir R, Kaufmann P. Placental villous stroma as a model system for myofibroblast differentiation. Histochem Cell Biol. 1996;105(6):415–29.
36. Toda A, Okabe M, Yoshida T, Nikaido T. The potential of amniotic membrane/amnion-derived cells for regeneration of various tissues. J Pharm Sci. 2007;105(3):215–28.
37. Parolini O, Soncini M, Evangelista M, Schmidt D. Amniotic membrane and amniotic fluid-derived cells: potential tools for regenerative medicine? Regen Med. 2009;4(2):275–91.
38. Diaz-Prado S, Muiños-López E, Hermida-Gómez T, Rendal-Vázquez ME, Fuentes-Boquete I, de Toro FJ, Bianco FJ. Multilineage differentiation potential of cells isolated from the human amniotic membrane. J Cell Biochem. 2010;111(4):846–57.
39. Diaz-Prado S, Muiños-López E, Hermida-Gómez T, Rendal-Vázquez ME, Fuentes-Boquete I, de Toro FJ, Bianco FJ. Human amniotic membrane as an alternative source of stem cells for regenerative medicine. Diff Res Biol Divers. 2011;81(3):162–71.

40. Kang NH, Hwang KA, Kim SU, Kim YB, Hyun SH, Jeung EB, Choi KC. Potential antitumor therapeutic strategies of human amniotic membrane and amniotic fluid-derived stem cells. Cancer Gene Ther. 2012;19(8):517–22.
41. Uyeno D. The physical properties and chemical composition of human amniotic fluid. J Biol Chem. 1919;37:77–103.
42. Sachs L, Serr DM, Danon M. Analysis of amniotic fluid cells for diagnosis of foetal sex. Br Med J. 1956;2:795–8.
43. Nadler HL. Prenatal detection of genetic defects. J Pediatr. 1969;74(1):132–43.
44. Nelson MM, Emery AE. Amniotic fluid cell cultures. J Med Genet. 1973;10(1):19–22.
45. Kelley K. Amniocentesis prior to 1980. 2013. From The Embryo Project Encyclopedia, http://embryo.asu.edu/pages/amniocentesis-prior-1980. Accessed 11 Oct 2013
46. Chang HC, Jones OW, Masui H. Human amniotic fluid cells grown in a hormone-supplemented medium: suitability for prenatal diagnosis. Proc Natl Acad Sci U S A. 1982;79(15):4795–9.
47. Martin AO. Characteristics of amniotic fluid cells in vitro and attempts to improve culture techniques. Clin Obstet Gynecol. 1980;7(1):143–73.
48. Porreco RP, Bradshaw C, Sarkar S, Jones OW. Enhanced growth of amniotic fluid cells in presence of fibroblast growth factor. Obstet Gynecol. 1980;55(1):55–9.
49. Rudiger HW, Wolff R, Wendel U, Passarge E. Enhancement of amniotic fluid cell growth in culture. Humangenetik. 1974;22(1):81–4.
50. Prusa AR, Hengstschläger M. Amniotic fluid cells and human stem cell research – a new connection. Med Sci Monit. 2002;8(11):RA253–7.
51. Hoehn H, Bryant EM, Karp LE, Martin GM. Cultivated cells from diagnostic amniocentesis in second trimester pregnancies. I. Clonal morphology and growth potential. Pediatr Res. 1974;8(8):746–54.
52. Sutherland GR, Bauld R, Bain AD. Observations on human amniotic fluid cell strains in serial culture. J Med Genet. 1974;11(2):190–5.
53. Virtanen I, von Koskull H, Lehto VP, Vartio T, Aula P. Cultured human amniotic fluid cells characterized with antibodies against intermediate filaments in indirect immunofluorescence microscopy. J Clin Invest. 1981;68(5):1348–55.
54. Schrage R, Bögelspacher HR, Wurster KG. Amniotic fluid cells in the second trimester of pregnancy. Acta Cytol. 1982;26(4):407–16.
55. Sutherland GR, Brock DJ, Scrimgeour JB. Amniotic-fluid macrophages and anencephaly. Lancet. 1973;10(2):1098–9.
56. Gosden C, Brock DJ. Amniotic fluid cell morphology in early antenatal prediction of abortion and low birth weight. Br Med J. 1978;2:1186–9.
57. Sutherland GR, Brock DJ, Scrimgeour JB. Amniotic fluid macrophages and the antenatal diagnosis of anencephaly and spina bifuda. J Med Genet. 1975;12(2):135–7.
58. Aula P, von Koskull H, Teramo K, Karjalainen O, Virtanen I, Lehto VP, Dahl D. Glial origin of rapidly adhering amniotic fluid cells. Br Med J. 1980;281:1456–7.
59. von Koskull H. Rapid identification of glial cells in human amniotic fluid with indirect immunofluorescence. Acta Cytol. 1984;28(4):393–400.
60. Prusa A, Marton E, Rosner M, Bernaschek G, Hengstschläger M. Oct-4 expressing cells in human amniotic fluid: a new source for stem cell research? Hum Reprod. 2003;18(7):1489–93.
61. Cananzi M, Atala A, De Coppi P. Stem cells derived from amniotic fluid: new potentials in regenerative medicine. Reprod Biomed Online. 2009;18 Suppl 1:17–27.
62. Rosner M, Schipany K, Shanmugasundaram B, Lubec G, Hengstchläger M. Amniotic fluid stem cells: future perspectives. Stem Cells Int. 2012;2012:741810.
63. Da Sacco S, De Filippo RE, Perin L. Amniotic fluid progenitor cells and their use in regenerative medicine. Adv Regen Med. 2011;Chapter 8:165–78.
64. Torricelli F, Brizzi L, Bernabei PA, Gheri G, Di Lollo S, Nutini L, Lisi E, Di Tommaso M, Cariati E. Identification of hematopoietic progenitor cells in human amniotic fluid before the 12th week of gestation. Ital J Anat Embryol. 1993;98(2):119–26.

65. Streubel B, Martucci-Ivessa G, Fleck T, Bittner RE. In vitro transformation of amniotic cells to muscle cells—background and outlook. Wiener Med Wochensch. 1996;146(9–10):2167–217 [Article in German].
66. Mosquera A, Fernandez JL, Campos A, Goyanes VJ, Ramiro-Diaz JR, Gosalvez J. Simultaneous decrease of telomere length and telomerase activity with ageing of human amniotic fluid cells. J Med Genet. 1999;36(6):494–6.
67. Kaviani A, Perry TE, Dzakovic A, Jennings RW, Ziegler MM, Fauza DO. The amniotic fluid as a source of cells for fetal tissue engineering. J Pediatr Surg. 2001;36(11):1662–5.
68. Bongso A, Richards M. History and perspective of stem cell research. Best Pract Res Clin Obstet Gynaecol. 2004;18(6):827–42.
69. Kaviani A, Guleserian K, Perry TE, Jennings RW, Ziegler MM, Fauza DO. Fetal tissue engineering from amniotic fluid. J Am Coll Surg. 2003;196(4):592–7.
70. Fuchs JR, Kaviani A, Oh JT, LaVan D, Udagawa T, Jennings RW, et al. Diaphragmatic reconstruction with autologous tendon engineered from mesenchymal amniocytes. J Pediatr Surg. 2004;39(6):834–8.
71. Cananzi M, De Coppi P. $CD117^+$ amniotic fluid stem cells: state of the art and future perspectives. Organogenesis. 2012;8(3):77–88.
72. Tsai M, Lee J, Chang Y, Hwang S. Isolation of human multipotent mesenchymal stem cells from second-trimester amniotic fluid using a novel two-stage culture protocol. Hum Reprod. 2004;19(6):1450–6.
73. Ryan JM, Pettit AR, Guillot PV, Chan JK, Fisk NM. Unraveling the pluripotency paradox in fetal and placental mesenchymal stem cells: Oct-4 expression and the case of The Emperor's New Clothes. Stem Cell Rev. 2013;9(4):408–21.
74. Turner CG, Klein JD, Wang J, Thakor D, Benedict D, Ahmed A, et al. The amniotic fluid as a source of neural stem cells in the setting of experimental neural tube defects. Stem Cells Dev. 2013;22(4):548–53.
75. Klein JD, Turner CG, Steigman SA, Ahmed A, Zurakowski D, Eriksson E, et al. Amniotic mesenchymal stem cells enhance normal fetal wound healing. Stem Cells Dev. 2011;20(6):969–76.

Chapter 8
Amniotic Membrane Stem Cell Populations

Rebecca Lim, Jean Tan, Ryan J. Hodges, and Euan M. Wallace

Introduction

Amniotic membrane, the amnion, has a long history in regenerative medicine as a wound or burns dressing and in ophthalmic surgery. However, it is really only over the past decade or so that it has been increasingly recognized as a potential source of cells that may have much broader applications in regenerative medicine and stem cell biology. In this chapter, we will describe the embryological derivation of the cells in the amnion, their characteristics and their stem-cell like properties, highlighting possible clinical applications as suggested by current experimental studies.

Derivation of Fetal Membranes

The process of forming the blastocyst is known as blastulation whereby the cells of the very early embryo, or morula, differentiate into an outer layer of cells, known as the trophoblast, and an inner cell mass or embryoblast. These two distinct cell populations are separated by a fluid filled cavity or blastocoele. While the trophoblast layer of cells will give rise to the definitive placenta, the inner cell mass differentiates into the epiblast and hypoblast, the latter of which migrates out along the trophoblast to form the primary yolk sac and give rise to primitive endodermal structures. The epiblast forms the definitive embryo by giving rise to ectoderm and invaginating mesoderm. The

R. Lim, Ph.D. • J. Tan, B.Sc. • R.J. Hodges, M.B.B.S., Ph.D.
E.M. Wallace, M.B.Ch.B., M.D. (✉)
The Ritchie Centre, Department of Obstetrics and Gynaecology, Monash University,
Level 5, Monash Medical Centre, 246 Clayton Road, Clayton, VIC, Australia
e-mail: rebecca.lim@monash.edu; jean.tan@monash.edu;
ryan.hodges@monash.edu; euan.wallace@monash.edu

D.O. Fauza, M. Bani (eds.), *Fetal Stem Cells in Regenerative Medicine*,
Stem Cell Biology and Regenerative Medicine,
DOI 10.1007/978-1-4939-3483-6_8

resulting trilaminar structure—ectoderm, mesoderm, and endoderm—is known as the gastrula. Immediately prior to formation of the gastrula, a stage called gastrulation, the epiblast gives off a layer of ectodermal cells that form the amniotic membrane or amnion. Given that these amnion cells are derived from the epiblast immediately prior to gastrulation it has long been thought that they might retain multipotent memory or plasticity and as such possess stem cell-like properties.

Amnion in Regenerative Therapy

The use of human amnion as biological dressings has been used in the fields of ophthalmology and wound care for decades owing to their bacteriostatic, antiphlogistic, pro-regenerative and scar-reducing properties (Fig. 8.1). In particular, in a comprehensive review, Kesting and colleagues describe the decades of amniotic membrane usage for burn treatment [1], highlighting that randomised clinical trials have demonstrated the benefits of amnion in the treatment of burns, promoting wound healing, improving patient comfort and reducing the need for dressing changes. Indeed, processed and dehydrated amnion/chorion allografts have been applied to other wound types, including as skin grafts where they have also been

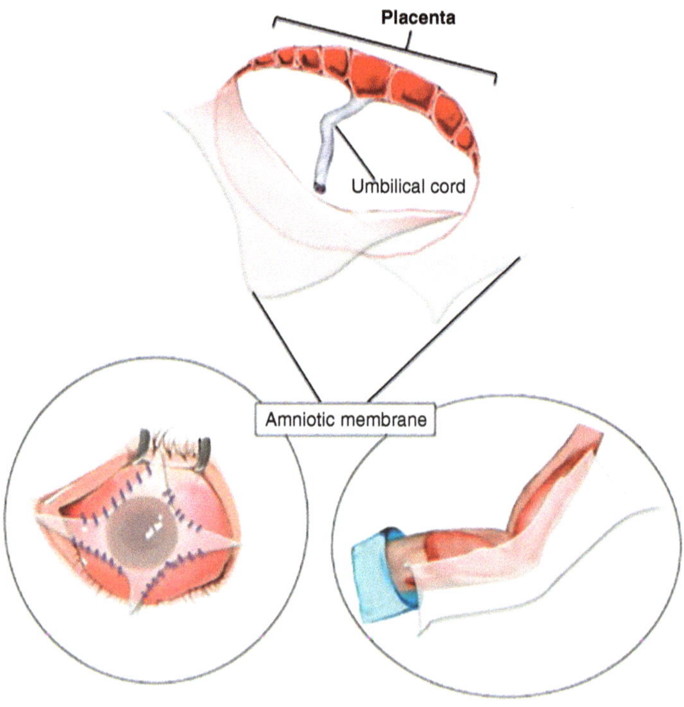

Fig. 8.1 The amniotic membrane can be physically separated from the chorionic membrane for ophthalmological applications and as biological dressings

shown to be of greater benefit than traditional bandages [2]. Similarly, amnion membrane and its derivatives have been shown to accelerate repair and reduce scarring in ophthalmic surgery, including as a treatment for corneal ulcer [3]. Indeed, such is the success of amnion and amnion-derived products as an aid to wound healing that allograft preparations are commercially available, such as PURION® and EpiFix®. It is claimed that these products have been specifically processed to retain the inherent biological properties of amnion, including the promotion of cell proliferation, modulation of inflammation, maintenance of metalloproteinase activity and recruitment of endogenous progenitors to facilitate wound repair [4]. Specifically, EpiFix® has been shown to promote resolution of refractory non-healing wounds [5]. That allogeneic amnion preparations can be successfully used without fear of tissue rejection most likely relates to the biological function of the cells at the maternal-fetal interface where they are believed to have critical roles in maternal immune re-programming necessary for maternal tolerance of the fetal allograft and a healthy pregnancy [6, 7].

It is this immune privileged property of fetal membranes that, more recently, has been thought to be fundamental to their reparative/regenerative properties. While the human fetal membranes, comprising the chorion (trophoblast derived) and amnion (epiblast derived), have both been shown to contain cells with stem-like cells properties (Table 8.1), it is the immunomodulatory properties of amnion derived stem-like cells in particular that have been exploited in a number of experimental animal models demonstrating pro-regenerative, and most recently, anti-tumor properties.

Table 8.1 Stem cell and stem cell-like properties of cells isolated from human fetal membrane

Cell type	Immune suppression	Self-renewal and pluripotent	*In vitro* differentiation potential	*In vivo* differentiation potential
Human amnion epithelial cells	Express very low levels of human leukocyte antigen (HLA)-A, -B, -C, and do not express HLA-DR [8–11]	Expression of Oct-4, Nanog and SOX-2 [12]	Adipogenic, chondrogenic, osteogenic, cardiomyocytic, hepatic, pancreatic and neurogenic differentiation [13–15]	Hepatic [16], pancreatic [17] and neurogenic [18, 19], differentiation
Human amnion mesenchymal stem/stromal cells		Expression of Oct-4, SSEA-3 and SSEA-4 [11, 20–22]	Adipogenic, chondrogenic, osteogenic, neurogenic, skeletal myogenic, pancreatic and cardiomyogenic differentiation [8, 11, 12, 20–24]	Cardiomyogenic [22] and chondrogenic [25] differentiation
Human chorion mesenchymal stem/stromal cells			Adipogenic, chondrogenic, osteogenic, neurogenic, skeletal myogenic differentiation [26, 27]	Chondrogenic [28]

Fetal Membrane Stem Cell Populations

Amniotic and Chorionic Derived Mesenchymal Stromal/Stem Cells

The most well described population of mesenchymal stromal/stem cells (MSCs) are derived from adult bone marrow, so called bone marrow mesenchymal stem cells (BM-MSCs). However, many other tissues harbor MSCs including adipose tissue, umbilical cord blood, umbilical cord tissue (Wharton's Jelly) and fetal membranes. Common features of these MSCs include multipotential differentiation ability and immunomodulatory properties. Unsurprisingly, MSCs obtained from the amnion and chorion bear similar properties. Human fetal membranes from normal placentae can be physically separated such that pure populations of human chorionic MSCs (hCMSCs) and amniotic MSCs (hAMSCs), and human amnion epithelial cells (hAECs), can be obtained by simple enzymatic digestion. Indeed, MSCs have been successfully isolated from first, second and third trimester placentae, including from the amnion, chorion, decidua parietalis and decidua basalis [8–11]. They represent about 1 % of all cells present in the placenta [12, 13]. Both hCMSCs and hAMSCs express CD73, CD90, and CD105, but not CD14, CD19, CD34, CD45, or HLA-DR [14]. The average yield of MSCs from a typical chorion or amnion is about 1–10 million cells per gram of tissue. Given that any clinical application is likely to require 100s millions of cells there is likely to be a need for considerable *ex vivo* expansion of MSCs. This is not a trivial consideration given the risks of karyotypic abnormalities and epigenetic changes that may accompany serial passaging of MSCs [15, 16]. In contrast to MSCs, much larger numbers of epithelial cells may be derived from the amnion layer. These are called amnion epithelial cells.

Amnion Epithelial Cells

Human amnion epithelial cells (hAECs) can be isolated in large numbers (150–200 million) from term placentae thereby circumventing the need for serial expansion. Further a xeno-free protocol has been developed to meet the requirements of clinical use in patients [17]. These cells express the epithelial cell surface marker, EpCAM, but do not express MSC markers such as CD105 and CD90. They share some stem cell properties such as multipotent differentiation potential and expression of pluripotent stem cell specific transcription factors including Oct4 and Nanog [18]. Unlike MSCs, which are plastic adherent, primary hAECs are composed of subpopulations of cells that are adherent, loosely adherent and free-floating. Of these, the adherent subpopulation of hAECs has the lowest expression of Oct4 and Nanog. Notably, most *in vivo* studies use the entire hAEC fraction rather than a specific subpopulation of hAECs.

Stem Cell Properties and Mechanisms of Action

Immunosuppressive Properties

Regardless of their source, MSCs have potent immunosuppressive properties and MSCs obtained from the fetal membranes are no exception. Both hCMSCs and hAMSCs suppress T cell proliferation to the same extent as BM-MSCs [14]. Interestingly, the production of the immunosuppressive prostaglandin E_2 (PGE_2) is higher from hAMSCs compared to hCMSCs. While PGE_2 is often considered a potent pro-inflammatory mediator, more recently it has become apparent that it plays important roles in limiting inflammatory processes and in regulating tissue remodelling after injury, particularly in the lung [19]. Specifically, PGE_2 modulates both macrophage and T cell functions during repair, polarizing M1 macrophages to the reparative M2 phenotype, thereby facilitating wound healing by increased macrophage phagocytosis [20]. PGE_2 also promotes the maturation of $CD4^+$ T cells to $CD4^+CD25^+FoxP3^+$ Tregs cells [21], enhancing the immunosuppressive capacity of existing Tregs and inducing *de novo* Treg function in $CD4^+CD25^-$ T cells with concurrent acquisition of $FoxP3^+$ expression akin to that in naturally occurring Tregs [29]. These "adaptive Tregs" both inhibit effector T cells by a contact-independent mechanism and express cyclooxygenase-2 (COX-2) thereby increasing PGE_2 further, enhancing T-cell inhibition in a paracrine manner [22]. In short, PGE_2 inhibits T-cell function by several modes of action, both directly as an inhibitory paracrine cytokine and by inducing and/or enhancing Treg function. It is, therefore, no surprise that PGE_2 has been recently identified as a key factor in MSC mediated immunosuppression [23].

Amnion epithelial cells (hAECs) share many common immunosuppressive properties with MSCs. They express potential immunosuppressive factors such as HLA-G [24] and Fas ligand [25] to mediate repair of injury. The histocompatibility leukocyte antigen (HLA)-G molecule is an MHC Class II fragment, which plays a major role in immune tolerance during pregnancy. At the maternal-fetal interphase, HLA-G secreted by the placenta inhibits migration and proliferation of maternal effector immune cells. A recent study showed that hAECs isolated from preterm amnion are not as effective at mitigating acute lung injury as term hAECs, a difference thought to be due to less HLA-G being secreted by preterm hAECs [24].

Another mechanism by which hAEC might mediate repair is via the apoptotic Fas-ligand (Fas-L), which is expressed by amniotic membrane and epithelial cells [25, 26]. Inhibition of hAEC Fas-L reduces the apoptotic activity of T cells by 50 % [25]. T cell-mediated immune suppression during wound healing is important and so the ability of hAECs to regulate T cell survival is an important mechanism by which they might mediate repair. Additionally, similar to MSCs, hAECs are able to suppress concanavalin A-activated splenocyte proliferation through prostaglandin E_2 (PGE_2). Inhibition of PGE_2 production using indomethacin, reversed the suppression of splenocyte proliferation *in vitro* [27]. hAECs are also able to mediated polarisation and limit recruitment of macrophages during injury, promoting a pro-reparative microenvironment predominated by highly phagocytic macrophages

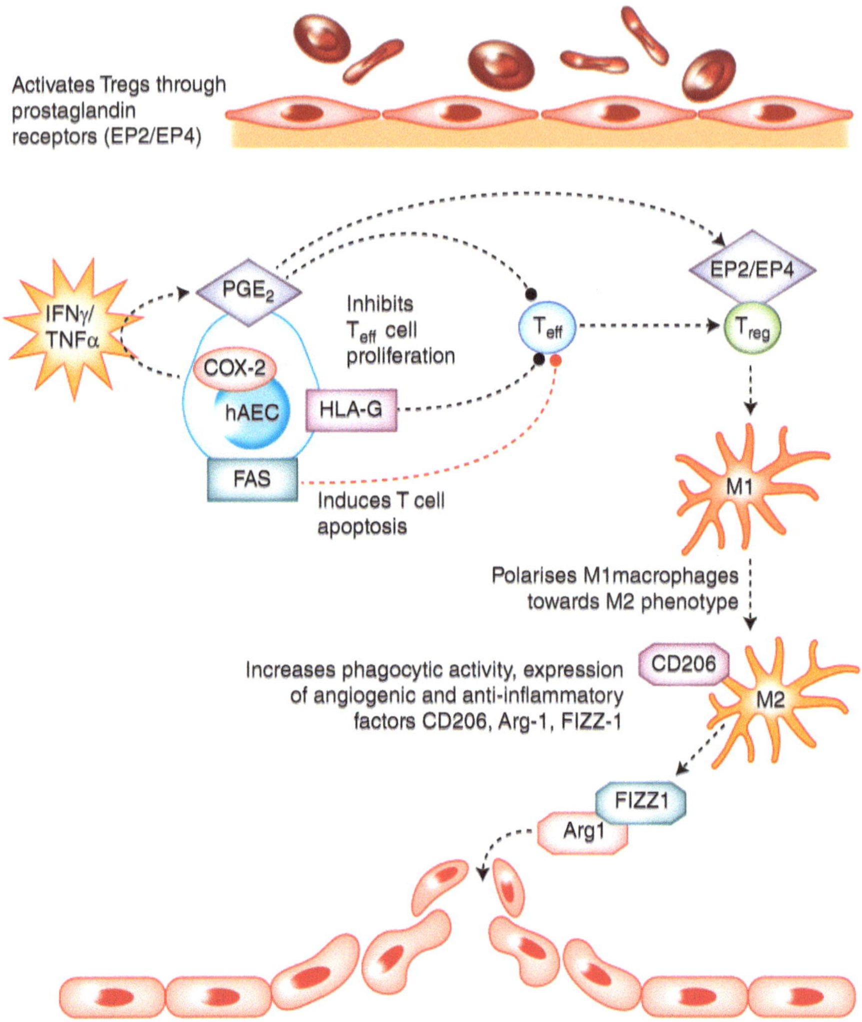

Fig. 8.2 Dynamic immunomodulatory properties of hAECs include reduction of effector T cell numbers and polarisation of macrophages towards an M2 phenotype

[28]. The dynamic capacity of hAECs to modulate host immune cell response in disease certainly warrants further study. Figure 8.2 illustrates the possible mechanisms and pathways by which hAECs might mediate and facilitate repair of injury.

In Vitro Characteristics

Amniotic and chorionic MSCs are described as being plastic adherent, with typical fibroblast-like morphology. They are able to form colonies and can be serially expanded [10]. Amniotic and chorionic MSCs express similar cell surface markers to BM-MSCs, including CD44, CD90, CD105, CD13, CD29, CD54, CD73 and CD166. They also retain the ability to differentiate into adipocytes, chondrocytes and osteocytes, indicative of a multipotent differentiation potential. However, the colony forming ability of hAMSCs is superior to that of hCMSCs, where the colony forming capacity of both MSC populations is enhanced by enriching for $CD271^+$ cells [10]. The proliferation rate of hCMSCs and hAMSCs remain steady even after >10 serial passages. However, HLA-DR expression diminishes in both types of MSCs after early passaging [30, 31]. This may explain their ability to avoid immune surveillance *in vivo.*

When cultured in semi-solid media hAMSCs spontaneously form capillary-like structures, even in the absence of VEGF, and they express VEGF receptor 1 and 2 (Flt-1 and KDR) [13]. This suggests that a further mechanism by which they may augment wound repair is by supporting neovascularization *in vivo.* On a similar note, a subtype of $CD200^+$ hC MSCs support proliferation of primary human hepatocytes *in vitro* [32].

In Vivo Characteristics

The pro-regenerative properties of CMSCs and AMSCs have been explored across a number of animal models. Since fetal membranes can be technically challenging to separate in rodents, allogeneic studies in rodent models of injury have largely relied on isolating MSCs from both fetal membranes together and therefore reported the effects of this mixed population of MSCs—both chorion and amnion MSCs. For example, mixed MSCs obtained from both fetal membranes of rats demonstrated therapeutic benefit when administered allogeneically in models of hindlimb ischaemia [33], autoimmune myocarditis [34], glomerulonephritis [35], renal ischaemia-reperfusion injury [36] and myocardial infarction [37]. In contrast, human fetal membranes are easily separated. The chorion and amnion can be physically pulled apart into discrete sheets, washed and processed for separate cell isolation. A recent study performed by Yamahara and colleagues showed that conditioned media from hAMSCs and hCMSCs were also able to inhibit cardiomyocyte and endothelial cell death triggered by hypoxia or serum starvation [14]. Both cell types secreted significant amounts of pro-angiogenic soluble factors including hepatocyte growth factor (HGF), insulin growth factor-1 (IGF-1), vascular endothelial growth factor (VEGF) and basic fibroblast growth factor (bFGF). Interestingly, the hAMSCs produced significantly higher levels of HGF and bFGF compared to hCMSCs. However, these researchers did not report significant differences in therapeutic benefit between the two MSC types when they were used xenogeneically in a rat model of hindlimb ischaemia and mouse model of acute graft versus host disease (GVHD) [14]. As such, hCMSCs and hAMSCs have been reported to exert similar therapeutic benefits even when administered xenogeneically across a number of animal models of acute and chronic diseases. These are described in the following sections.

Potential Clinical Applications

Graft-Versus-Host Disease

As mentioned above, both hCMSCs and hAMSCs have been shown to exert therapeutic benefit when used in an animal model of GVHD. In a study conducted by Yamahara and colleagues, four intravenous doses of 1×10^5 amniotic or chorionic MSCs reduced weight loss associated with GVHD [14]. It should be noted, however, that this study did not assess the extent of successful donor engraftment. In contrast, *ex vivo* expanded placental MSCs have been shown to enhance haematopoietic stem cell engraftment [29] such that a placental MSC adjuvant therapy is moving towards clinical use [38]. To date, the therapeutic application of hAECs in preclinical models of GVHD has not yet been reported but, as already described, hAECS possess many of the features and functions of amnion MSCs that it would be worthwhile assessing this more abundant cell type in this setting.

Wound Healing

The wound healing capabilities of the fetal membranes are well described. Amniotic MSCs have been reported to enhance wound healing in diabetic NOD/SCID mice [39]. When injected intra-dermally around full-thickness excision wounds, hAMSCs showed functional engraftment where transdifferentiation into keratinocytes was accompanied by accelerated re-epithelialisation. These appeared to be direct effects of the hAMSCs which promoted tube formation by human umbilical vein endothelial cells (HUVECs) on matrigel and cell migration in a scratch test assay. Interestingly, it appears that hAMSCs have superior wound healing capabilities than adipose derived MSCs [39].

Amniotic MSCs also promote angiogenesis *in vitro* and when injected into ischaemic hindlimbs of mice they spontaneously differentiated into vessel-like structures that expressed endothelial specific proteins including Von Willebrand's Factor and VEGF receptors Flt-1 and KDR [40]. Implantation of hAMSCs also improved blood perfusion and capillary density in these mice affected by hindlimb ischaemia, indicative of the vasculogenic potential of hAMSCs [40]. Given historical studies on scarless fetal wound healing [41, 42], it is thus tempting to speculate that the fetal origins of these MSCs may confer superior wound healing properties to adult derived MSCs such as BM-MSCs and/or adipose MSCs.

Cardiovascular Disease

Amniotic MSCs can differentiate *in vitro* into spontaneously beating cardiomyocytes, a differentiation significantly accelerated by the administration of interleukin-10 (IL-10) or progesterone [43]. *In vivo,* when transplanted into nude rats 2 weeks after an induced myocardial infarct, hAMSCs reduced myocardial fibrosis and significantly improved left ventricular function. Importantly, when transplanted into immune competent Wistar rats the amniotic MSCs survived for more than 4 weeks post-transplantation, differentiating into cardiomyocytes *in situ* without immunosuppression [43]. The survival of the hAMSCs in the immune competent rats was thought to be due to the secretion of HLA-G and the activation of regulatory T cells [43]. Similarly, the direct transplantation of hAMSCs into the border regions of induced ischaemic heart tissue increased survival in a NOD/SCID mouse model of myocardial infarction [44]. In that model, as with the previous report [43], hAMSC administration improved left ventricular function. This was thought to be effected, at least in part, by increased capillary density and increased levels of the pro-angiogenic factors angiopoetin-1 and vascular endothelial growth factor (VEGFA) [44]. In light of these reports for hAMSCs, perhaps not surprisingly hAECs have also been shown to reduce myocardial infarct size and improve ventricular contractility when directly transplanted into areas of infarction in athymic nude mice [45]. These improvements were associated with *in vivo* engraftment and differentiation of hAECs, where 3 % of the surviving hAECs expressed myosin heavy chain 4 weeks following the infarct event. The involvement of HLA-G, macrophage and/or T cell modulation by hAECs in models of myocardial ischaemia have not yet been reported. Further, to date, the impact of chorionic MSCs on cardiovascular disease has not been reported.

Liver Fibrosis

Both hAMSCs and hAECs have been assessed as possible cell therapies for acute and chronic liver disease. Carbon tetrachloride infusion induced liver fibrosis in immune competent C57Bl6 mice was reportedly ameliorated by hAMSC treatment [46]. Specifically, a single dose of 1×10^5 amniotic MSCs introduced intrasplenically 4 weeks following the commencement of carbon tetrachloride infusion effectively suppressed stellate cell activation and consequently reduced liver fibrosis. Hepatocyte apoptosis was also reduced while hepatocyte proliferation was increased, probably by reducing hepatocyte senescence. These cellular changes were associated with improved liver function, as evidenced by reduced serum levels of alanine aminotransferase and aspartate transaminase, and with engraftment and differentiation of hAMSCs into hepatocytes in the recipient livers. Anti-fibrotic properties of hCMSCs have not been assessed in liver fibrosis models to date however *in vitro* findings, as described earlier in this chapter, on the protective properties of hCMSCs on hepatocytes indicate that they may be similarly beneficial [32]. In contrast, hAECs have been shown to be an effective therapy for liver fibrosis, exerting anti-fibrotic effects when administered intravenously in a similar carbon tetrachloride induced liver fibrosis mouse model [47]. In that study, in addition to the suppression of stellate cell activation the authors showed that hAEC administration increased gelatinase activity. This effect may underlie the accelerated resolution of liver fibrosis that is observed following either hAEC or MSC administration [46, 47].

Intriguingly, as has been shown in ophthalmic surgery, the cells themselves may not be required for the reparative effects but rather secreted factors, in hAEC-conditioned media, can directly suppress stellate cell activation [48]. Further, as has been shown *in vitro* [28], hAECs are likely to exert their anti-fibrotic and anti-inflammatory effects in the livers of the carbon tetrachloride challenged animals inducing the polarisation of hepatic macrophages from the classically activated M1 phenotype to the alternatively activated M2 phenotype [49]. Since a direct comparison between hAECs and hAMSCs has not been performed to date it is not yet clear whether the different cell types work via different mechanisms and/or pathways. However, taken together, the findings from the various studies of hAMSCs and hAECs in liver damage suggest that that injury resolution and liver repair is most likely mediated by the cells via soluble factors acting upon endogenous host cell types such as the hepatic stellate cells and/or macrophages to resolve fibrosis, rather than through engraftment, differention and functional replacement of damaged hepatocytes.

Acute and Chronic Lung Disease

Stem cells derived from fetal membranes have been applied across a number of neonatal and adult respiratory diseases. In an immune competent mouse model of bleomycin induced lung injury (inflammation and fibrosis), a mixture of cells isolated from the human chorionic and amnion (50 % MSC:50 % hAECs mixture) were found to be as effective as allogeneic cells isolated from murine fetal membranes in preventing injury [50]. Indeed, the anti-inflammatory and anti-fibrotic effects of fetal membrane MSCs and AECs in lung injury were similar whether the cells were allogeneic or xenogeneic (human) and irrespective of route of administration, whether intravenous, intraperitoneal or intratracheal [50]. In accord with the studies in liver injury models, the administration of MSCs/AECs was associated with a profound reduction in pulmonary infiltration by all white cells: neutrophils, macrophages and lymphocytes [50]. That observation supported the notion that cells were most likely acting via modulation of the host immune response to injury rather than via engraftment and differentiation. Indeed, it was subsequently shown that, as in the liver [48], the cells themselves were not required for the protective effects in acute lung injury but that the same effects were achieved by the administration of hAMSC-conditioned media [51]. Of course, these studies were unable to unravel whether it the protective effects were being exerted by hAMSCs or by AECs, or by both.

In that regard, remarkably similar results in the lung have been achieved with pure populations of hAECs only [24, 52], whereby hAECs prevent inflammation and fibrosis following bleomycin-induced injury. As with MSCs it would appear that this effect may be exerted via macrophages. Macrophage recruitment, both *in vitro* and *in vivo* in the injured lung, is inhibited by hAECs and hAECs are able to induce macrophage polarisation from the classical, pro-inflammatory M1 phenotype to the alternative, pro-reparative M2 phenotype [28]. Indeed, the central role that macrophages play in the mechanisms of action of hAECs in mitigating acute lung injury, and possibly in other tissue injury repair, is illustrated by the inability of hAECs to be protective in a mouse model that has functionally deficient macrophages [53]. In that model, a *Sftpc*$^{-/-}$ mouse, hAECs were unable to prevent macrophage influx into the lung following bleomycin administration and unable to mitigate the subsequent fibrosis [53]. Since regulatory T cells are known to be involved in macrophage polarization and function, and play important roles in resolution of lung injury [23], it is likely that hAECs will mediate at least some of their effects via Tregs. Indeed, it is feasible that hAECs actions on macrophages are actually exerted, at least in part, via Tregs. This is certainly worth examination as it might offer new therapeutic insights into both hAEC and MSC mechanisms of action. Irrespective of the precise effects of hAECs on host immune responses, the very low engraftment rates of hAECs into the injured lung [52] suggests that, perhaps unlike heart [45] and liver [46], they are likely to augment the repair process by replacing injured lung epithelium through engraftment and differentiation [52]. It is also important to note that almost all the studies of hAECs and fetal membrane MSCs in lung injury explore the utility of the cell therapy given at the time or immediately after the insult [50–53]. Unlike the studies in liver and heart injury where cell therapy has been shown to improve established injury there is essentially only one study in the lung that has explored whether hAECs could accelerate repair of fibrotic injury [54]. In that study hAECs were administered either a week after bleomycin administration, when pulmonary inflammation is maximum, or 2 weeks after bleomycin, when fibrosis is maximum. At peak inflammation hAECs did not mitigate lung injury suggesting that either hAECS are inactive during active inflammation or that the cell dosage was inadequate [54]. In contrast, when administered at the time of peak fibrosis but after acute inflammation, the administration of hAECs accelerated the resolution of fibrosis [54]. These finding not only usually inform the design of future clinical trials they also provide further insights into likely mechanisms of action.

It is not only in adults that lung disease has been explored as a possible target. Bronchopulmonary dysplasia (BPD) is a chronic lung disease of the preterm infant resulting in alveolar maturational arrest, fibrosis and a disordered vasculature. There is no current treatment for BPD and, as such, it is an ideal disease target for cell therapies. Using a variety of insults to mimic those that are thought to be important in the development of BPD, such as hyperoxia, infection, and barotrauma, a series of neonatal mouse and fetal sheep experiments have shown that hAECs can reduce the BPD-like injury, normalizing the tissue:airspace ratio and lung architecture, reducing the inflammatory response, and reducing resultant fibrosis and scarring [55–57]. Interestingly, in the fetal sheep model of ventilation induced injury, mimicking barotrauma, the administration of hAECs both intravenously and intratracheally appeared to confer better protection/repair than when cells were administered by either route alone [57].

Neurological Diseases

Stroke is one of the leading causes of death and disability worldwide, a major burden of disease. Despite improvements in emergency therapies, such as thrombolysis, and rehabilitation medicine, treatment options for patients with profound ischaemic and haemorrhagic stroke remain limited. Not surprisingly, cell therapies, including fetal membrane cells have been assessed as possible treatment options. Broadly, two approaches have been taken: the administration of either partially differentiated neural cells derived from hAMSCs or the administration of primary undifferentiated amnion cells. With regard to the first approach, the direct administration, into the striatum of the brain, of partially differentiated neural cells from hAMSCs has been shown to improve neurological function in a rat stroke model where focal cerebral ischaemia is induced by middle cerebral artery occlusion [58]. Administration of the neurally differentiated MSCs 2 weeks after the induced stroke improved recovery of neurological function, as assessed by a test of balance test and other motor functional deficits (flexion of forelimb, circling towards the paretic side, falling down, etc.). These outcomes were coincident with an accumulation of hAMSCs within the ischaemic lesion and reduced pyknosis of pyramidal neurons [58]. Similarly, hAMSCs transfected with the brain derived neurotrophic factor (BDNF), as an approach to deliver gene therapy, when administered directly into the brain reduced infarct size and improved functional outcome in rats undergoing middle cerebral artery occlusion-induced ischaemic stroke [59]. That BDNF-hAMSCs were more effective than hAMSCs transfected with just the green fluorescent protein, EGFP (EGFP-hAMSCs) suggests that either the delivery of BDNF locally by the hAMSCs was important or that the BDNF transfection altered the hAMSCs in a manner to assist their function. In that regard, 3 weeks after transplantation the BDNF-hAMSCs could still be located within the brain and some of the cells expressed the neuronal markers MAP2 and nestin, consistent with *in vivo* neural differentiation [59].

Both amniotic fluid cells [60] and hAECs [61] have been assessed in the same models of ischaemic stroke. When delivered directly into the brain 24 h after middle cerebral artery occlusion, hAECs tracked to the infarct site, reduced cell death and infact size and improved neurological function 2 weeks later [61]. As with the hAMSCs, transfection with GDNF appeared to improve the effects of hAECs but GDNF was not required for an effect [61]. hAECs have also been administered to a rat model of intracerebral haemorrhage. When delivered by intraventricular injection, directly into the brain, hAECs were shown to reduce brain oedema and improve motor deficit following intracerebral haemorrhage [62]. Consistent with the observations in liver and lung where hAECs reduced macrophage number [28, 49], the administration of hAECs following intracerebral haemorrhage reduced the number of microglia suggesting a reduction in inflammatory response to the haemorrhage [62]. However, as has been highlighted [63], the direct administration of cells into the brain is not likely to find clinical favour. A more preferred, and safer, administration route such as intravenous delivery would be preferable. To date, there are no studies reporting the intravenous (iv) administration of either hAMSCs or hAECs for the treatment of stroke. However, amniotic fluid-derived cells have been successfully given iv with good effect [64], offering promise for hAECs and hAMSCs.

Indeed, in a fetal sheep model of white matter injury, to mimic the periventricular leucomalacia seen in preterm babies, the iv delivery of hAECs was associated with a significant reduction in microglial number and activation in the cortex, subcortical and periventricular white matter and with reduced apoptosis [65]. Such findings would be consistent with the hAECs exerting anti-inflammatory and neuroprotective effects in the fetal brain. Notably, while the hAECs were delivered intravenously they were detectable within the brain, confirming that they can cross the blood:brain barrier. However, unlike the hAMSCs [58, 59] there was no evidence of *in vivo* differentiation. Rather, it is likely that in the brain, as in lung and liver, hAECs exert their neuroprotective effects via modulation of the host inflammatory response. One additional mechanism of injury mitigation or facilitated repair in the brain is via melatonin [66]. hAECs express one of the melatonin receptors and the proliferative and neuroprotective properties of hAECs can be augmented by pretreatment with melatonin [66].

A number of animal models of other CNS disorders has been used to assess the therapeutic potential of hAECs. These include models of progressive degenerative diseases such as multiple sclerosis (MS), Parkinson's disease, amyotrophic lateral sclerosis (ALS) and traumatic injury such as spinal cord transections and peripheral nerve injury [63]. In an experimental autoimmune encephalomyelitis (EAE) mouse model of MS hAECs were shown to reduce proliferation of myelin oligodendrocyte glycoprotein-specific T cells and to decrease their secretion of pro-inflammatory cytokines, suggesting disease mitigating effects [67]. More recently, Liu and colleagues reported that iv administered hAECs mitigated demyelination in the EAE mouse model of MS, principally by reducing brain inflammation, an effect mediated by TGF-β and PGE_2 [27]. Importantly, there was no evidence of cell engraftment [27]. It is intriguing that the mechanisms of action of hAECs appear common across diseases and organs. That said, in a rat model of Parkinson's disease, when hAECs are transplanted directly into the striatum of the rat brain they survive and appear to differentiate, secreting dopamine [68]. Whether this is the primary mechanism of action in ameliorating this disease model or whether hAECs prevent further loss of host dopamine-secreting neurons [69], most likely though an anti-inflammatory action, remains to be further resolved.

Amyotrophic lateral sclerosis (ALS) is an adult onset progressive neurodegenerative disease where promising results have been achieved with multiple systemic transplantations of hAMSCs in a mouse model of the disease [70]. Treatment prevented the loss of motor neurons and reduced neuroinflammation While some hAMSCs were detected in the spinal cords of animals at the final stage of the disease, these were negative for β-tubulin III or glial fibrillary acidic protein, indicating that beneficial effects were not likely due to engraftment and differentiation of the hAMSCs to replace damaged neurons.

With regard to spinal cord transection and facilitated axonal regrowth more generally, hAECs appear to be able to survive for months in the spinal cord without inducing an inflammatory response and may be able to improve motor function consistent with facilitated cord repair [71]. The precise mechanisms of action of hAECs in spinal cord injury repair remain unclear but they appear to accelerate glial scar resolution facilitating axonal migration and penetration [72]. Such effects would certainly be consistent with the effects on macrophage polarisation previously discussed. It is also likely that cell therapy will assist with neovascularisation. For example, hAMSCs, when transplanted locally to the site of injured nerves improved vascularisation of the nerves and increased blood [73]. In that mouse model of peripheral nerve injury, the hAMSCs also appeared to engraft. Interestingly, the degree of neurovascular tropism and rescue from neuropathy *in vivo* afforded by hAMSCs was far greater than that seen by adipose tissue derived MSCs.

Conclusions

It is clear that the amniotic membrane harbours two populations of cells—hAMSCs and hAECs—that have special properties that make them an attractive and readily available cell therapy. While there is still much to be understood about how the cells exert their protective and regenerative effects, and therefore how the cells would be best used clinically, the long safety record of the amnion in wound healing and ophthalmic surgery makes these cells a relatively safe therapeutic avenue to translate to clinical care sooner rather than later. While it is possible that in some disease states, particularly where the cells are transplanted directly into the injured tissue, both hAMSCs and hAECs are able to engraft and differentiate into the host tissue cells the mounting evidence is that the cells most likely exert the majority of their effects on modulating the host immune response and driving host-led repair. In that regard, the identification of the secreted factors whereby the cells modulate host immune offer opportunities for novel therapeutics. Further, the ability of amnion cells to stimulate endogenous progenitor cells to accelerate repair is an area yet to be addressed and one that is likely to lead to important insights into how endogenous repair can be augmented. Whatever the mechanisms by which amnion cells work these are a most astonishing population of cells that have much to teach us.

References

1. Kesting MR, Wolff KD, Hohlweg-Majert B, Steinstraesser L. The role of allogenic amniotic membrane in burn treatment. J Burn Care Res. 2008;29:907–16.
2. Gruss JS, Jirsch DW. Human amniotic membrane: a versatile wound dressing. Can Med Assoc J. 1978;118:1237–46.
3. Ricciardelli G, Ceccuzzi R, Raneri M, Scalisi A, Bianchi PE. Management of recurrent corneal ulcers: use of amniotic membrane. Eur J Ophthalmol. 2014;24:793–6.

4. Koob TJ, Rennert R, Zabek N, Massee M, Lim JJ, Temenoff JS, Li WW, Gurtner G. Biological properties of dehydrated human amnion/chorion composite graft: implications for chronic wound healing. Int Wound J. 2013;10:493–500.
5. Sheikh ES, Sheikh ES, Fetterolf DE. Use of dehydrated human amniotic membrane allografts to promote healing in patients with refractory non healing wounds. Int Wound J. 2014;11:711–7.
6. de Moraes-Pinto MI, Vince GS, Flanagan BF, Hart CA, Johnson PM. Localization of IL-4 and IL-4 receptors in the human term placenta, decidua and amniochorionic membranes. Immunology. 1997;90:87–94.
7. Lefebvre S, Adrian F, Moreau P, Gourand L, Dausset J, Berrih-Aknin S, Carosella ED, Paul P. Modulation of HLA-G expression in human thymic and amniotic epithelial cells. Hum Immunol. 2000;61:1095–101.
8. In 't Anker PS, Scherjon SA, Kleijburg-van der C, de Groot-Swings GM, Claas FH, Fibbe WE, Kanhai HH. Isolation of mesenchymal stem cells of fetal or maternal origin from human placenta. Stem Cells. 2004;22:1338–45.
9. Portmann-Lanz CB, Schoeberlein A, Huber A, Sager R, Malek A, Holzgreve W, Surbek DV. Placental mesenchymal stem cells as potential autologous graft for pre- and perinatal neuroregeneration. Am J Obstet Gynecol. 2006;194:664–73.
10. Soncini M, Vertua E, Gibelli L, Zorzi F, Denegri M, Albertini A, Wengler GS, Parolini O. Isolation and characterization of mesenchymal cells from human fetal membranes. J Tissue Eng Regen Med. 2007;1:296–305.
11. Poloni A, Rosini V, Mondini E, Maurizi G, Mancini S, Discepoli G, Biasio S, Battaglini G, Berardinelli E, Serrani F, Leoni P. Characterization and expansion of mesenchymal progenitor cells from first-trimester chorionic villi of human placenta. Cytotherapy. 2008;10:690–7.
12. Zhang Y, Li CD, Jiang XX, Li HL, Tang PH, Mao N. Comparison of mesenchymal stem cells from human placenta and bone marrow. Chin Med J (Engl). 2004;117:882–7.
13. Alviano F, Fossati V, Marchionni C, Arpinati M, Bonsi L, Franchina M, Lanzoni G, Cantoni S, Cavallini C, Bianchi F, Tazzari PL, Pasquinelli G, Foroni L, Ventura C, Grossi A, Bagnara GP. Term amniotic membrane is a high throughput source for multipotent mesenchymal stem cells with the ability to differentiate into endothelial cells in vitro. BMC Dev Biol. 2007;7:11.
14. Yamahara K, Harada K, Ohshima M, Ishikane S, Ohnishi S, Tsuda H, Otani K, Taguchi A, Soma T, Ogawa H, Katsuragi S, Yoshimatsu J, Harada-Shiba M, Kangawa K, Ikeda T. Comparison of angiogenic, cytoprotective, and immunosuppressive properties of human amnion- and chorion-derived mesenchymal stem cells. PLoS One. 2014;9, e88319.
15. Schellenberg A, Lin Q, Schüler H, Koch CM, Joussen S, Denecke B, Walenda G, Pallua N, Suschek CV, Zenke M, Wagner W. Replicative senescence of mesenchymal stem cells causes DNA-methylation changes which correlate with repressive histone marks. Aging (Albany, NY). 2011;3:873–88.
16. Wang Y, Zhang Z, Chi Y, Zhang Q, Xu F, Yang Z, Meng L, Yang S, Yan S, Mao A, Zhang J, Yang Y, Wang S, Cui J, Liang L, Ji Y, Han ZB, Fang X, Han ZC. Long-term cultured mesenchymal stem cells frequently develop genomic mutations but do not undergo malignant transformation. Cell Death Dis. 2013;4, e950.
17. Murphy S, Rosli S, Acharya R, Mathias L, Lim R, Wallace E, Jenkin G. Amnion epithelial cell isolation and characterization for clinical use. Curr Protoc Stem Cell Biol. 2010;Chapter 1:Unit 1E.6. doi:10.1002/9780470151808.
18. Miki T, Lehmann T, Cai H, Stolz DB, Strom SC. Stem cell characteristics of amniotic epithelial cells. Stem Cells. 2005;23:1549–59.
19. Vancheri C, Mastruzzo C, Sortino MA, Crimi N. The lung as a privileged site for the beneficial actions of PGE2. Trends Immunol. 2004;25:40–6.
20. Prescott D, McKay DM. Aspirin-triggered lipoxin enhances macrophage phagocytosis of bacteria while inhibiting inflammatory cytokine production. Am J Physiol Gastrointest Liver Physiol. 2011;301:G487–97.
21. Mandapathil M, Szczepanski MJ, Szajnik M, Ren J, Jackson EK, Johnson JT, Gorelik E, Lang S, Whiteside TL. Adenosine and prostaglandin E2 cooperate in the suppression of immune responses mediated by adaptive regulatory T cells. J Biol Chem. 2010;285:27571–80.

22. Mahic M, Yaqub S, Johansson CC, Taskén K, Aandahl EM. FOXP3+CD4+CD25+ adaptive regulatory T cells express cyclooxygenase-2 and suppress effector T cells by a prostaglandin E2-dependent mechanism. J Immunol. 2006;177(1):246–54.
23. Aggarwal NR, D'Alessio FR, Tsushima K, Sidhaye VK, Cheadle C, Grigoryev DN, Barnes KC, King LS. Regulatory T cell-mediated resolution of lung injury: identification of potential target genes via expression profiling. Physiol Genomics. 2010;41:109–19.
24. Lim R, Chan ST, Tan JL, Mockler JC, Murphy SV, Wallace EM. Preterm human amnion epithelial cells have limited reparative potential. Placenta. 2013;34:486–9.
25. Li H, Niederkorn JY, Neelam S, Mayhew E, Word RA, McCulley JP, Alizadeh H. Immunosuppressive factors secreted by human amniotic epithelial cells. Invest Ophthalmol Vis Sci. 2005;46:900–7.
26. Kubo M, Sonoda Y, Muramatsu R, Usui M. Immunogenicity of human amniotic membrane in experimental xenotransplantation. Invest Ophthalmol Vis Sci. 2001;42:1539–46.
27. Liu YH, Vaghjiani V, Tee JY, To K, Cui P, Oh DY, Manuelpillai U, Toh BH, Chan J. Amniotic epithelial cells from the human placenta potently suppress a mouse model of multiple sclerosis. PLoS One. 2012;7, e35758.
28. Tan JL, Chan ST, Wallace EM, Lim R. Human amnion epithelial cells mediate lung repair by directly modulating macrophage recruitment and polarization. Cell Transplant. 2014;23:319–28.
29. Jang MJ, Kim HS, Lee HG, Kim GJ, Jeon HG, Shin HS, Chang SK, Hur GH, Chong SY, Oh D, Chung HM. Placenta-derived mesenchymal stem cells have an immunomodulatory effect that can control acute graft-versus-host disease in mice. Acta Haematol. 2013;129:197–206.
30. Kim J, Kang HM, Kim H, Kim MR, Kwon HC, Gye MC, Kang SG, Yang HS, You J. Ex vivo characteristics of human amniotic membrane-derived stem cells. Cloning Stem Cells. 2007;9:581–94.
31. Koo BK, Park IY, Kim J, Kim JH, Kwon A, Kim Y, Shin JC, Kim JH. Isolation and characterization of chorionic mesenchymal stromal cells from human full term placenta. J Korean Med Sci. 2012;27:857–63.
32. Wang J, Zhu Z, Huang Y, Wang P, Luo Y, Gao Y, Du Z. The subtype CD200-positive, chorionic mesenchymal stem cells from the placenta promote regeneration of human hepatocytes. Biotechnol Lett. 2014;36:1335–41.
33. Ishikane S, Ohnishi S, Yamahara K, Sada M, Harada K, Mishima K, Iwasaki K, Fujiwara M, Kitamura S, Nagaya N, Ikeda T. Allogeneic injection of fetal membrane-derived mesenchymal stem cells induces therapeutic angiogenesis in a rat model of hind limb ischemia. Stem Cells. 2008;26:2625–33.
34. Ohshima M, Yamahara K, Ishikane S, Harada K, Tsuda H, Otani K, Taguchi A, Miyazato M, Katsuragi S, Yoshimatsu J, Kodama M, Kangawa K, Ikeda T. Systemic transplantation of allogenic fetal membrane-derived mesenchymal stem cells suppresses Th1 and Th17 T cell responses in experimental autoimmune myocarditis. J Mol Cell Cardiol. 2012;53:420–8.
35. Tsuda H, Yamahara K, Ishikane S, Otani K, Nakamura A, Sawai K, Ichimaru N, Sada M, Taguchi A, Hosoda H, Tsuji M, Kawachi H, Horio M, Isaka Y, Kangawa K, Takahara S, Ikeda T. Allogenic fetal membrane-derived mesenchymal stem cells contribute to renal repair in experimental glomerulonephritis. Am J Physiol Renal Physiol. 2010;299:F1004–13.
36. Tsuda H, Yamahara K, Otani K, Okumi M, Yazawa K, Kaimori JY, Taguchi A, Kangawa K, Ikeda T, Takahara S, Isaka Y. Transplantation of allogenic fetal membrane-derived mesenchymal stem cells protects against ischemia/reperfusion-induced acute kidney injury. Cell Transplant. 2014;23:889–99.
37. Ishikane S, Hosoda H, Yamahara K, Akitake Y, Kyoungsook J, Mishima K, Iwasaki K, Fujiwara M, Miyazato M, Kangawa K, Ikeda T. Allogeneic transplantation of fetal membrane-derived mesenchymal stem cell sheets increases neovascularization and improves cardiac function after myocardial infarction in rats. Transplantation. 2013;96:697–706.
38. Prather WR, Toren A, Meiron M. Placental-derived and expanded mesenchymal stromal cells (PLX-I) to enhance the engraftment of hematopoietic stem cells derived from umbilical cord blood. Expert Opin Biol Ther. 2008;8:1241–50.

39. Kim SW, Zhang HZ, Guo L, Kim JM, Kim MH. Amniotic mesenchymal stem cells enhance wound healing in diabetic NOD/SCID mice through high angiogenic and engraftment capabilities. PLoS One. 2012;7, e41105.
40. Kim SW, Zhang HZ, Kim CE, An HS, Kim JM, Kim MH. Amniotic mesenchymal stem cells have robust angiogenic properties and are effective in treating hindlimb ischaemia. Cardiovasc Res. 2012;93:525–34.
41. Adzick NS, Longaker MT. Scarless fetal healing. Therapeutic implications. Ann Surg. 1992;215:3–7.
42. Lin RY, Sullivan KM, Argenta PA, Peter Lorenz H, Scott Adzick N. Scarless human fetal skin repair is intrinsic to the fetal fibroblast and occurs in the absence of an inflammatory response. Wound Repair Regen. 1994;2:297–305.
43. Tsuji H, Miyoshi S, Ikegami Y, Hida N, Asada H, Togashi I, Suzuki J, Satake M, Nakamizo H, Tanaka M, Mori T, Segawa K, Nishiyama N, Inoue J, Makino H, Miyado K, Ogawa S, Yoshimura Y, Umezawa A. Xenografted human amniotic membrane-derived mesenchymal stem cells are immunologically tolerated and transdifferentiated into cardiomyocytes. Circ Res. 2010;106:1613–23.
44. Kim SW, Zhang HZ, Kim CE, Kim JM, Kim MH. Amniotic mesenchymal stem cells with robust chemotactic properties are effective in the treatment of a myocardial infarction model. Int J Cardiol. 2013;168:1062–9.
45. Fang CH, Jin J, Joe JH, Song YS, So BI, Lim SM, Cheon GJ, Woo SK, Ra JC, Lee YY, Kim KS. In vivo differentiation of human amniotic epithelial cells into cardiomyocyte-like cells and cell transplantation effect on myocardial infarction in rats: comparison with cord blood and adipose tissue-derived mesenchymal stem cells. Cell Transplant. 2012;21:1687–96.
46. Zhang D, Jiang M, Miao D. Transplanted human amniotic membrane-derived mesenchymal stem cells ameliorate carbon tetrachloride-induced liver cirrhosis in mouse. PLoS One. 2011;6, e16789.
47. Manuelpillai U, Tchongue J, Lourensz D, Vaghjiani V, Samuel CS, Liu A, Williams ED, Sievert W. Transplantation of human amnion epithelial cells reduces hepatic fibrosis in immunocompetent CCl_4-treated mice. Cell Transplant. 2010;19:1157–68.
48. Hodge A, Lourensz D, Vaghjiani V, Nguyen H, Tchongue J, Wang B, Murthi P, Sievert W, Manuelpillai U. Soluble factors derived from human amniotic epithelial cells suppress collagen production in human hepatic stellate cells. Cytotherapy. 2014;16:1132–44.
49. Manuelpillai U, Lourensz D, Vaghjiani V, Tchongue J, Lacey D, Tee JY, Murthi P, Chan J, Hodge A, Sievert W. Human amniotic epithelial cell transplantation induces markers of alternative macrophage activation and reduces established hepatic fibrosis. PLoS One. 2012;7, e38631.
50. Cargnoni A, Gibelli L, Tosini A, Signoroni PB, Nassuato C, Arienti D, Lombardi G, Albertini A, Wengler GS, Parolini O. Transplantation of allogeneic and xenogeneic placenta-derived cells reduces bleomycin-induced lung fibrosis. Cell Transplant. 2009;18:405–22.
51. Cargnoni A, Ressel L, Rossi D, Poli A, Arienti D, Lombardi G, Parolini O. Conditioned medium from amniotic mesenchymal tissue cells reduces progression of bleomycin-induced lung fibrosis. Cytotherapy. 2012;14:153–61.
52. Murphy S, Lim R, Dickinson H, Acharya R, Rosli S, Jenkin G, Wallace E. Human amnion epithelial cells prevent bleomycin-induced lung injury and preserve lung function. Cell Transplant. 2011;20:909–23.
53. Murphy SV, Shiyun SC, Tan JL, Chan S, Jenkin G, Wallace EM, Lim R. Human amnion epithelial cells do not abrogate pulmonary fibrosis in mice with impaired macrophage function. Cell Transplant. 2012;21:1477–92.
54. Vosdoganes P, Wallace EM, Chan ST, Acharya R, Moss TJ, Lim R. Human amnion epithelial cells repair established lung injury. Cell Transplant. 2013;22:1337–49.
55. Vosdoganes P, Lim R, Koulaeva E, Chan ST, Acharya R, Moss TJ, Wallace EM. Human amnion epithelial cells modulate hyperoxia-induced neonatal lung injury in mice. Cytotherapy. 2013;15:1021–9.
56. Vosdoganes P, Hodges RJ, Lim R, Westover AJ, Acharya RY, Wallace EM, Moss TJ. Human amnion epithelial cells as a treatment for inflammation-induced fetal lung injury in sheep. Am J Obstet Gynecol. 2011;205:156.e26–33.

57. Hodges RJ, Jenkin G, Hooper SB, Allison B, Lim R, Dickinson H, Miller SL, Vosdoganes P, Wallace EM. Human amnion epithelial cells reduce ventilation-induced preterm lung injury in fetal sheep. Am J Obstet Gynecol. 2012;206:448.e8–15.
58. Li F, Miao ZN, Xu YY, Zheng SY, Qin MD, Gu YZ, Zhang XG. Transplantation of human amniotic mesenchymal stem cells in the treatment of focal cerebral ischemia. Mol Med Rep. 2012;6:625–30.
59. Tao J, Ji F, Liu B, Wang F, Dong F, Zhu Y. Improvement of deficits by transplantation of lentiviral vector-modified human amniotic mesenchymal cells after cerebral ischemia in rats. Brain Res. 2012;1448:1–10.
60. Rehni AK, Singh N, Jaggi AS, Singh M. Amniotic fluid derived stem cells ameliorate focal cerebral ischaemia-reperfusion injury induced behavioural deficits in mice. Behav Brain Res. 2007;183:95–100.
61. Liu T, Wu J, Huang Q, Hou Y, Jiang Z, Zang S, Guo L. Human amniotic epithelial cells ameliorate behavioral dysfunction and reduce infarct size in the rat middle cerebral artery occlusion model. Shock. 2008;29:603–11.
62. Dong W, Chen H, Yang X, Guo L, Hui G. Treatment of intracerebral haemorrhage in rats with intraventricular transplantation of human amniotic epithelial cells. Cell Biol Int. 2010;34:573–7.
63. Broughton BRS, Rebecca Lim R, Thiruma V, Arumugam TV, Grant R, Drummond GR, Euan M, Wallace EM, Sobey CG. Post-stroke inflammation and the potential efficacy of novel stem cell therapies: focus on amnion epithelial cells. Front Cell Neurosci. 2013;6:66.
64. Tajiri N, Acosta S, Glover LE, Bickford PC, Jacotte Simancas A, Yasuhara T, Date I, Solomita MA, Antonucci I, Stuppia L, Kaneko Y, Borlongan CV. Intravenous grafts of amniotic fluid-derived stem cells induce endogenous cell proliferation and attenuate behavioral deficits in ischemic stroke rats. PLoS One. 2012;7, e43779.
65. Yawno T, Schuilwerve J, Moss TJ, Vosdoganes P, Westover AJ, Afandi E, Jenkin G, Wallace EM, Miller SL. Human amnion epithelial cells reduce fetal brain injury in response to intrauterine inflammation. Dev Neurosci. 2013;35:272–82.
66. Kaneko Y, Hayashi T, Yu S, Tajiri N, Bae EC, Solomita MA, Chheda SH, Weinbren NL, Parolini O, Borlongan CV. Human amniotic epithelial cells express melatonin receptor MT1, but not melatonin receptor MT2: a new perspective to neuroprotection. J Pineal Res. 2011;50:272–80.
67. McDonald C, Siatskas C, Bernard CCA. The emergence of amnion epithelial stem cells for the treatment of multiple sclerosis. Inflamm Regen. 2011;31:256–71.
68. Kakishita K, Elwan MA, Nakao N, Itakura T, Sakuragawa N. Human amniotic epithelial cells produce dopamine and survive after implantation into the striatum of a rat model of Parkinson's disease: a potential source of donor for transplantation therapy. Exp Neurol. 2000;165:27–34.
69. Kakishita K, Nakao N, Sakuragawa N, Itakura T. Implantation of human amniotic epithelial cells prevents the degeneration of nigral dopamine neurons in rats with 6-hydroxydopamine lesions. Brain Res. 2003;980:48–56.
70. Sun H, Hou Z, Yang H, Meng M, Li P, Zou Q, Yang L, Chen Y, Chai H, Zhong H, Yang ZZ, Zhao J, Lai L, Jiang X, Xiao Z. Multiple systemic transplantations of human amniotic mesenchymal stem cells exert therapeutic effects in an ALS mouse model. Cell Tissue Res. 2014;357:571–82.
71. Sankar V, Muthusamy R. Role of human amniotic epithelial cell transplantation in spinal cord injury repair research. Neuroscience. 2033;118:11–7.
72. Wu ZY, Hui GZ. Materials for neuro-transplantation and the amnion. Chin Med J (Engl). 2006;119:1323–6.
73. Li Y, Guo L, Ahn HS, Kim MH, Kim SW. Amniotic mesenchymal stem cells display neurovascular tropism and aid in the recovery of injured peripheral nerves. J Cell Mol Med. 2014;18:1028–34.

Chapter 9
Amniotic Fluid Stem Cell Populations

Julie Di Bernardo and Shaun M. Kunisaki

Introduction

Amniotic fluid is the mildly acidic (pH 7.0) liquid surrounding the fetus that is essential for normal development in utero. It is entirely contained within the amniotic sac and protects the developing baby by cushioning against minor trauma to the maternal abdomen, allowing for easier fetal movement, and promoting musculoskeletal development. Amniotic fluid is "inhaled" and "exhaled" during fetal breathing movements as part of normal lung morphogenesis. Starting at the eighth week of gestation, swallowed amniotic fluid also creates urine that dramatically increases amniotic fluid volume over time, reaching a peak of 800 mL at 28 weeks gestation [7].

Amniotic fluid is not just fetal urine. It contains a heterogeneous population of cells as well as unique electrolyte, protein, carbohydrate, and lipid profiles useful for many diagnostic purposes, such as measuring fetal lung maturity late in gestation. The cells within amniotic fluid are of fetal origin. For this reason, ultrasound-guided amniocentesis using a fine needle and syringe has been widely adopted to evaluate the fetus for possible chromosomal anomalies. The procedure is associated with a low maternal/fetal morbidity rate in expert hands (i.e., less than 0.5 % complication rate). In the past decade, the therapeutic potential of amniotic fluid has been explored by a number of investigators pointing out how amniotic fluid is an abundant ethical autologous source of cells for regenerative medicine.

J. Di Bernardo, Ph.D. • S.M. Kunisaki, M.D., M.Sc. (✉)
Department of Surgery, C.S. Mott Children's Hospital and Von Voigtlander Women's Hospital, University of Michigan, 1540 E. Hospital Drive, SPC 4211, Ann Arbor, MI 48109, USA
e-mail: juliedib@med.umich.edu; shaunkun@umich.edu

D.O. Fauza, M. Bani (eds.), *Fetal Stem Cells in Regenerative Medicine*, Stem Cell Biology and Regenerative Medicine,
DOI 10.1007/978-1-4939-3483-6_9

Amniotic Fluid: A Dynamic Milieu

The amniotic cavity first appears between 7 and 8 days after fertilization. Shortly thereafter, amniotic fluid originates mostly from osmotic exchange with maternal plasma across the fetal membranes. Between 8 and 11 weeks gestation, the fetus starts to produce urine that rapidly becomes the predominant contributor to amniotic fluid, together with cells shed from the surrounding amniotic membranes and the fetus itself. Throughout gestation the cellular component changes dynamically in response to normal development of the fetus: in particular, cells from the fetus skin, urinary, gastrointestinal, respiratory tracts, and extra-embryonic membranes can be found [77]. This leads to an inherently heterogeneous cell population that expresses markers derived from all three germ layers, making difficult to establish its ontology [19, 24, 25, 46, 59]. Gestational age and the fetus health status represent additional sources of variability of the amniotic fluid cellular content. For example, in the specific case of open neural tube defects, neural cells presumably originating from exposed neural elements may also be found and have been used as a potential diagnostic tool in experimental models [52, 53]. In regards to extrinsic sources of cell heterogeneity, technical procedures commonly used to isolate and expand amniotic fluid cells *in vitro* can also introduce bias towards particular subpopulations of cells [12, 61, 74].

Characterization of Amniotic Fluid Cell Populations

Several pioneering studies from the late seventies attempted to classify the different amniocyte populations observed *in vitro*. Based on their morphological, growth, and biochemical characteristics, amniotic fluid cells that attach and form colonies under routine culture conditions can be classified as E-type (epithelioid), F-type (fibroblastic), and AF-type (amniotic fluid-specific). E- and AF-types coexist in the early stages of cultivation, whereas F-type appears later and not in all samples. F-type cells also replicate at a high rate and have a characteristic elongated shape. E-type cells tend to reach senescence earlier and dramatically decrease in number after few passages.

The ontology of these different cell populations has never been carefully studied. One hypothesis, based on hormone production, is that E-type cells derive from fetal skin and urine, AF-type from fetal membranes and trophoblast, and F-type from fibrous connective tissue and dermal fibroblasts [24, 25]. More recently, investigators have described two distinct populations within amniotic fluid (Fig. 9.1): one being round-shaped and slow growing (E-type, Fig. 9.1a), and the other stromal-like and spindle-shaped with a higher replication rate (F- or AF-type, Fig. 9.1b–d) [1, 5, 6, 65, 81, 85, 86].

In biochemical terms, the wide heterogeneity of stromal-like amniocytes is reflected by global gene expression [45, 85] and proteomic profiles [64, 76]. In a recent paper, Maguire and coworkers described how amniocytes can be considered transitioning cell types that co-express markers for both undifferentiated and dif-

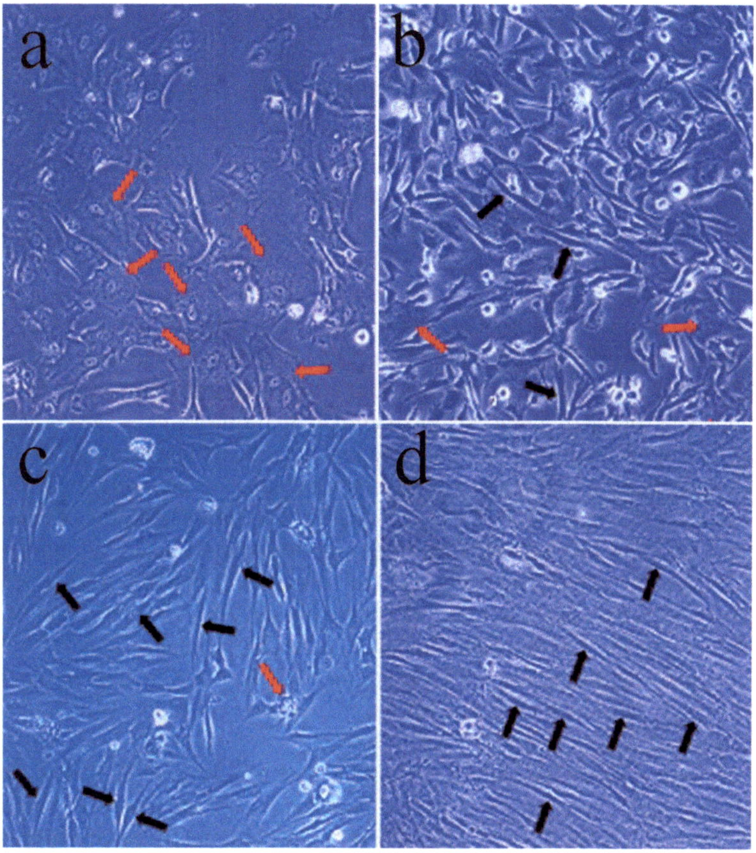

Fig. 9.1 Amniotic fluid-derived cell morphology *in vitro* is heterogeneous: some populations look mainly epithelial-like (**a**, *red arrows*), some fibroblast-like (**b–d**, *black arrows*). As a result of c-kit+ sorting, AFSCs have been isolated from population c and show a more homogeneous fibroblastoid shape. 200× magnification. Adapted from Bai et al. [2]

ferentiated derivatives, since they exist in an intermediate state between pluripotency and lineage-specific restriction. Amniocytes express many of the same genes as undifferentiated pluripotent cells, but core pluripotency genes are expressed at much lower levels and tend to further decrease with gestational age and time in culture. Other investigators have similarly shown that unselected amniotic cells express genes and proteins specific to pluripotent, committed progenitors and fully differentiated cells [5, 60, 76]. Because of their embryonic origin and nascent epigenetic background, the use of various amniotic fluid-derived cell populations has been fertile ground for iPSC (Induced Pluripotent Stem Cell) generation in recent years [28]. Several reports have shown that amniotic fluid cells can be rapidly and more efficiently reprogrammed into iPSCs than adult somatic cells [21, 39, 42], as well as reprogrammed directly into particular cell lineages [23].

Amniotic Fluid Progenitors and Stem Cells

Amniotic Fluid Mesenchymal Stromal Cells (AF-MSCs)

The availability, lack of ethical issues, and simplicity of isolation and cultivation of stromal-like amniocytes for potentially useful for therapeutic regenerative medicine purposes was formally proposed by Fauza and colleagues [31, 32]. However, the first published report highlighting the multi/pluripotent properties of amniotic fluid-derived cells was by Prusa and coworkers in July 2003 [60]. These investigators showed that second-trimester human amniotic fluid contains cells expressing the transcription factor OCT4, a marker of pluripotent stem cells. The following month, In't Anker and colleagues demonstrated that second-trimester human amniotic fluid is an abundant source of amniotic fluid mesenchymal stromal cells (AF-MSCs) with similar phenotype and differentiation potential than the bone marrow-derived counterpart. After these initial reports, a plethora of studies on amniotic fluid stem-like cells followed [5, 30, 61, 74–76]. Despite some overlapping patterns of expression of surface and pluripotency markers, it remains unproven whether all these studies describe the same cells. Discrepancies could also arise from technical issues more than tissue sources, gestational age, or culture conditions. A recent report by Ryan et al. systematically reviewed the literature on OCT4 in fetal or adnexal MSCs to show that most studies report OCT4 messenger RNA or protein expression, but no study provides definitive evidence for the true expression of OCT4A, the isoform associated with pluripotency [66].

Since most amniocenteses are performed between 15 and 22 weeks of pregnancy, the majority of amniotic fluid studies pertain to second trimester amniocytes. Nevertheless, it has been shown that MSCs can be consistently isolated from amniotic fluid later in gestation [6, 38, 83]. Klein and Fauza describe a simple and reproducible method to isolate, expand, and freeze AF-MSCs that has been widely adopted in the field. Isolation of AF-MSCs is achieved through a mechanical separation followed by natural selection by the culture medium [33].

AF-MSCs are considered to be *bona fide* mesenchymal stromal cells based on a number of factors, including [17]:

1. Plastic-adherence of cells when maintained in culture.
2. Expression of CD105, CD73 and CD90, and lack of expression of CD45, CD34, CD14, CD11b, CD79alpha, CD19, and HLA-DR surface molecules [26, 64, 80]. Beside this minimal set of markers, Amniotic Fluid and Bone Marrow Mesenchymal Stomal Cells (BM-MSCs) are positive for other surface markers such as CD44, CD29, CD49e, CD13, and CD166 [64].
3. By definition, AF-MSCs can differentiate *in vitro* in multiple mesodermal lineages such as osteoblasts, adipocytes, and chondroblasts under appropriate conditions [26, 37].

In addition to satisfying these criteria, AF-MSCs are considered immunologically privileged since they express lower levels of HLA-ABC compared to their bone marrow counterpart. In general, when compared to BM-MSCs, AF-MSCs are characterized by a more primitive and self-renewing phenotype given the expression of detectable amounts of OCT4, NANOG, and SSEA4, as well as some telomerase activity [50, 64]. AF-MSCs replicate at a higher rate than BM-MSCs, have longer telomeres, retain a normal karyotype, and do not display tumorigenic potential even after extensive expansion in culture [26, 32, 64, 69, 83].

Moreover, unlike BM-MSCs, some investigators have suggested that AF-MSCs may have greater plasticity and can differentiate into cell types derived from endoderm and ectoderm (Table 9.1). For example, AF-MSCs have been successfully differentiated into hepatocytes-like cells *in vitro* [64, 86] and *in vivo* [40] as well as into neural-like cells *in vitro* [27, 61, 64, 75]. These latter findings have been primarily based on the observation of the morphology and the expression of markers for neuronal progenitors in a subset of cells. It is still under considerable debate if this neurogenic differentiation potential is a result of a small fraction of pluripotent cells or if neuronal progenitors are already present in the cultured amniotic fluid [5, 27, 45]. To date there is no evidence that shows differentiation toward the ectodermal lineage *in vivo*. In several rodent models of neurodegeneration, injected AF-MSCs seem to ameliorate the outcome through a neuroprotective/neurotrophic pathway rather than through an actual differentiation into neurons [11, 51, 62, 72, 73].

One major drawback of studying AF-MSCs is related to the intrinsic nature of this cell population. Even though the isolation method illustrated by Klein and

Table 9.1 Main characteristics of AF-MSCs and AFSCs compared to other pluri/multipotent cell types

	ESC	iPSC	AFSC	AF-MSC	BM-MSC
Plasticity	Pluripotent	Pluripotent	Broadly multipotent	Multipotent	Multipotent
Source	Early stage embryo	Somatic cells	Amniotic fluid	Amniotic fluid	Bone marrow
Feeder cells	Required	Required	Not required	Not required	Not required
Markers	SOX2 SSEA4 OCT4	SOX2 SSEA4 OCT4	c-kit+++ SSEA4++ OCT4++ CD73 CD90 CD105	c-kit+/− SSEA4+/− OCT4+/− CD73 CD90 CD105	CD73 CD90 CD105
Teratoma formation	Yes	Yes	No	No	No
Lifespan *in vitro*	Infinite	Infinite	>250 doublings	30–50 doublings	30–50 doublings
Ethical issues	Yes	No	No	No	No
Clinical trials	Yes	No	No	No	Yes

Adapted from [29]

Fauza is straightforward and adopted by several labs, comparative analyses of the different studies remain difficult because of different protocols [74], medium compositions [12], periods of time in culture [56, 83], and time points in gestation have been used [6, 12, 27]. All these factors bring further variability to an already heterogeneous cell population and complicate the interpretation of findings. Some groups point out how the results from unfractionated AF-MSCs and single colonies derived from the same primary culture can be very different, most of all in terms of differentiation and self-renewal capabilities [27, 63, 65, 75, 81, 85].

On the other hand, from a clinical point of view, AF-MSCs have several ideal features that make them perfect candidates for autologous tissue engineering applications. First, these cells can be easily collected and isolated as a secondary product of routine prenatal testing with no ethical issues. Second, they can be expanded into large numbers in a short amount of time. Third, unlike human embryonic stem cells (ESCs) and iPSCs, AF-MSCs are not tumorigenic. All of these factors suggest that many autologous cell-based therapies, including implantable tissue-engineered grafts, could benefit from harvested AF-MSCs (Fig. 9.2). Depending of the urgency of the clinical problem, AF-MSCs could be manipulated for use before birth or immediately after delivery for surgical reconstruction [35]. So far this model has been successfully applied and validated in large animal models of congenital anomalies involving diaphragmatic, tracheal, cardiac valve, and sternal repairs [20, 34, 36, 37, 67, 71]. In view of prospective clinical trials, the Fauza group has shown that AF-MSCs can be isolated and expanded avoiding the use of animal products [38] and submitted to staged cell manufacturing within a Good Manufacturing Practice facility [70].

Amniotic Fluid-Derived Stem Cells (AFSCs)

In an attempt to work with a more defined and multipotent population, the Atala group isolated cells from amniotic fluid based on positive c-kit (CD117) expression and showed that a clonal population of amniocytes, termed amniotic fluid-derived stem cells (AFSC), could differentiate into six different lineages representing all the three germ layers [14]. Since AFSCs did not form teratomas *in vivo* and do not require feeder layers for propagation in culture, they are considered to be an intermediate stage between ESCs and lineage-restricted adult progenitor cells (Table 9.1).

In contrast to AF-MSCs, AFSCs represent a rare population that accounts for approximately 0.8–1.4 % of all the cells found in the amniotic fluid, appearing as early as the seventh week of gestation, peaking at the twentieth week, and decreasing dramatically to undetectable levels in the third trimester [2, 9, 12]. AFSCs can be sorted by positive selection for the expression of c-kit, a tyrosine-kinase receptor that specifically binds to the ligand stem cell factor (SCF). C-kit can be found not only in human ESCs and primordial germ cells, but also in hematopoietic stem cells and some somatic cells, making it difficult to speculate about the original source of

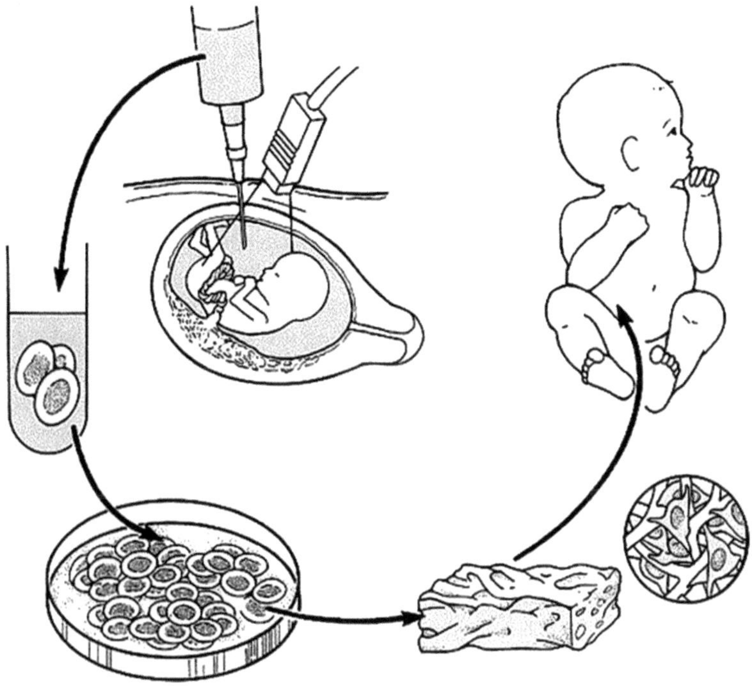

Fig. 9.2 Fetal tissue engineering from amniotic fluid: fetal cells derived from amniocentesis are expanded *in vitro* in parallel to the remainder of gestation. The resultant autologous tissue is ready for implantation in the newborn, if necessary, shortly after birth. From Kaviani et al. [32]

AFSCs in the developing fetus. A detailed protocol using immunomagnetic microspheres has been recently described [57].

Unlike ESCs and iPSCs, AFSCs do not require feeder layers to grow, which is an advantageous feature in terms of facilitating quality control and xeno-free culture conditions for eventual cell transplantation. AFSCs have a fibroblastoid shape (Fig. 9.1d), replicate at a rate comparable to ESCs (doubling time is about 36 h), retain long telomeres, and maintain a normal karyotype for over 250 population doublings [14]. Under specific conditions, AFSCs can differentiate into tissues from all three embryonic germ layers and are able to form embryoid bodies when cultured in suspension [78]. Even though broadly multipotent, they cannot be considered fully pluripotent as ESCs and iPSCs since they do not form teratomas when injected in nude mice. Their phenotype stands in between MSCs and ESCs. Similar to MSCs, they are positive for markers as CD29, CD44, CD73, CD90, CD105, and HLA-ABC. AFSCs stain negative for hematopoietic and endothelial markers like CD14, CD34, CD45, CD133, and CD31. Although about 90 % of the population expresses SSEA4, OCT4, and SOX2 [2, 14], there is no expression of other distinctive ESC markers. According to Moschidou and colleagues, first trimester amniotic fluid provides a more ES-like cell population expressing SSEA3, TRA-1-60, and TRA-1-81 [49] that can be induced towards full pluripotency with valproic acid [48].

Starting from single cell-derived clones under specific differentiation conditions, AFSCs have been shown to express genes specific to six distinct lineages: adipogenic, osteogenic, myogenic, endothelial, neurogenic, and hepatic [14]. In the wake of these promising results, several preclinical studies underscore the potential use of these cells in regenerative medicine. Ditadi and coworkers have demonstrated how AFSCs can differentiate into all three hematopoietic lineages (erythroid, myeloid and lymphoid) both *in vitro* and *in vivo* [16]. AFSCs can differentiate into cardiogenic, endothelial, and myogenic lineages, and have shown cardioprotective effects post myocardial infarction and muscle regeneration in animal models [3, 4, 22, 44, 82]. When injected into mouse embryonic lungs, AFSCs have been shown to integrate into the epithelium, express the early lung marker, TTF1 [10], and are able to migrate to the site of injury where they secrete factors to help wound repair [8]. Although De Coppi and coworkers first reported that c-kit+ cells are able to differentiate towards the neurogenic lineage based on morphology and expression of specific markers [14], evidence of their capacity to produce mature, functional neurons is still lacking. In a rodent model of Parkinson's disease, AFSCs fail to differentiate into dopamine neurons both *in vitro* and *in vivo* [18].

Aside from the differentiation potential that both AFSCs and AF-MSCs display, it is worthwhile to mention that both cell populations have more convincingly shown an important paracrine role as immunomodulators *in vitro* and may therefore be best utilized in immune-mediated disorders as well as in the treatment of graft-versus-host disease [15, 43, 47, 69]. Based in a chick embryo injury model, injected AFSCs act in a paracrine fashion reducing hemorrhage and improving overall survival [58]. Differentiation into renal cells has been studied by Perin and colleagues, proving an additional paracrine immunomodulatory effect in several animal injury models [54, 55, 68]. Recent reports have shown how AFSCs have beneficial effects in terms of survival and repair when injected in a rat model of necrotizing enterocolitis [84]. AFSCs protect and increase regeneration of damaged β-cells in a mouse model of type I diabetes mellitus [79].

A major drawback of working with AFSCs comes from their rarity. Unlike AF-MSCs, which are abundant and easily obtained until term, AFSCs cannot be reliably isolated from late second trimester and third trimester samples [2, 12]. These later time points are typically more clinically relevant since they correspond to the time at which most congenital anomalies are diagnosed by prenatal ultrasound. In general, AFSC isolation methods are also more laborious, time-consuming, and costly. For these reasons, AFSCs may be best suited as an "off-the-shelf" product amendable towards allogeneic cell transplantation applications.

Conclusions

Since their first isolation just over one decade ago, amniotic fluid-derived progenitor and stem cells have been proven to be one of the most interesting cells for regenerative medicine research. They are easily accessible from routine prenatal exams, thereby eliminating ethical issues that other sources like human ESCs raise. Both AF-MSCs and AFSCs have varying differentiation potential, in a spectrum between those of truly pluripotent stem cells and adult stem cells. Finally, they are non-tumorigenic, can duplicate at a high rate without feeder layers, and can be cryopreserved for cell banking and future use.

Despite the exciting findings in the field of amniocyte research, there is much that remains to be explored. We are far from characterizing the full gamut of amniotic fluid-derived cell populations. For example, little is known about the large portion of amniotic fluid cells that are viable yet do not adhere to treated or untreated tissue culture plastic *in vitro*. Furthermore, there are those subpopulations that adhere but do not replicate in routinely used culture conditions [59]. Sporadic reports describe amniotic fluid subpopulations of progenitor cells committed to specific fates and abilities to differentiate under particular conditions, like $CD133^+$ cells that express markers for neuronal progenitor/stem cells [61], $CD44^+$ cells that can be induced to become functional dopaminergic neuronal-like cells [41] or pancreatic beta-cells-like cells [87], and $CD24^+$OB-cadherin$^+$ cells that in turn include several kidney progenitor subpopulations [12, 13]. Moreover, given the challenges of understanding such a heterogeneous cell population, more standardized methods for cell isolation, expansion and characterization would be prudent prior to application in the clinical arena.

Abbreviations

AF-MSC	Amniotic fluid mesenchymal stromal cell
AFSC	Amniotic fluid stem cell
BM-MSC	Bone marrow mesenchymal stromal cell
ESC	Embryonic stem cell
iPSC	Induced pluripotent stem cell

References

1. Arnhold S, Gluer S, Hartmann K, Raabe O, Addicks K, Wenisch S, et al. Amniotic-fluid stem cells: growth dynamics and differentiation potential after a CD-117-based selection procedure. Stem Cells Int. 2011;2011:715341.
2. Bai J, Wang Y, Liu L, Chen J, Yang W, Gao L, et al. Human amniotic fluid-derived c-kit(+) and c-kit (−) stem cells: growth characteristics and some differentiation potential capacities comparison. Cytotechnology. 2012;64(5):577–89.
3. Bollini S, Cheung KK, Riegler J, Dong X, Smart N, Ghionzoli M, et al. Amniotic fluid stem cells are cardioprotective following acute myocardial infarction. Stem Cells Dev. 2011;20(11):1985–94.
4. Bollini S, Pozzobon M, Nobles M, Riegler J, Dong X, Piccoli M, et al. *In vitro* and *in vivo* cardiomyogenic differentiation of amniotic fluid stem cells. Stem Cell Rev. 2011; 7(2):364–80.

5. Bossolasco P, Montemurro T, Cova L, Zangrossi S, Calzarossa C, Buiatiotis S, et al. Molecular and phenotypic characterization of human amniotic fluid cells and their differentiation potential. Cell Res. 2006;16(4):329–36.
6. Bottai D, Cigognini D, Nicora E, Moro M, Grimoldi MG, Adami R, et al. Third trimester amniotic fluid cells with the capacity to develop neural phenotypes and with heterogeneity among sub-populations. Restor Neurol Neurosci. 2012;30(1):55–68.
7. Brace RA, Wolf EJ. Normal amniotic fluid volume changes throughout pregnancy. Am J Obstet Gynecol. 1989;161(2):382–8.
8. Buckley S, Shi W, Carraro G, Sedrakyan S, Da Sacco S, Driscoll BA, et al. The milieu of damaged alveolar epithelial type 2 cells stimulates alveolar wound repair by endogenous and exogenous progenitors. Am J Respir Cell Mol Biol. 2011;45(6):1212–21.
9. Cananzi M, De Coppi P. CD117(+) amniotic fluid stem cells: state of the art and future perspectives. Organogenesis. 2012;8(3):77–88.
10. Carraro G, Perin L, Sedrakyan S, Giuliani S, Tiozzo C, Lee J, et al. Human amniotic fluid stem cells can integrate and differentiate into epithelial lung lineages. Stem Cells. 2008;26(11):2902–11.
11. Cipriani S, Bonini D, Marchina E, Balgkouranidou I, Caimi L, Grassi Zucconi G, et al. Mesenchymal cells from human amniotic fluid survive and migrate after transplantation into adult rat brain. Cell Biol Int. 2007;31(8):845–50.
12. Da Sacco S, Sedrakyan S, Boldrin F, Giuliani S, Parnigotto P, Habibian R, et al. Human amniotic fluid as a potential new source of organ specific precursor cells for future regenerative medicine applications. J Urol. 2010;183(3):1193–200.
13. Da Sacco S, Lemley KV, Sedrakyan S, Zanusso I, Petrosyan A, Peti-Peterdi J, et al. A novel source of cultured podocytes. PLoS One. 2013;8(12), e81812.
14. De Coppi P, Bartsch Jr G, Siddiqui MM, Xu T, Santos CC, Perin L, et al. Isolation of amniotic stem cell lines with potential for therapy. Nat Biotechnol. 2007;25(1):100–6.
15. Di Trapani M, Bassi G, Fontana E, Giacomello L, Pozzobon M, Guillot PV, et al. Immune regulatory properties of CD117 amniotic fluid stem cells vary according to gestational age. Stem Cells Dev. 2014.
16. Ditadi A, de Coppi P, Picone O, Gautreau L, Smati R, Six E, et al. Human and murine amniotic fluid c-Kit+Lin– cells display hematopoietic activity. Blood. 2009;113(17):3953–60.
17. Dominici M, Le Blanc K, Mueller I, Slaper-Cortenbach I, Marini F, Krause D, et al. Minimal criteria for defining multipotent mesenchymal stromal cells. The International Society for Cellular Therapy position statement. Cytotherapy. 2006;8(4):315–7.
18. Donaldson AE, Cai J, Yang M, Iacovitti L. Human amniotic fluid stem cells do not differentiate into dopamine neurons *in vitro* or after transplantation *in vivo*. Stem Cells Dev. 2009;18(7):1003–12.
19. Fauza D. Amniotic fluid and placental stem cells. Best Pract Res Clin Obstet Gynaecol. 2004;18(6):877–91.
20. Fuchs JR, Kaviani A, Oh JT, LaVan D, Udagawa T, Jennings RW, et al. Diaphragmatic reconstruction with autologous tendon engineered from mesenchymal amniocytes. J Pediatr Surg. 2004;39(6):834–8. discussion 8.
21. Galende E, Karakikes I, Edelmann L, Desnick RJ, Kerenyi T, Khoueiry G, et al. Amniotic fluid cells are more efficiently reprogrammed to pluripotency than adult cells. Cell Reprogram. 2010;12(2):117–25.
22. Gekas J, Walther G, Skuk D, Bujold E, Harvey I, Bertrand OF. *In vitro* and *in vivo* study of human amniotic fluid-derived stem cell differentiation into myogenic lineage. Clin Exp Med. 2010;10(1):1–6.
23. Ginsberg M, James D, Ding BS, Nolan D, Geng F, Butler JM, et al. Efficient direct reprogramming of mature amniotic cells into endothelial cells by ETS factors and TGFbeta suppression. Cell. 2012;151(3):559–75.
24. Gosden CM. Amniotic fluid cell types and culture. Br Med Bull. 1983;39(4):348–54.
25. Hoehn H, Salk D. Morphological and biochemical heterogeneity of amniotic-fluid cells in culture. Method Cell Biol. 1982;26:11–34.

26. In 't Anker PS, Scherjon SA, Kleijburg-van der Keur C, Noort WA, Claas FH, Willemze R, et al. Amniotic fluid as a novel source of mesenchymal stem cells for therapeutic transplantation. Blood. 2003;102(4):1548–9.
27. Jezierski A, Gruslin A, Tremblay R, Ly D, Smith C, Turksen K, et al. Probing stemness and neural commitment in human amniotic fluid cells. Stem Cell Rev. 2010;6(2):199–214.
28. Jiang G, Di Bernardo J, Maiden MM, Villa-Diaz LG, Mabrouk OS, Krebsbach PH, et al. Human transgene-free amniotic-fluid-derived induced pluripotent stem cells for autologous cell therapy. Stem Cells Dev. 2014;23(21):2613–25.
29. Joo S, Ko IK, Atala A, Yoo JJ, Lee SJ. Amniotic fluid-derived stem cells in regenerative medicine research. Arch Pharm Res. 2012;35(2):271–80.
30. Karlmark KR, Freilinger A, Marton E, Rosner M, Lubec G, Hengstschlager M. Activation of ectopic Oct-4 and Rex-1 promoters in human amniotic fluid cells. Int J Mol Med. 2005;16(6):987–92.
31. Kaviani A, Perry TE, Dzakovic A, Jennings RW, Ziegler MM, Fauza DO. The amniotic fluid as a source of cells for fetal tissue engineering. J Pediatr Surg. 2001;36(11):1662–5.
32. Kaviani A, Guleserian K, Perry TE, Jennings RW, Ziegler MM, Fauza DO. Fetal tissue engineering from amniotic fluid. J Am Coll Surg. 2003;196(4):592–7.
33. Klein JD, Fauza DO. Amniotic and placental mesenchymal stem cell isolation and culture. Methods Mol Biol. 2011;698:75–88.
34. Klein JD, Turner CG, Steigman SA, Ahmed A, Zurakowski D, Eriksson E, et al. Amniotic mesenchymal stem cells enhance normal fetal wound healing. Stem Cells Dev. 2011;20(6):969–76.
35. Kunisaki SM. Congenital anomalies: treatment options based on amniotic fluid-derived stem cells. Organogenesis. 2012;8(3):89–95.
36. Kunisaki SM, Freedman DA, Fauza DO. Fetal tracheal reconstruction with cartilaginous grafts engineered from mesenchymal amniocytes. J Pediatr Surg. 2006;41(4):675–82. discussion 82.
37. Kunisaki SM, Jennings RW, Fauza DO. Fetal cartilage engineering from amniotic mesenchymal progenitor cells. Stem Cells Dev. 2006;15(2):245–53.
38. Kunisaki SM, Armant M, Kao GS, Stevenson K, Kim H, Fauza DO. Tissue engineering from human mesenchymal amniocytes: a prelude to clinical trials. J Pediatr Surg. 2007;42(6):974–9. discussion 9-80.
39. Li C, Zhou J, Shi G, Ma Y, Yang Y, Gu J, et al. Pluripotency can be rapidly and efficiently induced in human amniotic fluid-derived cells. Hum Mol Genet. 2009;18(22):4340–9.
40. Liu H, Liu DQ, Li BW, Guan LD, Yan ZF, Li YL, et al. Human amniotic fluid-derived stem cells can differentiate into hepatocyte-like cells *in vitro* and *in vivo*. In Vitro Cell Dev Biol Anim. 2011;47(9):601–8.
41. Liu T, Guo L, Liu Z, Huang Y, Cheng W. Induction of dopaminergic neuronal-like cells from CD44+ human amniotic fluids that are ameliorative to behavioral recovery in a Parkinson's disease rat model. Int J Mol Med. 2011;28(5):745–52.
42. Liu T, Zou G, Gao Y, Zhao X, Wang H, Huang Q, et al. High efficiency of reprogramming CD34(+) cells derived from human amniotic fluid into induced pluripotent stem cells with Oct4. Stem Cells Dev. 2012;21(12):2322–32.
43. Luo C, Jia W, Wang K, Chi F, Gu Y, Yan X, et al. Human amniotic fluid stem cells suppress PBMC proliferation through IDO and IL-10-dependent pathways. Curr Stem Cell Res Ther. 2014;9(1):36–45.
44. Ma X, Zhang S, Zhou J, Chen B, Shang Y, Gao T, et al. Clone-derived human AF-amniotic fluid stem cells are capable of skeletal myogenic differentiation *in vitro* and *in vivo*. J Tissue Eng Regen Med. 2012;6(8):598–613.
45. Maguire CT, Demarest BL, Hill JT, Palmer JD, Brothman AR, Yost HJ, et al. Genome-wide analysis reveals the unique stem cell identity of human amniocytes. PLoS One. 2013;8(1), e53372.
46. Milunsky A. Amniotic fluid cell types and culture. In: Milunsky A, editor. Genetic disorders and the fetus: diagnosis, prevention, and treatment. New York, NY: Plenum; 1979.
47. Moorefield EC, McKee EE, Solchaga L, Orlando G, Yoo JJ, Walker S, et al. Cloned, CD117 selected human amniotic fluid stem cells are capable of modulating the immune response. PLoS One. 2011;6(10), e26535.

48. Moschidou D, Mukherjee S, Blundell MP, Drews K, Jones GN, Abdulrazzak H, et al. Valproic acid confers functional pluripotency to human amniotic fluid stem cells in a transgene-free approach. Mol Ther. 2012;20(10):1953–67.
49. Moschidou D, Drews K, Eddaoudi A, Adjaye J, De Coppi P, Guillot PV. Molecular signature of human amniotic fluid stem cells during fetal development. Curr Stem Cell Res Ther. 2013;8(1):73–81.
50. Mosquera A, Fernandez JL, Campos A, Goyanes VJ, Ramiro-Diaz J, Gosalvez J. Simultaneous decrease of telomere length and telomerase activity with ageing of human amniotic fluid cells. J Med Genet. 1999;36(6):494–6.
51. Pan HC, Yang DY, Chiu YT, Lai SZ, Wang YC, Chang MH, et al. Enhanced regeneration in injured sciatic nerve by human amniotic mesenchymal stem cell. J Clin Neurosci. 2006;13(5):570–5.
52. Pennington EC, Gray FL, Ahmed A, Zurakowski D, Fauza DO. Targeted quantitative amniotic cell profiling: a potential diagnostic tool in the prenatal management of neural tube defects. J Pediatr Surg. 2013;48(6):1205–10.
53. Pennington EC, Rialon KL, Dionigi B, Ahmed A, Zurakowski D, Fauza DO. The impact of gestational age on targeted amniotic cell profiling in experimental neural tube defects. Fetal Diagn Ther. 2014.
54. Perin L, Giuliani S, Jin D, Sedrakyan S, Carraro G, Habibian R, et al. Renal differentiation of amniotic fluid stem cells. Cell Prolif. 2007;40(6):936–48.
55. Perin L, Sedrakyan S, Giuliani S, Da Sacco S, Carraro G, Shiri L, et al. Protective effect of human amniotic fluid stem cells in an immunodeficient mouse model of acute tubular necrosis. PLoS One. 2010;5(2), e9357.
56. Phermthai T, Suksompong S, Tirawanchai N, Issaragrisil S, Julavijitphong S, Wichitwiengrat S, et al. Epigenetic analysis and suitability of amniotic fluid stem cells for research and therapeutic purposes. Stem Cells Dev. 2013;22(9):1319–28.
57. Pozzobon M, Piccoli M, Schiavo AA, Atala A, De Coppi P. Isolation of c-Kit+ human amniotic fluid stem cells from second trimester. Methods Mol Biol. 2013;1035:191–8.
58. Prasongchean W, Bagni M, Calzarossa C, De Coppi P, Ferretti P. Amniotic fluid stem cells increase embryo survival following injury. Stem Cells Dev. 2012;21(5):675–88.
59. Prusa AR, Hengstschlager M. Amniotic fluid cells and human stem cell research: a new connection. Med Sci Monit. 2002;8(11):RA253–7.
60. Prusa AR, Marton E, Rosner M, Bernaschek G, Hengstschlager M. Oct-4-expressing cells in human amniotic fluid: a new source for stem cell research? Hum Reprod. 2003;18(7): 1489–93.
61. Prusa AR, Marton E, Rosner M, Bettelheim D, Lubec G, Pollack A, et al. Neurogenic cells in human amniotic fluid. Am J Obstet Gynecol. 2004;191(1):309–14.
62. Rehni AK, Singh N, Jaggi AS, Singh M. Amniotic fluid derived stem cells ameliorate focal cerebral ischaemia-reperfusion injury induced behavioural deficits in mice. Behav Brain Res. 2007;183(1):95–100.
63. Rosner M, Schipany K, Shanmugasundaram B, Lubec G, Hengstschlager M. Amniotic fluid stem cells: future perspectives. Stem Cells Int. 2012;2012:741810.
64. Roubelakis MG, Pappa KI, Bitsika V, Zagoura D, Vlahou A, Papadaki HA, et al. Molecular and proteomic characterization of human mesenchymal stem cells derived from amniotic fluid: comparison to bone marrow mesenchymal stem cells. Stem Cells Dev. 2007;16(6):931–52.
65. Roubelakis MG, Bitsika V, Zagoura D, Trohatou O, Pappa KI, Makridakis M, et al. *In vitro* and *in vivo* properties of distinct populations of amniotic fluid mesenchymal progenitor cells. J Cell Mol Med. 2011;15(9):1896–913.
66. Ryan JM, Pettit AR, Guillot PV, Chan JK, Fisk NM. Unravelling the pluripotency paradox in fetal and placental mesenchymal stem cells: Oct-4 expression and the case of The Emperor's New Clothes. Stem Cell Rev. 2013;9(4):408–21.
67. Schmidt D, Achermann J, Odermatt B, Breymann C, Mol A, Genoni M, et al. Prenatally fabricated autologous human living heart valves based on amniotic fluid derived progenitor cells as single cell source. Circulation. 2007;116(11 Suppl):I64–70.

68. Sedrakyan S, Da Sacco S, Milanesi A, Shiri L, Petrosyan A, Varimezova R, et al. Injection of amniotic fluid stem cells delays progression of renal fibrosis. J Am Soc Nephrol. 2012;23(4):661–73.
69. Sessarego N, Parodi A, Podesta M, Benvenuto F, Mogni M, Raviolo V, et al. Multipotent mesenchymal stromal cells from amniotic fluid: solid perspectives for clinical application. Haematologica. 2008;93(3):339–46.
70. Steigman SA, Armant M, Bayer-Zwirello L, Kao GS, Silberstein L, Ritz J, et al. Preclinical regulatory validation of a 3-stage amniotic mesenchymal stem cell manufacturing protocol. J Pediatr Surg. 2008;43(6):1164–9.
71. Steigman SA, Ahmed A, Shanti RM, Tuan RS, Valim C, Fauza DO. Sternal repair with bone grafts engineered from amniotic mesenchymal stem cells. J Pediatr Surg. 2009;44(6):1120–6. discussion 6.
72. Tajiri N, Acosta S, Glover LE, Bickford PC, Jacotte Simancas A, Yasuhara T, et al. Intravenous grafts of amniotic fluid-derived stem cells induce endogenous cell proliferation and attenuate behavioral deficits in ischemic stroke rats. PLoS One. 2012;7(8), e43779.
73. Tajiri N, Acosta S, Portillo-Gonzales GS, Aguirre D, Reyes S, Lozano D, et al. Therapeutic outcomes of transplantation of amniotic fluid-derived stem cells in experimental ischemic stroke. Front Cell Neurosci. 2014;8:227.
74. Tsai MS, Lee JL, Chang YJ, Hwang SM. Isolation of human multipotent mesenchymal stem cells from second-trimester amniotic fluid using a novel two-stage culture protocol. Hum Reprod. 2004;19(6):1450–6.
75. Tsai MS, Hwang SM, Tsai YL, Cheng FC, Lee JL, Chang YJ. Clonal amniotic fluid-derived stem cells express characteristics of both mesenchymal and neural stem cells. Biol Reprod. 2006;74(3):545–51.
76. Tsangaris G, Weitzdorfer R, Pollak D, Lubec G, Fountoulakis M. The amniotic fluid cell proteome. Electrophoresis. 2005;26(6):1168–73.
77. Underwood MA, Gilbert WM, Sherman MP. Amniotic fluid: not just fetal urine anymore. J Perinatol. 2005;25(5):341–8.
78. Valli A, Rosner M, Fuchs C, Siegel N, Bishop CE, Dolznig H, et al. Embryoid body formation of human amniotic fluid stem cells depends on mTOR. Oncogene. 2010;29(7):966–77.
79. Villani V, Milanesi A, Sedrakyan S, Da Sacco S, Angelow S, Conconi MT, et al. Amniotic fluid stem cells prevent beta-cell injury. Cytotherapy. 2014;16(1):41–55.
80. Whitsett CF, Priest JH, Priest RE, Marion J. HLA typing of cultured amniotic fluid cells. Am J Clin Pathol. 1983;79(2):186–94.
81. Wilson PG, Devkota L, Payne T, Crisp L, Winter A, Wang Z. Clonal populations of amniotic cells by dilution and direct plating: evidence for hidden diversity. Stem Cells Int. 2012;2012:485950.
82. Yeh YC, Wei HJ, Lee WY, Yu CL, Chang Y, Hsu LW, et al. Cellular cardiomyoplasty with human amniotic fluid stem cells: *in vitro* and *in vivo* studies. Tissue Eng Part A. 2010;16(6):1925–36.
83. You Q, Tong X, Guan Y, Zhang D, Huang M, Zhang Y, et al. The biological characteristics of human third trimester amniotic fluid stem cells. J Int Med Res. 2009;37(1):105–12.
84. Zani A, Cananzi M, Fascetti-Leon F, Lauriti G, Smith VV, Bollini S, et al. Amniotic fluid stem cells improve survival and enhance repair of damaged intestine in necrotising enterocolitis via a COX-2 dependent mechanism. Gut. 2014;63(2):300–9.
85. Zhang S, Geng H, Xie H, Wu Q, Ma X, Zhou J, et al. The heterogeneity of cell subtypes from a primary culture of human amniotic fluid. Cell Mol Biol Lett. 2010;15(3):424–39.
86. Zheng YB, Gao ZL, Xie C, Zhu HP, Peng L, Chen JH, et al. Characterization and hepatogenic differentiation of mesenchymal stem cells from human amniotic fluid and human bone marrow: a comparative study. Cell Biol Int. 2008;32(11):1439–48.
87. Zou G, Liu T, Zhang L, Liu Y, Li M, Du X, et al. Induction of pancreatic beta-cell-like cells from CD44+/CD105+ human amniotic fluids via epigenetic regulation of the pancreatic and duodenal homeobox factor 1 promoter. DNA Cell Biol. 2011;30(9):739–48.

Chapter 10
Amniotic Fluid Stem Cell Culture Methods

Dema Najem, Maria Ribecco-Lutkiewicz, Caroline Sodja, Felipe Mangoni Moretti, Danica Stanimirovic, and Mahmud Bani-Yaghoub

Introduction

Amniotic fluid (AF) is a dynamic, nourishing and protective milieu contributing to fetal development during pregnancy [1, 2]. Ultrasound and other live imaging methods in humans and animal models show that AF can be detected from the very beginning of formation of the gestational sac, even before the embryo is recognizable [3–5]. The production of AF begins with concomitant movement of water from maternal plasma through the fetal membranes based on hydrostatic and osmotic forces [6]. As a result, AF predominantly contains water and electrolytes during the very early stages of fetal development.

Amniotic Fluid: Cellular Composition and Function during Development

At the 8th week of gestation, the fetal kidneys begin to function and urine is present in the amniotic fluid. At approximately the 10th week of gestation, fetal breathing and swallowing begin; however, these functions, as well as urination, do not

D. Najem, M.Sc. • M. Ribecco-Lutkiewicz, Ph.D. • C. Sodja, M.Sc.
Translational Bioscience, Human Health Therapeutics Portfolio, National Research Council Canada, Bldg. M54, 1200 Montreal Road, Ottawa, ON, Canada, K1A 0R6
e-mail: Dema.Najem@nrc-cnrc.gc.ca; Maria.Ribecco-Lutkiewicz@nrc-cnrc.gc.ca; Caroline.Sodja@nrc-cnrc.gc.ca

F.M. Moretti, M.D.
Division of Maternal Fetal Medicine, Department of Obstetrics, Gynecology and Newborn Care, The Ottawa Hospital, 501 Smyth Road, Room 8472, Ottawa, ON, Canada, K1H 8L6
e-mail: fmoretti@toh.on.ca

D.O. Fauza, M. Bani (eds.), *Fetal Stem Cells in Regenerative Medicine*, Stem Cell Biology and Regenerative Medicine,
DOI 10.1007/978-1-4939-3483-6_10

contribute significantly to the volume or content of AF until the second half of pregnancy [7, 8]. By about the 12th week, a number of proteins, carbohydrates, lipids and phospholipids, and urea are present in AF, contributing to fetal growth. Since amniocentesis is used as a prenatal screening tool as early as 12th week of gestation, this time point can be also used to collect AF for cell culture (Fig. 10.1). Together, the removal of AF cells from the uterus followed by their expansion *in vitro* allows detecting chromosomal abnormalities, fetal infections as well as determining the sex of the fetus [9–11]. The AF volume increases significantly from 25 to 400 ml from the 10th to 20th weeks in correlation with fetal size (Fig. 10.1). During this period, there is still a rapid bi-directional diffusion between the fetus and the AF across the non-keratinized fetal skin as well as through the permeable surfaces of the amnion, placenta, and umbilical cord [9–13].

Among the developmental milestones contributing to changes in AF volume and composition is keratinization of fetal skin, a process starting around the 19th week and continuing up to approximately the 25th week of gestation. Following keratinization, AF composition changes by excretion of fetal urine (about 300 ml/kg fetal weight/day and 600–1200 ml/day around term) and the secretion of oral, nasal, tracheal, and pulmonary fluids (60–100 ml/kg fetal weight/day) [7, 8]. Fetal breathing movements contribute to the secretion of lung fluid into the AF, but nearly half of the effluent is swallowed rather than being expelled into the AF, keeping the AF volume changes to less than 5 ml per fetal breath (occurring for 20–30 min/h in late gestation) [8]. On the other hand, AF is mainly removed by fetal swallowing (200–250 ml/kg fetal weight/day), and fluid and solutes are also transferred from the amniotic cavity to the fetal circulation across the amniotic membranes (200–500 ml/day) [8, 14]. The transfer of AF across the fetal membranes and into the maternal circulation within the lining of the uterus is estimated to be only about 10 ml/day at term [8, 12]. AF reaches a volume of about 800 ml by the 28th week, plateaus around term and declines to about 400 ml at 42 weeks of pregnancy [8, 13]. These dynamic changes lead to the contribution of the cells from fetal skin, respiratory, digestive and urinary tracts along with proteins, carbohydrates, amino acids, enzymes, hormones and pigments to the amniotic fluid [15, 16]. Furthermore, the presence of a Y chromosome in AF cells derived from cases in which the pregnant women carried a male child confirms the fetal origin of these cells [15, 16].

The early reports by Serr et al., and Fuchs and Riis led to the successful cultivation of AF cells in culture a few years later by Steele and Breg, as well as Jacobson and Barter [17–20]. AF cells were generally cultured in 15–30 % fetal calf serum or human

D. Stanimirovic, M.D., Ph.D.
Translational Bioscience, Human Health Therapeutics Portfolio, National Research Council Canada, Bldg. M54, 1200 Montreal Road, Ottawa, ON, Canada, K1A 0R6

Department of Cellular and Molecular Medicine, Faculty of Medicine, University of Ottawa, 451 Smyth Rd., Ottawa, ON, Canada, K1H 8M5
e-mail: Danica.Stanimirovic@nrc-cnrc.gc.ca

M. Bani-Yaghoub, Ph.D. (✉)
Department of Cellular and Molecular Medicine,
University of Ottawa, Faculty of Medicine, Ottawa, ON, Canada
e-mail: Mahmud.Bani@nrc-cnrc.gc.ca; mBani@uOttawa.ca

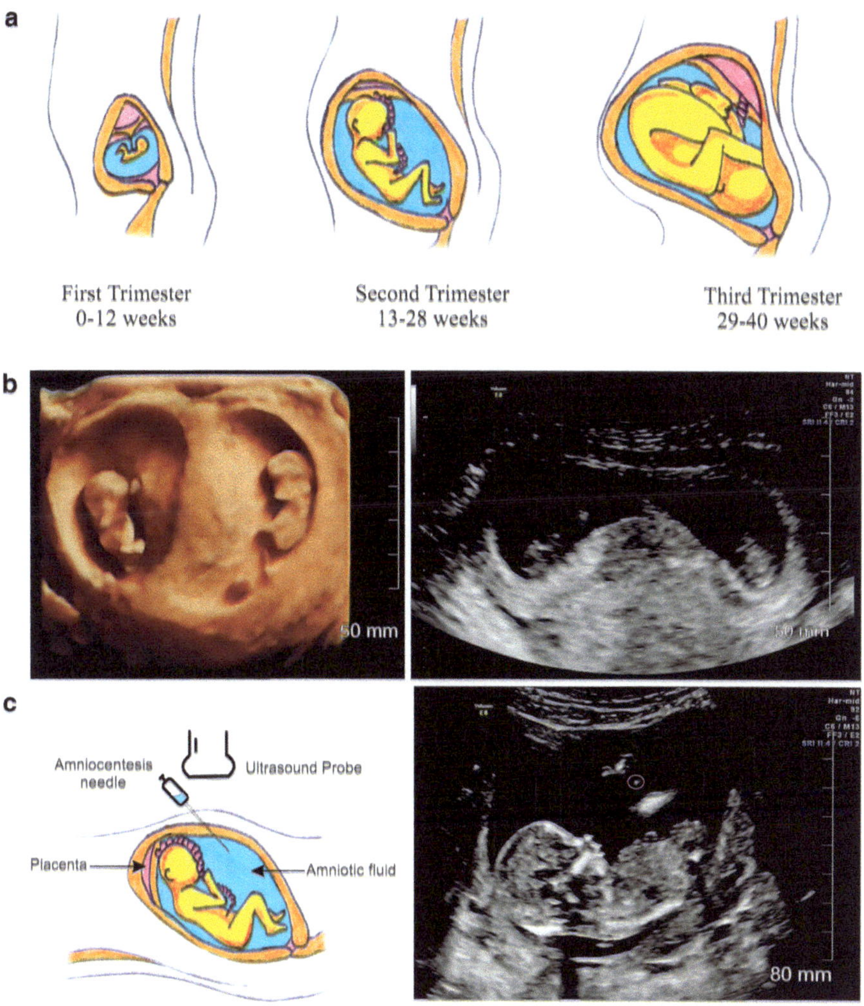

Fig. 10.1 Amniotic fluid can be collected by amniocentesis at different stages of pregnancy. (**a**) An illustration of the three trimesters of pregnancy in human; (**b**) The ultrasound images of twin pregnancy at 9 weeks of gestation. Similar to singleton pregnancies, the amniotic fluid volume in twin pregnancies can be estimated by different techniques, including subjective assessment, amniotic fluid index, single deepest pocket and two diameter pocket; (**c**) Depiction of ultrasound-guided amniocentesis and an ultrasound image of fetal head at 21.5 weeks of gestation. The red circle shows the tip of needle during amniocentesis

AB serum mainly to study chromosome abnormalities [21–23]. In the 1980s, efforts were made to optimize the quality of AF culture medium either by enrichment with growth and cell attachment factors or by using Amniomax and Amniochrome culture media [24, 25]. However, it took a few years before AF culturing techniques found their place in hospitals and affiliated research institutes. In addition, a quick and reliable live assay was required to determine the viability of AF cells in culture (Fig. 10.2).

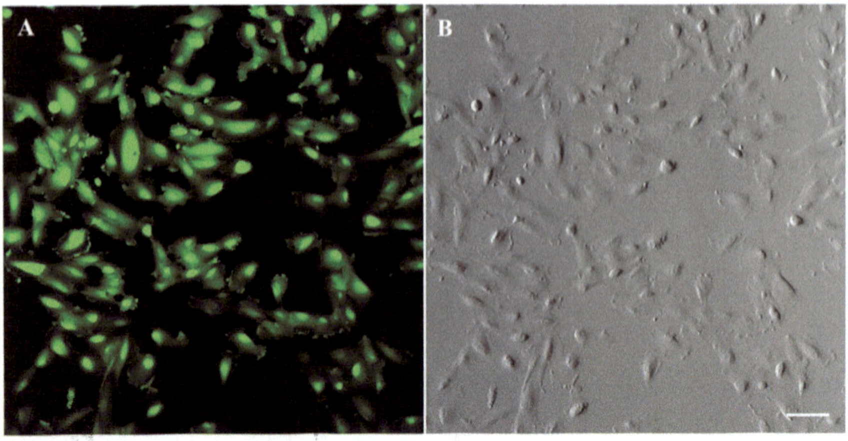

Fig. 10.2 The quality of amniotic fluid cultures can be determined by a simple and reliable cell viability assay. (**a**) Live human amniotic fluid cells have been labeled with carboxyfluorescein diacetate (CFDA, *green*); (**b**) The corresponding phase contrast image. Scale bar: 50 μm

Using conventional culturing methods and morphological classifications, AF cells have been grouped into adherent and non-adherent cells [26]. The adherent cells, which attach and form colonies *in vitro*, have been further classified into three main groups [9, 26]: epithelial-like cells (E-type) and amniotic fluid-specific cells (AF-type), both of which are observed in early AF cultures, and fibroblast-like (F-type) cells, which can be easily identified in cultures after many passages due to their adherence properties [9, 27]. However, the tissue-specific origin of these cells has yet to be determined [9, 26, 28]. AF has been reported to contain heterogeneous cell populations, representing stem cells (AFSCs), progenitors and differentiated cells (Fig. 10.3) [9, 16, 26, 29]. However, the identification, isolation, and purification of each cell type require well-standardized *in vitro* methods to ensure the reproducibility and consistency of data reported by different laboratories. Table 10.1 summarizes the use of AF culture for clinical and research applications throughout time. We have dedicated the following section to the progress made in this area since the discovery of amniotic fluid stem cells more than a decade ago.

Amniotic Fluid Stem Cells (AFSCs)

The potential applications of amniotic fluid cells in regenerative medicine and tissue engineering were first reported by Fauza's group in 2001 [30]. The authors identified a subpopulation of mesenchymal cells in AF and demonstrated that these cells can grow on polyglycolic acid polymer and acellular human dermis [30, 31]. Shortly after this initial report, Hengstschläger's group provided a connection between human AF and stem cells by showing OCT4 expression (Fig. 10.4) in a small subpopulation of AF cells [26, 28]. The work by these groups set the stage for the detection and isolation of different subpopulations of AF cells in culture.

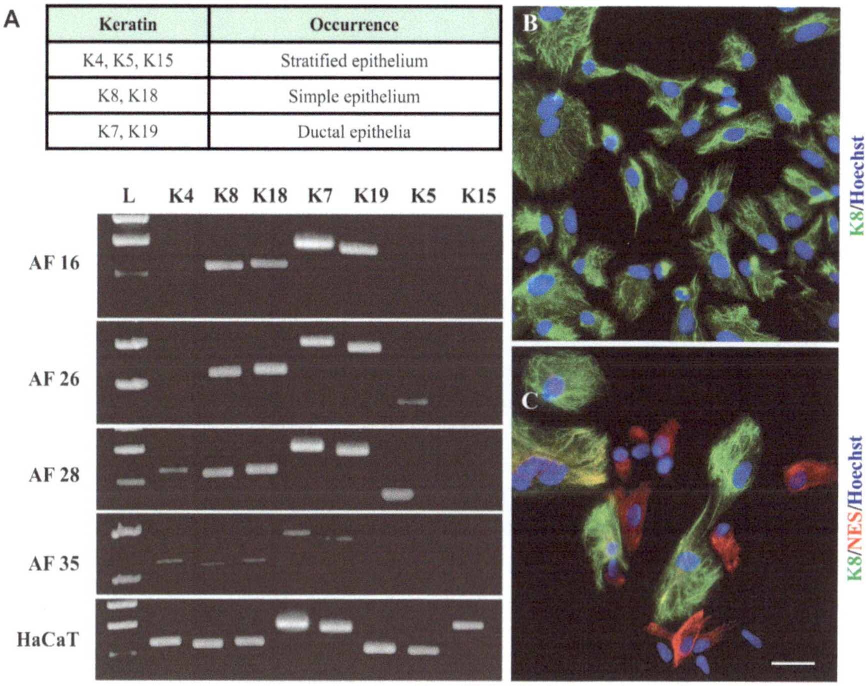

Fig. 10.3 Human amniotic fluid cultures contain heterogeneous cell populations. (**a**) The expression profile of *KERATIN* genes (*K4*, *K5*, *K7*, *K8*, *K15*, *K18* and *K19*) in AF cells at different gestational ages. AF16, AF26, AF28 and AF35 represent the AF samples collected at 16, 26, 28 or 35 weeks of gestation; (**b**, **c**) Double immunostaining with KERATIN 8 (K8, *green*) and NESTIN (NES, *red*) shows that some AF cultures only contain K8 positive cells (**b**), whereas others contain a heterogeneous population of K8 positive (*green*), NES positive (*red*) and cells positive for both markers (*yellow*) (**c**). Hoechst (*blue*) has been used as a counterstain. Scale bar: 50 μm

Table 10.1 Summary of the field progression in culturing and utilization of amniotic fluid cells

1950
1955—Cells from AF were used to diagnose sex before birth for the first time [18, 19].
1956—Antenatal sex determination [18].
1960
1966—Chromosome analysis of AF cells [20].
1967—Cells from AF first cultured in F10+30 % fetal calf serum for 18–25 days [17].
1968—AF cells were utilized to study the development of human fetal enzymes *in vitro* [132].
1969—First study utilizing AF cultured cells to explore prenatal genetic diagnosis [133].
1970
1970—First viability assessment of amniotic cells at different stages of gestation [134], First study to explore different culturing methods for AF cells [135].
1971—First comparison of different culture media for AF cells [136, 137].
1972–2012—Modifications in culturing media and methods for AF cells [138–149].
1974—Identification of F and AF cells type in amniotic fluid cell cultures [150].
1977–2012—Heterogeneity in AF cell cultures was reported [151–161].

(continued)

Table 10.1 (continued)

1980	
1980–Present—Cryopreservation to store AF cells was described [124–126, 131, 162–165].	
1990	
1990—Factors which affect growth of AF cells were further investigated [166].	
1990—Evaluation of surface bound HLA antigen of AF cells [167].	
1991—AF cells from different weeks of gestation are beginning to be compared [168].	
1993—Identification of hematopoietic progenitor cells in AF [27].	
1995—Examination of factors related to AF cell culture failure [146, 169].	
1999—Analysis of telomere length and telomerase in aging AF cells [170].	
2000	
2001—AF cells recognized as a source for potential fetal tissue engineering [30].	
2003—Stem cells marker expression in AF cells were discovered [28].	
2004—Early isolation methods for multipotent MSCs from AF [35].	
2004–Present—Regeneration and tissue engineering constructs from AFSCs [9, 66, 91, 171–208].	
2005—Proteomic analysis of AF cells [209].	
2006–Present—Isolation, characterization and differentiation methods of progenitor and stem cells in AF [54, 56, 71, 75, 80, 81, 99, 105, 148–172, 210, 211].	
• Adipogenic [212]	• Endothelial [119, 231, 232]
• Neural [111, 112, 115, 213–220]	• Cardiac [233–238]
• Renal [95]	• Osteogenic [239–251]
• Chondrogenic [117, 221, 222]	• Retinal [252]
• Lung [223]	• Smooth muscle [253]
• Pancreatic [224–230]	• Epithelial [254]
	• Urothelial [255]
2007—First comparison of MSCs from AF to MSCs from bone marrow [85].	
2007–Present—AFSCs as a new tool to study human genetic diseases [104, 106, 256, 257].	
2007–Present—The potential of AFSCs for cell-based therapies [96, 258–278].	
2007—First isolation of AFSCs based on CD117 selection method [62].	
2008—Introduction of a 3-stage amniotic mesenchymal stem cell manufacturing protocol in accordance with FDA mandates for clinical use [211].	
2008—Investigation of culturing media to isolate and differentiate AFSCs [279].	
2008–Present—Using reprogramming techniques to generate AF-iPSCs [70, 97, 98, 100, 103, 280–283].	
2008—International non-profit alliance to create a repository of stem cells from surplus cells founded [284].	
2009—Microfluid devices for separation of MSCs from amniotic fluid [285, 286].	
2009—Development of cloned embryos from AFSCs [287].	
2010	
2012—Compilation of markers in AFSCs [47].	
2013—Investigating the amniotic fluid as a source of autologous stem cells in the context of disease [275].	
2013—Culture media effects on expression of pluripotency markers in AF cells [288].	
2013—Use of AF-MSCs as a feeder layer for ESCs [289].	
2013—Proteome differences between male and female fetal cells in AF [290].	
2013—AF allograft used to treat 20 foot and ankle wounds in humans [291].	

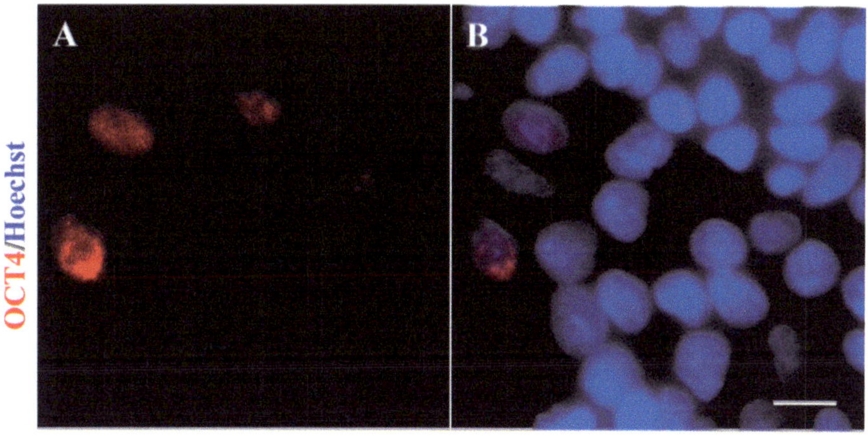

Fig. 10.4 OCT4 is used as a reliable marker to identify human amniotic fluid stem cells. (**a**) Immunostaining shows nuclear localization of OCT4 protein (*red*). (**b**) Hoechst (*blue*) has been used as a counterstain. Scale bar: 20 μm

In 2003–2004, our group made several attempts to isolate neural stem cells from AF samples based on NESTIN promoter activity and immunostaining. However, these attempts were not successful, partly because the majority of NESTIN positive subpopulations did not differentiate into neurons. Using our expertise in genomics and antibody production, we made an antibody against SOX2 and tested the expression of this protein in AF cells along with mouse brain sections and human NT2/D1 cell line. Only a very small number of AF cells were positive for SOX2 and those cultured in neuronal differentiation media did not produce substantial number of neurons for further experiments (unpublished observations). Fortunately, other laboratories were able to isolate a different subpopulation of AF cells. The term AF mesenchymal stem cells (MSCs) was introduced by In't Anker et al. based on relevant antigen expression and differentiation potential *in vitro* that was similar to that of bone marrow derived MSCs [29]. These observations have been further confirmed by a number of laboratories based on a set of standards to define MSCs proposed by the International Society for Cellular Therapy (ISCT) in 2006 [32, 33]. According to these standards (Table 10.2), AF-MSCs share several important features observed in other mesenchymal stem cells such as plastic adherence properties *in vitro*, fibroblast-like morphology and differentiation into osteogenic, adipogenic and chondrogenic cell lineages [1, 34]. AF-MSCs express cell surface markers such as CD44, CD73, CD90, CD117, and CD105, but lack the expression of CD11b, CD14, CD19, CD34, CD45, CD79a and HLA-DR cell surface markers [32]. AF-MSCs can be expanded, using a two-step culture protocol, while maintaining their differentiation capacity [35]. Using cell sorting methods, it was further determined that approximately 1 % of the AF cells isolated during the second trimester express C-KIT (CD117), a protein tyrosine kinase receptor with binding capability for the ligand stem cell factor (SCF) as well as important roles in cell survival, proliferation and differentiation [36]. Other cell surface antigens such

Table 10.2 ISCT standards used to define MSCs

Mesenchymal stem cells standard characteristics	
Adherence	Adherence to plastic *in vitro*
Morphology	Fibroblast-like morphology
Differentiation potential	Osteogenic, Adipogenic and Chondrogenic differentiation
Surface markers	
CD73	+
CD90	+
CD105	+
CD44	+
CD117	+
CD34	–
CD11b	–
CD14	–
CD19	–
CD45	–
CD79a	–
HLA-DR (MHC class II)	–

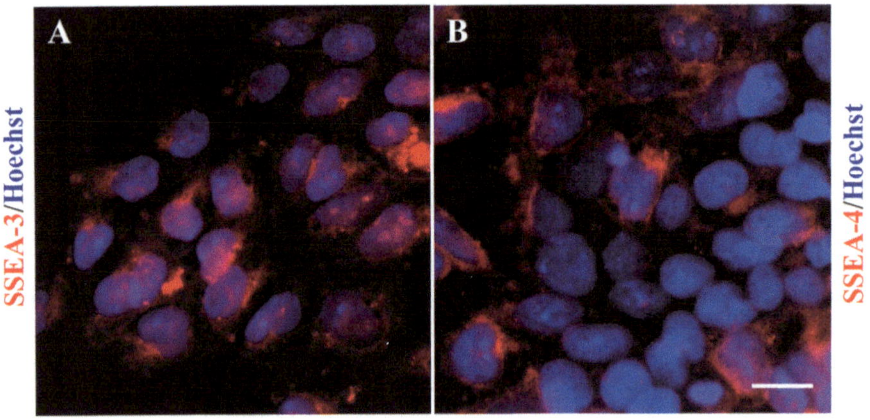

Fig. 10.5 Stage specific embryonic antigens (SSEAs) are commonly used as cell surface markers in human amniotic fluid cultures. Immunostaining with SSEA-3 (**a**) and SSEA-4 (**b**) show the presence of these proteins on cell surface. Hoechst has been used for nuclear staining. Scale bar: 20 μm

as the stage specific embryonic antigens-3 and -4 (SSEA-3, SSEA-4; Fig. 10.5) and more recently TRA-1-60 and TRA-1-81 have been extensively used to identify stem cells in amniotic fluid cultures (Table 10.3) [37–44]. Clonal expansion, protein expression profiling and comparative analyses of AFSCs with other stem cells have provided a better understanding of the morphological, biochemical and functional characteristics of AFSCs [44–50]. More detailed analyses focusing on

Table 10.3 Suggested profile of specific markers expression in AFSCs [61, 62, 65, 182, 292–295]

Features		
Cell phenotype	Mesenchymal/Amniotic fluid stem cells	
Potency	Pluripotent/Multipotent	
Doubling time/Population doubling	25–38 h (second trimester MSCs), 36 h (AFSCs), more than 250 doublings (AFSCs)	
Feeder layer/Matrigel requirement	–	
Teratoma formation in immunocompromised mice	–	
Commonly used markers		
Transcription factors	*Cell surface antigens*	*Enzymes and proteins*
OCT4	SSEA-3	Telomerase
NANOG	SSEA-4	KLF4
SOX2	TRA-1-60	
	TRA-1-81	
	C-KIT (CD117)	

transcriptomics, proteomics, and metabolomics of AFSCs have further proven the suitability of these cells for research in the field of regenerative medicine [44, 47, 51]. Interestingly, regardless of the cell culture methods used (see the following sections), OCT4 (the earliest stem cell marker detected in AF cells) still remains as a key determinant of cell fate in AFSCs (Fig. 10.4) [16, 52–59]. However, the lack of standardized methods remains to be addressed to further resolve the variability reported among amniotic fluid cell cultures from different laboratories.

Isolation and Culture of AFSCs

To optimize AFSC isolation and culturing techniques, several protocols have been discussed and their merits and caveats have been compared and summarized.

Early Methods for Isolation

Following the pioneering work by Fauza and Hengstschläger [26, 28, 30, 31], a double step cell culture protocol was developed by Tsai et al. to isolate and expand MSCs from human amniotic fluid. In the first step, AFCs were plated in Chang medium.[1] Non-adhering AFCs were collected, centrifuged, and plated in the

[1] Chang medium was developed for the primary culture of human AF cells for use in karyotyping and other antenatal genetic testing. The formula consists of a two part system: a liquid basal medium and supplement. The medium contains salts (7900 mg/L), dextrose (1400 mg/L), amino acids (1000 mg/L), polypeptides (66 mg/L), vitamins (22 mg/L), deoxyribonucleosides (21 mg/L),

medium supplemented with 20 % FBS, and fibroblast growth factor 2 (FGF2) [35]. The second step (AF-MSC culturing) included the collection of non-adhering amniotic fluid cells in the supernatant. The authors reported that cells with MSC phenotype appeared in culture 5 days after plating the non-adhering AFCs. These cells were grown to 90 % confluence within 3–7 days of culture, while maintaining a normal karyotype for at least 12 passages. In addition, they were positive for SH2, SH3, SH4, CD29, CD44, and HLA-ABC, low positive for CD99 and CD105 and negative for CD10, CD11b, CD14, CD34, CD117, EMA and HLA-DR, DP, DQ. Subcultures of the isolated human AF-MSCs were able to differentiate into mesoderm (adipocytes, osteocytes) and ectoderm (neuronal cells) lineages under inducible conditions [35]. Collectively, these data confirmed the MSC characteristics of these cells, based on the ISCT standards (Table 10.3).

Using this two-stage culture method, Tsai et al. further established a purification protocol by constructing clonal human AFSCs from single AF stem cells [60]. Single-cell derived AFSC clones were obtained by plating cells onto 96 well plates through limiting dilution in the medium supplemented with 20 % FBS and FGF2. The authors demonstrated NANOG and OCT4 expression in the expanded single cell-derived human AFSCs. Also, under appropriate conditions, the clonal cells maintained the capacity to differentiate into multiple cell types such as adipocytes, osteocytes, neurons and glial cells. Clonal cells differentiated into NES, TUBB3, NEFH, NEUNA60, GALC, and GFAP positive cells through neural induction. HPLC analysis showed evidence of dopamine release in the extract of dopaminergic induced clonal AFSCs. These results suggested that the AF contains subpopulation(s) of stem cells with potential to commit to mesenchymal and neural lineages, depending on culture conditions [60].

Kim et al. developed another technique for isolating human AF-MSCs, which involves prolonged culture of human AFSCs in the medium supplemented with EGF and 10 % FBS until a stem cell population with a homogenous morphology is obtained [61]. These cells were capable of differentiating into adipocytes, osteocytes, chondrocytes and neuronal cells; however, they became senescent after 27 passages, while undergoing 66 population doublings. *OCT4* gene expression was observed up to passage 19 with decreased telomerase activity by the 21st passage [61], supporting the notion that the stemness properties of AFSCs need to be evaluated over long term cultures.

Using Specific Markers to Isolate AFSCs

In addition to using OCT4 and other transcription factors to detect stem cells in amniotic fluid cell cultures, several attempts have been made to identify an applicable cell surface marker to select AFSCs and follow their fate both *in vitro* and *in vivo*. In particular, De Coppi et al. reported the suitability of C-KIT (CD117) to

---

ribonucleosides (20 mg/L), sodium pyruvate (110 mg/L), newborn calf serum (6 % v/v), fetal bovine serum (6 % v/v) as well as steroid hormones (0.0013 mg/L), hormones and trace element (0.0025 mg/L) and other components (8 mg/L).

sort AFSCs [62]. The cells were simply selected through incubation with a CD117 antibody on microbeads and cultured in the medium containing 15 % FBS, 1 % glutamine and supplemented with 18 % Chang B and 2 % Chang C media [62]. CD117 positive AFSCs, representing approximately 1 % of the population, were then expanded by sub-culturing at 70 % of confluence and clonal cell lines were generated by the limiting dilution methods [62]. The established clones expressed OCT4, maintained a constant telomere length after expansion to 250 population doublings and displayed a spindle-shaped fibroblast-like morphology similar to that of other MSC populations, as also reported by other laboratories [26, 62–68]. The presence of several other markers, including SSEA-3, SSEA-4, OCT4, NANOG, TRA-1-60, and TRA-1-81, has been confirmed in CD117 positive AFSCs [48, 69–72]. Furthermore, the expression of CD73, CD90, CD105, MHC-I and lack of MHC-II, CD40, CD80, CD86 in these cells suggest that they may have immunomodulatory function(s) [73, 74]. The immunological properties of AFSCs from different gestational ages were also evaluated in co-cultures with T, B and natural killer (NK) cells [75, 76]. As expected, AFSCs from the first trimester showed lower expression of HLA class-I molecules and NK-activating ligands than those obtained from the second and third trimesters [75]. Moreover, first trimester AFSCs significantly inhibited T and NK cell proliferation, while second and third trimester AFSCs were less efficient, suggestive of differences among the samples obtained at different gestational ages [75, 76].

Efforts to Standardize AFSC Culturing Methods

In an attempt to develop a standard isolation and culturing method, several CD117 positive AFSC clones (i.e., Q1, CB3 and CD117/2) were established [72]. These clones maintained normal karyotypes and did not show spontaneous differentiation or apoptosis [77]. The authors used complementary recipes (medium plus 15 % FBS, 1 % glutamine, 18 % Chang B and 2 % Chang C or medium containing 15 % FBS, 2-mercaptoethanol and glutamine) to cultivate these cell lines [78, 79]. The cells showed optimal proliferation in both media and formed embryoid bodies (EB) [77]. Furthermore, they maintained consistent morphology, doubling time, apoptosis rate, cell-cycle distribution and marker expression up to 25 passages [77, 80]. However, significant fluctuations were observed by proteomics over several passages (i.e., 5, 7, 11 and 25) for signaling, antioxidant, proteasomal and cytoskeletal proteins [80]. These observations warrant further standardization of AFSC culturing methods to advance the applications of these cells for drug screening and transplantation studies [68].

Table 10.4 CD117 positive AFSCs selection, application and differentiation

Isolation methods	Using a polyclonal CD117 antibody specific for amino acid 23–322, followed by magnetic microsphere selection and sorting through MACS
	Microbead conjugated with CD117 monoclonal antibody, followed by MACS
Culture media	Medium containing 15 % FBS, 1 % glutamine, 18 % Chang B, and 2 % Chang C
	Medium containing 2-mercaptoethanol, glutamine, and 15 % FBS
Characteristics	
Doubling time	~36 h
Morphology	Fibroblast morphology, similar to MSCs
Markers expressed by CD117 positive AFSCs	
CD73	+
CD90	+
CD105	+
MHC-I	+
CD40	–
CD80	–
CD86	–
MHC-II	–
Reported pluripotent marker expression	OCT4
	SSEA-4
	TRA-1-60
Advantages	*Other considerations*
Phenotypically and genetically stable up to passage 25	Reported cell volume increase
Self-renewal capabilities	Reported involvement in cell signaling, antioxidant, proteosomal, cytoskeleton, connective tissue and chaperone networks
No need for feeder layers, non-tumorigenic	CD117 selection may exclude other AFSCs
Best suited for osteogenic, chondrogenic, and adipogenic differentiation	CD117 negative AFSCs offer better neuronal differentiation

CD117 Positive Versus CD117 Negative AFSCs

Although, CD117 has been used to isolate AFSCs by many laboratories since the initial report by De Coppi et al., the suitability of this antibody as a *bona fide* stem cell marker in amniotic fluid cultures has been re-visited (Table 10.4). Arnhold et al. compared the growth dynamics and differentiation potential of human CD117 positive AFSCs to the CD117 negative cell fraction and to the total cell population [81]. All three cell populations showed similar growth characteristics with an average doubling time of 30–40 h [81]. The CD117 positive cell fraction revealed fibroblast morphology similar to mesenchymal stem cells, while the CD117 negative population seemed to have an epithelial morphology. Next, the ability of the three populations to differentiate towards osteogenic, chondrogenic, adipogenic and neuronal lineages was investigated. Osteogenic differentiation was compared between the fractions, using alkaline phosphatase (ALP) staining. CD117 positive cells were intensely labelled, unlike the CD117 negative and total cell populations. Similarly, Alcian blue staining, indicative of chondrogenic differentiation, and Oil Red O staining, indicative of adipogenic differentiation, were both present at high levels in CD117 positive cells, but not in the other two populations. Evaluation of HNK-1 expression revealed a significantly stronger level in the negative cell fraction than those of the total cell and CD117 positive cell populations. These results suggested that CD117 positive AFSCs may offer better adipogenic, osteogenic and chondrogenic differentiation, whereas the CD117 negative AFSCs may offer higher capacities to differentiate into other cell lineages [81]. This notion has been further emphasized in a paper by Rafii's group [82], in which CD117 negative amniotic cells were efficiently reprogrammed into vascular endothelial cells without transitioning through a pluripotent state. Furthermore, De Coppi's group have recently reported a role for CD117 to "enrich" for the stem cell and progenitor sub-population(s) in amniotic membrane cultures [83].

Isolation of AFSCs Based on Morphology

Based on morphological features, combined with antigen expression levels and functional properties, two different types of human AF-MSCs have been recently reported [73]. The AF-MSCs were named according to their morphology as spindle-shaped (SS) and round-shaped (RS) [84]. Human AF samples from the second trimester were utilized to obtain AFSCs, that were subsequently cultured in the medium supplemented with 20 % FBS for approximately 20 days until the first colonies appeared. Colony forming unit-fibroblast (CFU-F) were selected mechanically and sub-cultured separately. It was noted that RS-AF-MSCs were obtained more readily and represented 94 % of the MSCs obtained, while the SS-AF-MSCs type represented only 6 % of the MSCs. The SS-AF-MSCs exhibited high proliferative capacity and were expanded up to 30–50 passages with normal karyotype, whereas the RS-AF-MSCs exhibited a significantly lower proliferative potential and reached only passage 4–7. Interestingly, the RS cells exhibited a statistically significant increase in proliferation when cultured in conditioned media derived from SS-AF-MSCs, which suggested that paracrine factors derived from the SS cells may stimulate growth of the cultures. The cell surface antigens of these cells were examined by FACS and both types were negative for CD34, CD133, CD31, CD45, CD14 and HLA-DR. Both types were positive for MSC markers CD73, CD105, CD166, adherent molecules CD29, CD44, CD49e and HLA-ABC, consistent with previous reports [29, 62, 85–87]. RS-AF-MSCs shared similar morphological characteristics to human amniotic membrane epithelial cells (AECs), but they expressed very low

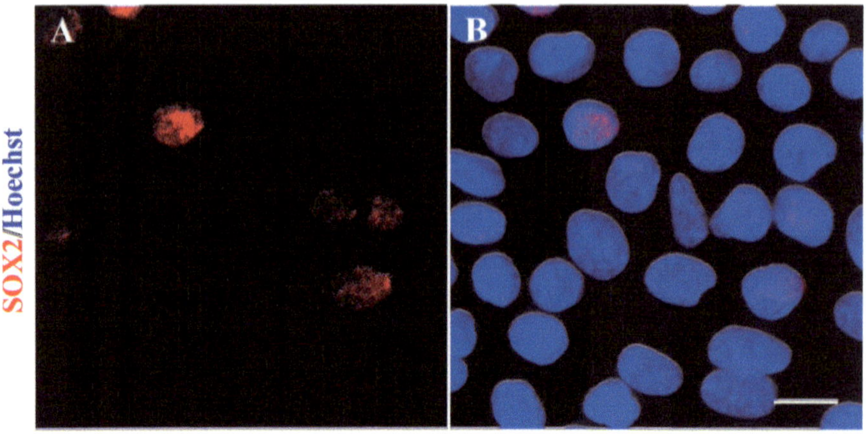

Fig. 10.6 SOX2 can be used to identify human amniotic fluid stem cells, including the subpopulation of cells that acquire a neurogenic potential. (**a**) Immunostaining shows nuclear SOX2 localization (*red*) in amniotic fluid culture. (**b**) Hoechst (*blue*) has been used as a counterstain. Scale bar: 20 μm

levels of CD90 and did not successfully differentiate to hepatocytes [88]. The SS-AF-MSCs shared the rapid expansion and the multi-lineage differentiation potential of human umbilical cord perivascular cells and they exhibited low expression of CD146 (marker for endothelial cells) [89]. Both subpopulations of AF-MSCs expressed very low/undetectable levels of CD117, which points to the existence of a wide variety of AF stem cells that can be isolated by their morphology or cell surface antigen(s) expression. The RS population expressed high levels of CD146, whereas the SS population expressed high levels of CD90, which decreased when the proliferation rate of SS was reduced due to temperature change. This alluded to the possibility that CD90 antigen expression could be related to the growth rate of AF-MSCs and could explain the difference of proliferation between the SS and the RS subpopulations. Both cell types expressed the pluripotency markers OCT4 and SOX2 (Figs. 10.4 and 10.6). The authors generated a comparative proteomic map of SS and RS-AF-MSCs, identifying 25 proteins that were differentially expressed and an additional ten proteins that showed unique expression in RS cells. Overall, SS-AF-MSCs exhibited significantly higher migration ability on extracellular matrices (fibronectin and laminin) compared to RS. Moreover, SS-AF-MSCs isolated based on their colony morphology and CD90 expression represented the population that can be expanded easily in culture, have differentiation capacity, lentiviral transduction efficiency and long-term survival *in vivo* and could be used in future *in vivo* therapeutic applications [84].

Another widely used method for the isolation of human AF-MSCs was developed by Klein and Fauza [90], based on the fact that MSCs are present throughout gestation and they can be easily identified and isolated. The MSC medium used was dubbed as mesenchymal-20 medium consisting of Dublecco's Modified Eagle's Medium (DMEM), L-glutamine, FBS, and recombinant human basic-fibroblast growth factor (rhFGF). Detailed step by step protocol, including freezing and thawing of AF-derived MSCs have been well described in this paper [90].

The "Xeno-Free" AF Cell Culture Conditions

Fauza's group demonstrated the feasibility of tissue engineering, using AF-derived mesenchymal cells expanded *ex vivo* in the absence of animal serum in 2007 [91]. Human AF-MSCs were obtained from third trimester samples and their phenotypic profile and cell proliferation rates were compared during expansion in two different media, containing either FBS or allogeneic human serum. There were no significant differences in the overall proliferation rates, based on serum type, and the cells remained positive for markers used for mesenchymal progenitors [91]. This work was further followed by Phermthai et al., suggesting that the techniques used for isolation and production of AFSCs might not be suitable for clinical purposes due to the length of time required for stem cell production (~2 months), possible heterogeneity of the cell culture populations, and xeno-contamination introduced to human cells by initial culturing of the cells in the presence of animal-based products such as culture media and antibodies [92]. The technique developed by this group selected fibroblast-like, adherent stem cells that were actively proliferating from a primary human AF culture and those selected were termed "starter cell". The starter cell was used for generating a clonal AFSC line, which showed high proliferation rate with 0.8 day population doubling time and maintained a normal karyotype over 20 subculture passages. The cells also expressed several stem cell markers, including OCT4, SSEA-4, CD29, CD73, CD90, CD105, and CD133, and differentiation capacity [92]. However, culture media of these cells were composed of 15 % FBS, glutamine, 18 % Chang B and 2 % Chang C, and FGF2 [92]. The fact that FBS was present in the culture medium contradicts the xeno-free premise the author had claimed. In a true xeno-free system, human cells do not come into contact with any animal-derived substances, including biological factors derived from serum [93], feeder cells, substrates used to coat tissue culture plates such as gelatin, and other supplements used for isolation, culturing, subculturing, expansion and cryopreservation. This notion has been well discussed in a recent review by Wang et al., elaborating on expansion of human stem cells under clinically compliant settings [94]. There is still a need in the scientific community to reach a concise definition of "xeno-free" to provide consistency across the literature.

Differentiation Potential of AFSCs in Culture

The capability of AFSCs to differentiate into multiple lineages of all three embryonic germ layers, including ectoderm (neurogenic), mesoderm (adipogenic, osteogenic, myogenic and hematopoietic), and endoderm (endothelial and hepatic) has been demonstrated by several laboratories (Table 10.5) [62, 95]. Furthermore, our

Table 10.5 Summary of media for differentiation techniques used for amniotic fluid stem cells

Coating	Differentiation media	Validation
Neurogenic differentiation		
–	Media low in glucose, NGF, BHA, DMSO	TUBBIII
		NEUN
		MAP2
		NSE
		Neurofilaments
		Dopaminergic induction
		Secretion of L-glutamate
		Expression of K^+ channels (Barium sensitive)
		Expression of Na^+ channels (Tetrodotoxin sensitive)
		Generation of action potential
Hepatogenic differentiation		
Matrigel	Media containing HGF, insulin, Oncostatin M, dexamethasone, FGF-4	Expression of albumin
		HNF4-α expression
		c-met receptor expression
		MDR membrane transporter
		α-fetoprotein
Adipogenic differentiation		
–	Media with low glucose, IBMX, insulin, indomethacin, 10 % FBS	Formation of intracellular lipid droplets
Osteogenic differentiation		
–	Media with 10 % FBS, dexamethasone, β-glycerophosphate, ascorbic-2 phosphate	Calcium precipitation
		Production of ALP
Chondrogenic differentiation		
Alginate	Media with dexamethasone, ascorbic acid-2 phosphate, sodium pyruvate, TGF-β-1	Production of sGAG
		Type II collagen secretion
Endothelial differentiation		
Gelatin	Endothelial basal media with 10 % FBS, EGF, VEGF, FGF-2, IGF-I, hydrocortisone, heparin, ascorbic acid	Cobblestone and capillary structures
Matrigel		Expression of Von Willebrand factor
		Endothelial nitric oxide synthase
		CD31
		VE-cadherin,
		VEGF receptor 2

group as well as others have used induced pluripotent stem cell (iPSC) techniques to generate various cell types from amniotic fluid cells (Fig. 10.6) [70, 71, 96–106]. Interestingly, iPSC technology is being currently used in a clinical trial for degenerative eye diseases in Japan [107], emphasizing the significance of research in this area to properly gauge the safety of such procedures in humans and extending their applications to other organs. While Table 10.5 provides a summary of AFSC differentiation into various cell types, we also describe the culturing conditions used for some of these cell lineages in the following sections.

Neurogenic cell lineage: Neuronal morphology has been observed in AFSC cultures treated with nerve-growth factor (NGF), DMSO, and/or butylated hydroxyanisole (BHA) [108]. In some cases, the cells expressed neuronal markers, including TUBB3, NEUN (FOX3), MAP2, neurofilaments and neuron-specific enolase (NSE) [35, 60, 61, 109]. In another study, serum withdrawal and DMSO addition have been used to induce neuronal differentiation in AFSC cultures [110]. Based on our experience, the efficient production of functional neurons with this method seems to be challenging. A few studies have demonstrated functional properties for AFSCs induced with neuronal differentiation media. For instance, the presence of dopaminergic neuronal markers [60], barium-sensitive K^+ channel [62], tetrodotoxin-sensitive voltage-gated sodium channel [111], and secretion of L-glutamate [62], although not necessarily exclusive to neurons [112], have been reported. It is now well-accepted that the ability of AFSCs to form functional mature neurons should be evaluated based on generating action potential and functional synapses [112]. These characteristics seem to be more achievable when AFSCs obtain a neural progenitor state as a key step towards neuronal differentiation. We have previously shown that SOX2 can be used for sorting and enriching neural cell populations in amniotic fluid cell cultures [58, 113, 114]. Since the level of SOX2 expression may vary from one cell to another, this selection method can be complemented with single cell cloning to further reduce heterogeneity in AFSCs [113, 114]. Using similar approaches, another recent study has shown that SOX9 (Fig. 10.7) can be used as a predictive neurogenic marker for AFSCs [115]. Notably, SOX9 also plays a critical role in chondrogenesis and the use of this marker for selecting neural progenitors may require tight culturing conditions to avoid AFSC differentiation to non-neuronal cell types [116].

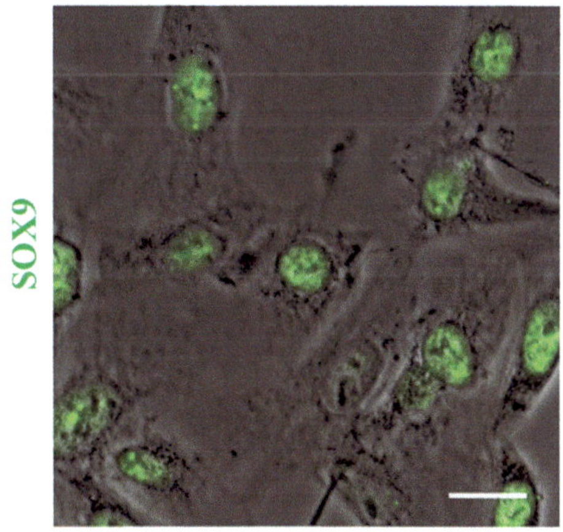

Fig. 10.7 SOX9 expression allows further evaluation of amniotic fluid cells. A representative image of SOX9 nuclear staining (*green*) and phase contrast. Scale bar: 20 μm

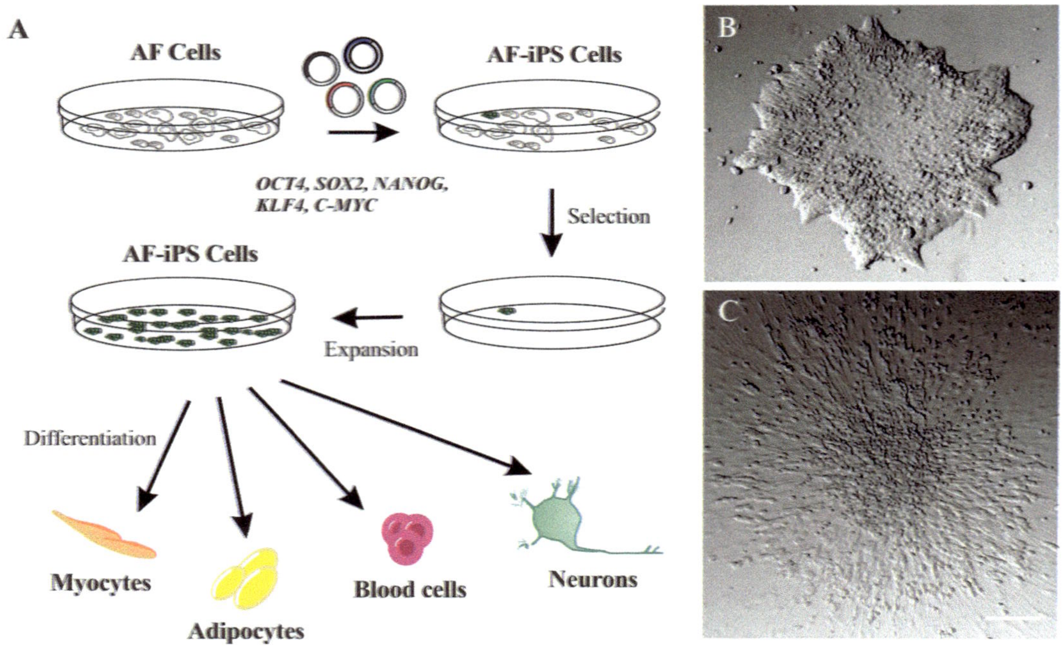

Fig. 10.8 Cell reprogramming techniques can be used to generate amniotic fluid induced pluripotent stem cells (AF-iPSCs), followed by their differentiation into various cell types. (**a**) An illustration of cell reprogramming in amniotic fluid cultures; (**b**) A colony of AF-iPSCs produced by *OCT4*, *SOX2*, *NANOG*, *KLF4* and *cMYC* nucleofection; (**c**) AF-iPSCs can readily differentiate into neurons, using appropriate neuronal induction media. Scale bar: 200 μm

In addition to these methods, several groups, including our laboratory, have used induced pluripotent stem cell (iPSC) techniques to optimize the generation of neuronal cultures from amniotic fluid cells [48, 96, 99, 100, 104]. Amniotic fluid-induced pluripotent stem cells (AF-iPSCs) can be obtained by introducing *OCT4*, *SOX2*, *NANOG*, *KLF4*, and/or *c-MYC* into AF cultures (Fig. 10.8). The AF-iPSC colonies are typically expanded in Matrigel-coated plates in mTeSR1 medium and further assessed for the absence of transgenes, optimal epigenetic state (i.e., reduced or lack of DNA methylation sites) and expression of endogenous stem cell markers. Subsequently, the fully-reprogrammed AF-iPSCs are treated with neural induction medium and assessed for formation of neural rosettes, followed by characterization of neurons at the molecular, cellular and functional levels. We find it essential to further confirm neural identity by loss of OCT4 and other pluripotent markers, while cells maintain SOX2 and acquire the expression of markers such as PAX6. Furthermore, neuronal phenotype and function need to be validated by combined immuncytochemistry and electrophysiology. Taking a step further, we have demonstrated that AF-iPSC-derived neurons can be used as an important tool for pharmacological assays such as neurotoxicity (Fig. 10.9) as well as transplantation studies to facilitate their applications in regenerative medicine.

Adipogenic cell lineage: Differentiation of AFSCs into adipogenic lineage has been observed in media containing 10 % FBS, 3-isobutyl-1-methyl-xanthine (IBMX), insulin and indomethacin. Adipogenic features are generally confirmed within 3 weeks by observing the formation of intracellular lipid droplets and Oil Red O staining [62, 84, 108].

Osteogenic cell lineage: This method generally involves culturing AFSCs in media containing 10 % FBS, dexamethasone, β-glycerophosphate and ascorbic acid-2-phosphate. The differentiation can be confirmed by observation of calcium precipitation and production of ALP [62, 84, 108].

Chondrogenic cell lineage: Placing AFSCs in an alginate hydrogel and culturing in media containing dexamethasone, ascorbic acid-2-phosphate, sodium pyruvate, proline and transforming growth factor-β1 (TGF-β1) appear to induce these cells to differentiate into chondrocytes. Cells are usually maintained for about 20 days and their differentiation can be confirmed by production of sulfated glycosaminoglycan (sGAG) and type II collagen [84, 117].

Hematopoietic cell lineage: Under the presumption that AFSCs have the potential to give rise to hematopoietic cells, Moorefield et al. isolated CD117-positive/Lineage (Lin)-negative (KL) population from human AFSCs and stained them with an APC-conjugated CD117 antibody and a cocktail of Lin PE-conjugated antibodies (containing anti-CD3, CD4, CD8, CD13, CD16, CD19, CD20, CD33, CD56 and CD235a) [67, 73]. The isolated KL cells demonstrated the presence of multipotent hematopoietic progenitors and colony-forming unit-granulocytes, erythrocytes, monocytes and megakaryocytes. KL cultures were able to generate CD13+ CD33+ cells, confirming their myeloid differentiation potential. Furthermore, they displayed an *in vitro* multilineage hematopoietic potential where KL cultures gave rise to natural killer cells (CD56+ CD16+) and T-cell precursors. The authors concluded

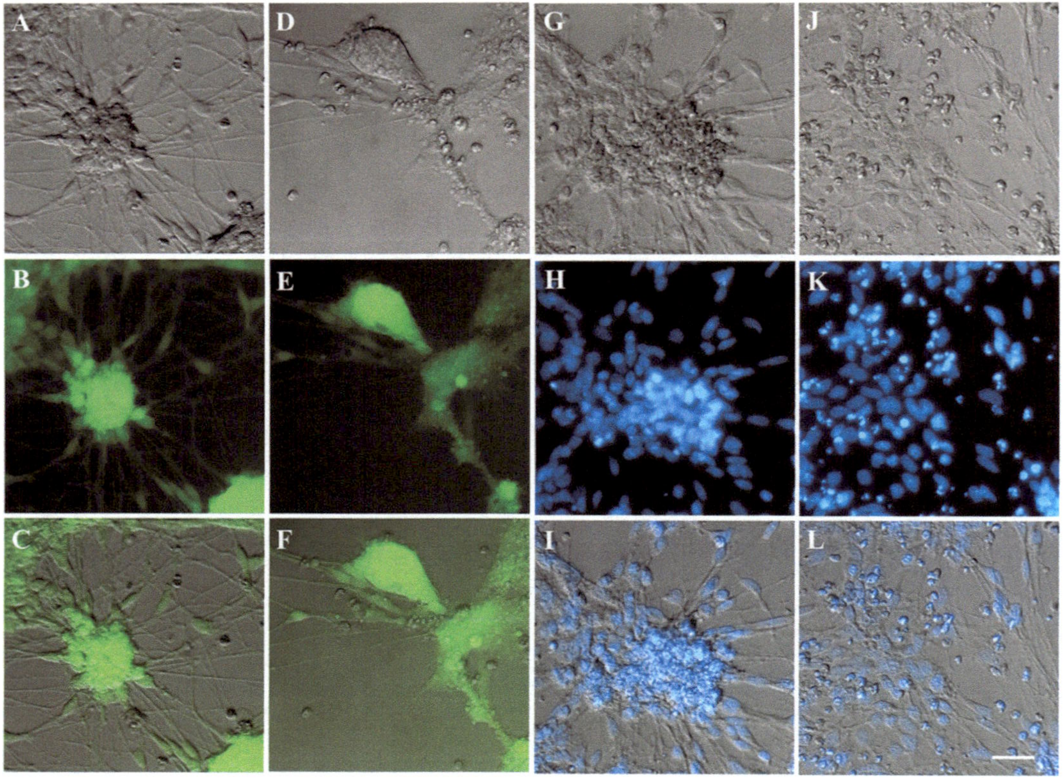

Fig. 10.9 The neurons derived from human amniotic fluid induced pluripotent stem cells (AF-iPSCs) can serve as a tool for neurotoxicity assays. Neuronal viability has been determined by CFDA fluorescence (*green*) in the absence (**A–C**) or presence of glutamate (**D–F**) in culture. A decrease in cell viability is observed in glutamate treated cultures (**D–F**). This assay has been complemented by Hoechst staining (*blue*, **G–I**), further confirming the morphological changes in glutamate-treated (**J–L**) versus control (**G–I**) cultures. Phase contrast images have been shown without (**A, D, G** and **J**) and in combination with Hoechst (**C, F, I** and **L**). Scale bar: 125 μm

that human AF contains cells with hematopoietic potential, as demonstrated by the generation of erythroid, myeloid, and lymphoid cells *in vitro* [67].

Hepatogenic cell lineage: To induce hepatogenic cell differentiation, AFSCs are cultured on Matrigel coated vessels in medium containing hepatocyte growth factor (HGF), insulin, Oncostatin M, dexamethasone, and fibroblast growth factor 4 (FGF-4). The cells differentiate into hepatocytes, as evident by expression of albumin, transcription factor HNF4α, C-MET receptor, MDR membrane transporter and α-fetoprotein. The formation of hepatocytes is usually observed within 2–3 weeks under appropriate differentiation conditions [62, 84, 108, 118].

Endothelial cell lineage: The differentiation of AFSCs into endothelial cells can be induced by culturing cells in the gelatin- or Matrigel-coated plates in endothelial basal medium containing 10 % FBS, epithelial growth factor (EGF), vascular endothelial growth factor (VEGF), fibroblast growth factor 2 (FGF-2), insulin-like growth factor (IGF-1), hydrocortisone, heparin and ascorbic acid [62, 108]. After 2 weeks, the AFSC-derived endothelial cells show morphological characteristics such as cobblestone and capillary structures on 2 and 3 dimensional culture substrates, respectively [62]. The cells also express von Willebrand factor (VWF), endothelial nitric oxide synthase, CD31, VE-cadherin, and VEGF receptor 2 [108]. Functionally, AFSC-derived endothelial cells form networks and metabolize acetylated low density lipoprotein [119]. Proteomic analysis, including proteome profiling arrays, can provide a better understanding of the proteins involved in differentiation of AFSCs into various endothelial cell types such as vascular and lymphatic endothelial cells [120].

Cryopreservation of AFSCs

Since AFSCs are proving to be of immense potential for future diagnostic and therapeutic applications, a long-term banking system of these cells is necessary. Currently, a major obstacle exists in producing clinical grade AFSCs due to the lack of good manufacturing practices in cell processing, cryopreservation, storage and distribution. Developing effective techniques for cryopreservation of AFSCs is an important step in the banking of stem cells. During this process, cells need to remain healthy, as they go through freeze and thaw cycles [121]. Similar to other cells, cooling AFSCs at a slow controlled rate avoids intracellular ice buildup, which can cause the cell membrane to rupture. However, even slow freezing can result in dehydration of the cells by formation of extracellular ice, and for this reason a cryoprotective agent (CPA) is usually added to the freezing medium [122]. CPAs decrease the freezing temperature and increase viscosity as the freezing solution becomes amorphous ice [123]. Currently, the cryopreservation method most commonly used for AFSCs includes a freezing medium consisting of 10 % v/v of DMSO and up to 90 % v/v of animal or human serum. DMSO is a hygroscopic polar compound and can be toxic to cells, however, several studies have illustrated that cryopreservation of AFSCs using DMSO does not appear to affect the biological properties of the cells, including morphology, proliferation rate, viability, cell cycle, karyotype, gene expression and differentiation potential [124–126]. Studies that investigate the clinical effects of the presence of DMSO in the freezing media used for AFSCs are lacking. However, several reports about transplantation of hematopoietic stem cell products preserved in DMSO show that there are serious side effects such as vomiting, hypotension, acute abdominal pain, dyspnea, cardiac arrhythmia, and hemoglobinuria associated with using this compound [127–129]. Efforts have been focusing on eliminating or reducing DMSO concentration and introducing a xeno-free cryopreservation solution. Natural disaccharides including trehalose and sucrose have been investigated as non-toxic and non-permeable CPAs [121, 130]. Recent studies showed an increase in post-thaw cell viability of AFSCs cryopreserved in freezing solutions containing trehalose, catalase, and caspase inhibitor (zVAD-fmk) with 5 % or 2.5 % DMSO compared to freezing solution containing 10 % DMSO and 30 % FBS [131]. While the data showed the ability to preserve AFSCs using natural CPAs, 20 % FBS containing media with either 5 %, or 10 % DMSO or glycerol showed workable cell viability, but not sucrose or trehalose after 6 months of storage, despite the maintenance of differentiation capacity [126]. It becomes apparent that animal and human studies are needed to confirm the safety and efficacy in AFSC banking. Developing CPA-free media or non-toxic CPAs for the cryopreservation and storage of AFSCs is imperative for long term preservation and maintenance of their viability and biological functions for future therapeutic applications. These issues have been discussed by Albanna and Woods in a separate chapter in this book.

References

1. Fauza D. Amniotic fluid and placental stem cells. Best Pract Res Clin Obstet Gynaecol. 2004;18(6):877–91.
2. Klemmt PA, Vafaizadeh V, Groner B. The potential of amniotic fluid stem cells for cellular therapy and tissue engineering. Expert Opin Biol Ther. 2011;11(10):1297–314.
3. Downs KM, Harmann C. Developmental potency of the murine allantois. Development. 1997;124(14):2769–80.
4. Snow MH, Bennett D. Gastrulation in the mouse: assessment of cell populations in the epiblast of tw18/tw18 embryos. J Embryol Exp Morphol. 1978;47:39–52.
5. Ichikawa T, et al. Live imaging of whole mouse embryos during gastrulation: migration analyses of epiblast and mesodermal cells. PLoS One. 2013;8(7):e64506.
6. Pappa KI, Anagnou NP. Novel sources of fetal stem cells: where do they fit on the developmental continuum? Regen Med. 2009;4(3):423–33.
7. Gilbert WM, Brace RA. Amniotic fluid volume and normal flows to and from the amniotic cavity. Semin Perinatol. 1993;17(3):150–7.
8. Underwood MA, Gilbert WM, Sherman MP. Amniotic fluid: not just fetal urine anymore. J Perinatol. 2005;25(5):341–8.
9. Parolini O, et al. Amniotic membrane and amniotic fluid-derived cells: potential tools for regenerative medicine? Regen Med. 2009;4(2):275–91.
10. Bartha JL, et al. Alpha-fetoprotein and hematopoietic growth factors in amniotic fluid. Obstet Gynecol. 2000;96(4):588–92.

11. Heidari Z, et al. Characterization of the growth factor activity of amniotic fluid on cells from hematopoietic and lymphoid organs of different life stages. Microbiol Immunol. 1996;40(8):583–9.
12. Sherer DM. A review of amniotic fluid dynamics and the enigma of isolated oligohydramnios. Am J Perinatol. 2002;19(5):253–66.
13. Brace RA, Wolf EJ. Normal amniotic fluid volume changes throughout pregnancy. Am J Obstet Gynecol. 1989;161(2):382–8.
14. Gilbert WM, et al. Technetium Tc 99m rapidly crosses the ovine placenta and intramembranous pathway. Am J Obstet Gynecol. 1996;175(6):1557–62.
15. Medina-Gomez P, del Valle M. The culture of amniotic fluid cells. An analysis of the colonies, metaphase and mitotic index for the purpose of ruling out maternal cell contamination. Ginecol Obstet Mex. 1988;56:122–6.
16. Murphy SV, Atala A. Amniotic fluid stem cells. In: Perinatal stem cells. New York, NY: John Wiley & Sons; 2013. p. 1–15.
17. Jacobson CB, Barter RH. Intrauterine diagnosis and management of genetic defects. Am J Obstet Gynecol. 1967;99(6):796–807.
18. Fuchs F, Riis P. Antenatal sex determination. Nature. 1956;177(4503):330.
19. Serr DM, Sachs L, Danon M. The diagnosis of sex before birth using cells from the amniotic fluid (a preliminary report). Bull Res Counc Isr. 1955;5B(2):137–8.
20. Steele MW, Breg Jr WR. Chromosome analysis of human amniotic-fluid cells. Lancet. 1966;1(7434):383–5.
21. Nelson MM, Emery AE. Amniotic fluid cells; prenatal sex prediction and culture. Br Med J. 1970;1(5695):523–6.
22. Shih VE, Schulman JD. Ornithine-ketoacid transaminase activity in human skin and amniotic fluid cell culture. Clin Chim Acta. 1970;27(1):73–5.
23. Chitham RG, Quayle SJ, Hill L. The selection of a method for the culture of cells from amniotic fluid. J Clin Pathol. 1973;26(9):721–3.
24. Chang HC, Jones OW. Reduction of sera requirements in amniotic fluid cell culture. Prenat Diagn. 1985;5(5):305–12.
25. Chang HC, Jones OW. Enhancement of amniocyte growth on a precoated surface. Prenat Diagn. 1991;11(6):357–70.
26. Prusa AR, Hengstschlager M. Amniotic fluid cells and human stem cell research: a new connection. Med Sci Monit. 2002;8(11):RA253–7.
27. Torricelli F, et al. Identification of hematopoietic progenitor cells in human amniotic fluid before the 12th week of gestation. Ital J Anat Embryol. 1993;98(2):119–26.
28. Prusa AR, et al. Oct-4-expressing cells in human amniotic fluid: a new source for stem cell research? Hum Reprod. 2003;18(7):1489–93.
29. In 't Anker PS, et al. Amniotic fluid as a novel source of mesenchymal stem cells for therapeutic transplantation. Blood. 2003;102(4):1548–9.
30. Kaviani A, et al. The amniotic fluid as a source of cells for fetal tissue engineering. J Pediatr Surg. 2001;36(11):1662–5.
31. Kaviani A, et al. Fetal tissue engineering from amniotic fluid. J Am Coll Surg. 2003;196(4):592–7.
32. Dominici M, et al. Minimal criteria for defining multipotent mesenchymal stromal cells. The International Society for Cellular Therapy position statement. Cytotherapy. 2006;8(4):315–7.
33. Keating A. Mesenchymal stromal cells: new directions. Cell Stem Cell. 2012;10(6):709–16.
34. Zhou J, et al. Amniotic fluid-derived mesenchymal stem cells: characteristics and therapeutic applications. Arch Gynecol Obstet. 2014;290(2):223–31.
35. Tsai MS, et al. Isolation of human multipotent mesenchymal stem cells from second-trimester amniotic fluid using a novel two-stage culture protocol. Hum Reprod. 2004;19(6):1450–6.
36. Fleischman RA. From white spots to stem cells: the role of the Kit receptor in mammalian development. Trends Genet. 1993;9(8):285–90.

37. Andrews PW, et al. Embryonic stem (ES) cells and embryonal carcinoma (EC) cells: opposite sides of the same coin. Biochem Soc Trans. 2005;33(Pt 6):1526–30.
38. Andrews PW, et al. Three monoclonal antibodies defining distinct differentiation antigens associated with different high molecular weight polypeptides on the surface of human embryonal carcinoma cells. Hybridoma. 1984;3(4):347–61.
39. Inniss K, Moore H. Mediation of apoptosis and proliferation of human embryonic stem cells by sphingosine-1-phosphate. Stem Cells Dev. 2006;15(6):789–96.
40. Henderson JK, et al. Preimplantation human embryos and embryonic stem cells show comparable expression of stage-specific embryonic antigens. Stem Cells. 2002;20(4):329–37.
41. Shevinsky LH, et al. Monoclonal antibody to murine embryos defines a stage-specific embryonic antigen expressed on mouse embryos and human teratocarcinoma cells. Cell. 1982;30(3):697–705.
42. Fox N, et al. Distribution of murine stage-specific embryonic antigens in the kidneys of three rodent species. Exp Cell Res. 1982;140(2):331–9.
43. Shamblott MJ, et al. Derivation of pluripotent stem cells from cultured human primordial germ cells. Proc Natl Acad Sci U S A. 1998;95(23):13726–31.
44. Maguire CT, et al. Genome-wide analysis reveals the unique stem cell identity of human amniocytes. PLoS One. 2013;8(1):e53372.
45. Zia S, et al. Routine clonal expansion of mesenchymal stem cells derived from amniotic fluid for perinatal applications. Prenat Diagn. 2013;33(10):921–8.
46. Petsche Connell J, et al. Amniotic fluid-derived stem cells for cardiovascular tissue engineering applications. Tissue Eng Part B Rev. 2013;19(4):368–79.
47. Roubelakis MG, Trohatou O, Anagnou NP. Amniotic fluid and amniotic membrane stem cells: marker discovery. Stem Cells Int. 2012;2012:107836.
48. Moschidou D, et al. Valproic acid confers functional pluripotency to human amniotic fluid stem cells in a transgene-free approach. Mol Ther. 2012;20(10):1953–67.
49. Bani-Yaghoub M, et al. Amniotic stem cells: potential in regenerative medicine. Stem Cells Int. 2012;2012:530674.
50. Dionigi B, et al. A comparative analysis of human mesenchymal stem cell response to hypoxia in vitro: implications to translational strategies. J Pediatr Surg. 2014;49(6):915–8.
51. Romani R, et al. Comparative proteomic analysis of two distinct stem-cell populations from human amniotic fluid. Mol Biosyst. 2015;11(6):1622–32.
52. Nichols J, et al. Formation of pluripotent stem cells in the mammalian embryo depends on the POU transcription factor Oct4. Cell. 1998;95(3):379–91.
53. Takeda J, Seino S, Bell GI. Human Oct3 gene family: cDNA sequences, alternative splicing, gene organization, chromosomal location, and expression at low levels in adult tissues. Nucleic Acids Res. 1992;20(17):4613–20.
54. Hatano SY, et al. Pluripotential competence of cells associated with Nanog activity. Mech Dev. 2005;122(1):67–79.
55. Rizzino A. Sox2 and Oct-3/4: a versatile pair of master regulators that orchestrate the self-renewal and pluripotency of embryonic stem cells. Wiley Interdiscip Rev Syst Biol Med. 2009;1(2):228–36.
56. Chambers I, Tomlinson SR. The transcriptional foundation of pluripotency. Development. 2009;136(14):2311–22.
57. Ghaleb AM, et al. Kruppel-like factors 4 and 5: the yin and yang regulators of cellular proliferation. Cell Res. 2005;15(2):92–6.
58. Jezierski A, et al. Probing stemness and neural commitment in human amniotic fluid cells. Stem Cell Rev. 2010;6(2):199–214.
59. Wang KH, et al. Upregulation of nanog and sox-2 genes following ectopic expression of Oct-4 in amniotic fluid mesenchymal stem cells. Biotechnol Appl Biochem. 2015;62:591.
60. Tsai MS, et al. Clonal amniotic fluid-derived stem cells express characteristics of both mesenchymal and neural stem cells. Biol Reprod. 2006;74(3):545–51.
61. Kim J, et al. Human amniotic fluid-derived stem cells have characteristics of multipotent stem cells. Cell Prolif. 2007;40(1):75–90.

62. De Coppi P, et al. Isolation of amniotic stem cell lines with potential for therapy. Nat Biotechnol. 2007;25(1):100–6.
63. Dobreva MP, et al. On the origin of amniotic stem cells: of mice and men. Int J Dev Biol. 2010;54(5):761–77.
64. Pozzobon M, Ghionzoli M, De Coppi P. ES, iPS, MSC, and AFS cells. Stem cells exploitation for Pediatric Surgery: current research and perspective. Pediatr Surg Int. 2010;26(1):3–10.
65. De Coppi P, et al. Amniotic fluid and bone marrow derived mesenchymal stem cells can be converted to smooth muscle cells in the cryo-injured rat bladder and prevent compensatory hypertrophy of surviving smooth muscle cells. J Urol. 2007;177(1):369–76.
66. Piccoli M, et al. Amniotic fluid stem cells restore the muscle cell niche in a HSA-Cre, Smn(F7/F7) mouse model. Stem Cells. 2012;30(8):1675–84.
67. Ditadi A, et al. Human and murine amniotic fluid c-Kit+Lin- cells display hematopoietic activity. Blood. 2009;113(17):3953–60.
68. Cananzi M, De Coppi P. CD117(+) amniotic fluid stem cells: state of the art and future perspectives. Organogenesis. 2012;8(3):77–88.
69. Zani A, et al. Amniotic fluid stem cells improve survival and enhance repair of damaged intestine in necrotising enterocolitis via a COX-2 dependent mechanism. Gut. 2014;63(2):300–9.
70. Wolfrum K, et al. The LARGE principle of cellular reprogramming: lost, acquired and retained gene expression in foreskin and amniotic fluid-derived human iPS cells. PLoS One. 2010;5(10):e13703.
71. Moschidou D, et al. Human mid-trimester amniotic fluid stem cells cultured under embryonic stem cell conditions with valproic acid acquire pluripotent characteristics. Stem Cells Dev. 2013;22(3):444–58.
72. De Coppi P. C-kit/CD117 cells from amniotic fluid and membranes and their cardiomyogenic potential. Circ Res. 2010;107(6):e11. author reply e12.
73. Moorefield EC, et al. Cloned, CD117 selected human amniotic fluid stem cells are capable of modulating the immune response. PLoS One. 2011;6(10):e26535.
74. Romani R, et al. Stem cells from human amniotic fluid exert immunoregulatory function via secreted indoleamine 2,3-dioxygenase1. J Cell Mol Med. 2015;19:1593.
75. Di Trapani M, et al. Immune regulatory properties of CD117(pos) amniotic fluid stem cells vary according to gestational age. Stem Cells Dev. 2015;24(1):132–43.
76. Di Trapani M, et al. Comparative study of immune regulatory properties of stem cells derived from different tissues. Stem Cells Dev. 2013;22(22):2990–3002.
77. Valli A, et al. Embryoid body formation of human amniotic fluid stem cells depends on mTOR. Oncogene. 2010;29(7):966–77.
78. Kurosawa H. Methods for inducing embryoid body formation: in vitro differentiation system of embryonic stem cells. J Biosci Bioeng. 2007;103(5):389–98.
79. Narita N, Bielinska M, Wilson DB. Cardiomyocyte differentiation by GATA-4-deficient embryonic stem cells. Development. 1997;124(19):3755–64.
80. Chen WQ, et al. Variations of protein levels in human amniotic fluid stem cells CD117/2 over passages 5-25. J Proteome Res. 2009;8(11):5285–95.
81. Arnhold S, et al. Amniotic-fluid stem cells: growth dynamics and differentiation potential after a CD-117-based selection procedure. Stem Cells Int. 2011;2011:715341.
82. Ginsberg M, et al. Efficient direct reprogramming of mature amniotic cells into endothelial cells by ETS factors and TGFbeta suppression. Cell. 2012;151(3):559–75.
83. Resca E, et al. Enrichment in c-Kit improved differentiation potential of amniotic membrane progenitor/stem cells. Placenta. 2015;36(1):18–26.
84. Roubelakis MG, et al. In vitro and in vivo properties of distinct populations of amniotic fluid mesenchymal progenitor cells. J Cell Mol Med. 2011;15(9):1896–913.
85. Roubelakis MG, et al. Molecular and proteomic characterization of human mesenchymal stem cells derived from amniotic fluid: comparison to bone marrow mesenchymal stem cells. Stem Cells Dev. 2007;16(6):931–52.

86. Goodwin HS, et al. Multilineage differentiation activity by cells isolated from umbilical cord blood: expression of bone, fat, and neural markers. Biol Blood Marrow Transplant. 2001;7(11):581–8.
87. Pittenger MF, et al. Multilineage potential of adult human mesenchymal stem cells. Science. 1999;284(5411):143–7.
88. Parolini O, et al. Concise review: isolation and characterization of cells from human term placenta: outcome of the first international Workshop on Placenta Derived Stem Cells. Stem Cells. 2008;26(2):300–11.
89. Sarugaser R, et al. Human umbilical cord perivascular (HUCPV) cells: a source of mesenchymal progenitors. Stem Cells. 2005;23(2):220–9.
90. Klein JD, Fauza DO. Amniotic and placental mesenchymal stem cell isolation and culture. Methods Mol Biol. 2011;698:75–88.
91. Kunisaki SM, et al. Tissue engineering from human mesenchymal amniocytes: a prelude to clinical trials. J Pediatr Surg. 2007;42(6):974–9. discussion 979–80.
92. Phermthai T, et al. A novel method to derive amniotic fluid stem cells for therapeutic purposes. BMC Cell Biol. 2010;11:79.
93. Summers MC, Biggers JD. Chemically defined media and the culture of mammalian preimplantation embryos: historical perspective and current issues. Hum Reprod Update. 2003;9(6):557–82.
94. Wang Y, Cheng L, Gerecht S. Efficient and scalable expansion of human pluripotent stem cells under clinically compliant settings: a view in 2013. Ann Biomed Eng. 2014;42(7): 1357–72.
95. Perin L, et al. Renal differentiation of amniotic fluid stem cells. Cell Prolif. 2007;40(6): 936–48.
96. Jiang G, et al. Human transgene-free amniotic-fluid-derived induced pluripotent stem cells for autologous cell therapy. Stem Cells Dev. 2014;23(21):2613–25.
97. Li C, et al. Pluripotency can be rapidly and efficiently induced in human amniotic fluid-derived cells. Hum Mol Genet. 2009;18(22):4340–9.
98. Galende E, et al. Amniotic fluid cells are more efficiently reprogrammed to pluripotency than adult cells. Cell Reprogram. 2010;12(2):117–25.
99. Lu HE, et al. Selection of alkaline phosphatase-positive induced pluripotent stem cells from human amniotic fluid-derived cells by feeder-free system. Exp Cell Res. 2011;317(13):1895–903.
100. Liu T, et al. High efficiency of reprogramming CD34(+) cells derived from human amniotic fluid into induced pluripotent stem cells with Oct4. Stem Cells Dev. 2012;21(12):2322–32.
101. Fan Y, et al. Generation of human beta-thalassemia induced pluripotent stem cells from amniotic fluid cells using a single excisable lentiviral stem cell cassette. J Reprod Dev. 2012;58(4):404–9.
102. Kunisaki SM. Congenital anomalies: treatment options based on amniotic fluid-derived stem cells. Organogenesis. 2012;8(3):89–95.
103. Diecke S, Wu JC. Pushing the reset button: chemical-induced conversion of amniotic fluid stem cells into a pluripotent state. Mol Ther. 2012;20(10):1839–41.
104. Lu HE, et al. Modeling neurogenesis impairment in Down syndrome with induced pluripotent stem cells from Trisomy 21 amniotic fluid cells. Exp Cell Res. 2013;319(4):498–505.
105. Li Q, et al. Generation of induced pluripotent stem cells from human amniotic fluid cells by reprogramming with two factors in feeder-free conditions. J Reprod Dev. 2013;59(1):72–7.
106. Pipino C, et al. Trisomy 21 mid-trimester amniotic fluid induced pluripotent stem cells maintain genetic signatures during reprogramming: implications for disease modeling and cryobanking. Cell Reprogram. 2014;16(5):331–44.
107. Kanemura H, et al. Tumorigenicity studies of induced pluripotent stem cell (iPSC)-derived retinal pigment epithelium (RPE) for the treatment of age-related macular degeneration. PLoS One. 2014;9(1):e85336.

108. Joo S, et al. Amniotic fluid-derived stem cells in regenerative medicine research. Arch Pharm Res. 2012;35(2):271–80.
109. Prusa AR, et al. Neurogenic cells in human amniotic fluid. Am J Obstet Gynecol. 2004;191(1):309–14.
110. Croft AP, Przyborski SA. Formation of neurons by non-neural adult stem cells: potential mechanism implicates an artifact of growth in culture. Stem Cells. 2006;24(8):1841–51.
111. Mareschi K, et al. Multipotent mesenchymal stem cells from amniotic fluid originate neural precursors with functional voltage-gated sodium channels. Cytotherapy. 2009;11(5):534–47.
112. Toselli M, et al. Do amniotic fluid-derived stem cells differentiate into neurons in vitro? Nat Biotechnol. 2008;26(3):269–70. author reply 270–1.
113. Rennie K, et al. Therapeutic potential of amniotic fluid-derived cells for treating the injured nervous system. Biochem Cell Biol. 2013;91(5):271–86.
114. Zhang S, Cui W. Sox2, a key factor in the regulation of pluripotency and neural differentiation. World J Stem Cells. 2014;6(3):305–11.
115. Wei PC, et al. SOX9 as a predictor for neurogenesis potentiality of amniotic fluid stem cells. Stem Cells Transl Med. 2014;3(10):1138–47.
116. Akiyama H, Lefebvre V. Unraveling the transcriptional regulatory machinery in chondrogenesis. J Bone Miner Metab. 2011;29(4):390–5.
117. Kolambkar YM, et al. Chondrogenic differentiation of amniotic fluid-derived stem cells. J Mol Histol. 2007;38(5):405–13.
118. Atala A. Amniotic fluid-derived pluripotential cells. Chapter 16. In: Essentials of stem cell biology. Amsterdam: Elsevier Inc.; 2009.
119. Benavides OM, et al. Evaluation of endothelial cells differentiated from amniotic fluid-derived stem cells. Tissue Eng Part A. 2012;18(11–12):1123–31.
120. Roubelakis MG, et al. Spindle shaped human mesenchymal stem/stromal cells from amniotic fluid promote neovascularization. PLoS One. 2013;8(1):e54747.
121. Buchanan SS, et al. Cryopreservation of stem cells using trehalose: evaluation of the method using a human hematopoietic cell line. Stem Cells Dev. 2004;13(3):295–305.
122. Pegg DE. Long-term preservation of cells and tissues: a review. J Clin Pathol. 1976;29(4):271–85.
123. Rall WF, Fahy GM. Ice-free cryopreservation of mouse embryos at -196 degrees C by vitrification. Nature. 1985;313(6003):573–5.
124. Angelo PC, et al. Cryopreservation does not alter karyotype, multipotency, or NANOG/SOX2 gene expression of amniotic fluid mesenchymal stem cells. Genet Mol Res. 2012; 11(2):1002–12.
125. Miranda-Sayago JM, et al. Evaluation of a low cost cryopreservation system on the biology of human amniotic fluid-derived mesenchymal stromal cells. Cryobiology. 2012;64(3):160–6.
126. Janz Fde L, et al. Evaluation of distinct freezing methods and cryoprotectants for human amniotic fluid stem cells cryopreservation. J Biomed Biotechnol. 2012;2012:649353.
127. Zambelli A, et al. Clinical toxicity of cryopreserved circulating progenitor cells infusion. Anticancer Res. 1998;18(6B):4705–8.
128. Alessandrino P, et al. Adverse events occurring during bone marrow or peripheral blood progenitor cell infusion: analysis of 126 cases. Bone Marrow Transplant. 1999;23(6):533–7.
129. Davis J, Rowley SD, Santos GW. Toxicity of autologous bone marrow graft infusion. Prog Clin Biol Res. 1990;333:531–40.
130. Rodrigues JP, et al. Evaluation of trehalose and sucrose as cryoprotectants for hematopoietic stem cells of umbilical cord blood. Cryobiology. 2008;56(2):144–51.
131. Seo JM, et al. Cryopreservation of amniotic fluid-derived stem cells using natural cryoprotectants and low concentrations of dimethylsulfoxide. Cryobiology. 2011;62(3):167–73.
132. Nadler HL. Patterns of enzyme development utilizing cultivated human fetal cells derived from amniotic fluid. Biochem Genet. 1968;2(2):119–26.

133. DeMars R, et al. Lesch-Nyhan mutation: prenatal detection with amniotic fluid cells. Science. 1969;164(3885):1303–5.
134. Wahlstrom J, Brosset A, Bartsch F. Viability of amniotic cells at different stages of gestation. Lancet. 1970;2(7681):1037.
135. Gregson NM. A technique for culturing cells from amniotic fluid. Lancet. 1970;1(7637):84.
136. Gray C, Davidson RG, Cohen MM. A simplified technique for the culture of amniotic fluid cells. J Pediatr. 1971;79(1):119–22.
137. Marchant GS. Evaluation of methods of amniotic fluid cell culture. Am J Med Technol. 1971;37(10):391–6.
138. Knorr-Gartner H, Harle I. A modified method of culturing human amniotic fluid cells for prenatal detection of genetic disorders. Humangenetik. 1972;14(4):333–4.
139. Felix JS, et al. Amniotic fluid cell culture. I. Experimental design for evaluating cell culture variables: determination of optimal fetal calf serum concentration. Pediatr Res. 1974;8(11):870–4.
140. Butterworth J, et al. Effect of serum concentration, type of culture medium and pH on the lysosomal enzyme activity of cultured human amniotic fluid cells. Clin Chim Acta. 1974;53(2):239–46.
141. Rudiger HW, et al. Enhancement of amniotic fluid cell growth in culture. Humangenetik. 1974;22(1):81–4.
142. Huddleston JF, Carney Jr FE. Midtrimester amniocentesis and culture of amniotic fluid cells. Clin Obstet Gynecol. 1976;19(4):941–56.
143. Chettur L, Christensen E, Philip J. Stimulation of amniotic fluid cells by fibroblast growth factor. Clin Genet. 1978;14(4):223–8.
144. Zhou XT, et al. A simple method of culturing amniotic fluid cells. Chin Med J (Engl). 1980;93(8):553–5.
145. Martin AO. Characteristics of amniotic fluid cells in vitro and attempts to improve culture techniques. Clin Obstet Gynaecol. 1980;7(1):143–73.
146. Gosden CM. Amniotic fluid cell types and culture. Br Med Bull. 1983;39(4):348–54.
147. Seguin LR, Palmer CG. Variables influencing growth and morphology of colonies of cells from human amniotic fluid. Prenat Diagn. 1983;3(2):107–16.
148. Brackertz M, et al. Decreased oxygen supply enhances growth in culture of human mid-trimester amniotic fluid cells. Hum Genet. 1983;64(4):334–8.
149. Miron PM. Preparation, culture, and analysis of amniotic fluid samples. Curr Protoc Hum Genet. 2012; Chapter 8: Unit 8.4.
150. Hoehn H, et al. Cultivated cells from diagnostic amniocentesis in second trimester pregnancies. I. Clonal morphology and growth potential. Pediatr Res. 1974;8(8):746–54.
151. Priest RE, et al. Differentiation in human amniotic fluid cell cultures: I: Collagen production. J Med Genet. 1977;14(3):157–62.
152. Megaw JM, et al. Differentiation in human amniotic fluid cell cultures: II: Secretion of an epithelial basement membrane glycoprotein. J Med Genet. 1977;14(3):163–7.
153. Aula P, et al. Glial origin of rapidly adhering amniotic fluid cells. Br Med J. 1980;281(6253):1456–7.
154. Priest RE, et al. Differentiation in human amniotic fluid cell cultures: chorionic gonadotropin production. In Vitro. 1979;15(2):142–7.
155. Sarkar S, et al. Neural origin of cells in amniotic fluid. Am J Obstet Gynecol. 1980;136(1):67–72.
156. Cremer M, et al. Demonstration of astrocytes in cultured amniotic fluid cells of three cases with neural-tube defect. Hum Genet. 1981;56(3):365–70.
157. Harris A. Glycoproteins that distinguish different cell types found in amniotic fluid. Hum Genet. 1982;62(3):188–92.
158. Hoehn H, Salk D. Morphological and biochemical heterogeneity of amniotic fluid cells in culture. Methods Cell Biol. 1982;26:11–34.

159. Zhang S, et al. The heterogeneity of cell subtypes from a primary culture of human amniotic fluid. Cell Mol Biol Lett. 2010;15(3):424–39.
160. Wilson PG, et al. Clonal populations of amniotic cells by dilution and direct plating: evidence for hidden diversity. Stem Cells Int. 2012;2012:485950.
161. Bottai D, et al. Third trimester amniotic fluid cells with the capacity to develop neural phenotypes and with heterogeneity among sub-populations. Restor Neurol Neurosci. 2012;30(1):55–68.
162. Pentz S, Horler H. A cryopreservative procedure for storing cultivated and uncultivated amniotic fluid cells in liquid nitrogen. J Med Genet. 1980;17(6):472–5.
163. Schmidt D, et al. Cryopreserved amniotic fluid-derived cells: a lifelong autologous fetal stem cell source for heart valve tissue engineering. J Heart Valve Dis. 2008;17(4):446–55. discussion 455.
164. Niknejad H, et al. Serum-free cryopreservation of human amniotic epithelial cells before and after isolation from their natural scaffold. Cryobiology. 2013;67(1):56–63.
165. Cho HJ, et al. Evaluation of cell viability and apoptosis in human amniotic fluid-derived stem cells with natural cryoprotectants. Cryobiology. 2014;68(2):244–50.
166. Mathew S, et al. Factors which affect growth of amniotic fluid cells. Indian J Med Res. 1990;92:86–8.
167. Bishara A, et al. Determination of surface bound and shed HLA antigens of amniotic cells by ELISA. Isr J Med Sci. 1990;26(1):34–8.
168. Byrne D, Azar G, Nicolaides K. Why cell culture is successful after early amniocentesis. Fetal Diagn Ther. 1991;6(1–2):84–6.
169. Persutte WH, Lenke RR. Failure of amniotic-fluid-cell growth: is it related to fetal aneuploidy? Lancet. 1995;345(8942):96–7.
170. Mosquera A, et al. Simultaneous decrease of telomere length and telomerase activity with ageing of human amniotic fluid cells. J Med Genet. 1999;36(6):494–6.
171. Fuchs JR, et al. Diaphragmatic reconstruction with autologous tendon engineered from mesenchymal amniocytes. J Pediatr Surg. 2004;39(6):834–8. discussion 834–8.
172. Kunisaki SM, Freedman DA, Fauza DO. Fetal tracheal reconstruction with cartilaginous grafts engineered from mesenchymal amniocytes. J Pediatr Surg. 2006;41(4):675–82. discussion 675–82.
173. Pan HC, et al. Post-injury regeneration in rat sciatic nerve facilitated by neurotrophic factors secreted by amniotic fluid mesenchymal stem cells. J Clin Neurosci. 2007;14(11):1089–98.
174. Schmidt D, et al. Prenatally fabricated autologous human living heart valves based on amniotic fluid derived progenitor cells as single cell source. Circulation. 2007;116(11 Suppl):164–70.
175. Dai W, Kloner RA. Myocardial regeneration by human amniotic fluid stem cells: challenges to be overcome. J Mol Cell Cardiol. 2007;42(4):730–2.
176. Kunisaki SM, et al. A comparative analysis of cartilage engineered from different perinatal mesenchymal progenitor cells. Tissue Eng. 2007;13(11):2633–44.
177. Steigman SA, et al. Sternal repair with bone grafts engineered from amniotic mesenchymal stem cells. J Pediatr Surg. 2009;44(6):1120–6. discussion 1126.
178. Peister A, et al. Amniotic fluid stem cells produce robust mineral deposits on biodegradable scaffolds. Tissue Eng Part A. 2009;15(10):3129–38.
179. Pan HC, et al. Combination of G-CSF administration and human amniotic fluid mesenchymal stem cell transplantation promotes peripheral nerve regeneration. Neurochem Res. 2009;34(3):518–27.
180. Donaldson AE, et al. Human amniotic fluid stem cells do not differentiate into dopamine neurons in vitro or after transplantation in vivo. Stem Cells Dev. 2009;18(7):1003–12.
181. Pan HC, et al. Human amniotic fluid mesenchymal stem cells in combination with hyperbaric oxygen augment peripheral nerve regeneration. Neurochem Res. 2009;34(7):1304–16.
182. Cananzi M, Atala A, De Coppi P. Stem cells derived from amniotic fluid: new potentials in regenerative medicine. Reprod Biomed Online. 2009;18 Suppl 1:17–27.

183. Liu YW, et al. Xenografted human amniotic fluid-derived stem cell as a cell source in therapeutic angiogenesis. Int J Cardiol. 2013;168(1):66–75.
184. Turner CG, et al. Preclinical regulatory validation of an engineered diaphragmatic tendon made with amniotic mesenchymal stem cells. J Pediatr Surg. 2011;46(1):57–61.
185. Da Sacco S, De Filippo RE, Perin L. Amniotic fluid as a source of pluripotent and multipotent stem cells for organ regeneration. Curr Opin Organ Transplant. 2011;16(1):101–5.
186. Antonucci I, et al. Amniotic fluid as a rich source of mesenchymal stromal cells for transplantation therapy. Cell Transplant. 2011;20(6):789–95.
187. Weber B, Zeisberger SM, Hoerstrup SP. Prenatally harvested cells for cardiovascular tissue engineering: fabrication of autologous implants prior to birth. Placenta. 2011;32 Suppl 4:S316–9.
188. Mirabella T, et al. Recruitment of host's progenitor cells to sites of human amniotic fluid stem cells implantation. Biomaterials. 2011;32(18):4218–27.
189. Skardal A, et al. Bioprinted amniotic fluid-derived stem cells accelerate healing of large skin wounds. Stem Cells Transl Med. 2012;1(11):792–802.
190. Turner CG, et al. Craniofacial repair with fetal bone grafts engineered from amniotic mesenchymal stem cells. J Surg Res. 2012;178(2):785–90.
191. Klemmt P. Application of amniotic fluid stem cells in basic science and tissue regeneration. Organogenesis. 2012;8(3):76.
192. Santana AC, et al. Protective effects of human amniotic fluid stem cells in a model of aorta allograft vasculopathy in rats. Transplant Proc. 2012;44(8):2490–4.
193. Tajiri N, et al. Intravenous grafts of amniotic fluid-derived stem cells induce endogenous cell proliferation and attenuate behavioral deficits in ischemic stroke rats. PLoS One. 2012;7(8):e43779.
194. Kim BS, et al. Human amniotic fluid stem cell injection therapy for urethral sphincter regeneration in an animal model. BMC Med. 2012;10:94.
195. Rodrigues MT, et al. Amniotic fluid-derived stem cells as a cell source for bone tissue engineering. Tissue Eng Part A. 2012;18(23–24):2518–27.
196. Yang DY, et al. Dual regeneration of muscle and nerve by intravenous administration of human amniotic fluid-derived mesenchymal stem cells regulated by stromal cell-derived factor-1 alpha in a sciatic nerve injury model. J Neurosurg. 2012;116(6):1357–67.
197. Weber B, et al. Prenatally engineered autologous amniotic fluid stem cell-based heart valves in the fetal circulation. Biomaterials. 2012;33(16):4031–43.
198. Vadasz S, et al. Second and third trimester amniotic fluid mesenchymal stem cells can repopulate a de-cellularized lung scaffold and express lung markers. J Pediatr Surg. 2014;49(11):1554–63.
199. Zavatti M, et al. Critical-size bone defect repair using amniotic fluid stem cell/collagen constructs: effect of oral ferutinin treatment in rats. Life Sci. 2015;121:174–83.
200. Payton S. Basic research: urethral sphincter regeneration using stem cells from amniotic fluid. Nat Rev Urol. 2014;11(7):365.
201. Kim JA, et al. MYOD mediates skeletal myogenic differentiation of human amniotic fluid stem cells and regeneration of muscle injury. Stem Cell Res Ther. 2013;4(6):147.
202. Yang JD, et al. Effect of amniotic fluid stem cells and amniotic fluid cells on the wound healing process in a white rat model. Arch Plast Surg. 2013;40(5):496–504.
203. Yang CM, et al. Engraftment of genetically modified human amniotic fluid-derived progenitor cells to produce coagulation factor IX after in utero transplantation in mice. Cell Biol Int. 2013;37(5):420–9.
204. Barboni B, et al. Synthetic bone substitute engineered with amniotic epithelial cells enhances bone regeneration after maxillary sinus augmentation. PLoS One. 2013;8(5):e63256.
205. Kim J, et al. In vitro osteogenic differentiation of human amniotic fluid-derived stem cells on a poly(lactide-co-glycolide) (PLGA)-bladder submucosa matrix (BSM) composite scaffold for bone tissue engineering. Biomed Mater. 2013;8(1):014107.

206. Roh DH, et al. Transplantation of human umbilical cord blood or amniotic epithelial stem cells alleviates mechanical allodynia after spinal cord injury in rats. Cell Transplant. 2013;22(9):1577–90.
207. Hartmann-Fritsch F, et al. Human amniotic fluid derived cells can competently substitute dermal fibroblasts in a tissue-engineered dermo-epidermal skin analog. Pediatr Surg Int. 2013;29(1):61–9.
208. Berardinelli P, et al. Role of amniotic fluid mesenchymal cells engineered on MgHA/collagen-based scaffold allotransplanted on an experimental animal study of sinus augmentation. Clin Oral Investig. 2013;17(7):1661–75.
209. Tsangaris G, et al. The amniotic fluid cell proteome. Electrophoresis. 2005;26(6):1168–73.
210. Skardal A, et al. Substrate elasticity controls cell proliferation, surface marker expression and motile phenotype in amniotic fluid-derived stem cells. J Mech Behav Biomed Mater. 2013;17:307–16.
211. Steigman SA, et al. Preclinical regulatory validation of a 3-stage amniotic mesenchymal stem cell manufacturing protocol. J Pediatr Surg. 2008;43(6):1164–9.
212. De Gemmis P, et al. A real-time PCR approach to evaluate adipogenic potential of amniotic fluid-derived human mesenchymal stem cells. Stem Cells Dev. 2006;15(5):719–28.
213. McLaughlin D, et al. Stable expression of a neuronal dopaminergic progenitor phenotype in cell lines derived from human amniotic fluid cells. J Neurosci Res. 2006;83(7):1190–200.
214. Jiang TM, et al. Schwann-like cells can be induction from human nestin-positive amniotic fluid mesenchymal stem cells. In Vitro Cell Dev Biol Anim. 2010;46(9):793–800.
215. Pfeiffer S, McLaughlin D. In vitro differentiation of human amniotic fluid-derived cells: augmentation towards a neuronal dopaminergic phenotype. Cell Biol Int. 2010;34(9):959–67.
216. Orciani M, et al. Neurogenic potential of mesenchymal-like stem cells from human amniotic fluid: the influence of extracellular growth factors. J Biol Regul Homeost Agents. 2011;25(1):115–30.
217. Jezierski A, et al. Human amniotic fluid cells form functional gap junctions with cortical cells. Stem Cells Int. 2012;2012:607161.
218. Preitschopf A, et al. mTORC1 is essential for early steps during Schwann cell differentiation of amniotic fluid stem cells and regulates lipogenic gene expression. PLoS One. 2014;9(9):e107004.
219. Maraldi T, et al. Human amniotic fluid stem cells: neural differentiation in vitro and in vivo. Cell Tissue Res. 2014;357(1):1–13.
220. Hartmann K, et al. Amniotic fluid derived stem cells give rise to neuron-like cells without a further differentiation potential into retina-like cells. Am J Stem Cells. 2013;2(2):108–18.
221. Park JS, et al. Chondrogenic potential of stem cells derived from amniotic fluid, adipose tissue, or bone marrow encapsulated in fibrin gels containing TGF-beta3. Biomaterials. 2011;32(32):8139–49.
222. Preitschopf A, et al. Chondrogenic differentiation of amniotic fluid stem cells and their potential for regenerative therapy. Stem Cell Rev. 2012;8(4):1267–74.
223. Carraro G, et al. Human amniotic fluid stem cells can integrate and differentiate into epithelial lung lineages. Stem Cells. 2008;26(11):2902–11.
224. Trovato L, et al. Pluripotent stem cells isolated from human amniotic fluid and differentiation into pancreatic beta-cells. J Endocrinol Invest. 2009;32(11):873–6.
225. Zou G, et al. Induction of pancreatic beta-cell-like cells from CD44+/CD105+ human amniotic fluids via epigenetic regulation of the pancreatic and duodenal homeobox factor 1 promoter. DNA Cell Biol. 2011;30(9):739–48.
226. Li B, et al. Neuronal restrictive silencing factor silencing induces human amniotic fluid-derived stem cells differentiation into insulin-producing cells. Stem Cells Dev. 2011;20(7):1223–31.
227. Chun SY, et al. Pdx1 and controlled culture conditions induced differentiation of human amniotic fluid-derived stem cells to insulin-producing clusters. J Tissue Eng Regen Med. 2015;9(5):540–9.

228. Carnevale G, et al. In vitro differentiation into insulin-producing beta-cells of stem cells isolated from human amniotic fluid and dental pulp. Dig Liver Dis. 2013;45(8):669–76.
229. Zhou Y, et al. Genetic modification of primate amniotic fluid-derived stem cells produces pancreatic progenitor cells in vitro. Cells Tissues Organs. 2013;197(4):269–82.
230. Gage BK, et al. Cellular reprogramming of human amniotic fluid cells to express insulin. Differentiation. 2010;80(2–3):130–9.
231. Zhang P, et al. Endothelial differentiation of amniotic fluid-derived stem cells: synergism of biochemical and shear force stimuli. Stem Cells Dev. 2009;18(9):1299–308.
232. Connell JP, et al. Formation of functional gap junctions in amniotic fluid-derived stem cells induced by transmembrane co-culture with neonatal rat cardiomyocytes. J Cell Mol Med. 2013;17(6):774–81.
233. Walther G, Gekas J, Bertrand OF. Amniotic stem cells for cellular cardiomyoplasty: promises and premises. Catheter Cardiovasc Interv. 2009;73(7):917–24.
234. Bollini S, et al. In vitro and in vivo cardiomyogenic differentiation of amniotic fluid stem cells. Stem Cell Rev. 2011;7(2):364–80.
235. Guan X, et al. In vitro cardiomyogenic potential of human amniotic fluid stem cells. J Tissue Eng Regen Med. 2011;5(3):220–8.
236. Ma X, et al. Clone-derived human AF-amniotic fluid stem cells are capable of skeletal myogenic differentiation in vitro and in vivo. J Tissue Eng Regen Med. 2012;6(8):598–613.
237. Maioli M, et al. Amniotic fluid stem cells morph into a cardiovascular lineage: analysis of a chemically induced cardiac and vascular commitment. Drug Des Devel Ther. 2013;7: 1063–73.
238. Gao Y, et al. Amniotic fluid-derived stem cells demonstrated cardiogenic potential in indirect co-culture with human cardiac cells. Ann Biomed Eng. 2014;42(12):2490–500.
239. Antonucci I, et al. Isolation of osteogenic progenitors from human amniotic fluid using a single step culture protocol. BMC Biotechnol. 2009;9:9.
240. Chen Q, et al. AFM studies of cellular mechanics during osteogenic differentiation of human amniotic fluid-derived stem cells. Anal Sci. 2010;26(10):1033–7.
241. Hipp JA, et al. Ethanol alters the osteogenic differentiation of amniotic fluid-derived stem cells. Alcohol Clin Exp Res. 2010;34(10):1714–22.
242. Sun H, et al. Osteogenic differentiation of human amniotic fluid-derived stem cells induced by bone morphogenetic protein-7 and enhanced by nanofibrous scaffolds. Biomaterials. 2010;31(6):1133–9.
243. Peister A, et al. Osteogenic differentiation of amniotic fluid stem cells. Biomed Mater Eng. 2008;18(4–5):241–6.
244. Higuchi A, et al. Osteoblast differentiation of amniotic fluid-derived stem cells irradiated with visible light. Tissue Eng Part A. 2011;17(21–22):2593–602.
245. De Rosa A, et al. Amniotic fluid-derived mesenchymal stem cells lead to bone differentiation when cocultured with dental pulp stem cells. Tissue Eng Part A. 2011;17(5–6):645–53.
246. Barba M, et al. Lim mineralization protein 3 induces the osteogenic differentiation of human amniotic fluid stromal cells through Kruppel-like factor-4 downregulation and further bone-specific gene expression. J Biomed Biotechnol. 2012;2012:813894.
247. Rodrigues MT, et al. The effect of differentiation stage of amniotic fluid stem cells on bone regeneration. Biomaterials. 2012;33(26):6069–78.
248. Liu M, Li Y, Yang ST. Effects of naringin on the proliferation and osteogenic differentiation of human amniotic fluid-derived stem cells. J Tissue Eng Regen Med. 2014. doi:10.1002/term.1911.
249. Liu M, Li Y, Yang ST. Curculigoside improves osteogenesis of human amniotic fluid-derived stem cells. Stem Cells Dev. 2014;23(2):146–54.
250. de Lara Janz F, et al. Simvastatin induces osteogenic differentiation in human amniotic fluid mesenchymal stem cells (AFMSC). Fundam Clin Pharmacol. 2014;28(2):211–6.
251. Zavatti M, et al. Ferutinin promotes proliferation and osteoblastic differentiation in human amniotic fluid and dental pulp stem cells. Life Sci. 2013;92(20–21):993–1003.
252. Decembrini S, et al. Comparative analysis of the retinal potential of embryonic stem cells and amniotic fluid-derived stem cells. Stem Cells Dev. 2011;20(5):851–63.

253. Ghionzoli M, et al. Human amniotic fluid stem cell differentiation along smooth muscle lineage. FASEB J. 2013;27(12):4853–65.
254. Li Y, et al. Differentiation of human amniotic fluid-derived mesenchymal stem cells into type II alveolar epithelial cells in vitro. Int J Mol Med. 2014;33(6):1507–13.
255. Kang HH, et al. Urothelial differentiation of human amniotic fluid stem cells by urothelium specific conditioned medium. Cell Biol Int. 2014;38(4):531–7.
256. Siegel N, et al. Stem cells in amniotic fluid as new tools to study human genetic diseases. Stem Cell Rev. 2007;3(4):256–64.
257. Rosner M, et al. Human amniotic fluid stem cells as a model for functional studies of genes involved in human genetic diseases or oncogenesis. Oncotarget. 2011;2(9):705–12.
258. Rehni AK, et al. Amniotic fluid derived stem cells ameliorate focal cerebral ischaemia-reperfusion injury induced behavioural deficits in mice. Behav Brain Res. 2007;183(1): 95–100.
259. Chun SY, et al. Human amniotic fluid stem cell-derived muscle progenitor cell therapy for stress urinary incontinence. J Korean Med Sci. 2012;27(11):1300–7.
260. Liu T, et al. CD44+/CD105+ human amniotic fluid mesenchymal stem cells survive and proliferate in the ovary long-term in a mouse model of chemotherapy-induced premature ovarian failure. Int J Med Sci. 2012;9(7):592–602.
261. Zheng YB, et al. Amniotic-fluid-derived mesenchymal stem cells overexpressing interleukin-1 receptor antagonist improve fulminant hepatic failure. PLoS One. 2012;7(7):e41392.
262. Sedrakyan S, et al. Injection of amniotic fluid stem cells delays progression of renal fibrosis. J Am Soc Nephrol. 2012;23(4):661–73.
263. Zagoura DS, et al. Therapeutic potential of a distinct population of human amniotic fluid mesenchymal stem cells and their secreted molecules in mice with acute hepatic failure. Gut. 2012;61(6):894–906.
264. Peng SY, et al. Therapeutic potential of amniotic-fluid-derived stem cells on liver fibrosis model in mice. Taiwan J Obstet Gynecol. 2014;53(2):151–7.
265. Morigi M, De Coppi P. Cell therapy for kidney injury: different options and mechanisms--mesenchymal and amniotic fluid stem cells. Nephron Exp Nephrol. 2014;126(2):59.
266. Jun EK, et al. Hypoxic conditioned medium from human amniotic fluid-derived mesenchymal stem cells accelerates skin wound healing through TGF-beta/SMAD2 and PI3K/Akt pathways. Int J Mol Sci. 2014;15(1):605–28.
267. Villani V, et al. Amniotic fluid stem cells prevent beta-cell injury. Cytotherapy. 2014;16(1):41–55.
268. Stenson WF. Preventing necrotising enterocolitis with amniotic fluid stem cells. Gut. 2014;63(2):218–9.
269. Li Y, et al. Therapeutic effects of amniotic fluid-derived mesenchymal stromal cells on lung injury in rats with emphysema. Respir Res. 2014;15:120.
270. Xiao GY, et al. Amniotic fluid stem cells prevent follicle atresia and rescue fertility of mice with premature ovarian failure induced by chemotherapy. PLoS One. 2014;9(9):e106538.
271. Bottai D, et al. Third trimester NG2-positive amniotic fluid cells are effective in improving repair in spinal cord injury. Exp Neurol. 2014;254:121–33.
272. Chang YJ, et al. Amniotic fluid stem cells with low gamma-interferon response showed behavioral improvement in Parkinsonism rat model. PLoS One. 2013;8(9):e76118.
273. Lai D, et al. Human amniotic fluid stem cells have a potential to recover ovarian function in mice with chemotherapy-induced sterility. BMC Dev Biol. 2013;13:34.
274. Sun D, et al. Therapeutic effects of human amniotic fluid-derived stem cells on renal interstitial fibrosis in a murine model of unilateral ureteral obstruction. PLoS One. 2013;8(5):e65042.
275. Turner CG, et al. Intra-amniotic delivery of amniotic-derived neural stem cells in a syngeneic model of spina bifida. Fetal Diagn Ther. 2013;34(1):38–43.
276. Grisafi D, et al. Human amniotic fluid stem cells protect rat lungs exposed to moderate hyperoxia. Pediatr Pulmonol. 2013;48(11):1070–80.

277. Kim SW, et al. Amniotic mesenchymal stem cells with robust chemotactic properties are effective in the treatment of a myocardial infarction model. Int J Cardiol. 2013;168(2): 1062–9.
278. Pederiva F, et al. Amniotic fluid stem cells rescue both in vitro and in vivo growth, innervation, and motility in nitrofen-exposed hypoplastic rat lungs through paracrine effects. Cell Transplant. 2013;22(9):1683–94.
279. Orciani M, et al. Potential role of culture mediums for successful isolation and neuronal differentiation of amniotic fluid stem cells. Int J Immunopathol Pharmacol. 2008;21(3): 595–602.
280. Tweedell KS. New paths to pluripotent stem cells. Curr Stem Cell Res Ther. 2008;3(3): 151–62.
281. Giuliani M, et al. Human mesenchymal stem cells derived from induced pluripotent stem cells down-regulate NK-cell cytolytic machinery. Blood. 2011;118(12):3254–62.
282. Kobari L, et al. Human induced pluripotent stem cells can reach complete terminal maturation: in vivo and in vitro evidence in the erythropoietic differentiation model. Haematologica. 2012;97(12):1795–803.
283. Spinelli V, Guillot PV, De Coppi P. Induced pluripotent stem (iPS) cells from human fetal stem cells (hFSCs). Organogenesis. 2013;9(2):101–10.
284. Simantov R. Amniotic stem cell international. Reprod Biomed Online. 2008;16(4):597–8.
285. Wu HW, et al. A microfluidic device for separation of amniotic fluid mesenchymal stem cells utilizing louver-array structures. Biomed Microdevices. 2009;11(6):1297–307.
286. Wu HW, et al. The culture and differentiation of amniotic stem cells using a microfluidic system. Biomed Microdevices. 2009;11(4):869–81.
287. Zheng YM, et al. Development of cloned embryos from porcine neural stem cells and amniotic fluid-derived stem cells transfected with enhanced green fluorescence protein gene. Reproduction. 2009;137(5):793–801.
288. Bryzek A, et al. Expression and co-expression of surface markers of pluripotency on human amniotic cells cultured in different growth media. Ginekol Pol. 2013;84(12):1012–24.
289. Soong YK, et al. The use of human amniotic fluid mesenchymal stem cells as the feeder layer to establish human embryonic stem cell lines. J Tissue Eng Regen Med. 2013. doi:10.1002/term.1702.
290. Chen CP, et al. Proteome differences between male and female fetal cells in amniotic fluid. OMICS. 2013;17(1):16–26.
291. Werber B, Martin E. A prospective study of 20 foot and ankle wounds treated with cryopreserved amniotic membrane and fluid allograft. J Foot Ankle Surg. 2013;52(5):615–21.
292. Sessarego N, et al. Multipotent mesenchymal stromal cells from amniotic fluid: solid perspectives for clinical application. Haematologica. 2008;93(3):339–46.
293. Horwitz EM. Stem cell plasticity: the growing potential of cellular therapy. Arch Med Res. 2003;34(6):600–6.
294. Bajek A, et al. Human amniotic-fluid-derived stem cells: a unique source for regenerative medicine. Expert Opin Biol Ther. 2014;14(6):831–9.
295. Pozzobon M, Piccoli M, De Coppi P. Stem cells from fetal membranes and amniotic fluid: markers for cell isolation and therapy. Cell Tissue Bank. 2014;15(2):199–211.

Chapter 11
MicroRNA Expression in Amniotic Fluid Cells

Maria Ribecco-Lutkiewicz, Dao Ly, Caroline Sodja, Julie Haukenfrers, Brandon Smith, Qing Yan Liu, Marianna Sikorska, and Mahmud Bani-Yaghoub

Introduction

MicroRNAs (miRNAs) are small non-coding functional RNAs of approximately 19–22 nucleotides that downregulate protein expression by base-pairing with complementary sequences within messenger RNA molecules [1–4]. The first miRNA was discovered in 1993 during a study of *lin-4*, a gene known for controlling the timing of larval development in *C. elegans* [5]. This pioneering report has led to the systematic discovery of hundreds of miRNAs, using innovative computational analysis of small RNA sequencing data [See [6] for a review]. To date, about 2800 human miRNA candidates have been reported [1], of which more than 1100 have been validated, and recorded in the miRBase (http://www.mirbase.org/). In addition, Mirror Suite has been developed as a miR target site (http://www.mirrorsuite.cs.huji.ac.il/). MiRNA genes can encode individual miRNAs (monocistronic) or miRNA clusters (polycistronic). Polycistronic miRNAs often share sequence similarity, but can be from different families [7]. Using mathematical modeling of

M. Ribecco-Lutkiewicz, Ph.D. • D. Ly, B.Sc. • C. Sodja, M.Sc. • J. Haukenfrers, B.Sc.
B. Smith, B.Sc., M.Sc. • Q.Y. Liu, Ph.D. • M. Sikorska, Ph.D.
Translational Bioscience, Human Health Therapeutics Portfolio, National Research Council Canada, Bldg. M54, 1200 Montreal Road, Ottawa, ON, Canada, K1A 0R6
e-mail: Maria.Ribecco-Lutkiewicz@nrc-cnrc.gc.ca; Dao.Ly@nrc-cnrc.gc.ca; Caroline.Sodja@nrc-cnrc.gc.ca; Julie.Haukenfrers@nrc.gc.ca; Brandon.Smith@nrc-cnrc.gc.ca; Qing.Liu@nrc-cnrc.gc.ca; Marianna.sikorska@bell.net

M. Bani-Yaghoub, Ph.D. (✉)
Department of Cellular and Molecular Medicine, University of Ottawa, Faculty of Medicine, Ottawa, ON, Canada
e-mail: Mahmud.Bani@nrc-cnrc.gc.ca; mBani@uOttawa.ca

D.O. Fauza, M. Bani (eds.), *Fetal Stem Cells in Regenerative Medicine*, Stem Cell Biology and Regenerative Medicine,
DOI 10.1007/978-1-4939-3483-6_11

miRNA turnover, a half-life of 119 h has been estimated for miRNAs, suggesting that they are much more stable than messenger RNA [8].

How Do MicroRNAs Work?

Transcription of endogenous miRNA genes by RNA polymerase II generates primary miRNAs (pri-miRNAs) that are processed in the nucleus into pre-miRNAs (Fig. 11.1; also see [3] for a review). This processing is performed by Drosha RNase III endonuclease. Pre-miRNAs are exported to the cytoplasm by Exportin 5 and are processed by another RNAse III endonuclease (Dicer) into mature miRNAs. Subsequently, the mature miRNA is incorporated into the RNA-induced silencing complex (RISC). RISC was first defined as a large RNA–protein complex with sequence-specific RNA cleavage activity. MiRNAs can direct RISC to downregulate gene expression by either of two post-transcriptional mechanisms: mRNA cleavage or translational repression [3]. In cases where there is a high degree of

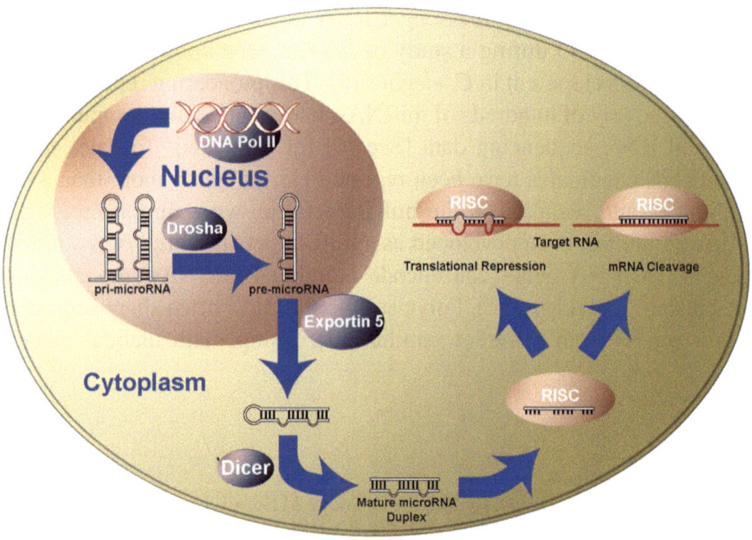

Fig. 11.1 Silencing of protein expression by miRNAs. MiRNAs control gene expression by binding to partially or fully complementary sequences in target mRNAs, resulting in translational repression or mRNA degradation. MiRNAs genes are transcribed by RNA polymerase II, producing long RNA primary precursors (pri-miRNAs). The pri-miRNAs are processed in the nucleus by an RNase III enzyme, Drosha, to stem-loop structures of approximately 70-nucleotides (pre-miRNAs). The pre-miRNAs are exported into the cytoplasm by exportin 5 where they are cleaved by another RNAse III enzyme, Dicer, generating mature miRNA duplexes. The duplexes consist of two imperfect complementary RNA strands; one of these strands is preferentially loaded onto the RNA-induced silencing complex (RISC). RISC-loaded miRNAs are single-stranded and can affect translation by binding to imperfect complementary sites typically located in the 3′ untranslated regions (3′-UTRs) of their mRNA targets. The resulting miRNA:mRNA duplex leads to repressed translation or degredation of mRNA

complementarity between the microRNA and its mRNA target, RISC can stabilize the miRNA target strand, guide it to the target mRNA, and activate endonuclease cleavage. RISC can also inhibit productive translation, if the miRNA does not have sufficient base-pairing to activate endonuclease cleavage of its target mRNA [3].

The interaction between a miRNA and its target mRNA does not require perfect sequence complementarity, permitting a high degree of variability in the mRNA target sequence. Thus, a single miRNA has the potential to regulate many mRNAs. Furthermore, mRNAs can contain numerous miRNA target sites and can be regulated by multiple miRNAs. It has been suggested that the majority of protein coding genes are regulated by miRNAs, depending on the developmental, cellular and physiological context [9]. Because of their ability to simultaneously control the expression of a large number of genes, miRNAs are perfect candidates as master regulators to maintain or switch cell fate by regulating biological processes such as cell survival, proliferation and differentiation [10].

Extracellular and Intracellular MicroRNAs

Both cell contact-dependent and -independent transfer of functional miRNAs appear to play essential roles in regulating protein expression in various biological processes. Data from several laboratories confirm that miRNAs can be shuttled between cells by microvesicles (MVs), plasma membrane fragments of 0.1–1 μm in diameter [11]. In particular, MVs released from donor bone marrow-derived mesenchymal cells can deliver miRNAs into recipient cells, reducing target-specific protein levels [12]. Extracellular transfer of functional miRNAs between cells has been also reported for exosomes, extracellular vesicles of up to 100 nm in diameter that are released from a late endosomal cellular compartment [2, 13–15]. Extracellular miRNAs have been identified in various human body fluids, including amniotic fluid, plasma, cerebrospinal fluid and saliva [16–18]. An important issue regarding extracellular miRNAs is their cellular origin. This is particularly important for amniotic fluid, which contains cells originating from different fetal tissues. One way to address this question is to obtain amniotic fluid at different stages of pregnancy and establish miRNA profiles for each stage by comparing extracellular and cellular miRNAs. This approach may also provide an opportunity to use miRNA profiles as diagnostic biomarkers to identify and monitor developmental, physiological and pathological conditions.

In addition to microvesicles and exosomes, microRNAs can be also transferred through gap junctions in a cell contact-dependent manner (Fig. 11.2), leading to down-regulation of target-specific proteins [14, 19–22]. Gap junctions are intercellular channels of 1.5–2 nm diameter formed by hemi-channels (connexons), which are in turn formed by six connexin monomers. Gap junctions permit direct cell–cell transfer of small ions and molecules such as ATP, amino acids, glucose, glutathione, small interfering (siRNAs) and miRNAs among neighbouring cells [23]. They facilitate miRNA transfer from one cell to another cell, coordinating the regulation of various biological processes such as cell survival, proliferation and differentiation [20, 21, 24]. This transport may occur as a single or double stranded mature miRNA

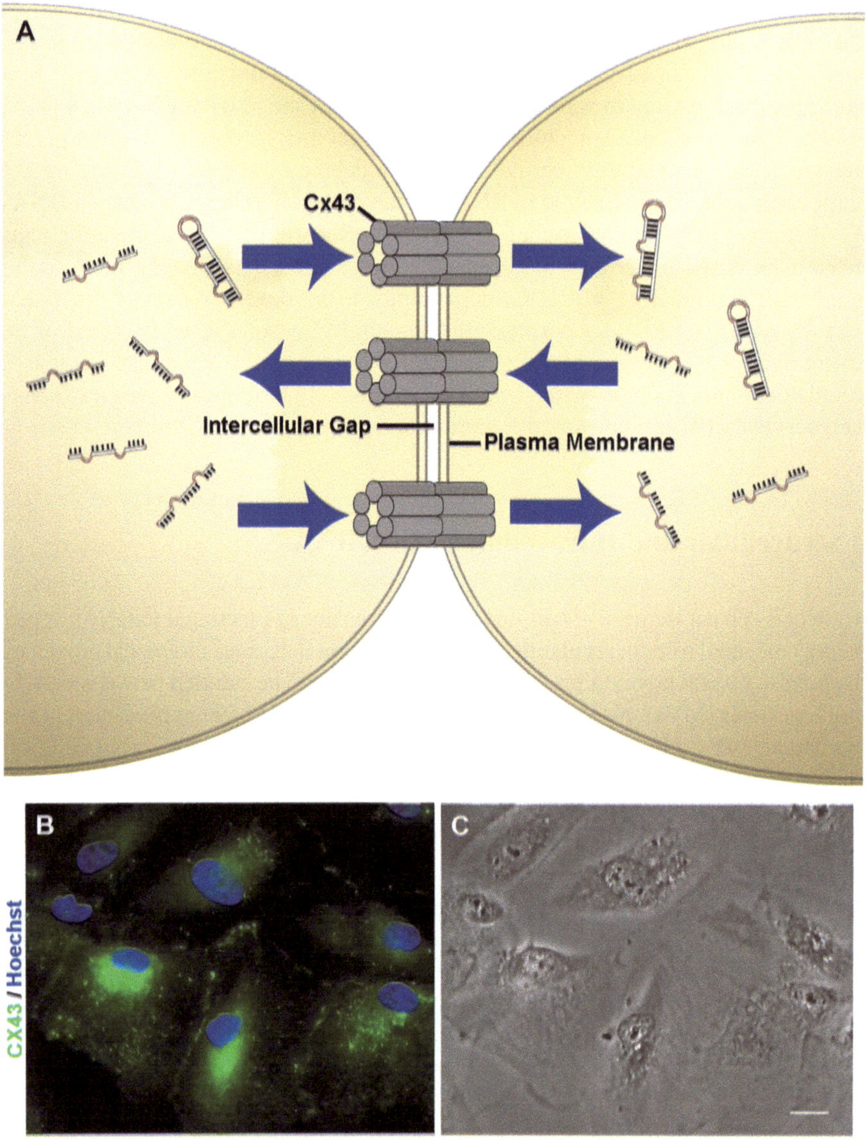

Fig. 11.2 Intercellular transfer of miRNAs through gap junctions. (**A**) In addition to their effects at the intracellular level, miRNAs can be transported through exosomes or gap junction channels to neighbouring cells where they can affect expression of their target mRNAs. (**B**) Amniotic fluid cells (isolated at 26 weeks of gestation) express abundant levels of Connexin43 (Cx43, green), a protein widely reported for its involvement in the formation of functional gap junctions. Nuclei have been stained with Hoechst (blue). (**C**) Phase contrast. Scale bar: 10 μm

or its precursor (pre-miRNA). The transfer of miRNAs through gap junctions is possibly facilitated by RNA-binding proteins (RBPs) [14].

Several lines of evidence confirm that miRNAs have the capability to regulate the establishment and maintenance of pluripotent stem cells [25, 26]. More specifically, miRNAs can be used to directly silence pluripotency by repressing the related transcription factors such as OCT4, SOX2, KLF4, MYC and NANOG, and to promote differentiation [27]. Since miRNAs and their targets are expected to have correlated expression patterns, computational analyses comparing stem cells with differentiated cells have further resulted in the identification of miRNAs that repress pluripotency as well as the ones that are involved in cell differentiation [27–29]. For instance, miRNA profiling of iPSC-derived endothelial cells has led to the identification of 376 miRNAs, including miR-20a, -20b, -222 and -210, which are known to be involved in endothelial cell differentiation [30]. MiR519c-3p, miR-193a-5p, MiR-650, miR-215 and most significantly miR-590-5p and miR-611 had the highest expression in iPSC-derived endothelial cells compared with their expression in human umbilical vein endothelial cells (HUVEC) [30].

About 360 miRNAs have been identified in amniotic fluid, with miR-518e, miR-335, miR-302c, miR515-3p, miR-452, miR-892a, miR671-5p, miR-515-5p, miR-137, and miR-593 as the most abundantly expressed species [16]. In another study, focusing on miRNA expression in amniotic fluid cells [31], two morphologically different subpopulations of fetal mesenchymal stem cells (MSC) were isolated from amniotic fluid cells: spindle-shaped (SS) and round-shaped (RS). While both subpopulations expressed mesenchymal stem cell markers CD73, CD105, CD166, and integrins, CD29 and CD49e at similar levels, they showed differences in pluripotency, proliferation, differentiation and protein expression. A total of 32 miRNAs were differentially expressed between the SS-AF-MSCs and RS-AFMSCs [31]. In particular, miR-21 was expressed at higher levels in RS-AF-MSCs compared with SS-AF-MSCs. Interestingly, SS-AF-MSCs exhibited higher expression levels of SOX2 compared with RS-AF-MSCs, and the induction of miR-21 downregulated SOX2 expression in SS-AF-MSCs, resulting in reduced clonogenic and proliferative potential. The opposite effect was observed upon miR-21 inhibition in RS-AF-MSCs, which led to an enhanced proliferation rate. Furthermore, miR-21 induction accelerated osteogenesis and impaired adipogenesis and chondrogenesis in SS-AF-MSCs. These results indicate that miR-21 may serve as a key regulator of proliferation and differentiation in amniotic fluid MSCs by controlling SOX2 expression. Using a fluorescently-labeled morpholino mimic of miR-21 (with an approximate molecular weight, width and length of 2 kDa, 1 nm and 7.5 nm, respectively), our laboratory has shown that amniotic fluid cells expressing abundant levels of Connexin43 and with functional gap junctions, have the capability to transfer miRNA-like molecules from one cell to another [32, 33]. In the work presented in this chapter, we have selected several key miRNAs following a microarray analysis, and have evaluated their expression in amniotic fluid cells (Fig. 11.3).

MiR-7e: Lethal-7 (also known as let-7) was the first human miRNA discovered. Let-7 family members are known for their role(s) in controlling stem cell proliferation, maintenance and differentiation (See [34] for a review). There are 12 let-7

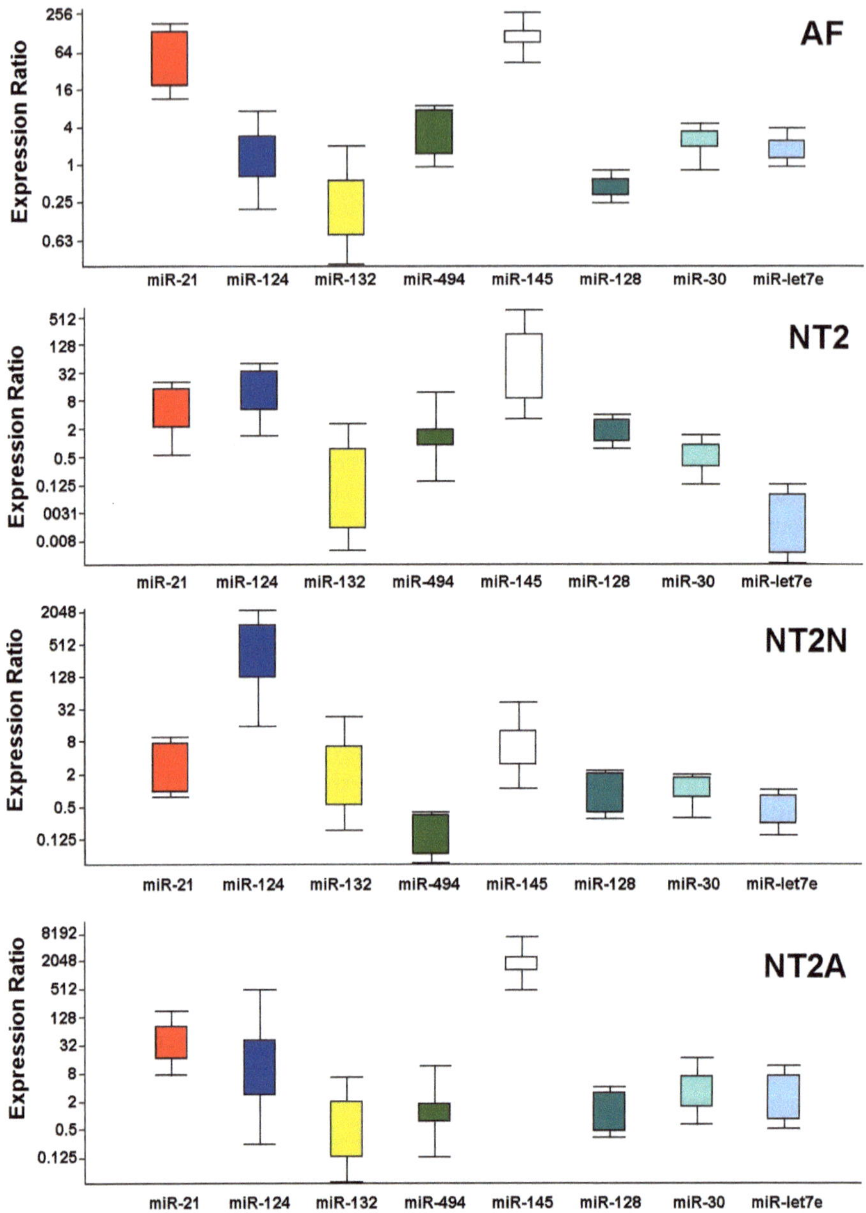

Fig. 11.3 Differential expression of miRNA in amniotic fluid cells. The expression of a selected group of miRNAs was evaluated in amniotic fluid (AF) cells, NT2 cells, NT2 neurons (NT2N) and NT2 astrocytes (NT2A) and scored against those of HEK293 cells as reference*. Among the miRNAs tested, miR-21 and miR-145 were the most abundant in AF cells, whereas miR-128 and miR-132 showed the lowest expression levels. Similar to AF cells, other cell types maintained low levels of miR-132. In contrast, the highest expression of miR-145 was observed in NT2A cells

family members in human [35], many of which are encoded in polycistronic clusters. The levels of let-7 family members rise during embryogenesis. In particular, pri-let7a and pri-let7e are upregulated during mouse brain development [36].

MiR-21: There is abundant expression of miR-21 in MSC-derived cells [37] and the depletion of miR-21 enhances reprogramming efficiency in mouse embryonic fibroblasts [54]. In contrast, miR-21 overexpression in VEGF-treated iPSCs leads to upregulation of *VE-cadherin* and *Flk1* via targeting PTEN [38]. In addition to its expression in MSC-derived cells, miR-21 is highly expressed during fetal brain development [39]. It is significantly upregulated following spinal cord injury and there is a potential role for miR-21 in glial scar progression [40]. MiR-21 also protects neurons following ischemia [41], and inhibits apoptosis and promotes angiogenesis after traumatic brain injury in rats [42]. These results suggest that the molecular signature of miR-21 can be used as a biomarker for therapeutic strategies [37].

MiR-30: The miR-30 family is essential for the maintenance of epithelial state. In particular, the expression of miR-30 family members is reduced during epithelial-mesenchymal transition (EMT) in human pancreatic cells, a phenomenon recognized in organogenesis [43]. In addition, overexpression of miR-30b in endothelial cells has been shown to increase vessel number and length by targeting Delta-Like 4 (DLL4), a membrane-bound ligand belonging to the Notch signaling family [44].

MiR-124: As one of the most abundant miRNAs in the brain, miR-124 demonstrates a spatiotemporal expression pattern in various cell types and it regulates a wide range of biological functions [45]. MiR-124, along with miR-7, has been shown to regulate the differentiation of neural stem and progenitor cells into specific neural cell types [46].

MiR-128: MiR-128 is known as a brain-enriched miRNA with a different expression time-point, compared with miR-124, during neuronal development [47]. It is expressed in adult neurons, and it modulates neuronal excitability and motor activity by suppressing the expression of several ion channels and signaling components of the extracellular signal-regulated kinase 2 (ERK2) network. A reduction of miR-128 expression in postnatal neurons causes increased motor activity and epilepsy in mice. In contrast, miR-128 overexpression attenuates neuronal responsiveness, suppresses motor activity, and alleviates motor abnormalities associated with seizure and Parkinson's disease [48]. Transduction of human iPS cells with miR-124 and miR-128 leads to upregulation of MAP2, NSE, GFAP and BDNF [47].

MiR-132: Because of its anti-inflammatory effect via the targeting of acetylcholinesterase and the subsequent increase in acetylcholine levels, a therapeutic role has been considered for miR-132 [49]. More recently, nanoparticles have been used to deliver miR-132 to human umbilical vein endothelial cells and enhance vessel formation following subcutaneous transplantation in a mouse model [50]. This strategy has the potential to be expanded to other applications in the field of tissue engineering.

MiR-145: The iPS cells derived from human amniotic epithelial cells and transfected with a miR-145 mutant gene that interferes with endogenous miR-145 activity show abundant expression of SOX2 and other stem cell markers, suggesting that SOX2 expression may be regulated by miR-145 [51]. Given the reciprocal inhibitory effects between miR-145 and the pluripotency factors, OCT4, SOX2 and KLF4 [29], miR-145 can be used to optimize the generation of iPSC lines. Furthermore, Götte et al. have shown that overexpression of miR-145 leads to downregulation of junctional adhesion molecule A (JAM-A), and the actin bundling protein, fascin [52].

MiR-494: MiR-494 is highly expressed in decidua-derived MSCs (dMSCs), and it regulates proliferation and differentiation of mesenchymal stem cells [53]. More specifically, it inhibits the growth and paracrine function of MSCs by arresting cell cycle at G1/S transition and downregulating CDK6, CCND1 and VEGF [53].

Among the microRNAs examined, miR-21 and miR-145 had the highest expression levels in amniotic fluid cells (Fig. 11.3). Using human NT2 neural progenitors (NT2), neurons (NT2N) and astrocytes (NT2A) for comparative analysis, our results also show relatively higher expression levels for miR-21, miR-124 and miR-145 in NT2 cells. MiR-124 maintained abundant expression in NT2N neurons, whereas miR-21 and more significantly miR-145 were highly expressed in NT2A astrocytes. These results support the presence of cell type-specific expression patterns for miR-21, miR-124 and miR-145 throughout differentiation. Furthermore, recent data from other laboratories also show that Let7, miR-21 and miR-29a serve as barriers to the initial stage of cell reprogramming, a process challenged by C-MYC [54, 55]. MiR-7 family members are known for regulating cell size in human embryonic stem cell-derived cardiac myocytes [56]. Together, these results prompted to determine whether the expression of microRNAs change during reprogramming and differentiation in amniotic fluid cells, which are further associated with changes in cell size and shape. Similar to other induced pluripotent cells, amniotic fluid cells demonstrated significant reduction in cell size during reprogramming (Fig. 11.4 A-D). The induction of pluripotency was confirmed by staining cells with TRA1-81 (Fig. 11.4 E, F), OCT4, KLF4, SOX2 and NANOG (Fig. 11.5). Parallel cultures of amniotic induced pluripotent stem cells were primed with neuronal, cardiac and endothelial differentiation media for 1 week and subsequently evaluated for miRNA expression. Our data show that the expression of miR-145 is reduced upon reprogramming in amniotic fluid cells, whereas various clones of differentiated AF-iPS cells demonstrated miR-145 upregulation within the early stage of differentiation (Fig. 11.6). Interestingly, there were no significant changes in miR-21 expression during cell reprogramming or differentiation and coincident morphological changes.

As a major contributor to the early stage of reprogramming, C-MYC plays an essential role in regulating miR-21, miR-29a and miR-7, which serve as barriers to the initial stage of reprogramming [54, 55]. Furthermore, depletion of miR-21 and miR-29a enhances reprogramming efficiency in mouse embryonic fibroblasts, suggesting that MEF-enriched miRNAs also function as reprogramming barriers [54]. On the other hand, our results show an increase in the level of miR-145 following differentiation of AF-iPS cells into neuronal (AF-iPSC-N2), mesenchymal/cardiac (AF-iPSC-M1) or endothelial cells (AF-iPSC-E1 and E2). These observations are in agreement with previous reports in which miR-145 was shown to inhibit self-renewal and pluripotency in human embryonic stem cells and its loss impaired

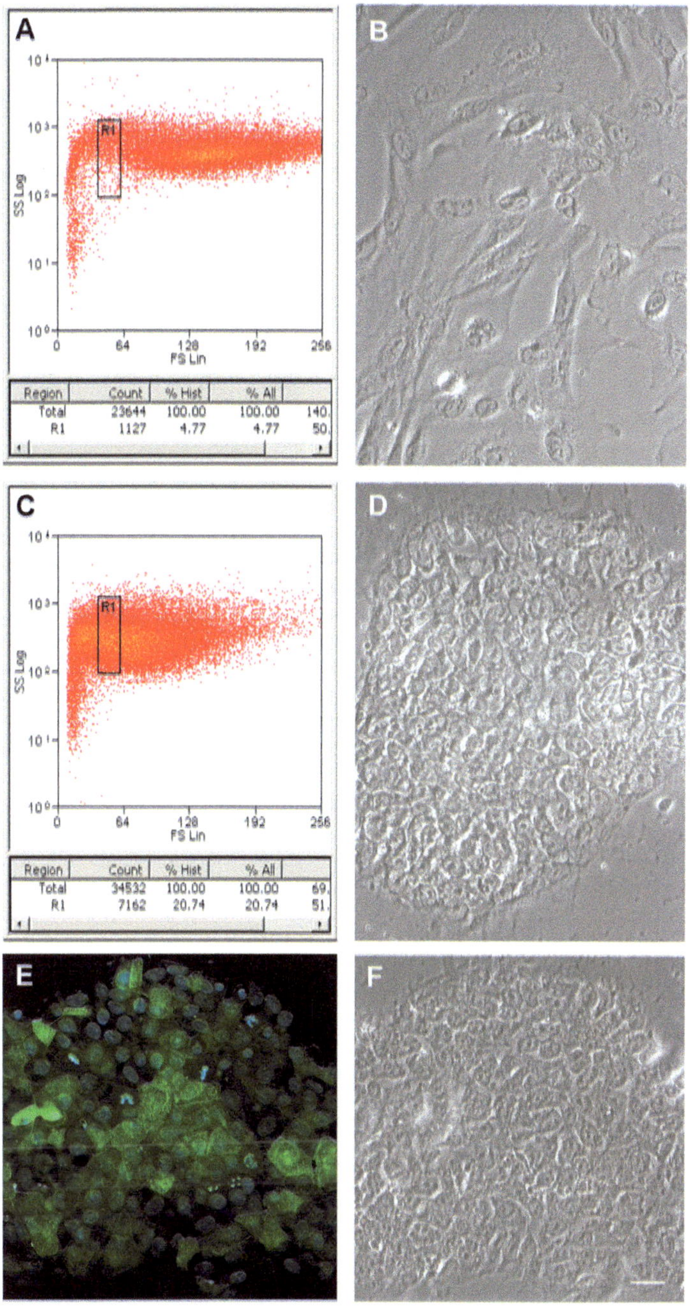

Fig. 11.4 Morphological changes in amniotic fluid cells after reprogramming. (**a**) FACS analysis indicates a small population (4.77%) of single amniotic fluid cells within gate R1. (**b**) Corresponding phase contrast image prior to sorting shows the cells with average dimensions of 47 μm x 83 μm at 80% confluency. (**c**) There is a significant increase in the number of single cells witinn gate R1 (20%) following reprogramming. (**d**) Corresponding phase contrast image of amniotic fluid induced pluripotent stem cells shows compact colonies of smaller cells with an approximate diameter of 15-20 μm and a reduced cytoplasmic to nuclear ratio. (**e**) The induced pluripotent state of reprogrammed cells was confirmed by live TRA1-81 staining. (**f**) The corresponding phase contrast image of panel (**e**) Scale bar: 15 μm (**b**), 50 μm (**d** and **f**)

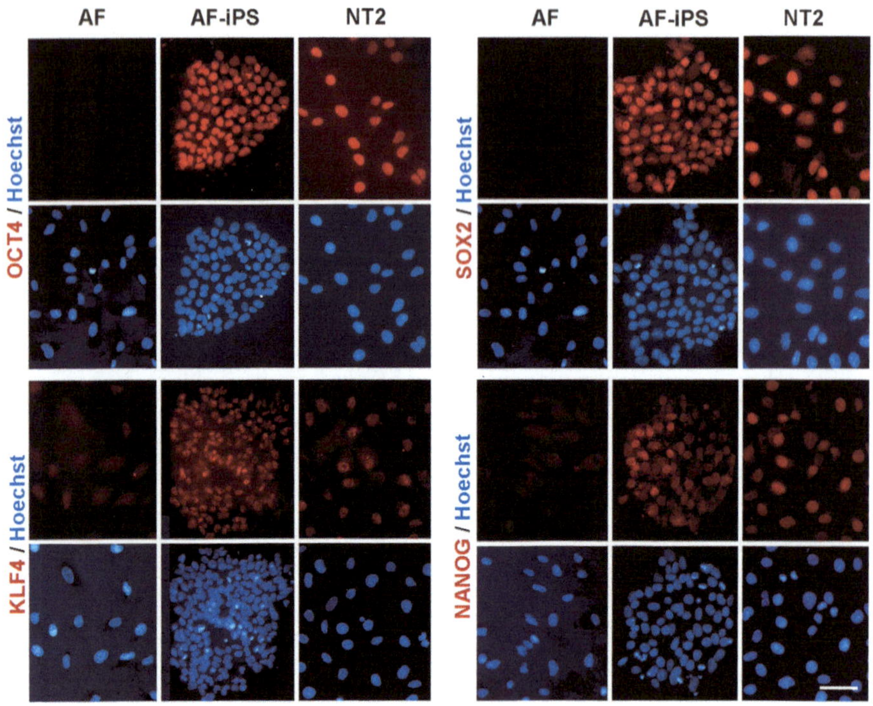

Fig. 11.5 Validation of reprogramming in amniotic fluid induced pluripotent stem cells. Immunocytochemistry with pluripotency markers confirms efficient reprogramming of amniotic fluid cells. Abundant levels of OCT4, KLF4, SOX2 and NANOG are expressed in amniotic fluid induced pluripotent stem cells (AF-iPSCs), compared with amniotic fluid cells (AF, isolated at 26 weeks of gestation). Human NT2/D1 cells were used as a positive control. Hoechst: Nuclear counterstain. Scale bar: 75 μm (AF, NT2), 100 μm (AF-iPS)

differentiation by maintaining high levels of OCT4, SOX2 and KLF4 [29, 55]. Together, these results suggest the presence of a negative feedback loop involving OCT4, SOX2, KLF4 and miR-145.

Recent advances in identifying and validating miRNA functions have encouraged the field to design miRNA delivery systems for research and therapeutic applications. For instance, miR-132 has been shown to play multiple roles in promoting neuronal outgrowth, regulating synaptic transmission, limiting inflammation and inducing angiogenesis. Considering the low levels of endogenous miR-132 in amniotic fluid cells (Fig. 11.3), we designed a vector that allows enriching and monitoring miR-132 in these cells (Fig. 11.7) as well as other cell systems. MiR-132-GFP amniotic fluid cells can be used as a model to test the role of miR-132 in neuronal and endothelial differentiation *in vitro* and following transplantation in animal models. MiR-132-GFP amniotic fluid cells can be also used to screen drugs that have the potential to target miR-132. Furthermore, amniotic fluid cells provide an excellent opportunity to study miRNA-target interactions, regulatory mechanisms and crosstalk between miRNA targets.

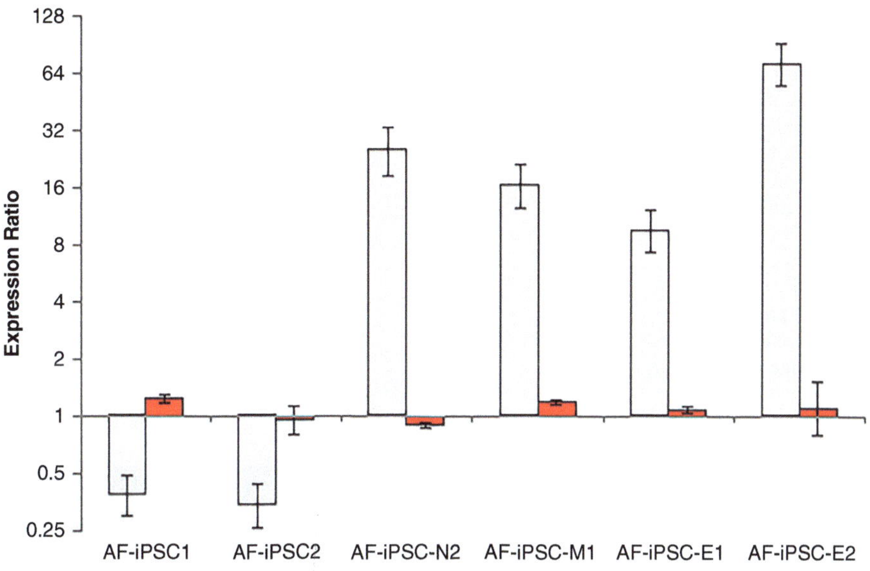

Fig. 11.6 Changes in miR-145 expression following reprogramming and differentiation of amniotic fluid cells. Real time PCR analysis shows a decrease in miR-145 expression (*gray bars*) in amniotic fluid induced pluripotent stem cells (AF-iPSCs) compared with amniotic fluid (AF) cells. AF-iPSCs were differentiated into neuronal, mesodermal and endodermal cell lineages and further assessed for miR-21 and miR-145 expression. In contrast to AF and AF-iPSCs, miR-145 expression was increased in neuronal (AF-iPSC-N2), mesodermal (AF-iPSC-M1) and endothelial cell populations (AF-iPSC-E1 and E2) derived from AF-iPSCs, a phenomenon not observed for miR-21 (*red bars*)

Endnote

Immunostaining and FACS analysis were performed, as previously described [32, 33, 57].

Total RNA was isolated from each cell type in triplicate and used for microRNA analysis following well-established methods to ensure stability [17, 58, 59]. In brief, Real-time PCR was performed, using the TaqMan Human MicroRNA Assay and the expression ratio for each cell type was scored against that of HEK-293 cells as a reference. AF cells were reprogrammed into induced pluripotent stem cells, using an episomal gene delivery system, and differentiated into neuronal, mesenchymal and endodermal cell lineages following methods established in our laboratory.

MiR-132 was first PCR amplified and then subcloned at *XhoI* sites of pMXs-miR-EGFP vector (Cell Biolabs, Inc.). This retroviral vector allows for the expression of the miRNA precursor in its native context, while preserving putative hairpin structures to ensure biologically relevant interactions with endogenous processing machinery and regulatory partners. Thus, it leads to properly cleaved miRNAs. The presence of GFP allows monitoring and selecting miR-132 positive cells.

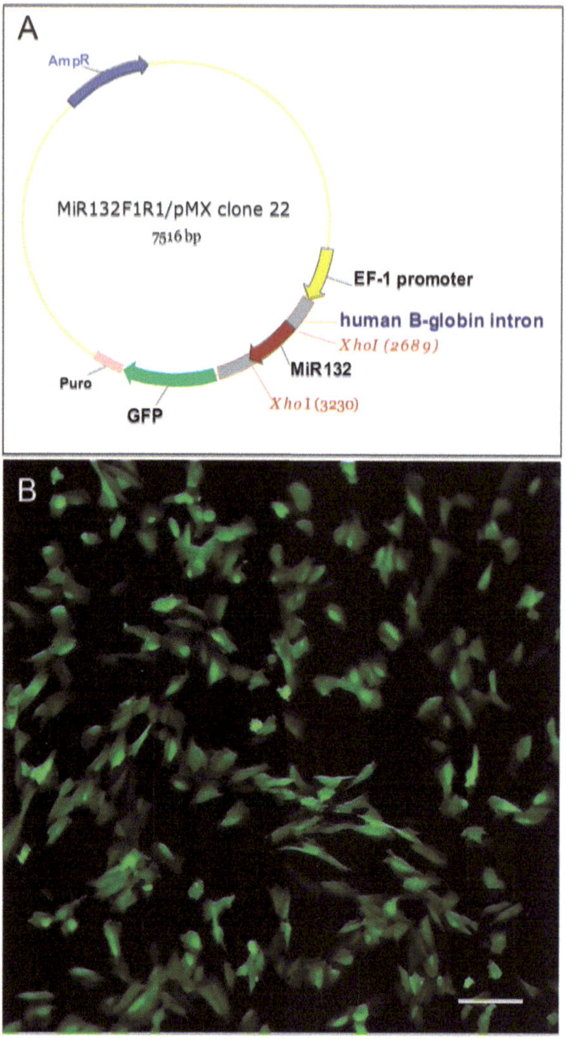

Fig. 11.7 Amniotic fluid cells as miRNA delivery vehicles. (**a**) Schematic diagram of pMXs-miR132-EGFP vector. MiR-132 was PCR amplified and subcloned into pMXs-miR-EGFP retroviral vector at XhoI sites. This vector provides cells with the capability to express the miRNA precursor in its native context, while preserving putative hairpin structures. The latter allows biologically relevant interactions with endogenous processing machinery and regulatory partners, leading to properly cleaved miRNAs. The use of EF-1α promoter in this vector facilitates high expression levels in mammalian cells. Furthermore, GFP-Puro fusion marker allows monitoring of cells positive for miR-132 expression and stable selection with either GFP or puromycin resistance. (**b**) Amniotic fluid cells readily express miR-132-GFP after electroporation. Scale bar: 150 μm

References

1. Londin E, et al. Analysis of 13 cell types reveals evidence for the expression of numerous novel primate- and tissue-specific microRNAs. PNAS. 2015; E1106–E1115 www.pnas.org/cgi/doi/10.1073/pnas.1420955112.
2. Zhang J, et al. Exosome and exosomal microRNA: trafficking, sorting, and function. Genomics Proteomics Bioinformatics. 2015;13(1):17–24.
3. Li Z, Rana TM. Therapeutic targeting of microRNAs: current status and future challenges. Nat Rev Drug Discov. 2014;13(8):622–38.
4. Jonas S, Izaurralde E. Towards a molecular understanding of microRNA-mediated gene silencing. Nat Rev Genet. 2015;16(7):421–33.

5. Lee RC, Feinbaum RL, Ambros V. The C. elegans heterochronic gene lin-4 encodes small RNAs with antisense complementarity to lin-14. Cell. 1993;75(5):843–54.
6. Friedlander MR, et al. Evidence for the biogenesis of more than 1,000 novel human microRNAs. Genome Biol. 2014;15(4):R57.
7. Zeng Y. Principles of micro-RNA production and maturation. Oncogene. 2006;25(46): 6156–62.
8. Gantier MP, et al. Analysis of microRNA turnover in mammalian cells following Dicer1 ablation. Nucleic Acids Res. 2011;39(13):5692–703.
9. Friedman RC, et al. Most mammalian mRNAs are conserved targets of microRNAs. Genome Res. 2009;19(1):92–105.
10. Guo WT, Wang XW, Wang Y. Micro-management of pluripotent stem cells. Protein Cell. 2014;5(1):36–47.
11. Javidi MA, et al. Cell-free microRNAs as cancer biomarkers: the odyssey of miRNAs through body fluids. Med Oncol. 2014;31(12):295.
12. Collino F, et al. Microvesicles derived from adult human bone marrow and tissue specific mesenchymal stem cells shuttle selected pattern of miRNAs. PLoS One. 2010;5(7):e11803.
13. Simpson RJ, et al. Exosomes: proteomic insights and diagnostic potential. Expert Rev Proteomics. 2009;6(3):267–83.
14. Higa GS, et al. MicroRNAs in neuronal communication. Mol Neurobiol. 2014;49(3): 1309–26.
15. Qin J, Xu Q. Functions and application of exosomes. Acta Pol Pharm. 2014;71(4):537–43.
16. Weber JA, et al. The microRNA spectrum in 12 body fluids. Clin Chem. 2010;56(11):1733–41.
17. Cortez MA, et al. MicroRNAs in body fluids--the mix of hormones and biomarkers. Nat Rev Clin Oncol. 2011;8(8):467–77.
18. Keller S, et al. Body fluid derived exosomes as a novel template for clinical diagnostics. J Transl Med. 2011;9:86.
19. Lim PK, et al. Gap junction-mediated import of microRNA from bone marrow stromal cells can elicit cell cycle quiescence in breast cancer cells. Cancer Res. 2011;71(5):1550–60.
20. Hosoda T. The mircrine mechanism controlling cardiac stem cell fate. Front Genet. 2013;4:204.
21. Greco SJ, Rameshwar P. Analysis of the transfer of circulating microRNA between cells mediated by gap junction. Methods Mol Biol. 2013;1024:87–96.
22. Hong X, et al. Gap junctions modulate glioma invasion by direct transfer of microRNA. Oncotarget. 2015;6:15566.
23. Katakowski M, et al. Functional microRNA is transferred between glioma cells. Cancer Res. 2010;70(21):8259–63.
24. Nielsen MS, et al. Gap junctions. Compr Physiol. 2012;2(3):1981–2035.
25. Judson RL, et al. Embryonic stem cell-specific microRNAs promote induced pluripotency. Nat Biotechnol. 2009;27(5):459–61.
26. Wang Y, et al. miR-294/miR-302 promotes proliferation, suppresses G1-S restriction point, and inhibits ESC differentiation through separable mechanisms. Cell Rep. 2013;4(1):99–109.
27. Shenoy A, Blelloch RH. Regulation of microRNA function in somatic stem cell proliferation and differentiation. Nat Rev Mol Cell Biol. 2014;15(9):565–76.
28. Tay Y, et al. MicroRNAs to Nanog, Oct4 and Sox2 coding regions modulate embryonic stem cell differentiation. Nature. 2008;455(7216):1124–8.
29. Xu N, et al. MicroRNA-145 regulates OCT4, SOX2, and KLF4 and represses pluripotency in human embryonic stem cells. Cell. 2009;137(4):647–58.
30. Wang L, et al. Gene and MicroRNA profiling of human induced pluripotent stem cell-derived endothelial cells. Stem Cell Rev. 2015;11(2):219–27.
31. Trohatou O, et al. Sox2 suppression by miR-21 governs human mesenchymal stem cell properties. Stem Cells Transl Med. 2014;3(1):54–68.
32. Jezierski A, et al. Human amniotic fluid cells form functional gap junctions with cortical cells. Stem Cells Int. 2012;2012:607161.

33. Rennie K, et al. Therapeutic potential of amniotic fluid-derived cells for treating the injured nervous system. Biochem Cell Biol. 2013;91(5):271–86.
34. Roush S, Slack FJ. The let-7 family of microRNAs. Trends Cell Biol. 2008;18(10):505–16.
35. Barh D, et al. MicroRNA let-7: an emerging next-generation cancer therapeutic. Curr Oncol. 2010;17(1):70–80.
36. Sanek NA, Young WS. Investigating the in vivo expression patterns of miR-7 microRNA family members in the adult mouse brain. Microrna. 2012;1(1):11–8.
37. Sekar D, et al. Role of microRNA 21 in mesenchymal stem cell (MSC) differentiation: a powerful biomarker in MSCs derived cells. Curr Pharm Biotechnol. 2015;16(1):43–8.
38. Di Bernardini E, et al. Endothelial lineage differentiation from induced pluripotent stem cells is regulated by microRNA-21 and transforming growth factor beta2 (TGF-beta2) pathways. J Biol Chem. 2014;289(6):3383–93.
39. Polajeva J, et al. miRNA-21 is developmentally regulated in mouse brain and is co-expressed with SOX2 in glioma. BMC Cancer. 2012;12:378.
40. Bhalala OG, et al. microRNA-21 regulates astrocytic response following spinal cord injury. J Neurosci. 2012;32(50):17935–47.
41. Buller B, et al. MicroRNA-21 protects neurons from ischemic death. FEBS J. 2010;277(20):4299–307.
42. Ge XT, et al. miR-21 improves the neurological outcome after traumatic brain injury in rats. Sci Rep. 2014;4:6718.
43. Joglekar MV, et al. The miR-30 family microRNAs confer epithelial phenotype to human pancreatic cells. Islets. 2009;1(2):137–47.
44. Bridge G, et al. The microRNA-30 family targets DLL4 to modulate endothelial cell behavior during angiogenesis. Blood. 2012;120(25):5063–72.
45. Sun Y, et al. An updated role of microRNA-124 in central nervous system disorders: a review. Front Cell Neurosci. 2015;9:193.
46. Meza-Sosa KF, Pedraza-Alva G, Perez-Martinez L. microRNAs: key triggers of neuronal cell fate. Front Cell Neurosci. 2014;8:175.
47. Zare M, et al. A novel protocol to differentiate induced pluripotent stem cells by neuronal microRNAs to provide a suitable cellular model. Chem Biol Drug Des. 2015;86:232.
48. Tan CL, et al. MicroRNA-128 governs neuronal excitability and motor behavior in mice. Science. 2013;342(6163):1254–8.
49. O'Neill LA. Boosting the brain's ability to block inflammation via microRNA-132. Immunity. 2009;31(6):854–5.
50. Devalliere J, et al. Sustained delivery of proangiogenic microRNA-132 by nanoparticle transfection improves endothelial cell transplantation. FASEB J. 2014;28(2):908–22.
51. Liu T, et al. Human amniotic epithelial cell feeder layers maintain human iPS cell pluripotency via inhibited endogenous microRNA-145 and increased Sox2 expression. Exp Cell Res. 2012;318(4):424–34.
52. Götte M, et al. miR-145-dependent targeting of junctional adhesion molecule A and modulation of fascin expression are associated with reduced breast cancer cell motility and invasiveness. Oncogene. 2010;29(50):6569–80.
53. Chen S, et al. MicroRNA-494 inhibits the growth and angiogenesis-regulating potential of mesenchymal stem cells. FEBS Lett. 2015;589(6):710–7.
54. Yang CS, Li Z, Rana TM. microRNAs modulate iPS cell generation. RNA. 2011;17(8):1451–60.
55. Yang CS, Rana TM. Learning the molecular mechanisms of the reprogramming factors: let's start from microRNAs. Mol Biosyst. 2013;9(1):10–7.
56. Kuppusamy KT, et al. Let-7 family of microRNA is required for maturation and adult-like metabolism in stem cell-derived cardiomyocytes. Proc Natl Acad Sci U S A. 2015;112(21):E2785–94.
57. Jezierski A, et al. Neuroprotective effects of GDNF-expressing human amniotic fluid cells. Stem Cell Rev. 2014;10(2):251–68.
58. Smith B, et al. Large-scale expression analysis reveals distinct microRNA profiles at different stages of human neurodevelopment. PLoS One. 2010;5(6):e11109.
59. Davis CJ, Clinton JM, Krueger JM. MicroRNA 138, let-7b, and 125a inhibitors differentially alter sleep and EEG delta-wave activity in rats. J Appl Physiol (1985). 2012;113(11):1756–62.

Part III
Umbilical Cord and Placenta

Chapter 12
Historical Perspectives

Karen K. Ballen

Introduction

The first umbilical cord blood transplantation (UCBT) was performed in France in a child with Fanconi Anemia (FA) in 1988 by Dr. Eliane Gluckman and colleagues. The clinical success grew from scientific collaborations with Drs. Hal Broxmeyer, Arleen Auerbach and others. Over the last 25 years, UCB banking and transplantation have grown exponentially. Over 600,000 UCB units have been donated for public use worldwide, and over 30,000 UCBT have been performed. UCB serves as an alternative source of cells for hematopoietic cell transplantation (HCT), which is a curative therapy for patients with leukemia, lymphoma, myelodysplasia, and many genetic disorders. Only 30 % of patients will have a human leukocyte antigen (HLA) matched sibling. There are approximately 20 million adult volunteer donors in the National Marrow Donor Program and affiliated registries; however, only 60 % of Whites and 20 % of Blacks and other minorities will have a suitably matched unrelated volunteer donor identified in the required time period [1]. UCB collection is safe, readily available, and does not need to be as closely matched to the patient as traditional stem cell sources of bone marrow and peripheral blood stem cells. Therefore UCBT has extended access to transplantation, especially to patients of racial and ethnic minorities [2, 3].

K.K. Ballen, M.D. (✉)
Division of Hematology/Oncology, Massachusetts General Hospital Cancer Center, Zero Emerson, Suite 118, Boston, MA 02114, USA
e-mail: kballen@partners.org

D.O. Fauza, M. Bani (eds.), *Fetal Stem Cells in Regenerative Medicine*, Stem Cell Biology and Regenerative Medicine,
DOI 10.1007/978-1-4939-3483-6_12

The Need for Umbilical Cord Blood Transplantation

The need for using an additional stem cell source arose from the concern of many patients not fortunate to have an HLA identical sibling or unrelated donor to proceed to a potentially curative HCT. Thomas and colleagues in Seattle pioneered HCT in the 1960s, initially for children with relapsed leukemia and lymphoma, hematologic malignancies that often could not be cured by chemotherapy alone. The transplant consisted of high doses (myeloablative) chemotherapy or chemotherapy and radiation treatments to eradicate disease and immunosuppress the patient, followed by infusion of the donor cells through an intravenous line (the transplant itself takes less than 1 h), and intensive supportive care with antibiotics, transfusions, nutrition, graft vs. host disease (GVHD) prophylaxis medications for several months. A limiting factor was the availability of an HLA matched donor. Life threatening graft rejection or graft versus host disease could occur if the patient and the donor were not HLA matched. Graft versus host disease, an immune reaction between the donor and recipient lymphocytes, was characterized by skin rash, diarrhea, and liver problems, and could be fatal. Each sibling has a 25 % chance of matching the patient; given the size of most US families, only 30 % of patients have an HLA-matched sibling donor.

The National Marrow Donor Program (NMDP) was established in 1988 to assist patients who did not have a matched family donor [4, 5]. Although the NMDP and related registries have grown to over 20 million volunteer donors, it is particularly difficult for Black patients and racial/ethnic minority patients to find matched donors in the registry, within the required time frame to proceed to HCT. Despite intensive recruiting efforts, the majority of donors in the NMDP are of Northern European ancestry. An estimated 30–40 % of White patients and 60–70 % of Black and other minority patients will have no matched related or unrelated donor. UCB cells are immunologically more naïve, and therefore patient and UCB donor do not need to be perfectly matched for the HCT to be successful. This ability to transplant across HLA boundaries contributed to the growth of UCB field as an alternative stem cell source for patients with hematologic malignancies in need of a potentially curative HCT.

The Early Days of Umbilical Cord Blood Transplantation

The first UCBT were performed in children, as there was concern that the cell dose might be too low to successfully engraft the larger adult patient. The scientific basis of the UCBT was based on the work of Broxmeyer and colleagues at Indiana University. Broxmeyer established that the UCB cells had extensive proliferative capacity, could be stored for days at room temperature, and could be cryopreserved and thawed with no loss of cells [6]. In addition, murine data showed that UCB cells could engraft a lethally irradiated mouse [7].

Fanconi Anemia was selected as the first disease to try this new technology. HLA matched sibling HCT was an accepted treatment for FA, and there was the possibility of an HLA identical sibling who did not have the disease. Eliane Gluckman in Paris had designed an attenuated, safer conditioning regimen for children with FA undergoing allogeneic HCT [8]. At the same time, Arleen Auerbach from the Rockefeller University in New York described a method for prenatal diagnosis to determine before birth if a child (and potential donor) would be affected by FA [9]. Prior to the first UCBT, as UCB had never been used before in humans, the French National Ethics committee authorized the procedure.

The first UCBT was performed in Paris by Dr. Gluckman in a child with FA, whose normal sibling had UCB collected at birth [10]. The cryopreserved UCB unit was transported from Indiana to Paris in a dry shipper that maintained the temperature at −175 °C. The patient received a modified conditioning regimen and the UCB cells were infused into the patient without separation or washing. The first indication of neutrophil engraftment occurred on day 22 with subsequent complete donor derived hematological reconstitution. The patient is alive and well, now 25 years after UCBT.

The Growth of Umbilical Cord Blood Banking

Following the first successful UCBT, interest in the field grew, to extend the application to unrelated UCBT. Achievement of this goal required establishments of large repositories for pregnant women to donate their UCB for the future use of a patient in need of an HCT for a hematologic malignancy or genetic disorder. In the early 1990s the first UCB banks were established in Dusseldorf, Milan, London, Paris, St Louis, and New York. Dr. Pablo Rubinstein, at the New York Blood Center, established the largest unrelated UCB bank [11]. Dr. Rubinstein and others developed standards for UCB collection, processing, storage, and thawing [12]. The advantages of UCB were as follows: a single UCB unit contained enough progenitor cells to successfully engraft pediatric patients, the collection procedure was safe, and UCB could be easily frozen and then thawed when needed for use.

With the growth of the UCB banking industry came interest in private UCB banking: charging a fee for upfront collection and yearly storage in return for storing the UCB unit for personal or family use. Multiple private UCB banking companies have developed in the US although private UCB banking is illegal in several European countries, including France and Italy. The chance of using privately stored UCB is very low, estimated at about 0.01 %. Therefore, the American Society of Blood and Marrow Transplantation, the American Academy of Pediatrics, and the American College of Obstetrics and Gynecology have all issued position statements discouraging private UCB banking and encouraging donation to a public UCB bank whenever possible [13].

Pediatric Umbilical Cord Blood Transplantation

The first UCBT were performed in children, due to the concern that the cell dose might be too low for adult UCBT. Studies in related UCBT showed that UCBT was associated with similar survival to HLA-identical sibling bone marrow transplants, but with a lower rate of acute and chronic GVHD, one of the serious immune complications of HCT [14]. But, the real need for UCBT was for unrelated use. Using the UCB stored at the New York Blood Center, Joanne Kurtzberg at Duke University reported the first series of unrelated UCBT in 1995 [15].

In this series of 25 patients, the 100-day overall survival (OS) was 64 %, demonstrating the feasibility of unrelated mismatched UCBT. Since this initial series, multiple centers have reported successful engraftment and sustained disease free survival in pediatric patients with malignant and nonmalignant diseases [16, 17]. Retrospective comparisons of UCBT with traditional graft sources such as matched unrelated donor (MUD) HCT have shown comparable survival [18]. The European Eurocord group showed that UCBT were associated with delayed neutrophil and platelet engraftment and less acute and chronic GVHD, compared to MUD HCT. Mary Eapen and colleagues compared outcomes of 503 children with acute leukemia receiving unrelated mismatched UCBT to 282 children receiving a MUD HCT [19]. Leukemia free survival and overall survival were similar between BM and one or two HLA antigen mismatched UCBT; children who received an HLA matched UCBT had improved survival to children who received an HLA matched unrelated bone marrow HCT. An important finding was that children who received a UCBT with a higher cell dose (>3 × 10 [7] total nucleated cells (TNC)/kg) had improved survival. In an attempt to increase cell dose, double UCBT was studied in pediatric patients; however, a recent randomized study showed no benefit to two UCB units (double UCBT) versus single UCBT [20].

Excellent results have also been achieved with UCBT and non-malignant diseases. These include thalassemia, Fanconi anemia, and metabolic storage diseases such as Hunter's and Hurler's syndrome. The pediatric transplant program at Duke has specialized in these rare diseases. Results are improved when UCBT in performed in the first few months of life with OS of 58 % at 5 years [21].

Adult Umbilical Cord Blood Transplantation

After the encouraging results in children, investigators used a single UCBT for adults with hematologic malignancies who needed a HCT but did not have a matched related or unrelated donor. Unfortunately, the initial results were poor, with high toxicity from infection and organ damage and 40 % of patients dying before day 100 [22]. A critical observation was that a higher cell dose was associated with an improved survival. Therefore, in the last 10 years more stringent patient selection for HCT, selection of UCB units with higher cell doses, and improved supportive care have contributed to better outcome results. Current studies report disease free survival of 30–60 % [23–25]. Some of the most impressive results have been reported by the Japanese groups, with survivals of 60–70 %; these excellent results have been attributed to the smaller size of Japanese patients, genetic homogeneity between patient and UCB donor, and stricter patient selection [26]. A summary of selected adult UCBT series is presented in Table 12.1.

Table 12.1 Selected adult UCBT series

Author	Number of patients	Conditioning	Single or double UCBT	Median age (years)	DFS
Takahasi [55]	71	Myelo Ablative	Single UCBT	38	70 %
Eapen [35]	165	Myelo Ablative	Single UCBT	28	44 %
Chen [37]	64	RIC	Double UCBT	53	30 %
Brunstein [38]	50	RIC	Double UCBT	58	46 %

UCBT umbilical cord blood transplant, *DFS* disease free survival, *RIC* reduced intensity conditioning

Double Cord Blood Transplantation

Double UCBT, the infusion of two partially matched UCB units to the same patient, was initially pioneered by the Minnesota group in an effort to increase the cell dose infused [27]. Patients from the USA are 10 kg heavier than in Europe and 15 kg heavier than in Asia; double UCBT therefore is especially popular in the US. In addition, the use of less intensive or reduced intensity conditioning (RIC) regimens allowed older patients to be transplanted more safely, using related donor, matched unrelated donor, or UCBT [28]. Numerous adult UCBT series of double UCBT reported disease free survivals of 30–50 % [29, 30].

While a large randomized pediatric study showed no difference in survival between single and double UCBT, the adult data is less clear [20]. No randomized studies have been done to date. A Eurocord retrospective analysis reported similar survival to single UCBT with an adequate cell dose as compared with double UCBT using a myeloablative preparative regimen of thiotepa, busulfan, and fludarabine [31].

Which Graft Source Is Best?

Given the improving results in adults, a natural question was the effectiveness of UCBT compared to other graft sources. In the last 10 years, there has also been a growth in the use of mismatched related donor (haploidentical) transplants, pioneered by the group at Hopkins [32]. In addition, improvements in HLA typing and GVHD prophylaxis and treatment regimens have permitted HCT using HLA mismatched unrelated donors [33, 34]. No randomized prospective studies have been completed to delineate the best graft source for adults. Multiple retrospective studies have indicated comparable overall and disease free survivals among UCBT, matched unrelated donor, and mismatched unrelated donor. The risk of infection was higher after UCBT but GVHD and in some studies, relapse of disease was lower [35–37]. To answer these important questions, the National Institutes of Health funded Clinical Trials Network (CTN) compared RIC double UCBT and RIC haploidentical HCT in two parallel Phase 2 studies [38]. The 1 year overall and disease free survivals were comparable. The UCBT patients had a higher rate of transplant related complications (transplant related mortality), 24 % for UCBT vs. 7 % for haploidentical HCT, but the relapse rate at 1 year was lower after UCBT (31 %) vs. haplo (45 %). There was concern that the relapse rate might increase as patients are followed for longer periods of time post HCT. A large CTN randomized study is ongoing in the United States to compare long term outcomes of the UCBT and haploidentical HCT approaches.

Challenges of Umbilical Cord Blood Transplant

While there have been many advances over the last 25 years in the field of UCBT, multiple challenges remain. They include a high rate of infection due to poor immune reconstitution, cost, limited access for some populations, and relapsed disease. Although neutrophil engraftment has improved with the use of better UCB unit selection or, for adults, double UCBT, there is delayed immune reconstitution after UCBT. Jacobson and colleagues showed delayed T cell recovery, including CD4+ and CD8+ cells, in the first year post UCBT [39]. This poor immune recovery contributed to the increased and severe infections seen after UCBT, including cytomegalovirus reactivation, post transplant lymphoproliferative disease, aspergillus, and human herpes virus 6 encephalitis [40, 41]. Cost is another major concern after UCBT; the acquisition cost per UCB unit can be $40,000 US, with an additional $200,000 US for the transplant itself. The CIBMTR has recently shown that the length of stay is longer after UCBT compared to MUD transplantation [42].

While one of the goals of UCBT and banking was to collect UCB units from minorities and increase access to HCT, that goal has not been completely achieved. UCB units collected from the children of Black mothers has been shown to have lower CD34+ counts, an important marker for engraftment potential [43]. In addition, Black patients have lower overall survival after single UCBT than White patients, and Black patients on average received UCB units that were smaller and less well matched [44]. Relapsed disease remains a major cause of death after all HCT, and this problem also exists after UCBT.

Several strategies have been undertaken to address these important challenges. To improve engraftment and immune recovery, several centers are investigating intramarrow injection of the UCBT, UCB expansion, homing strategies, and combination of UCBT and haploidentical approaches. Frassoni and colleagues have delivered the UCBT directly into the bone marrow via a bone marrow aspiration needle, to bypass the need to "home" to the bone marrow [45]. intramarrow injection was associated with improved engraftment. Recently, this finding was confirmed in a Eurocord study comparing intramarrow injection to double UCBT [46].

Multiple centers have explored UCB expansion as a strategy to increase the low cell doses infused. One example, employed by the group at MD Anderson, is the use of mesenchymal stem cells [47]. Using a coculture *ex vivo* with mesenchymal progenitor cells in one of two UCB units in 31 patients, this group reported an improvement in engraftment of 9 days, compared to historical controls. A larger Phase III study is currently enrolling patients. Another platform to improve UCBT outcomes is to increase the homing to and nurturing of cells within the hematopoietic microenvironment. Our center has pretreated UCB cells with a modified Prostaglandin (PG) E molecule, as a means to upregulate the CXCR4 expression and increase marrow homing [48]. In a Phase I study in which one UCB in a double UCBT was incubated with PGE2, neutrophil engraftment improved by 3.5 days [49]. Finally, combining a haploidentical mismatched family member bone marrow with UCBT may speed engraftment, and this approach is being tested with both single and double UCBT [50].

The cost of UCBT is considerable, and this is related to the acquisition cost of the UCB product, a longer length of hospital stay, and the need for transfusions and often antiviral therapy. Less than 10 % of the UCB units in public inventory have been used for UCBT. As demand increases, the cost of the UCB unit is likely to decrease. In addition, attempts to speed engraftment as discussed above may decrease hospital length of stay [51].

Access to care for all populations remains a major impediment to the widespread success of HCT. Although the immunologic naivety of the UCB cells allows patients to proceed to UCBT without a perfectly matched donor, it is still difficult for Black and other minorities to find appropriately matched UCB units of sufficient size. However, UCBT has extended the access to HCT. Barker and colleagues have demonstrated that unrelated bone marrow grafts were identified for 53 % of patients with European ancestry but only 21 % of patients of non-European ancestry [2]. In contrast, 56 % of UCBT recipients were of non-European ancestry. Early recognition of this issue and prompt search strategies for both unrelated bone marrow and UCB donors are recommended by the authors. Relapse remains a major problem after all HCT. Strategies to overcome relapse include addition of post UCBT maintenance chemotherapy or targeted therapy, such as the use of a flt-3 inhibitor for patients with flt-3 mutated acute myeloid leukemia [52].

The Future of Cord Blood

Tremendous progress has been made in the field of UCBT. Over the next 10 years, there will be continued progress to reduce the risk of infection and control costs. Exciting work is being done outside of oncology. Preliminary results suggest that patients with cerebral palsy had decrease in pro-inflammatory factors such as interleukin-6 and tumor necrosis factor after allogeneic UCBT [53]. UCB derived mesenchymal stem cell have been used in the treatment of peripheral arterial occlusive disease [54].

Conclusion

Twenty-five years have passed since the first UCBT in France. Over 600,000 UCB units have been generously donated by pregnant women for public use. Approximately 30,000 UCBT have been performed and over 10,000 patients have been cured. Results continue to improve, and the next 25 years should be even more exciting.

References

1. Appelbaum FR. Pursuing the goal of a donor for everyone in need. N Engl J Med. 2012;367(16):1555–6.
2. Barker JN, Byam CE, Kernan NA, et al. Availability of cord blood extends allogeneic hematopoietic stem cell transplant access to racial and ethnic minorities. Biol Blood Marrow Transplant. 2010;16(11):1541–8.
3. Ballen KK, Gluckman E, Broxmeyer HE. Umbilical cord blood transplantation: the first 25 years and beyond. Blood. 2013;122(4):491–8.
4. Ballen KK, King R, Carston M, et al. Outcome of unrelated transplants in patients with multiple myeloma. Bone Marrow Transplant. 2005;35:675–81.
5. Ballen KK, King RJ, Chitphakdithai P, Bolan CD, Agora E, Hartzman RJ, Kernan NA. The National Marrow Donor Program experience: the first 20 years of unrelated donor stem cell transplantation. Biol Blood Marrow Transplant. 2008;14:2–7.
6. Broxmeyer HE, Gordon GW, Hangoc G, et al. Human umbilical cord blood as a potential source of transplantable hematopoietic stem/progenitor cells. Proc Natl Acad Sci U S A. 1989;86(10):3828–32.
7. Broxmeyer HE, Kurtzberg J, Gluckman E, et al. Umbilical cord blood hematopoietic stem and repopulating cells in human clinical transplantation. Blood Cells. 1991;17(2):313–29.
8. Gluckman E, Devergie A, Dutreix J. Radio sensitivity in Fanconi anemia: application to the conditioning regimen for bone marrow transplantation. Br J Haematol. 1983;54(3):431–40.
9. Auerbach AD, Liu O, Ghosh R, et al. Prenatal identification of potential donors for umbilical cord blood transplantation for Fanconi anemia. Transfusion. 1990;30(8):682–7.
10. Gluckman E, Broxmeyer HA, Auerbach AD, et al. Hematopoietic reconstitution in a patient with Fanconi's anemia by means of umbilical-cord blood from an HLA-identical sibling. N Engl J Med. 1989;321(17):1174–8.
11. Rubinstein P, Carrier C, Scaradavou A, et al. Outcomes among 562 recipients of placental-blood transplants from unrelated donors. N Engl J Med. 1998;339(22):1565–77.
12. Szczepiorkowski ZM, Alder BC, Allickson J, et al. Standards for cellular therapy product services. Bethesda, MD: American Association of Blood Banks Press; 2004.
13. Ballen KK, Barker JN, Stewart S, Greene M, Lane T. ASBMT committee report: collection and preservation of cord blood for personal use. Biol Blood Marrow Transplant. 2008;14:356–63.
14. Rocha V, Wagner JE, Sobocinski K, et al. Graft-versus-host disease in children who have received a cord blood or bone marrow transplant from an HLA identical sibling. N Engl J Med. 2000;342(22):1846–54.
15. Kurtzberg J, Laughlin M, Graham ML, et al. Placental blood as a source of hematopoietic stem cells for transplantation into unrelated recipients. N Engl J Med. 1996;335(3):157–66.
16. Locatelli F, Rocha V, Chastang C, et al. Factors associated with outcome after cord blood transplantation in children with acute leukemia. Blood. 1999;93(11):3662–7.
17. Wagner JE, Barker JN, DeFor TE, et al. Transplantation of unrelated donor umbilical cord blood in 102 patients with malignant and non malignant diseases: influence of CD34 cell dose and HLA disparity on treatment-related mortality and survival. Blood. 2002;100(5):1611–8.
18. Rocha V, Cornish J, Sievers E, et al. Comparison of outcomes of unrelated bone marrow and umbilical cord blood transplants in children with acute leukemia. Blood. 2001;97(10):2962–71.
19. Eapen M, Rubinstein P, Zhang MJ, et al. Outcomes of transplantation of unrelated donor umbilical cord blood and bone marrow in children with acute leukemia: a comparison study. Lancet. 2007;369(9577):1947–54.
20. Wagner JE, Eapen M, Carter SL, et al. One unit versus two unit cord blood transplantation for hematologic malignancies. N Engl J Med. 2014;371(18):1885–94.
21. Prasad V, Kurtzberg J. Cord blood and bone marrow transplantation in inherited metabolic diseases: scientific basis, current status, and future directions. Br J Hematol. 2010;148:356–72.

22. Laughlin MJ, Barker J, Bambach B, et al. Hematopoietic engraftment and survival in adult recipients of umbilical-cord blood from unrelated donors. N Engl J Med. 2001;344(24):1815–22.
23. Cohen YC, Scaradavou A, Stevens CE, et al. Factors affecting mortality following myeloablative cord blood transplantation in adults: a pooled analysis of three international registries. Bone Marrow Transplant. 2011;46(1):70–6.
24. Robin M, Sanz GF, Ionescu I, et al. Unrelated cord blood transplantation in adults with myelodysplasia or secondary acute myeloblastic leukemia: a survey on behalf of Eurocord and CLWP of EBMT. Leukemia. 2011;25(1):75–81.
25. Sanz J, Boluda JC, Martin C, et al. Single-unit umbilical cord blood transplantation from unrelated donors in patients with hematologic malignancy using busulfan, thiotepa, fludarabine, and ATG as myeloablative conditioning regimen. Bone Marrow Transplant. 2012;47(10):1287–93.
26. Ooi J, Takahashi S, Tomonari A, et al. Unrelated cord blood transplantation after myeloablative conditioning in adults with acute myelogenous leukemia. Biol Blood Marrow Transplant. 2008;14(12):1341–7.
27. Giralt S, Ballen K, Rizzo D, et al. Reduced intensity conditioning workshop: defining the dose spectrum. Biol Blood Marrow Transplant. 2009;15(3):367–9.
28. Barker JN, Weisdorf DJ, DeFor TE, et al. Rapid and complete donor chimerism in adult recipients of unrelated donor umbilical cord blood transplantation after reduced intensity conditioning. Blood. 2003;102(5):1915–9.
29. Ballen KK, Spitzer TR, Yeap BY, et al. Double unrelated reduced-intensity umbilical cord blood transplantation in adults. Biol Blood Marrow Transplant. 2007;13(1):82–9.
30. Cutler C, Stevenson K, Kim HT, et al. Double umbilical cord blood transplantation with reduced intensity conditioning and sirolimus based GVHD prophylaxis. Bone Marrow Transplant. 2011;46(5):273–7.
31. Ruggeri A, Sanz G, Bittencourt H, et al. Comparison of outcomes after single or double cord blood transplantation in adults with acute leukemia using different types of myeloablative conditioning regimen, a retrospective analysis on behalf of Eurocord and the acute leukemia working party of EBMT. Leukemia. 2014;28(4):779–86.
32. Luznik L, O'Donnell PV, Fuchs EJ. Post-transplantation cyclophosphamide for tolerance induction in HLA-haploidentical bone marrow transplantation. Semin Oncol. 2012;39:683–93.
33. Koreth J, Stevenson KE, Kim HT, et al. Bortezomib-based graft-versus-host disease prophylaxis in HLA-mismatched unrelated donor transplantation. J Clin Oncol. 2012;26:3202–8.
34. Ballen KK, Koreth J, Chen YB, Dey BR, Spitzer TR. Selection of optimal alternative graft source: mismatched unrelated donor, umbilical cord blood, or haploidentical transplant. Blood. 2012;119(9):1972–80.
35. Eapen M, Rocha V, Sanz G, et al. Effect of graft source on unrelated donor haemopoietic stem-cell transplantation in adults with acute leukaemia: a retrospective analysis. Lancet Oncol. 2010;11(7):653–60.
36. Ponce DM, Zheng J, Gonzales AM, et al. Reduced late mortality risk contributes to similar survival after double-unit cord blood transplantation compared with related and unrelated donor hematopoietic stem cell transplantation. Biol Blood Marrow Transplant. 2011;17(9):1316–26.
37. Chen YB, Aldridge J, Kim HT, et al. Reduced intensity conditioning stem cell transplantation: comparison of double umbilical cord blood and unrelated donor grafts. Biol Blood Marrow Transplant. 2012;18(5):805–12.
38. Brunstein CG, Fuchs EJ, Carter SL, et al. Alternative donor transplantation after reduced intensity conditioning: results of parallel phase 2 trials using partially HLA-mismatched related bone marrow or unrelated double umbilical cord blood grafts. Blood. 2011;118(2):282–8.
39. Jacobson CA, Turki AT, McDonough SM, et al. Immune reconstitution after double umbilical cord blood stem cell transplantation: comparison with unrelated peripheral blood stem cell transplantation. Biol Blood Marrow Transplant. 2012;18:565–74.
40. Kanda J, Chiou LW, Szalbolcs P, et al. Immune recovery in adult patients after myeloablative dual umbilical cord blood, matched sibling, and matched unrelated donor hematopoietic cell transplantation. Biol Blood Marrow Transplant. 2012;18:1664–76.

41. Ballen KK, Cutler C, Yeap BY, et al. Donor derived second hematologic malignancies after cord blood transplantation. Biol Blood Marrow Transplant. 2010;16:1025–31.
42. Ballen KK, Joffe S, Brazauskas R, et al. Hospital length of stay in the first 100 days after allogeneic hematopoietic cell transplantation for acute leukemia in complete remission: comparison among alternative graft sources. Biol Blood Marrow Transplant. 2014;20(11):1819–27.
43. Ballen KK, Hicks J, Dharan B, et al. Racial and ethnic composition of volunteer cord blood donors: comparison with volunteer unrelated marrow donors. Transfusion. 2002;42(10):1279–84.
44. Ballen KK, Klein JP, Pederson TL, et al. Relationship of race/ethnicity and survival after single umbilical cord blood transplantation for adults and children with leukemia and myelodysplastic syndromes. Biol Blood Marrow Transplant. 2012;18(6):903–12.
45. Frassoni F, Gualandi F, Podesta M, et al. Direct intrabone transplant of unrelated cord-blood cells in acute leukaemia: a phase I/II study. Lancet Oncol. 2008;9(9):831–9.
46. Rocha V, Labopin M, Ruggeri A, et al. Unrelated cord blood transplantation: outcomes after single-unit intrabone injection compared with double-unit intravenous injection in patients with hematologic malignancies. Transplantation. 2013;95(10):1284–91.
47. de Lima M, McNiece I, Robinson SN, et al. Cord-blood engraftment with ex vivo mesenchymal-cell coculture. N Engl J Med. 2012;367(24):2305–15.
48. Goessling W, Allen RS, Guan X, et al. Prostaglandin E2 enhances human cord blood stem cell xenotransplants and shows long-term safety in preclinical nonhuman primate transplant models. Cell Stem Cell. 2011;8(4):445–58.
49. Cutler C, Multani P, Robbins D, et al. Prostaglandin-modulated umbilical cord blood hematopoietic stem cell transplantation. Blood. 2013;122(17):3074–81.
50. Liu H, Rich ES, Godley L, et al. Reduced-intensity conditioning with combined haploidentical and cord blood transplantation results in rapid engraftment, low GVHD, and durable remissions. Blood. 2011;118(24):6438–45.
51. Khera N, Zeliadt SB, Lee SJ. Economics of hematopoietic cell transplantation. Blood. 2012;120(8):1545–51.
52. Chen YB, Li S, Lane AA, et al. Phase I study of maintenance sorafenib after allogeneic hematopoietic stem cell transplantation for Fms-like tyrosine kinase 3 internal tandem duplication acute myeloid leukemia. Biol Blood Marrow Transplant 2014;20:2042–8.
53. Bae SH, Lee HS, Kang MS, et al. The levels of pro-inflammatory factors are significantly decreased in cerebral palsy patients following an allogeneic umbilical cord blood cell transplant. Int J Stem Cells. 2012;5:31–8.
54. Yang SS, Kim NR, Park KB, et al. A Phase I study of human cord blood-derived mesenchymal stem cell therapy in patients with peripheral arterial occlusive disease. Int J Stem Cells. 2013;6:37–44.
55. Takahashi S, Ooi J, Tomonari A, et al. Comparative single-institute analysis of cord blood transplantation from unrelated donors with bone marrow or peripheral blood stem-cell transplants from related donors in adult patients with hematologic malignancies after myeloablative conditioning regimen. Blood. 2007;109(3):1322–30.

Chapter 13
Umbilical Cord Blood Stem Cell Populations

David T. Harris

Introduction

Not that long ago neither stem cells nor talk of their use in regenerative medicine was as commonplace as it seems today. Stem cells have gone from "cure for whatever ails you" when first discovered to "political hot potato" between 2004 and 2008 to "promising medical hope", all in the span of 10 years. One of the more promising and intriguing stem cell types has been those found in umbilical cord blood (CB) and umbilical cord tissue (CT).

Work from Boyse [1], Broxmeyer [2], Harris [3] and others showed that cord blood was comparable to bone marrow for use in stem cell transplantation [1–9]. Over the past 20 years, more than 30,000 transplants have been performed worldwide using cord blood stem cells [10]. However, stem cell transplantation for cancer and genetic blood disorders (such as sickle cell anemia) is an uncommon occurrence. Fortuitously, research performed by several laboratories including our own [11–16] has demonstrated that cord blood also contains a mixture of different stem cells capable of giving rise to cells derived from the endodermal, mesodermal, and ectodermal lineages. In addition, mesenchymal stem cells (MSC) can be isolated from the cord tissue (CT) and preserved for later use [17]. Thus, both CB and CT can be readily available for use in tissue engineering and regenerative medicine applications, which are hypothesized to be more frequent events than the need for a typical stem cell transplant. Clinical trials using cord blood stem cells to treat cerebral palsy and peripheral vascular disease among other indications have been ongoing for several years [18, 19]. Recent efforts have focused on the isolation,

D.T. Harris (✉)
Department Immunobiology, University of Arizona, P.O. Box 245221, AHSC 6122,
Tucson, AZ 85724, USA
e-mail: davidh@email.arizona.edu

D.O. Fauza, M. Bani (eds.), *Fetal Stem Cells in Regenerative Medicine*,
Stem Cell Biology and Regenerative Medicine,
DOI 10.1007/978-1-4939-3483-6_13

characterization and utilization of MSC found in CT. In fact, CT stem cells have also now made their way into clinical trials [20–24]. In terms of clinically available stem cell sources, CT represents an abundant source of MSC for various clinical uses. In the US there are more than four million babies born annually from whom such MSC may be collected. MSCs present in CT may be collected and banked economically, for less than $1500, and can either be expanded for immediate use or banked for future applications [25].

As both CB and CT collections are once in a lifetime opportunities, cord blood and tissue banks have been established during the past two decades to assist in the preservation of these tissues. These biobanks serve to harvest, process, evaluate and cryopreserve such biological specimens for both autologous and general public usage. Once banked the samples can then be made available when needed for whatever purpose has been developed. This review will highlight the processes involved in such endeavors as well as the recent and pending clinical applications utilizing such resources.

Stem Cell Sources

Stem cells can be found throughout the body, being present in many tissues and organs (e.g., heart, brain and muscle), throughout one's lifespan. Over time these stem cells age and can be impaired by chronic disease and other changes in health status. However, stem cells can also be obtained from the youngest and healthiest biological source available, the leftover CB and CT that is the byproduct of one's entry into the world. When considering the use of stem cells for regenerative medicine and tissue engineering, one must consider several aspects. Ideally, when considering a source of stem cells for use in therapy one would prefer a source of autologous tissue for regenerative medicine and tissue engineering applications (to avoid immune rejection issues) that can be readily and inexpensively accessed, and which contains large (or at least sufficient) numbers of stem cells (not requiring expensive, extensive and time-consuming *in vitro* expansion before clinical use). It is our belief that these constraints limit our ideal choices of stem cell sources to cord blood and tissue, and to adipose tissue. Adipose stem cell banking has been the subject of several recent papers and review [17, 26–29]. Therefore, this review will focus on banking stem cells collected from cord blood and tissue.

Cord blood is a rich source of hematopoietic stem cells (primarily). Although CB contains other types of stem cells, including endothelial stem cells, mesenchymal stem cells and some ESC-like stem cells [11–16], the numbers and reproducibility of these cell populations are such that CB is not a practical source of such cells. Genetic and molecular analyses have demonstrated that CB stem cells are among the purest ones available and developmentally lie somewhere between mature stem cells like those found in bone marrow, and fetal stem cells [30]. CB stem cells generally are present at approximately 1 ± 0.5 % of the mononuclear cell (MNC) fraction of cord blood, and have a phenotype of being positive for CD34+, CD45+, CD90+, and CD133+, while being negative for all other mature lineage markers

[31]. The primary cell populations found in CB are lymphocytes [3] as would be expected for any source of peripheral blood (note that cord blood is in fact the child's peripheral blood). The major lymphocyte population is the T cell, although there are differences in the ratio of $CD4^{+}$–$CD8^{+}$ T cells [3]. As expected, virtually 100 % of the lymphocytes are immature ($CD45RA^{+}$) in phenotype which is in agreement with a functional immaturity observed upon challenge with antigens or mitogens [3, 32–35].

Cord tissue is a viable source of MSC, containing cells comparable to MSC found in both bone marrow and adipose tissue. That is, the cells exhibit a $CD45^{-}$, $CD44^{+}$, $CD90^{+}$, $CD73^{+}$, $CD105^{+}$ phenotype and are capable of differentiating into adipogenic, chondrogenic, osteogenic, and neurogenic lineages [17]. A significant disadvantage of utilizing CT-MSC is that the length of umbilical cord that can be harvested is limited in size (generally less than 10 in.), which translates in fewer cells being harvested than found with adipose tissue. As CT-MSC are not efficiently harvested from the tissues, what cells can be obtained need to be expanded *in vitro* prior to clinical use. In addition, it is very difficult to bank CT intact and later recovery significant MSC numbers upon thawing [17, 28, 36]. Thus, for immediate applications, unlike CB, CT is not really a suitable stem cell source.

Cord Blood Banking

Cord blood banking involves collection, transport, processing, evaluation and cryopreservation of the stem cell samples. Of course, donor and physician education precedes these steps and is critical in determining the eventual success of a cord blood bank, but is a discussion more appropriate for a different publication. Most CB banks have adopted the use of small blood bags (approximately 250–300 cc in size) for CB collections. These collection bags can be economically purchased for approximately $25–30 from a variety of FDA-approved vendors. All but one cord blood bank generally utilizes bags that contain CPD as an anticoagulant, while the other CB bank utilizes lyophilized heparin. Both types of anticoagulants have been in use in this and other medical applications for more than 25 years without problems. Current arguments for one anticoagulant over another are little more than marketing hoopla, rather than significant scientific fact. In our experience we find minimal differences in use of either anticoagulant as long as samples are processed within 36 h [37, 38]. Bag collections have evolved over the years as the preferred approach to collection of the blood (and subsequent processing) because it involves a closed system (preferable for most regulatory guidelines). However, bag collections require active oversight in order to prevent unintended contamination or loss of blood flow from occurring. CB collections are easily completed within 5–10 min (prior to placental expulsion, after clamping and sectioning of the cord) by accessing the umbilical vein. Alternatively, one can wait for delivery of the placenta and collect the blood directly from the expelled placenta. Once collected the CB sample may be stored at room temperature for up to 48 h before needing to be processed [39, 40]. This large window of time before a CB sample needs to be processed allows for the collection of CB at distant sites and overnight shipment to a processing facility that may even be located in a different state (or country). Most public and private CB banks provide collectors with collection kits (containing everything needed to make the harvest) that also serve as a shipping container. These kits are generally insulated and crush-resistant packages that can be shipped back to the processing facility within 24–36 h, at a cost of $35–100 by a number of transport companies.

The vast majority of the cellular constituent in a cord blood collection is red blood cells (RBC), followed by neutrophils (making up 70–80 % of the leukocyte population). It is not thought that either of these cell populations is involved in the biological function of the CB collection when used for either stem cell transplant or for regenerative medicine. In reality, only the MNC fraction (approximately 20 % of the leukocyte population) which contains the stem cell population (about 1 % of the MNC fraction), is needed for banking. However, isolation and banking solely the stem cell population is impractical, expensive and requires special FDA oversight. CB has a very high hematocrit and RBC can make up more than half of the collection by volume. Thus, to facilitate the banking procedure, CB collections are routinely RBC-depleted or reduced prior to cryopreservation. The most commonly used methods to accomplish this task include Hespan sedimentation to obtain a modified buffy coat [41], density gradient centrifugation (Ficol method) to obtain enriched MNCs [42], and two automated processes (Sepax® from Biosafe SA, Eysins, Switzerland and the AutoXpress Platform® (AXP) from Thermogenesis, Rancho Cordova, CA; [43, 44]) that also result in a buffy coat product. The Hespan, Sepax, and AXP processing methods result in cord blood products containing all nucleated cell populations found in the original collection (MNC, neutrophils, some normal as well as nucleated RBC), while the Ficol method enriches for the stem cell-containing MNC subpopulation (generally greater than 85 % of the final cell composition are MNC with the remaining cells being neutrophils and nucleated RBC). Total cell counts obtained in the final Ficol product are generally 50 % or less of the cell counts found in the other processes for this reason, which may be deceiving at first, but absolute MNC and stem cell recovery is similar to the other processes. The AXP and Sepax devices are functionally closed systems, which are recommended under the current regulatory guidelines [34], although not the only closed processing systems available. There are also a few CB banks that perform plasma reduction as a means of sample volume reduction prior to banking. It is thought that there may be important components in the non-leukocyte fraction that would be important for clinical use. In addition, the RBC may always be removed later after thawing [45]. An average cord blood collection is generally 70–80 mL of blood from a typical full-term (40 weeks), live birth, containing an average of slightly more (or less) than $(850–1100) \times 10^6$ total nucleated cells. Our own experience has been that the AXP and plasma-reduction approaches to CB processing recover the highest percentage of total nucleated cells, although all approaches may be equivalent in terms of MNC and stem cell recovery. It has not been determined if samples processed by any one of these approaches is clinically superior to another approach. In general, it is desirable to recover 80 % or greater of the starting total MNC and stem cells to insure confidence in clinical utility.

The majority of CB banks currently store CB units in what is advertised as "multiple aliquots". This partitioning of the sample is usually accomplished by the use of a freezing bag divided into multiple compartments. Such freezing bags can be obtained from multiple vendors for approximately $25–30 per bag. In theory, cryopreservation of the CB sample in multiple aliquots would allow for future use of the stem cells in cell expansion, gene therapy, or for regenerative medicine uses, which may only require a fraction of the frozen unit. Thus, it would not be necessary to thaw the entire unit unless absolutely needed, avoiding the damaging effects of repeated episodes of freezing/thawing. The commonly available freezing bags generally provide for two aliquots comprising a 20 % and an 80 % fraction of the processed unit in separate compartments. In reality, the smaller fraction is almost always too small to be clinically useful without significant cell expansion beforehand, which currently is not feasible. Thus, storage of CB samples in multiple aliquots is more marketing hype than anything else. A better strategy would be to store the processed sample in two smaller bags of equal volume, either of which could be clinically useful. Alternatively, we have shown that CB samples may be thawed and refrozen several times before losing biological function [39, 40, 46].

Cryopreservation of the processed CB samples is generally performed with the use of a controlled-rate freezing device to avoid cell death during this process. Although there may be multiple cryoprotectant solutions advertised that have been "optimized" for this process, 40 years of extensive experience with dimethylsulfoxide (DMSO) as a cryoprotectant (and in our experience a better than 90 % recovery of nucleated and stem cells upon thawing) leave this author unconvinced that better products are needed or that significant improvements can be obtained. Further, the use of DMSO has the regulatory seal of approval for clinical use. Once frozen the samples are placed in liquid nitrogen-containing dewars at −196 °C for long term storage. This desired temperature range was once only accomplished by placing samples in the liquid phase of such dewars, but recent equipment innovations have now made this possible using vapor phase dewars, which avoids possible cross-sample contamination issues that may arise with sample immersion. We and others [31, 47] have shown that samples stored in this fashion remain viable and clinically useful for more than a decade after placement. In fact, the FDA and other regulatory agencies have noted that samples stored in this manner have no expiration date.

Cord Tissue and Other Cell Banking

An additional source of stem cells that can also be obtained at the time of birth is the umbilical cord tissue (CT) itself which is a source of MSCs [48–50]. These CT-MSCs appear to be identical in phenotype and function to MSC isolated from both bone marrow and adipose tissue [17], sources commonly used in the clinic. That is, the CT-MSCs are CD45−, CD44+, CD90+. CD73+, CD105+ and are capable of differentiating into adipogenic, chondrogenic and osteogenic lineages [17, 28]. CT can be collected and banked as a future source of stem cells for regenerative medicine and tissue engineering. In addition to MSCs, CT also contains endothelial and epithelial precursor cells that may be useful for these applications [51, 52]. CT is derived from the human umbilical cord that develops during gestation to support the development of the fetus. The average length of the cord itself is between 30 and 50 cm [53]. Stem cells have been identified in the Wharton's jelly and the perivascular space surrounding the blood vessels [54], which can be isolated by various methodologies [55–71]. Regardless of the isolation methodology used, one generally finds that the predominant cell population obtained is the MSC [54]. In addition, there are also small numbers of endothelial, epithelial and hematopoietic stem cells present depending on which isolation protocol is utilized.

Cord tissue is generally obtained from full term deliveries. Most collections are 5–8 in. tissue pieces cut from the umbilical cord with sterilized scissors and processed within 24 h. Many CB banks have elected to cryopreserve the intact CT rather than the isolated MSC in order to avoid regulatory restrictions and oversight that could prohibit this type of operation by for-profit facilities (due to the requirement for an IND or BLA application). Generally, the surface of the CT is cleaned and sterilized with alcohol and betadine by the collector after the cord blood has been collected and prior to expulsion of the placenta. It is then cut with sterile scissors and a 5–8 in. segment is placed into a sterile container containing a transport buffer (usually containing penicillin-streptomycin, gentamycin, and amphotericin) for shipment to the processing facility, utilizing a collection kit provided by the processing facility and specifically designed for this purpose. Research has shown that the sample may be held and transported at room temperatures for up to 48 h. Processing of the CT sample begins by washing the CT in isotonic saline followed by a 70 % ethanol wash, and a final sterile saline wash. The CT is then either cut into small 5 mm ringlets or minced into small pieces using a sterile scalpel. The CT pieces are then placed in isotonic saline (containing human serum albumin and 1.5 mol/L DMSO) for 30 min at 4C on a rocking platform. Samples are frozen in 4.5 cc cryovials (~1.0–1.5 g total/cryovial) using a controlled rate freezer similar to how the cord blood sample is frozen. Samples are stored in liquid nitrogen dewars much as cord blood samples are stored. Significant difficulties are encountered however, when the CT sample is thawed for MSC isolation and clinical utilization.

MSCs from cord tissue are best isolated after thawing using a non-enzymatic digestion procedure previously described [17, 26, 72–74]. Briefly, pieces of fresh (or thawed) cord tissue are extensively washed with PBS containing penicillin and streptomycin in a 100 mm petri-plate. The minced tissue is placed into a 25 cm^2 culture flask. After 4–6 days the pieces are removed and cultured in a new flask. In 10–14 days cell colonies can be observed in the culture flasks. The cells are then harvested using trypsin-EDTA and pooled. Our experience [17, 28, 36] has been that it is not possible to isolate significant numbers of MSC from frozen and thawed CT utilizing an enzymatic digestion approach [75]. The reason for this observation appears to be the difficulty of sufficiently removing the DMSO cryopreservation buffer from the intact tissue after thawing prior to placement at 37 °C for the enzymatic digestion process. If DMSO is present in the tissue when placed at room temperature (or above) it is toxic to the cells. Thus, banked CT is not a readily available or significant source of MSC for clinical use without extensive *in vitro* expansion beforehand. Even if the freshly harvested CT is enzymatically digested prior to cryopreservation, the total number of MSC that can be obtained is quite low [54, 56, 58, 71, 76, 77]. Best estimates are that if it were possible to obtain a 30 cm length of CT at birth [54], that the number of freshly isolated MSC would be in the range of millions of cells, assuming every cell isolated was actually a MSC (which is probably not realistic; [78]). Thus, even with freshly harvested CT only limited numbers of MSC can be obtained immediately upon collection. Again, *in vitro* culture expansion would generally be needed prior to clinical use which is a major disadvantage. In our experience cryopreservation reduces the numbers of MSC upon thawing by 75 % or greater [17, 28, 36]. However, once cultures are established the CT-MSC are capable of rapid expansion and continue to demonstrate the capacity to undergo multi-lineage differentiation [36]. Thus, it appears that both fresh and frozen CT requires extensive *in vitro* expansion before being able to be used in clinical applications. Clinical grade MSC expansion would require at a minimum a Good Tissue Practice (GTP) facility, and possibly a Good Manufacturing Practices (GMP)-qualified facility, and an investigational approval from the FDA (in terms of an Investigational New Drug (IND) protocol or Biologic License Agreement (BLA)). These requirements require additional resources (time and monies) in order to be accomplished successfully. Thus, currently there is no ready-to-use, out of the box, clinical trial-ready CT-MSC methodology available (in contrast to other adult sources of MSC as the number of CT-MSC that can routinely be obtained ranges from 250,000 to 10 million total cells). Thus, although CT must and can be collected at the time of birth and can be banked frozen for extended periods of time prior to expected use, it is not a practice that comes highly recommended (although it has become quite faddish).

Other Cell Populations

The majority of cells present in both CB and CT are not stem cells. In CB, as mentioned above, neutrophils are the cells in greatest number, followed by typical lymphocytes [3]. However, based on typical collection volumes of 80–100 cc, and a TNC of $12–16 \times 10^6$/cc, the absolute numbers of these cells in a typical cord blood collection pale in comparison to what can be obtained from typical peripheral blood harvests, with or without prior G-CSF mobilization. Thus, one would not elect to bank this tissue solely for the purpose of having access to either the PMNs or lymphocytes. Both CB and CT also contain some endothelial cell precursors as well as limited numbers of ES-like pluripotential cells [13]. Again, absolute numbers are too small to justify banking these tissues solely for this purpose. Although these cell populations could be useful for regenerative medicine and tissue engineering applications, it is difficult to justify banking these tissues solely for the purpose of having access to these populations. Fortuitously, banking CB for the hematopoietic stem cells and banking CT for the MSC also stores these other potentially useful cell populations.

Clinical Trials

Cord blood has been used to treat more than 80 malignant and non-malignant hematologic conditions requiring stem cell transplant, with more than 30,000 transplants performed to date [10]. However, the use of cord blood (or any other stem cells source) in this setting is thankfully uncommon. Thus, investigations have been initiated over the last decade to determine if these stem cells might be used in regenerative medicine and tissue engineering applications, which is predicted to afflict most individuals during their lifetime. In that regard, multiple studies [79–86] have shown that CB administration in animal stroke models have resulted in some degree of therapeutic benefit with no adverse effects. Neuroprotective effects [79–81, 84, 85, 87] as well as functional/behavioral improvements [80, 84, 85] have been widely reported. Beneficial effects have also been reported for CB use in traumatic brain injury [88] and spinal cord injury [83, 89, 90] models. Additional pre-clinical animal studies have also examined the use of CB in cardiovascular disease, including myocardial infarction (MI) with positive effects [91–100]. Finally, other animal model studies have demonstrated CB's utility in the regenerative medicine setting including orthopedics (for review please see reference [29]).

A recent review of the www.clinicaltrials.gov website (1 Apr 2014) revealed that there are currently 223 ongoing cord blood clinical trials registered. As might be expected the majority of the trials were concerned with stem cell transplant for cancer and other blood disorders. However, a total of 15 trials were currently investigating the use of CB stem cells for regenerative medicine. These trials included children with certain types of autism (Sutter Health, Sacramento, CA; NCT01638819), acute burns (China, NCT01443689), acquired hearing loss (Florida Hospital, NCT02038972), type 1 diabetes (China, NCT01996228), and cardiovascular disease (Mayo Clinic, NCT01883076). The area with the most interest has been the neurological arena with eight individual clinical trials for global development delay (Korea, NCT01769716), cerebral palsy (Duke University, NCT01147653; Korea, NCT02025972; Univ. Texas-Houston, NCT01988584), traumatic brain injury (Korea, NCT01451528; Singapore, NCT01649648), stroke (Hong Kong, NCT01673932), and hypoxic-ischemic encephalopathy (Duke University, NCT00593242). In total, these trials have or will treat more than 350 patients over the next several years.

CT has only recently been recognized as a potential stem cell source, and thus fewer studies have been performed to date. CT-MSCs have shown promising results in a variety of pre-clinical animal studies including intracerebral hemorrhage models [101], spinal cord injury models [102–105] and in animal models of Parkinson's disease [106, 107]. In clinical trials CT-MSC were first expanded *in vitro* under an IND before use. In the transplant setting CT-MSCs showed positive results in the treatment of graft versus host disease (GVHD) following stem cell transplantation. Wu et al. [81] found that CT-MSC had superior proliferative potential and increased immunosuppressive effects as compared to bone marrow MSC. Two pediatric patients with severe steroid-resistant GVHD were infused with *ex-vivo* expanded CT MSCs. The GVHD improved dramatically in both patients following infusion of

CT-MSCs, although one patient needed to receive multiple infusions of MSCs over the course of treatment [22]. CT-MSCs have also been evaluated for potential therapeutic benefits in autoimmune diseases. Liang et al. [23] reported that CT-MSCs stabilized the disease course of a patient with progressive multiple sclerosis that was not responsive to conventional treatment [23]. A subsequent study from the same group reported dramatic improvements in a patient with systemic lupus erythematosus following intravenous infusion of CT-MSCs [24]. A recent report from Xue et al. examined the use of CT MSCs in patients with non-healing bone fractures [21]. This study reported significant clinical benefit from intravenously infused MSC.

Examination of www.clinicaltrials.gov revealed an additional seven trials registered with the FDA including three additional trials for autoimmune diseases: lupus (China, NCT01741857), multiple sclerosis (NCT02034188) and rheumatoid arthritis (NCT01985464). Additionally there was listed one study using CT-MSC to induce tolerance for organ transplantation (China, NCT01690247), one to treat liver cirrhosis (China, NCT01220492), one for cardiovascular disease (China, NCT01946048), and one for osteoarthritis (NCT02003131). In total these trials have or are treating more than 250 patients.

Problems and Issues

If one has elected to bank stem cells, and/or desires to participate in any of the numerous stem cell therapies being touted, how does one go about doing that? Fortunately and unfortunately, stem cell clinics seem to be everywhere these days. A recent Google search turned up 21.7 million hits on the topic (search performed on 1 Apr 2014). Many of these clinics reside outside of the USA, and outside of any type of medical and regulatory oversight. It seems that many of these clinics are making unsubstantiated claims without any basis in fact, and really are doing nothing more than taking advantage of the hype surrounding the stem cell arena. Desperate patients are being preyed upon by individuals without the knowledge or training to offer safe and effective therapies, but merely false hope. One should be cautious not only of therapies offered only outside the USA in "second world" countries, but also any trial not registered with the www.clinicaltrials.gov website. If it seems to be too good to be true, it probably is. Although it seems that many patients have lost nothing more than money and time, there will eventually be a time that one or more will suffer injury or death in one of these unlicensed trials. Once that happens the field of regenerative medicine will suffer a significant setback, which is to no one's benefit. Thus, something must be done to restrict the practice of stem cell therapy to someone with more than a medical license or an overseas clinic. More must be done to educate patients about promising as well as fraudulent therapies. And something must really be done to streamline the process of bringing new therapies to fruition at more reasonable costs.

Although cord blood banking has been practiced for more than 20 years now (and CT banking approximately 5 years or so), a recent public survey ($N=1000$) that we conducted has shown that less than half of expectant mothers were aware of cord blood banking, most were not familiar to any extent with the different uses of cord blood stem cells, and most found it prohibitively expensive (almost regardless of income). At current costs of approximately $2000 more than half of all expectant families are economically excluded from considering family stem cell banking. And with few public stem cell banks available more than 90 % of all expectant families are not eligible to donate their stem cells if they should desire to do so. Obviously, the system needs changes.

Even fewer families are aware of CT banking and whether or not they should even consider it. We have also found that although a greater percentage of OBs and midwives are familiar with CB banking, most are not current with cord blood stem cell uses, especially in the fields of regenerative medicine and tissue engineering. Thus, education of both the public and the medical community remains a challenge. Without education families may miss the opportunity to donate or bank stem cells for their child, but more importantly because they or their caregivers are not aware of treatment breakthroughs they may miss the opportunity to participate in clinical trials that could save or significantly better their lives. We must not let that happen.

References

1. Broxmeyer HE, Gluckman E, Auerbach A, Douglas GW, Friedman H, Cooper S, Hangoc G, Kurtzberg J, Bard J, Boyse EA. Human umbilical cord blood: a clinically useful source of transplantable hematopoietic stem/progenitor cells. Int J Cell Cloning. 1990;8 Suppl 1:76–89. discussion 89-91.
2. Gluckman E, Broxmeyer HA, Auerbach AD, Friedman HS, Douglas GW, Devergie A, Esperou H, Thierry D, Socie G, Lehn P, et al. Hematopoietic reconstitution in a patient with Fanconi's anemia by means of umbilical-cord blood from an HLA-identical sibling. N Engl J Med. 1989;321(17):1174–8.
3. Harris D, Schumacher MJ, LoCacsio J, Besencon FJ, Olson G, DeLuca D, Bard J, Boyse E. Phenotypic and functional immaturity of human umbilical cord blood T lymphocytes. Proc Natl Acad Sci U S A. 1992;89:10006–10.
4. Gluckman E. Stem cell harvesting from cord blood: a new perspective. In: Wunder HA, editor. Peripheral blood stem cell autographs. Berlin: Springer; 1990.
5. Broxmeyer HA, Kurtzberg J, Gluckman E, Auerbach A, Douglas G, Cooper S, Falkenburg JH, Bard J, Boyse EA. Umbilical cord blood hematopoietic stem and repopulating cells in human clinical transplantation: an expanded role for cord blood transplantation. Blood Cells. 1991;17(2):330–7.
6. Broxmeyer HE, Kurtzberg J, Gluckman E, Auerbach AD, Douglas G, Cooper S, Falkenburg JH, Bard J, Boyse EA. Umbilical cord blood hematopoietic stem and repopulating cells in human clinical transplantation. Blood Cells. 1991;17(2):313–29.
7. Broxmeyer HE, Douglas GW, Hangoc G, Cooper S, Bard J, English D, Arny M, Thomas L, Boyse EA. Human umbilical cord blood as a potential source of transplantable hematopoietic stem/progenitor cells. Proc Natl Acad Sci U S A. 1989;86(10):3828–32.
8. Vilmer E, Sterkers G, Rahimy C, Denamur E, Elion J, Broyart A, Lescoeur B, Tiercy JM, Gerota J, Blot P. HLA-mismatched cord-blood transplantation in a patient with advanced leukemia. Transplantation. 1992;53(5):1155–7.
9. Wagner JE, Kernan NA, Steinbuch M, et al. Allogeneic sibling umbilical cord blood transplantation in children with malignant and nonmalignant disease. Lancet. 1995;346:214–9.

10. Butler MG, Menitove JE. Umbilical cord blood banking: an update. J Assist Reprod Genet. 2011;28(8):669–76.
11. McGuckin C, Forraz N, Baradez MO, et al. Production of stem cells with embryonic characteristics from human umbilical cord blood. Cell Prolif. 2005;38:245–55.
12. McGuckin CP, Forraz N, Allouard Q, Pettengell R. Umbilical cord blood stem cells can expand hematopoietic and neuroglial progenitors in vitro. Exp Cell Res. 2004;295:350–9.
13. Rogers I, Yamanaka N, Bielecki R, Wong CJ, Chua S, Yuen S, Casper RF. Identification and analysis of in vitro cultured CD45-positive cells capable of multi-lineage differentiation. Exp Cell Res. 2007;313:1839–52.
14. Kucia M, Halasa M, Wysoczynski M, et al. Morphological and molecular characterization of novel population of CXCR4+ SSEA-4+ Oct-4+ very small embryonic-like cells purified from human umbilical cord blood-preliminary report. Leukemia. 2007;21:297–303.
15. Harris DT, He X, Badowski M, Nichols JC. Regenerative medicine of the eye: a short review. In: Levicar N, Habib NA, Dimarakis I, Gordon MY, editors. Stem cell repair and regeneration, vol. 3. London: Imperial College Press; 2008. p. 211–25.
16. Sunkomat JNE, Goldman S, Harris DT. Cord blood-derived MNCs delivered intracoronary contribute differently to vascularization compared to CD34+ cells in the rat model of acute ischemia. J Mol Cell Cardiol. 2007;42(6 Suppl 1):97.
17. Choudhery MS, Badowski M, Muise A, Harris DT. Comparison of the regenerative potential of human adipose and cord tissue derived mesenchymal stem cells. Cytotherapy. 2013;15:330–43.
18. Harris DT, Rogers I. Umbilical cord blood: a unique source of pluripotent stem cells for regenerative medicine. Curr Stem Cell Res Ther. 2007;2:301–9.
19. www.clinicaltrials.gov
20. Harris DT, Badowski M, Ahmad N, Gaballa M. The potential of cord blood stem cells for use in regenerative medicine. Expert Opin Biol Ther. 2007;7(9):1311–22.
21. Xue G, He M, Zhao J, Chen Y, Tian Y, Zhao B. Intravenous umbilical cord mesenchymal stem cell infusion for the treatment of combined malnutrition nonunion of the humerus and radial nerve injury. Regen Med. 2011;6:733–41.
22. Wu KH, Chan CK, Tsai C, Chang YH, Sieber M, Chiu TH, Ho M, Peng CT, Wu HP, Huang JL. Effective treatment of severe steroid-resistant acute graft-versus-host disease with umbilical cord-derived mesenchymal stem cells. Transplantation. 2011;91(12):1412–6.
23. Liang J, Zhang H, Hua B, Wang H, Wang J, Han Z, Sun L. Allogeneic mesenchymal stem cells transplantation in treatment of multiple sclerosis. Mult Scler. 2009;15(5):644–6.
24. Liang J, Gu F, Wang H, Hua B, Hou Y, Shi S, Lu L, Sun L. Mesenchymal stem cell transplantation for diffuse alveolar hemorrhage in SLE. Nat Rev Rheumatol. 2010;6(8):486–9.
25. Reza HM, Ng B-Y, Gimeno FL, Phan TT, Ang LP-K. Umbilical cord lining stem cells as a novel and promising source for ocular surface regeneration. Stem Cell Rev Rep. 2011;7:935–47.
26. Choudhery MS, Badowski M, Muise A, Pierce J, David T, Harris DT. Cryopreservation of whole adipose tissue for future use in regenerative medicine. J Surg Res. 2014;187(1):24–35.
27. Choudhery MS, Badowski MS, Muise A, Pierce J, Harris DT. Donor age negatively impacts adipose tissue-derived mesenchymal stem cell expansion and differentiation. J Transl Med. 2014;12:8. doi:10.1186/1479-5876-12-8; URL:http://www.translational-medicine.com/content/12/1/8.
28. Choudhery MS, Badowski M, Muise A, Harris DT. Utility of cryopreserved umbilical cord tissue for regenerative medicine. Curr Stem Cell Res Ther. 2013;8(5):870–80.
29. Harris DT. Stem cell banking for regenerative and personalized medicine. Biomedicines. 2014;2:50–79. doi:10.3390/biomedicines2010050.
30. He X, Gonzalez V, Tsang A, Thompson J, Tsang T, Harris DT. Differential gene expression profiling of CD34+CD133+ umbilical cord blood hematopoietic stem-progenitor cells. Stem Cells Dev. 2005;14(2):188–98.
31. Harris DT, Schumacher MJ, Rychlik S, Booth A, Acevedo A, Rubinstein P, Bard J, Boyse EA. Collection, separation and cryopreservation of umbilical cord blood for use in transplantation. Bone Marrow Transplant. 1994;13:135–43.

32. Harris D, LoCascio J, Besencon F. Analysis of the alloreactive capacity of human umbilical cord blood: implications for graft-versus-host disease. Bone Marrow Transplant. 1994;14:545–53.
33. Harris DT. In vitro and in vivo assessment of the graft-versus-leukemia activity of cord blood. Bone Marrow Transplant. 1995;15:17–23.
34. Harris DT, Schumacher MJ, LoCascio L, Booth A, Bard J, Boyse EA. Immunoreactivity of umbilical cord blood and post-partum peripheral blood with regard to HLA-haploidentical transplantation. Bone Marrow Transplant. 1994;14:63–8.
35. Harris DT. GVL and GVHD implications of cord blood. Proceedings of the international conference/workshop on cord blood transplantation and biology/immunology. Blood Cells. 1994;20:560–5.
36. Badowski M, Muise A, Harris DT. Mixed effects of long term frozen storage on cord tissue stem cells. Cytotherapy. 2014;16(9):1313–21.
37. Badowski M, Shultz C, Harris DT. The effect of anti-coagulant choice on collection of cord blood. Transfusion. 2014;54(9):2364.
38. Badowski M, Shultz C, Harris DT. The effect of anti-coagulant choice on cord blood processing results. Cytotherapy. 2014;pii:S1465-3249(14)00600-8.
39. Harris DT, Wang J, He X, Brett SC, Moore ME, Brown H. Studies on practical issues for cord blood banking: effects of ionizing radiation and cryopreservation variables. Open Stem Cell J. 2010;2:37–44. doi:10.2174/1876893801002010037 (ISSN:1876–8938, Volume 2, 2010).
40. Harris DT. Optimizing cord blood sample cryopreservation. Cytotherapy. 2012;14(3):359–65.
41. Rubinstein P, Rosenfield RE, Adamson JW, Stevens CE. Stored placental blood for unrelated bone marrow reconstitution. Blood. 1993;81:1679–90.
42. Badowski MS, Harris DT. Collection, processing, and banking of umbilical cord blood stem cells for transplantation and regenerative medicine. In: Singh SR, editor. Somatic stem cells: methods and protocols, methods in molecular biology, vol. 879. New York, NY: Springer; 2011. p. 279–90.
43. Papassavas AC, Goika V, Chatzistamatiou T, et al. A strategy of splitting individual high volume cord blood units into two half subunits prior to processing increases the recovery of cells and facilitates ex vivo expansion of the infused hematopoietic progenitor cells in adults. Int J Lab Hematol. 2008;30(2):124–32.
44. Harris DT, McGaffey AP, Schwarz RH, et al. Comparing the mononuclear cell (MNC) recovery of AXP and Hespan. Obstet Gynecol. 2007;109(4):93S.
45. Chow R, Nademanee A, Rosenthal J, Karanes C, Jaing T-H, Graham ML, Tsukahara E, Wang B, Gjertson D, Tan P, Forman S, Petz LD. Analysis of hematopoietic cell transplants using plasma-depleted cord blood products that are not red blood cell reduced. Biol Blood Marrow Transplant. 2007;13:1346–57.
46. Harris DT, Pipes B, Wang J, He X, Brown H. Assessment of the effects of repeated freeze/thaw on the stem/progenitor cell capacity of cord blood samples. AABB annual meeting, Oct 2005, Seattle, WA
47. Broxmeyer HE, Srour EF, Hangoc G, Cooper S, Anderson SA, Bodine DM. High-efficiency recovery of functional hematopoietic progenitor and stem cells from human cord blood cryopreserved for 15 years. Proc Natl Acad Sci U S A. 2003;100(2):645–50.
48. Lee OK, Kuo TK, Chen WM, Lee KD, Hsieh SL, Chen TH. Isolation of multipotent mesenchymal stem cells from umbilical cord blood. Blood. 2004;103:1669–75.
49. Musina RA, Bekchanova ES, Belyavskii AV, Grinenko TS, Sukhikh GT. Umbilical cord blood mesenchymal stem cells. Bull Exp Biol Med. 2007;143:127–31.
50. Secco M, Zucconi E, Vieira NM, et al. Multipotent stem cells from umbilical cord: cord is richer than blood. Stem Cells. 2008;26:146–50.
51. Lindenmair A, Hatlapatka T, Kollwig G, et al. Mesenchymal stem or stromal cells from amnion and umbilical cord tissue and their potential for clinical applications. Cells. 2012;1:1061–88.
52. Huang Y-C, Parolini O, La Rocca G, Deng L. Umbilical cord versus bone marrow-derived mesenchymal stromal cells. Stem Cells Dev. 2012;21:2900–3.
53. William PL, Banister LH, Berry MM, et al. Grays anatomy. 38th ed. London: ELBS Churchill Livingstone; 1995.

54. Lu LL, Liu YJ, Yang SG, et al. Isolation and characterization of human umbilical cord mesenchymal stem cells with hematopoiesis-supportive function and other potentials. Haematologica. 2006;91:1017–26.
55. Schugar RC, Chirieleison SM, Wescoe KE. High harvest yield, high expansion, and phenotype stability of CD146 mesenchymal stromal cells from whole primitive human umbilical cord tissue. J Biomed Biotechnol. 2009;Article ID 789526:11 pages.
56. Capelli C, Gotti E, Morigi M, et al. Minimally manipulated whole human umbilical cord is a rich source of clinical grade human mesenchymal stromal cells expanded in human platelet lysate. Cytotherapy. 2011;13:786–801.
57. Bosch J, Houben AP, Radke TF, et al. Distinct differentiation potential of 'MSC' derived from cord blood and umbilical cord: are cord-derived cells true mesenchymal stromal cells? Stem Cells Dev. 2012;21(11):1977–88. doi:10.1089/scd.2011.0414.
58. Kita K, Gauglitz GG, Phan TT, Herndon DN, Jeschke MG. Isolation and characterization of mesenchymal stem cells from the sub amniotic human umbilical cord lining membrane. Stem Cells Dev. 2010;19:491–502.
59. Sarugaser R, Lickorish D, Baksh D, Hosseini MM, Davies JE. Human umbilical cord perivascular (HUCPV) cells: a source of mesenchymal progenitors. Stem Cells. 2005;23:220–9.
60. Romanov YA, Svintsitskaya VA, Smirnov VN. Searching for alternative sources of postnatal human mesenchymal stem cells: candidate MSC-like cells from umbilical cord. Stem Cells. 2003;21:105–10.
61. Karahuseyinoglu S, Cinar O, Kilic E, et al. Biology of stem cells in human umbilical cord stroma: in situ and in vitro surveys. Stem Cells. 2007;25:319–31.
62. Fu YS, Shih YT, Cheng YC, Min MY. Transformation of human umbilical mesenchymal cells into neurons in vitro. J Biomed Sci. 2004;11:652–60.
63. Tsagias N, Koliakos I, Karagiannis V, Eleftheriadou M, Koliakis GG. Isolation of mesenchymal stem cells using the total length of umbilical cord for transplantation purposes. Transfus Med. 2011;21:253–61.
64. Dominici M, Le Blanc K, Mueller I, et al. Minimal criteria for defining multipotent mesenchymal stromal cells. The International Society for Cellular Therapy position statement. Cytotherapy. 2006;8:315–7.
65. Abdallah BM, Kassem M. Human mesenchymal stem cells: from basic biology to clinical applications. Gene Ther. 2008;15(2):109–16.
66. Uccelli A, Moretta L, Pistoia V. Mesenchymal stem cells in health and disease. Nat Rev Immunol. 2008;8:726–36.
67. Friedenstein AJ, Chailakhjan RK, Lalykina KS. The development of fibroblast colonies in monolayer cultures of guinea-pig bone marrow and spleen cells. Cell Tissue Kinet. 1970;3:393–403.
68. Bieback K, Kern S, Kocaomer A, Ferlik K, Bugert P. Comparing mesenchymal stromal cells from different human tissues: bone marrow, adipose tissue and umbilical cord blood. Biomed Mater Eng. 2008;18(1 Suppl):S71–6.
69. Prather WR, Toren A, Meiron M, Ofir R, Tschope C, Horwitz EM. The role of placental-derived adherent stromal cell (PLX-PAD) in the treatment of critical limb ischemia. Cytotherapy. 2009;11(4):427–34.
70. Jo CH, Kim OS, Park EY, et al. Fetal mesenchymal stem cells derived from human umbilical cord sustain primitive characteristics during extensive expansion. Cell Tissue Res. 2008;3: 423–33.
71. Troyer DL, Weiss ML. Wharton's jelly-derived cells are a primitive stromal cell population. Stem Cells. 2008;26(3):591–9.
72. Marmotti A, Mattia S, Bruzzone M, Buttiglieri S, Risso A, Bonasia DE, Blonna D, Castoldi F, Rossi R, Zanini C, Ercole E, Defabiani E, Tarella C, Peretti GM. Minced umbilical cord fragments as a source of cells for orthopaedic tissue engineering: an in vitro study. Stem Cells Int. 2012;2012:326813. doi:10.1155/2012/326813.
73. Nekanti U, Mohanty L, Venugopal P, Balasubramanian S, Totey S, Ta M. Optimization and scale-up of Wharton's jelly-derived mesenchymal stem cells for clinical applications. Stem Cell Res. 2001;5:244–54.

74. Pilz GA, Ulrich C, Ruh M, Abele H, Schäfer R, Kluba T, Bühring HJ, Rolauffs B, Aicher WK. Human term placenta-derived mesenchymal stromal cells are less prone to osteogenic differentiation than bone marrow-derived mesenchymal stromal cells. Stem Cells Dev. 2011;20:635–46.
75. Kikuchi-Taura A, Taguchi A, Kanda T, et al. Human umbilical cord provides a significant source of unexpanded mesenchymal stromal cells. Cytotherapy. 2012;14:441–50.
76. Pappa KI, Anagnou NP. Novel sources of fetal stem cells: where do they fit on the developmental continuum? Regen Med. 2009;4(3):423–33.
77. Can A, Karahuseyinoglu S. Concise review: human umbilical cord stroma with regard to the source of fetus-derived stem cells. Stem Cells. 2007;25(11):2886–95.
78. Fan C-G, Zhang Q-J, Zhou J-R. Therapeutic potentials of mesenchymal stem cells derived from human umbilical cord. Stem Cell Rev Rep. 2011;7:195–207.
79. Xiao J, Nan Z, Motooka Y, Low WC. Transplantation of a novel cell line population of umbilical cord blood stem cells ameliorates neurological deficits associated with ischemic brain injury. Stem Cells Dev. 2005;14:722–33.
80. Newcomb JD, Ajrno CT, Sanberg CD, Sanberg PR, Pennypacker KR, Willing AE. Timing of cord blood treatment after experimental stroke determines therapeutic efficacy. Cell Transplant. 2006;15:213–23.
81. Vendrame M, Gemma C, Pennypacker KR, Bickford PC, Davis Sanberg C, Sanberg PR, Willing AE. Cord blood rescues stroke-induced changes in splenocyte phenotype and function. Exp Neurol. 2006;199(1):191–200.
82. Meier C, Middelanis J, Wasielewski B, Neuhoff S, Roth-Haerer A, Gantert M, Dinse HR, Dermietzel R, Jensen A. Spastic paresis after perinatal brain damage in rats is reduced by human cord blood mononuclear cells. Pediatr Res. 2006;59(2):244–9.
83. Chen SH, Chang FM, Tsai YC, Huang KF, Lin CL, Lin MT. Infusion of human umbilical cord blood cells protect against cerebral ischemia and damage during heatstroke in the rat. Exp Neurol. 2006;199(1):67–76.
84. Vendrame M, Gemma C, de Mesquita D, Collier L, Bickford PC, Sanberg CD, Sanberg PR, Pennypacker KR, Willing AE. Anti-inflammatory effects of human cord blood cells in a rat model of stroke. Stem Cells Dev. 2005;14(5):595–604.
85. Nan Z, Grande A, Sanberg CD, Sanberg PR, Low WC. Infusion of human umbilical cord blood ameliorates neurologic deficits in rats with hemorrhagic brain injury. Ann N Y Acad Sci. 2005;1049:84–96.
86. Nystedt J, Mäkinen S, Laine J, Jolkkonen J. Human cord blood CD34+ cells and behavioral recovery following focal cerebral ischemia in rats. Acta Neurobiol Exp (Wars). 2006;66(4):293–300.
87. Mäkinen S, Kekarainen T, Nystedt J, Liimatainen T, Huhtala T, Närvänen A, Laine J, Jolkkonen J. Human umbilical cord blood cells do not improve sensorimotor or cognitive outcome following transient middle cerebral artery occlusion in rats. Brain Res. 2006;1123(1):207–15.
88. Lu D, Sanberg PR, Mahmood A, Li Y, Wang L, Sanchez-Ramos J, Chopp M. Intravenous administration of human umbilical cord blood reduces neurological deficit in the rat after traumatic brain injury. Cell Transplant. 2002;11:275–81.
89. Kuh SU, Cho YE, Yoon DH, Kim KN, Ha Y. Functional recovery after human umbilical cord blood cells transplantation with brain derived-neurotropic factor into the spinal cord injured rats. Acta Neurochir (Wein). 2005;14:985–92.
90. Kang K-S, Kim SW, Oh YH, Yu JW, Kim K-Y, Park HK, Song C-H, Han H. A thirty-seven-year old spinal cord-injured female patient, transplanted of multipotent stem cells from human UC blood with improved sensory perception and mobility, both functionally and morphologically: a case study. Cytotherapy. 2005;7:368–73.
91. Copeland N, Harris D, Gaballa MA. Human umbilical cord blood stem cells, myocardial infarction and stroke. Clin Med. 2009;9(4):342–5.
92. Botta R, Gao E, Stassi G, et al. Heart infarct in NOD-SCID mice: therapeutic vasculogenesis by transplantation of human CD34+ cells ad low dose CD34+KDR+ cells. FASEB J. 2004;18:1392–4.

93. Henning RJ, Abu-Ali H, Balis JU, et al. Human umbilical cord blood mononuclear cells for treatment of acute myocardial infarction. Cell Transplant. 2004;13:729–39.
94. Chen HK, Hung HF, Shyu KG, et al. Combined cord blood cells and gene therapy enhances angiogenesis and improves cardiac performance in mouse after acute myocardial infarction. Eur J Clin Invest. 2005;35:677–86.
95. Hirata Y, Sata M, Motomura N, et al. Human umbilical cord blood cells improve cardiac function after myocardial infarction. Biochem Biophys Res Commun. 2005;327:609–14.
96. Kim BO, Tian H, Prasongsukarn K, et al. Cell transplantation improves ventricular function after a myocardial infarction: a preclinical study of human unrestricted somatic stem cells in a porcine model. Circulation. 2006;112(9 Suppl):196–204.
97. Leor J, Guetta E, Feinberg MS, et al. Human umbilical cord blood-derived CD133+ cells enhance function and repair of the infracted myocardium. Stem Cells. 2006;24(3):772–80.
98. Ma N, Stamm C, Kaminski A, et al. Human cord blood cells induce angiogenesis following myocardial infarction in NOD/scid mice. Cardiovasc Res. 2005;66:45–54.
99. Amado LC, Saliaris AP, Schuleri KH, et al. Cardiac repair with intramyocardial injection of mesenchymal stem cells after myocardial infarction. Proc Natl Acad Sci U S A. 2005;102:11474–9.
100. Bonnano G, Mariotti A, Procoli A, et al. Human cord blood CD133+ cells imunoselected by a clinical-grade apparatus differentiate in vitro into endothelial- and cardiomyocyte-like cells. Transfusion. 2007;47:280–9.
101. Liao W, Zhong J, Yu J, et al. Therapeutic benefit of human umbilical cord derived mesenchymal stromal cells in intracerebral hemorrhage rat: implications of anti-inflammation and angiogenesis. Cell Physiol Biochem. 2009;24(3–4):307–16.
102. Lim JH, Byeon YE, Ryu HH, et al. Transplantation of canine umbilical cord blood-derived mesenchymal stem cells in experimentally induced spinal cord injured dogs. J Vet Sci. 2007;8(3):275–82.
103. Yang CC, Shih YH, Ko MH, Hsu SY, Cheng H, Fu YS. Transplantation of human umbilical mesenchymal stem cells from Wharton's jelly after complete transection of the rat spinal cord. PLoS One. 2008;3(10), e3336.
104. Zhang L, Zhang HT, Hong SQ, Ma X, Jiang XD, Xu RX. Cografted Wharton's jelly cells-derived neurospheres and BDNF promote functional recovery after rat spinal cord transection. Neurochem Res. 2009;34(11):2030–9.
105. Hu S-L, Luo H-S, Li J-T, Xia Y-Z, Li L, Zhang L-J, Meng H, Cui G-Y, Chen Z, Wu N, Lin J-K, Zhu G, Feng H. Functional recovery in acute traumatic spinal cord injury after transplantation of human umbilical cord mesenchymal stem cells. Crit Care Med. 2010;38:2181–9.
106. Weiss ML, Medicetty S, Bledsoe AR, et al. Human umbilical cord matrix stem cells: preliminary characterization and effect of transplantation in a rodent model of Parkinson's disease. Stem Cells. 2006;24(3):781–92.
107. Fu YS, Cheng YC, Lin MY, et al. Conversion of human umbilical cord mesenchymal stem cells in Wharton's jelly to dopaminergic neurons in vitro: potential therapeutic application for Parkinsonism. Stem Cells. 2006;24(1):115–24.

Chapter 14
Wharton's Jelly Stem Cells

Marina Bastawrous, Mibel M. Pabón, Sandra Acosta, Ike de la Peña, Diana Hernandez-Ontiveros, Meaghan Staples, Kazutaka Shinozuka, Paolina Pantcheva, Naoki Tajiri, Yuji Kaneko, and Cesar V. Borlongan

Introduction

The field of stem cells has witnessed a fast pace in research and development as they offer a tool to decipher the cell's biology and also because of their vast applications in the area of tissue engineering and regenerative medicine. Wharton's Jelly (WJ), first described by Thomas Wharton in 1656 [1–5]. as the primitive connective tissue of the human umbilical cord (UC), is considered an essential source where populations of stem cells are derived [1]. Perhaps one of the most unique populations that display the stemness property is the mesenchymal stromal stem cells (MSCs). Other sources of MSCs are found in the adult's bone marrow and adipose tissue. Different characteristics, predominantly the stemness and immune properties, pertain to populations of MSCs depending on the source where they are derived from whether adult

M. Bastawrous, B.S. • M.M. Pabón, M.S., Ph.D. • S. Acosta, Ph.D. • M. Staples, B.S.
K. Shinozuka, Ph.D. • P. Pantcheva, B.S., M.S. • N. Tajiri, Ph.D. • Y. Kaneko, Ph.D.
C.V. Borlongan, B.S., M.S., Ph.D. (✉)
Center of Excellence for Aging and Brain Repair, Department of Neurosurgery and Brain Repair, Morsani College of Medicine, University of South Florida College of Medicine, 12901 Bruce B. Downs Blvd, Tampa, FL 33612, USA
e-mail: mbastawr@health.usf.edu; mpabon@health.usf.edu; sacosta@health.usf.edu; meaghans@mail.usf.edu; ppantche@health.usf.edu; ntajiri@health.usf.edu; ykaneko@health.usf.edu; cborlong@health.usf.edu

I. de la Peña, Ph.D.
Pharmaceutical and Administrative Sciences, Loma Linda University, 11175 Campus St., Loma Linda, CA, USA
e-mail: idelapena@llu.edu

D. Hernandez-Ontiveros, M.S., Ph.D.
Department of Physiology and Pharmacology, University of South Florida College of Medicine, 12901 Bruce B. Downs Blvd, Tampa, FL 33612, USA
e-mail: dhernan1@health.usf.edu

D.O. Fauza, M. Bani (eds.), *Fetal Stem Cells in Regenerative Medicine*, Stem Cell Biology and Regenerative Medicine,
DOI 10.1007/978-1-4939-3483-6_14

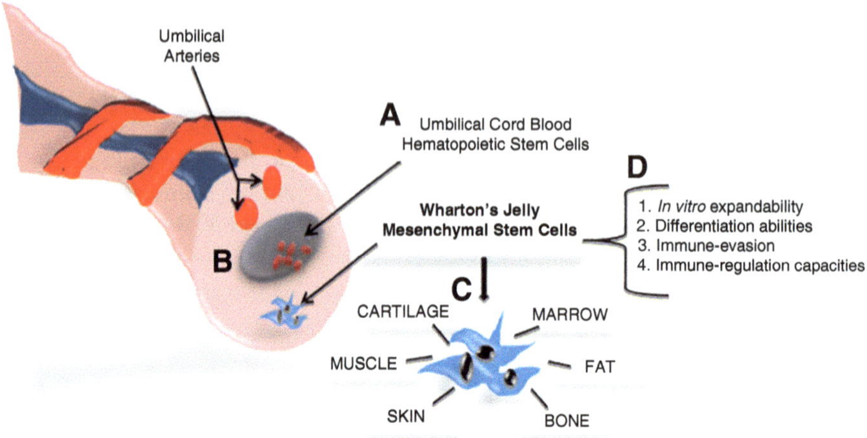

Fig. 14.1 Schematic cartoon of the umbilical cord anatomical structures and Wharton's jelly mesenchymal stem cells. Arteries and vein supply the umbilical cord where blood, enriched in hematopoietic stem cells, flows (**A**). These arteries and vein are embedded in a mucous proteoglycan-rich matrix, known as Wharton's Jelly (WJ) (**B**). WJ constitutes the major component of the umbilical cord tissue, and it is an essential source of mesenchymal stem cell (MSC) populations. WJ-MSCs are multipotent stem cells able to differentiate into cartilage, marrow, muscle, skin, fat, and bone (**C**). Their therapeutic potential includes ease of sourcing, *in vitro* expandability, differentiation abilities, immune-evasion and immune-regulation capacities (**D**)

or fetal. It was found that cell populations harvested from WJ display more robust proliferative, immunosuppressive and therapeutically potent stem cells than those populations harvested from adult bone marrow or adipose (Fig. 14.1). Wharton's Jelly MSCs display a defined set of surface markers which include CD90, CD73 and CD105. These surface markers attribute to the prototypical feature of MSCs which is their plastic adherence. Since McElreavey et al. [2] have first described the isolation and culture of stem cells from WJ; other studies have been focused on defining the optimum method for the isolation and differentiation of these cells. This work offers a compilation of the major discoveries on the properties of WJ-derived stem cells that should support further research and clinical application.

Wharton's Jelly and Other Anatomical Structures of the UC as Sources of MSCs

The umbilical cord encloses two arteries and one vein embedded in a mucous proteoglycan-rich matrix, known as WJ, which is then covered by amniotic epithelium [7–11]. Due to its elasticity, the UC functions to connect between the fetus and the placenta, thus, prevent the umbilical cord vessels from bending and compressing (Fig. 14.1). WJ-MSCs were previously mentioned in literature as "umbilical cord matrix stem cells (UCMSCs)" to distinguish them from endothelial cells isolated from umbilical vein (HUVEC) as well as MSCs isolated from UC blood (UCB-MSCs) [13, 14].

The isolation of the first fibroblast like MSC population from WJ was done by Kobayashi et al. [12] more than 10 years ago. Literature proposes two theories explaining how stem cells originated in WJ; the first postulates that two waves of fetal stem cell migration occurred which eventually ended with some of the MSCs being trapped in the WJ [15]. The second theory hypothesizes that MSCs evolved from cells of the mesenchyme of the UC matrix that function to form a gelatinous ground substance to prevent the strangulation of the UC vessels by secreting glycoproteins, mucopolysaccharides, glycosaminoglycans and extracellular matrix proteins [16]. Different compartments of the UC, identified as distinct regions by Jeschke et al. [17], are considered hosts for stem cell populations including the amniotic compartment (outer epithelial layer and inner subamniotic mesenchymal layer), the WJ compartment, the perivascular compartment surrounding the vessels, the media and adventitia compartment of the walls of UC blood vessels, the endothelial compartment (inner lining of the vein) and the vascular compartment (blood lying within the UC blood vessels) [16]. It is worth to mention that the nomenclature has not been standardized, with terms such as "subamnion", "cord lining (subamnio)", "intervascular", "perivascular" and "hUVEC" being used. There is no histological defined landmark that separates these compartments leading to difficulty in determining whether the WJ-MSCs populations isolated from in between the compartments have the same identity or not. Other factors contributing to the identification of WJ-MSCs populations include the absence of standardization for the region of interest and the method of isolation of MSCs from UC, and the absence of a consensus of the optimum derivation protocol [16]. Previous studies have reported isolation of WJ-MSCs from the intervascular and sub-amnion regions [18], while Troyer et al. [19] reported an additional third region which is the perivascular zone. Populations of cells isolated from the previously mentioned three regions display differences in the number, nature and properties of cells as indicated by the structural, immunohistochemical, and functional analysis performed *in vitro* [20, 21]. This may suggest that these populations originate from different preexisting structures [22]. In general, WJ-MSCs populations isolated from near the amniotic surface have more robust proliferative properties and less differentiation than population isolated from near the umbilical cord vessels [20, 21]. One of the cell populations isolated from around the umbilical cord vessels is the Human umbilical cord perivascular cells (HUCPVCs) [23, 24] also, other stem cell-like cells which show equal potency were isolated from the sub-amnion (cord lining; CL) [17, 25].

Characteristic Features of WJ-MSCs for Cell Therapy

Immunomodulatory Property of WJ

The most pivotal characteristics that WJ-MSCs display are the decreased immunogenicity and the high proliferative potential; both are required to make WJ-MSCs a good candidate for allogenic and xenogeneic transplants. Moreover, they are able to suppress the immune cells and avoid the immune response (Fig. 14.1). Prasanna et al. presented a review that described the different immunomodulatory molecules secreted by WJ-MSCs and data from *in vivo* and *in vitro* studies of their immuno-modulating activities [18]. WJ-MSCs express low levels of MHC class I (HLA-ABC) [18, 26–29] which may explain why they have the ability to escape natural killer cell-mediated cell lysis [18]. No expression of MHC class II (HLA-DR) and the co-stimulatory antigens CD80, CD86, which are involved in the activation of both T and B cell responses, was found [18, 26–29]. When comparing the immune responses and the immuno-modulating activities of WJ-MSCs and BM-MSCs, it is found that the former secrete high amounts of the anti-inflammatory IL-10 and TGF-β than BM-MSCs [30–43]. Also, WJ-MSCs secrete HLA-G, an important protein that functions to suppress the immune response against the fetus during pregnancy by upregulating the regulatory T cells [39, 40, 42–45]. HLA-G was found to be associated with lower immune rejection rate in blood, heart and liver/kidney grafts [46]. Another distinction between these two cell populations is their response to pro-inflammatory cytokines. Upon treatment with IFN-γ, HLA-DR show substantial induction in BM-MSCs, while in WJ-MSCs negligible induction is observed due to the absence of the targeted receptor [26, 30]. Moreover, WJ-MSCs secrete two essential proteins for their immunosuppressive capability which are IL-6 and VEGF [29, 34]. Recently, a study was done to further evaluate the immuno-moduatory properties of WJ-MSCs regarding their effects on T-cell proliferation and IFN-γ secretion and their therapeutic potential in diabetic rat models [47]. The study also investigated whether or not the transplanted WJ-MSCs will differentiate into pancreatic β-cells. Results confirmed the immunosuppressive property of WJ-MSCs as they did not induce the proliferation of allogenic T-cell response and they did not express CD40, CD40L, CD80 and CD86 which are involved in the immune response. In addition, when peripheral blood lymphocytes were co-cultured with WJ-MSCs, IFN-γ secretion was reduced remarkably. Their therapeutic effects were recognized when transplanted in type-1 diabetic rats, in that WJ-MSCs reduced hyperglycemia and pancreatic cell destruction [47]. The exact mechanism remains unknown, but most probably the mode of actions involves the immunosuppressive effects of WJ-MSCs rather than their ability to differentiate into pancreatic cells.

Although WJ-MSCs are less immunogenic than their fetal and bone marrow counterparts, they can produce immune response under certain circumstances. Evidence has shown that UCMSCs produced immune response when injected repeatedly in the same location, injected in an inflamed area or administered with preceding IFN-γ stimulation [35]. This must be put in consideration especially if multiple injections of these cells in the same location are required in case of regenerative therapy.

In general, MSCs have the ability to inhibit equally CD4(+), CD8(+), CD2(+) and CD3(+) subsets [37]. WJ-MSCS display more robust immuno-modulating activity than their bone marrow-derived counterparts as data show that they attenuate the T-cell response that is produced against mitogens, alloantigens or specific antigens in a dose dependent manner *in vitro* [36]. However, WJ-MSCS display significant suppression of mitogen induced CD3(+) T cell activity even at low doses when compared to the suppression activity of BM-MSCs [26, 38]. Also, when T-cells are stimulated allogeneically, WJ-MSCs have greater suppression activity than BM-MSCs or adipose-derived MSCs [18]. One advantage that WJ-MSCs have compared to fetal liver-derived MSCs is that the latter only attenuate lympho-proliferative activity that are mitogen driven only, while they do not display this activity towards allogeneic T-cell responses [39]. Also, WJ-MSCs exhibit the lympho-proliferative regulatory effect in a dose independent manner unlike fetal MSCs [18]. In addition to, WJ-MSCs have an indirect effect on T-cell allogeneic response through affecting the maturation and activation of dendritic cell (DC) precursors. When WJ-MSCs are cultured with (CD14+) monocytes, they failed to develop to mature dendritic cells in a contact dependent manner, rather results show that cells were halted as immature dendritic cells and the upregulation of co-stimulatory ligands was blocked [40]. Previous work has examined closely the kinetic secretion pattern of pro-inflammatory cytokines by PHA-activated lymphocytes co-cultured with WJ-MSCs and BM-MSCs [26]. Results showed that only in BM-MSCs culture a change in the threshold and kinetics of IL-2 secretion was observed [26]. In WJ-MSCs culture there was more evident activation of negative co-stimulatory ligands on peripheral blood lymphocytes compared to BM-MSCs [26]. There is a similarity of the secretion profile between different populations of MSCs, however, IL-12, IL-15 and Platelet-derived growth factor (PDGF) were found to be only secreted by WJ-MSCs and cord blood MSCs. To summarize, the immuno-modulatory properties of WJ-MSCs are attributed to mechanisms that includes secretion of immunosuppressive soluble factors, upregulation of negative co-stimulatory ligands, generation of memory cells, cell fusion to escape recognition, immune avoidance mechanisms specific to fetal-maternal interface, attenuation of antigen-presenting cell functions, altered migration of immune cells, and unresponsive T cell and apoptosis tolerance [18].

Phenotypic Characterization of WJ

The characterization of the WJ was laid out in 2011 by Conconi et al. where he provided an overview on the human UC [41]. This review illustrates the characterization of phenotypes of different UC cell populations. Due to the absence of standard extraction, culture and analysis methods, it is difficult to discern accurately the identity of UC stromal cells. However, cells from WJ were demonstrated to carry the mesenchymal phenotypic characteristics which include the expression of specific lineage cytoskeletal markers, such as SMA and vimentin. Also, some cell populations express the ESCs markers Oct-4, Stage-Specific Embryonic Antigen 4 (SSEA4), nucleostemin, SOX-2 and Nanog However, HUCPV cells were found to lack Oct-4 and Nanog expression as well as the expression of CD59, a receptor involved in preventing cell lysis by the regulation of complement system, and CD146, an endothelial specific cell molecule that is found also expressed on MSCs [41]. A study done by Karahuseyinoglu et al. hypothesized that HUCPV cells are more differentiated than WJ-MSCs as the results showed that HUCPV cells have stronger pan-cytokeratin staining than WJ-MSCs [20]. This finding can be an explanation why HUCPV cells fail to differentiate to neuronal cells. A recent study aimed to assess the role of the ESC marker SSEA4 in maintaining the multipotency of WJ-MSCs [48]. Results showed that the stemness of WJ-MSCs explants and their ability to differentiate into osteocytes and adipocytes did not vary in SSEA4+ and SSEA4− cells which suggests that SSEA4 is not an indispensible marker for maintaining cell multipotency. An important protein for promoting the proliferative activity of WJ-MSCs is non-muscle myosin II (NMII) which is a protein involved in cell division, migration and differentiation. Sharma et al. recently showed that the inhibition of NMII causes attenuation of the proliferative property of WJ-MSCs by locking the cell in the G_0/G_1 state [49]. In this study blebbistatin, which is a potent NMII adenosine triphosphate inhibitor, was added to the culture medium. Results from flow cytometry, proliferation kinetics, senescence assay and DNA microarray indicate decrease in the dividing activity of the cell in a dose-dependent manner [49]. An explanation for these findings can be that the G_0/G_1 cell arrest causes the up-regulation of CDKN1A, CDKN2A and CDKN2B, which are cell cycle inhibitory genes, and down-regulation of genes that promote S/M phase transition [49]. Another group recently examined the effect of prolonged cell culture on the gene expression of WJ-MSCs by means of microarray analysis of their transcriptome after 4 and 12 passages of cell culture [50]. Data analysis showed that after the 12th passage, there were up-regulation of 157 genes and down-regulation of 440 genes compared with the transcriptome after the fourth passage. Most of these genes are related to inflammatory and cell stress response, cell proliferation and maturation, and apoptosis. These results suggest that after multiple cycles of cell expansion, WJ-MSCs start to acquire features of aging cells such as limited proliferative ability and resistance to apoptosis [50].

A difference between WJ-MSCs and the cord lining membrane mesenchymal stem cells (CL-MSCs) receptor- expression profile is the macrophage marker CD14 which is found expressed only by CL-MSCs [25]. It was found that CD14 in its soluble form has a down-regulating activity on T-cell activation [42]. As mentioned previously, WJ-MSCs are adequate for cell-based therapy. This claim is strengthened by the fact that they express HLA-G6 isoform, a protein involved in immunomodulation. Cell populations with variable phenotypic profiles were found not only in different parts of the UC rather, within the same UC regions. This may infer that UCMSCs offer cell populations with different stemness degrees. Studies reported CD105(+)/CD31(−)/KDR(−) cell populations from WJ that showed the ability to differentiate *in vivo* towards myogenic lineage which was confirmed by *in vitro* assays, as well as to aid in the muscle regeneration [43]. Thus, MSCs isolated from WJ remain the most possibly applicable cells for therapeutic purposes. A study done by Kita et al. to isolate MSCs from the sub-amnion (cord lining) of the UC showed that CL-MSCs are expressing mesenchymal characteristics *in vitro* while they are significantly discrete from ESCs as they do not induce tumorigenicity *in vitro* [25]. After that, another attempt by Jeschke and colleagues specified regions of the sub-amnion and WJ where stem cell niches are mainly concentrated [17]. As mentioned previously, isolated multipotent cells from the UC show heterogeneity and this is attributed to the different anatomical compartments of the UC where the cells are isolated as well as the different isolation protocols used. Although CL-MSCs are promising to be used in cell-based therapy due to their multipotent and proliferative capacities [17], their use is limited by the time-consuming isolation process. On the other hand, WJ represents a generous source for MSCs but with variable cell quality. Thus, it is paramount to take into account the quality and quantity of stem cells required with respect to each application.

It is hypothesized that the perinatal environment has a significant impact on the biological characteristics of MSCs. Previous work has shown an increased probability of offspring acquiring diabetes and obesity when intrauterine metabolic disturbances occur during pregnancy as a result of maternal hyperglycemia [44–46]. This hypothesis is further confirmed by data from animal models suggesting that MSCs start the commitment process to pre-adipocytes and mature adipocytes during fetal development and perinatal life [51]. Also, the number of pre-adipocytes and mature adipocytes were found lower in normal subjects than in obese subjects [52]. Further studies on how the perinatal environment influence the differentiation of fetal MSCs, especially in unregulated gestational diabetes, is still needed. A recent study was done by Pierdomenico et al. to examine the underlying mechanism causing metabolic diseases in offspring of diabetic mothers [53]. Isolated WJ-MSCs from the UC of healthy and diabetic mothers showed similarity in the type of markers expressed, however; differences in their expression level between the two groups were observed which is possibly due to a difference in the functional characteristics of both groups. Particularly, CD90 levels were lower in WJ-MSCs from diabetic mothers which may explain the limited plasticity of these cells. However, CD44, CD29, CD73, CD166, SSEA4 and TERT were highly expressed in WJ-MSCs isolated from the diabetic group which may attribute to the proliferative characteristics of these cells. Moreover, WJ-MSCs obtained from the diabetic group showed higher capability to differentiate into adipocytes which may indicate pre-commitment to adipocyte lineage. In summary, many results have evidence that diabetic uterine environment can be the major cause of MSCs pre-commitment to the adipocyte lineage resulting in an increase in adipocytes production upon an incorrect diet style which can cause diabetes and obesity.

Clinical Applications of WJ-Derived Stem Cells

Cancer Therapy

Stem cells have been regarded as a prospective cell-based therapy to treat primary and metastatic cancers. Previous work reported that un-engineered human and rat UCMSCs caused significant suppression of amplification of multiple cancer cell lines *in vivo* and *in vitro* [54, 55]. The exact mechanism for this attenuation activity is not fully understood. Tamura et al. examined the intrinsic stem cell-dependent regulation of cancer growth, other potential underlying mechanisms, the use of stem cells to deliver exogenous anti-cancer agents and their potential clinical application [56]. There are two theories explaining the tumor suppression action by UCMSCs; the first theory is the secretion of multiple secretory proteins that causes the activation of caspases and eventually the induction of cell death and cell cycle arrest [30, 56, 57]. This hypothesis is strengthened by data from microarray expression profile of rat UCMSCs indicating over-expression of tumor suppressor genes [56]. The second theory of tumor suppression is via inducing an immune response against the tumor as shown by CD8(+) T cell infiltration of the tumor tissues in the immunohistochemistry analysis [55]. This theory can be considered conflicting with the low immunogenicity of UCMSCs, however; this can be due to the contribution of the UCMSC microenvironment as well as the tumor cells to their immunogenicity properties.

Tumor cells secrete various cytokines and growth factors in large amounts that interact with their target receptors on stem cells. This is considered the mechanism of migration of UCMSCs and other MSCs to tumor tissues by sensing the concentration gradient of these cytokines [56]. UCMSCs have higher levels of IL-8 receptor and CXCR than BM-MSCs, thus, the former show higher capability to migrate to the tumor site. This adds to UCMSCs clinical utility as they can be loaded with nanoparticles to be delivered to the tumor or they can be genetically manipulated to express cytotoxic cytokines upon reaching the tumor site [58, 59]. To test this, previous studies were done where UCMSCs with inserted IFN-β gene showed induction of tumor death in breast adenocarcinoma cells and bronchioloalveolar carcinoma cells *in vitro* and *in vivo* [28, 59]. Due to the homing ability to cancer tissue and to other inflammatory tissues and tumor suppression ability, UCMSCs are considered a potential therapeutic cancer treatment modality. In addition to their abundance, low immunogenicity, lack of CD34 and CD45 expression and the availability of simple methods for their isolation and *in vitro* expansion, UCMSCs are appropriate for allogenic transplantations.

Liver Disease

WJ-MSCs have been also considered for regenerating liver tissues instead of orthotopic liver transplantation based on their potential to differentiate into endodermal lineage including hepatocyte-like cells. Scheers et al. showed the possible use of UCMSCs in this purpose *in vivo* and *in vitro* [60]. Evidences that boost this finding is the expression of hepatic cell markers that correspond to the sequence of liver development and the ability to differentiate and express albumin and AFP after 2, 4 and 6 weeks following *in vivo* transplantation in SCID mice with partial hepatectomy [61]. In addition to that, undifferentiated UCMSCs were shown to decrease the rate of hepatic cell fibrosis and thus, help in rescuing the injured hepatocytes *in vivo*. In other words, even if transdifferentiation did not occur, UCMSCs are still beneficial by promoting the endogenous parenchymal cell differentiation and fibrous matrix degradation [62]. The anti-inflammatory and anti-fibrosis properties are attributed to secreted metalloproteinases. In order to enhance the differentiation ability of UCMSCs to hepatic lineage *in vivo* and *in vitro*, it is suggested to add hepatogenic factors to the culture medium. However, this differentiation process requires further research in order to translate it to clinical application.

Cardiovascular Diseases

Semenov et al. suggested the use of cells from WJ in engineering cultured tissues [63] which will constitute an alternative treatment modality to cardiovascular diseases instead of the non-autologous valves or conduits which may acquire obstructive tissue ingrowths and calcification [64, 65]. Thus, there are *in vitro* attempts to create autologous living tissue that can differentiate into cardiomyocytes by cardiovascular fetal tissue engineering [66]. Other studies reported the success of implantation of completely autologous tri-leaflet heart valves in sheep models for up to 20 weeks. These implants were derived from human WJ-MSCs and they displayed similar functional, biochemical and structural characteristics as those of the native semilunar heart valves. Kadner et al. reported that the UC, as a source of stem cells, is considered favorable for cardiovascular tissue engineering compared to other cell sources and thus, avoiding the invasive procedure for harvesting intact vascular structures [6]. A recent study was done aiming to develop a myocardial patch to be used to repair tissue of myocardial infarctions and boost the long-term heart function [67]. The study integrated two microporous tubes to develop a 3D aligned microfibrous myocardial tissue construct cultured under transient perfusion for 14 days. The biodegradable microporous tubes function to supply growth media to the cells within the construct and cell seeded, fiber mats around them. WJ-MSCs are embedded into the mat in a similar fashion to the parallel cell organization in native myocardium. Results showed that the constructs provided nutrient supply for the cells which lead to increased cell viability, uniform cell distribution and alignment in static and perfused cultures. Ample input of information is available about the characteristics of WJ-MSCs and their potential clinical application in cardiovascular tissue regeneration, however; their limitations are still emerging and require further research. To mention, implanted cells were unable to grow and differentiate, and had functional limitations when foreign, natural, synthetic or hybrid polymers are used in the scaffold which may also lead to infection and the formation of thromboembolus. Therefore, the use of biocompatible materials that have no effect on the regenerative and immuno-modulatory properties of WJ-MSCs in cardiovascular tissue engineered scaffolds is required [63]. Moreover, there should be a consensus on the criteria for WJ-MSCs isolation, characterization methods and long-term culturing to encounter side effects resulted from the long-term survival of stem cell in the host tissue and to implement an effective treatment regimen at the same time.

Cartilage Regeneration

Cartilages are known for their limited self-repair and regeneration after exposure to traumatic injury or autoimmune diseases which lead to cartilage damage eventually. Studies reported that WJ-MSCs have the ability to differentiate into chondrocyte-like cells *in vivo* and *in vitro* [68], which can place them to be used in regenerative repair for articular diseases. This finding is confirmed by Arufe et al. where WJ-MSCs differentiation potential analysis revealed their multipotency and chondrogenic capacity [69]. Also, it was found that WJ-MSCs up-regulated the production of hyaluronic acid and GAGs in addition to the expression of key genes including SOX9, COMP, Collagen type II and FMOD [70]. Another study showed the enhancement of WJ-MSCs biosynthesis and mechanical integrity to form fibrocartilage-poly-glycolic acid scaffolds is achieved when high tissue seeding density was used, specifically 25 million cells/mL [71]. When WJ-MSCs were cultured on nanofibrous substrates with a two sequential culture medium environment, their chondrogenic differentiation is promoted [70]. For a successful development of cartilage implants, there must be integration between bone and cartilage cells to mimic the native osteochondral tissue. This attempt was done by Wang et al. where the study managed to produce a supportive osteochondral tissue derived from WJ-MSCs by embedding the cells into two poly-L-lactic-acid (PLLA) scaffolds with chondrogenic and osteogeic media respectively for 3 weeks [72]. Then, suturing of the chondrogenic and osteogenic constructs is applied to form four different osteochondral assemblies where immunohistochemical assays for glycosaminoglycans, type I collagen and calcium showed improved integration and transition of the matrices between two layers in the composite group containing sandwiched cells as compared to other control composites [72]. Another study showed that WJ-MSCs can differentiate to form intervertebral disc (IVD)-like tissues that exhibit immature nucleus pulposus (NP) phenotype in a pseudo-three-dimensional culture system [73]. The same group has shown previously that the WJ-MSCs produced immature NP cells possessing specific laminin isoforms and laminin-binding receptors which may lead to the formation of NP-like cells eventually. All these results suggest the use of WJ-MSCs in repairing injured IVD. Thus, WJ-MSCs have a great promise to be used in osteochondral regeneration due to their ability to differentiate into chondrocyte-like cells. In addition, their low immunogenicity suggests their use in autoimmune diseases such as, osteoarthritis and rheumatoid arthritis. More research efforts are required to optimize methods of isolation of these cells from different sources, define the most suitable materials for scaffolds and matrices and determine the growth factors necessary for the cell growth and differentiation.

Peripheral Nerve Repair

One of the many therapeutic approaches for restoring neural function post peripheral nervous system (PNS) injury is tissue engineering. Attempts in this scope have developed bioartificial nerve conduits placed between neural gaps to guide the regrowth of axons [74, 75]. A caveat in these neural conduits is their limited growth-promoting action when the nerve gap is long. To yield better nerve growth, Schwann cells, cells essential in myelin formation, are preferentially added to the microenvironment of the damaged nerves [76]. To secure a suitable source for Schwann cells, MSCs from different tissues are considered instead of impractical isolation process from other peripheral nerve tissues. A study by Dezawa et al. in 2001 first reported the induction pathway of BM-MSCs to differentiate to Schwann cells [77]. After that, other studies focused on umbilical cord-derived mesenchymal stromal cells (UCMSCs) as their mesenchymal source, reported the derivation of Schwann cells, support of neural regeneration and myelin formation [78, 79]. Some evidences suggest that UC-Schwann cells are a better viable alternative than native Schwann cells for purposes of cell-based therapy. When human UC-Schwann cells were transplanted into a rat with transected sciatic nerve, they were found to keep their differentiation status *in vivo* and to contribute in the axonal regeneration and recovery. Also, secretion of neurotropic factors such as NGF and BDNF by UC-Schwann cells from WJ was reported by another groups [80, 81]. For spinal cord injury, Schwann cells play a role in axonal regeneration and myelin sheath construction. A comparative study of MSCs sources for the purpose of Schwann cells production was done by Kuroda et al. Data reported that WJ-MSCs have a high therapeutic property in spinal cord injury [79].

In addition, human umbilical cord-derived MSCs (hUCMSCs) have the ability to ameliorate neonatal hypoxic-ischemic encephalopathy (HIE) in rat models [82]. After HIE was induced, subjects were divided into two groups; the first received the transplanted cells 24 h post HIE and the second group received the cells 72 h post HIE. Results reported differentiation of the UCMSCs into neurons. The behavioral analysis showed better locomotor function in rats that received UCMSCs 24 h after HIE induction compared to rats that administered UCMSCs 72 h after HIE induction [82]. Another aim of this study is to test the efficiency of the intravenous route of cell administration versus the intraperitoneal route. It was shown that more homing of UCMSCs to the ischemic frontal cortex was observed in case of intravenous administration. These findings suggest that the intravenous administration of UCMSCs at an early stage after HIE can be considered a therapeutic modality for this condition. Another study by Hsieh et al. suggested that WJ-MSCs can be used in treatment of stroke due to their ability to promote neural generation and neuroprotection [83]. Secretome analysis of WJ-MSCs cultured in an oxygen-glucose deprivation culture model indicated the secretion of proteins that promote neural differentiation and cell migration, and reduce rate of apoptosis in the primary cortical cells of the model [83]. Paldino et al. examined recently the capability of WJ-MSCs to differentiate into dopaminergic neural-like cells in presence of forskolin which has an increasing effect on the levels of intracellular cAMP [84]. Microarray analysis reported the modulation of 1532 genes, almost all of them are required in the signaling pathways of the neurons and some of them are essential for the neuronal dopaminergic induction. To further confirm these results, immunohistochemistry and Western blot analysis were performed revealing the up-regulation of Nurr1, NeuroD1, and TH proteins which are all specific to the dopaminergic phenotype [84].

An attempt was made to maximize the yield and differentiation of isolated stem cells from WJ by adjusting oxygen concentration to 5 %, instead of 21 %, and using low plating density [85]. These conditions were suggested by previous work to produce a standardized isolation protocol that leads to increased proliferation rate of MSCs and permits the expansion and maintenance of colony-forming unit-fibroblast (CFU-F). For better understanding of the underlying mechanisms, a recent study investigated the effects of both oxygen concentration and plating density on WJ-MSCs [85]. Results have confirmed that decreasing the oxygen exposure from 21 % (room temperature) to 5 % during cell proliferation amplified the yield and maintained CFU-F which is hypothesized to be due to the up-regulation of hypoxia inducible factors (HIFs) by the reduced oxygen concentration which will lead eventually to these results by stimulating telomerase activity. Regarding plating density, reduction from 100 to 10 cells/cm^2 was found to cause an increase in CFU-F frequency. In summary, key factors for standardized stem cell isolation and cultivation protocols are oxygen concentration and plating density. These factors can be manipulated according to the targeted cell population for tissue engineering or cell-based therapy.

Cardiac Differentiation of Human WJ-Derived Stem Cells

The capacity of BM-MSCs to differentiate into multiple tissue types *in vivo* after transplantation is not limited by surrounding tissue of myocardial infarction which poses the risk of undesired cells types within the infracted area leading to life-threatening consequences [86]. Subsequent studies then proposed to impart a certain level of differentiation to these cells to form myocardial lineage cells before transplanting them [87, 88]. To achieve this, a defined culture medium of WJ-MSCs is treated with 5-azacytidine for 3 weeks. Results showed that the cells successfully acquired features of the myocyte morphology and also expressed cardiomyocyte markers, cardiac troponin I, connexin 43, and desmin [89]. Also, the addition of oxytocin, which was found highly expressed in fetal cardiac tissues compared to the adult, embryo-like aggregates and transforming growth factor-β1 (TGF-β1), PDGF and basic fibroblast growth factor (bFGF) enhanced the differentiation of stem cells to form myocytes [90–93]. Among the many published protocol for inducing cardiac differentiation, one study reported that the addition of oxytocin and 5-azacytidine and "embryoid bodies formation" is essential for the formation of cardiac differentiated UCMSC [92]. However, comparative morphological and immunocytochemical analyses of these cells indicated that oxytocin is the most important differentiation stimulator to form cardiomyocyte-like cells compared to 5-azacytidine and "embryoid bodies" formation. The long-term therapeutic effects of BM-MSCs and WJ-MSCs were compared recently by Lopez et al. where the cells were injected 24–48 h post MI in a rat model [94]. Results showed that animals received MSCs exhibited significant improvement in ejection fraction 25–31 weeks after treatment. Also, when WJ-MSCs were co-cultured with fetal MSCs, their cardiac differentiation potential was enhanced compared to when they are co-cultured with adult MSCs. This was indicated by the formation of myotube structures in two-three days as well as the observation of spontaneously contracting cells in 5–7 days. Therefore, UCMSCs and WJ-derived stem cells are considered a promising option for treatment and regeneration of cardiac tissue due to their accessibility and differentiation capability into cardiomyocyte-like cells. To be feasible for use in cardiac repair, the cardiac functional properties of UCMSCs need to be assessed.

The New Research Frontiers in WJ Research

Clonal MSCs

Sarugaser et al. managed to isolate a nonhematopoetic (CD45–, CD34–, SH2+, Thy-1+, CD44+) HUCPVC population [24]. This population of cells also displays a non-hematopoietic myofibroblastic MSC phenotype (CD45–, CD34–, CD105+, CD73+, CD90+, CD44+, CD106+, 3G5+, CD146+) and has similar immunological properties to BM-MSCs [95]. Moreover, HUCPVCs have robust self-renewing multi-potential differentiation capacity *in vitro* and are able to contribute to both musculo-skeletal and dermal wound healing *in vivo* [95]. The identification of the perivascular region of the human UC as a rich source of HUCPVCs led to the first single cell clonal confirmation of a hierarchy of MSC differentiation [24, 96–98]. HUCPVCs exhibit properties that enable them to be used for allogenic cell-based therapies such as their high MHC–/– phenotype, high frequencies of CFU-F and CFU-osteogenic subpopulation, and their rapid doubling time. More attempts are needed to study the clonal expansions of stem cells derived from the WJ similar to the case of HUCPVCs so that their therapeutic potential can be translated from bench to clinic.

Use of Magnetic Resonance Imaging in Contrast Labeled-UC Stem Cells

Although, magnetic resonance agents has not being used to directly examine the behavior of WJ, a subpopulation of WJ correspond to MSC are routinely tracked [99, 100]. A recent study examined the efficiency of labeling MSC using aminopropyltriethoxysilane-modified magnetic iron oxide nanoparticles (APTS-MNPs). Results demonstrated that MSC cells were efficiently labeled and MRI migration patterns were consistent with histological examinations. Also, it was reported that proliferation and differentiation patterns were not significantly affected [100]. Moreover, a different study was done to examine marker expression pattern and the proliferation and differentiation capacities of stem cell populations isolated from the intervascular and perivascular regions of the UCM [99]. As the cells isolated from the perivascular portion are more differentiated than those isolated from the intervascular region, the latter have faster doubling times. Both isolates were shown to express MSC mRNA markers (CD29, CD105, CD44, CD166) and were negative for CD34 and MHC-II. Also, staining and gene expression data confirmed osteogenic, adipogenic, chondrogenic and neurogenic differentiation of the isolated cell populations from both portions. Also, this study examined the *in vitro* labeling efficiency of MSC with the magnetic resonance agents particularly, superparamagnetic iron oxide particles (SPIO) and manganese chloride. Results reported more sensitive results for SPIO, but both agents showed simple, effective and safe labeling methods. Thus, magnetic resonance labeling by SPIO can be used in studying migration of stem cells to injured and non-injured tissues as well as their mechanism of action of cell therapy which requires further *in vivo* studies.

Conclusions

This review offers insight on the phenotypic and therapeutic properties of stem cells derived from the Wharton's Jelly as well as gaps in knowledge for other biological properties. In order to render them applicable for cell-based therapy and tissue engineering, further investigations must be done to determine the optimal protocols to achieve this.

Acknowledgments This work was supported by NIH NINDS RO1 1R01NS071956-01 (C.V.B.), James and Esther King Biomedical Research Program 09KB-01-23123 (C.V.B.) and 1KG01-33966 (C.V.B.).

References

1. La Rocca G. Connecting the dots: the promises of Wharton's Jelly stem cells for tissue repair and regeneration. Open Tissue Eng Regen Med J. 2011;4:3–5.
2. McElreavey KD, Irvine AI, Ennis KT, McLean WH. Isolation, culture and characterisation of fibroblast-like cells derived from the Wharton's Jelly portion of human umbilical cord. Biochem Soc Trans. 1991;19:29S.
3. Wharton T. Adenographia. Oxford: Oxford University Press; 1996. Transl. Freer S.
4. Chacko AW, Reynolds SRM. Architecture of distended and nondistended human umbilical cord tissues, with special reference to the arteries and veins. Contrib Embryol. 1954;35: 135–50.
5. Kadner A, Hoerstrup SP, Tracy J, Breymann C, Maurus CF, Melnitchouk S, Kadner G, Zund G, Turina M. Human umbilical cord cells: a new cell source for cardiovascular tissue engineering. Ann Thorac Surg. 2002;74:S1422–1428.
6. Kadner A, Zund G, Maurus C, Breymann C, Yakarisik S, Kadner G, Turina M, Hoerstrup SP. Human umbilical cord cells for cardiovascular tissue engineering: a comparative study. Eur J Cardiothorac Surg. 2004;25:635–41.
7. Mitchell KE, Weiss ML, Mitchell BM, Martin P, Davis D, Morales L, Helwig B, Beerenstrauch M, Abou-Easa K, Hildreth T, et al. Matrix cells from Wharton's Jelly form neurons and glia. Stem Cells. 2003;21:50–60.

8. Naughton BA, San Roman J, Liu K, Purchio A, Pavelec R, Rekettye L. Cells isolated from Wharton's Jelly of the human umbilical cord develop a cartilage phenotype when treated with tgf-β in vitro. FASEB J. 1997;11:A19.
9. Purchio AF, Naughton BA, Roman, JS. Production of cartilage tissue using cells isolated from Wharton's Jelly. U.S. Patent 5,919,702. 1999.
10. Romanov YA, Svintsitskaya VA, Smirnov VN. Searching for alternative sources of postnatal human mesenchymal stem cells: candidate msc-like cells from umbilical cord. Stem Cells. 2003;21:105–10.
11. Takechi K, Kuwabara Y, Mizuno M. Ultrastructural and immunohistochemical studies of Wharton's Jelly umbilical cord cells. Placenta. 1993;14:235–45.
12. Kobayashi K, Kubota T, Aso T. Study on myofibroblast differentiation in the stromal cells of Wharton's Jelly: expression and localization of alpha-smooth muscle actin. Early Hum Dev. 1998;51:223–33.
13. Markov V, Kusumi K, Tadesse MG, William DA, Hall DM, Lounev V, Carlton A, Leonard J, Cohen RI, Rappaport EF, et al. Identification of cord blood-derived mesenchymal stem/stromal cell populations with distinct growth kinetics, differentiation potentials, and gene expression profiles. Stem Cells Dev. 2007;16:53–73.
14. Baudin B, Bruneel A, Bosselut N, Vaubourdolle M. A protocol for isolation and culture of human umbilical vein endothelial cells. Nat Protoc. 2007;2:481–5.
15. Wang XY, Lan Y, He WY, Zhang L, Yao HY, Hou CM, Tong Y, Liu YL, Yang G, Liu XD, et al. Identification of mesenchymal stem cells in aorta-gonad-mesonephros and yolk sac of human embryos. Blood. 2008;111:2436–43.
16. Bongso A, Fong CY. The therapeutic potential, challenges and future clinical directions of stem cells from the Wharton's Jelly of the human umbilical cord. Stem Cell Rev. 2013;9:226–40.
17. Jeschke MG, Gauglitz GG, Phan TT, Herndon DN, Kita K. Umbilical cord lining membrane and Wharton's Jelly-derived mesenchymal stem cells: the similarities and differences. Open Tissue Eng Regen Med J. 2011;4:21–7.
18. Prasanna SJ, Jahnavi VS. Wharton's jelly mesenchymal stem cells as off-the -shelf cellular therapeutics: a closer look into their regenerative and immunomodulatory properties. Open Tissue Eng Regen Med J. 2011;4:28–38.
19. Troyer DL, Weiss ML. Wharton's jelly-derived cells are a primitive stromal cell population. Stem Cells. 2008;26:591–9.
20. Karahuseyinoglu S, Cinar O, Kilic E, Kara F, Akay GG, Demiralp DO, Tukun A, Uckan D, Can A. Biology of stem cells in human umbilical cord stroma: In situ and in vitro surveys. Stem Cells. 2007;25:319–31.
21. Nanaev AK, Kohnen G, Milovanov AP, Domogatsky SP, Kaufmann P. Stromal differentiation and architecture of the human umbilical cord. Placenta. 1997;18:53–64.
22. Can A, Karahuseyinoglu S. Concise review: human umbilical cord stroma with regard to the source of fetus-derived stem cells. Stem Cells. 2007;25:2886–95.
23. Baksh D, Yao R, Tuan RS. Comparison of proliferative and multilineage differentiation potential of human mesenchymal stem cells derived from umbilical cord and bone marrow. Stem Cells. 2007;25:1384–92.
24. Sarugaser R, Lickorish D, Baksh D, Hosseini MM, Davies JE. Human umbilical cord perivascular (hucpv) cells: a source of mesenchymal progenitors. Stem Cells. 2005;23:220–9.
25. Kita K, Gauglitz GG, Phan TT, Herndon DN, Jeschke MG. Isolation and characterization of mesenchymal stem cells from the sub-amniotic human umbilical cord lining membrane. Stem Cells Dev. 2010;19:491–502.
26. Prasanna SJ, Gopalakrishnan D, Shankar SR, Vasandan AB. Pro-inflammatory cytokines, ifngamma and tnfalpha, influence immune properties of human bone marrow and Wharton jelly mesenchymal stem cells differentially. PLoS One. 2010;5, e9016.
27. La Rocca G, Anzalone R, Corrao S, Magno F, Loria T, Lo Iacono M, Di Stefano A, Giannuzzi P, Marasa L, Cappello F, et al. Isolation and characterization of oct-4+/hla-g+ mesenchymal

stem cells from human umbilical cord matrix: Differentiation potential and detection of new markers. Histochem Cell Biol. 2009;131:267–82.
28. Rachakatla RS, Pyle MM, Ayuzawa R, Edwards SM, Marini FC, Weiss ML, Tamura M, Troyer D. Combination treatment of human umbilical cord matrix stem cell-based interferon-beta gene therapy and 5-fluorouracil significantly reduces growth of metastatic human breast cancer in scid mouse lungs. Cancer Invest. 2008;26:662–70.
29. Weiss ML, Anderson C, Medicetty S, Seshareddy KB, Weiss RJ, VanderWerff I, Troyer D, McIntosh KR. Immune properties of human umbilical cord Wharton's jelly-derived cells. Stem Cells. 2008;26:2865–74.
30. Deuse T, Stubbendorff M, Tang-Quan K, Phillips N, Kay MA, Eiermann T, Phan TT, Volk HD, Reichenspurner H, Robbins RC, et al. Immunogenicity and immunomodulatory properties of umbilical cord lining mesenchymal stem cells. Cell Transplant. 2011;20:655–67.
31. Selmani Z, Naji A, Zidi I, Favier B, Gaiffe E, Obert L, Borg C, Saas P, Tiberghien P, Rouas-Freiss N, et al. Human leukocyte antigen-g5 secretion by human mesenchymal stem cells is required to suppress t lymphocyte and natural killer function and to induce cd4+cd25highfoxp3+ regulatory t cells. Stem Cells. 2008;26:212–22.
32. Griffin MD, Ritter T, Mahon BP. Immunological aspects of allogeneic mesenchymal stem cell therapies. Hum Gene Ther. 2010;21:1641–55.
33. Zarkhin V, Talisetti A, Li L, Wozniak LJ, McDiarmid SV, Cox K, Esquivel C, Sarwal MM. Expression of soluble hla-g identifies favorable outcomes in liver transplant recipients. Transplantation. 2010;90:1000–5.
34. Djouad F, Charbonnier LM, Bouffi C, Louis-Plence P, Bony C, Apparailly F, Cantos C, Jorgensen C, Noel D. Mesenchymal stem cells inhibit the differentiation of dendritic cells through an interleukin-6-dependent mechanism. Stem Cells. 2007;2025–2032:25.
35. Cho PS, Messina DJ, Hirsh EL, Chi N, Goldman SN, Lo DP, Harris IR, Popma SH, Sachs DH, Huang CA. Immunogenicity of umbilical cord tissue derived cells. Blood. 2008;111:430–8.
36. Di Nicola M, Carlo-Stella C, Magni M, Milanesi M, Longoni PD, Matteucci P, Grisanti S, Gianni AM. Human bone marrow stromal cells suppress t-lymphocyte proliferation induced by cellular or nonspecific mitogenic stimuli. Blood. 2002;99:3838–43.
37. Aggarwal S, Pittenger MF. Human mesenchymal stem cells modulate allogeneic immune cell responses. Blood. 2005;105:1815–22.
38. Najar M, Rouas R, Raicevic G, Boufker HI, Lewalle P, Meuleman N, Bron D, Toungouz M, Martiat P, Lagneaux L. Mesenchymal stromal cells promote or suppress the proliferation of t lymphocytes from cord blood and peripheral blood: the importance of low cell ratio and role of interleukin-6. Cytotherapy. 2009;11:570–83.
39. Gotherstrom C, Ringden O, Westgren M, Tammik C, Le Blanc K. Immunomodulatory effects of human foetal liver-derived mesenchymal stem cells. Bone Marrow Transplant. 2003;32:265–72.
40. Tipnis S, Viswanathan C, Majumdar AS. Immunosuppressive properties of human umbilical cord-derived mesenchymal stem cells: role of b7-h1 and ido. Immunol Cell Biol. 2010;88:795–806.
41. Conconi MT, Di Liddo R, Tommasini M, Calore C, Parnigotto PP. Phenotype and differentiation potential of stromal populations obtained from various zones of human umbilical cord: an overview. Open Tissue Eng Regen Med J. 2011;4:6–20.
42. Rey Nores JE, Bensussan A, Vita N, Stelter F, Arias MA, Jones M, Lefort S, Borysiewicz LK, Ferrara P, Labeta MO. Soluble cd14 acts as a negative regulator of human t cell activation and function. Eur J Immunol. 1999;29:265–76.
43. Conconi MT, Burra P, Di Liddo R, Calore C, Turetta M, Bellini S, Bo P, Nussdorfer GG, Parnigotto PP. Cd105(+) cells from Wharton's jelly show in vitro and in vivo myogenic differentiative potential. Int J Mol Med. 2006;18:1089–96.
44. Dabelea D, Pettitt DJ. Intrauterine diabetic environment confers risks for type 2 diabetes mellitus and obesity in the offspring, in addition to genetic susceptibility. J Pediatr Endocrinol Metab. 2001;14:1085–91.

45. Gillman MW, Rifas-Shiman S, Berkey CS, Field AE, Colditz GA. Maternal gestational diabetes, birth weight, and adolescent obesity. Pediatrics. 2003;111:e221–226.
46. Clausen TD, Mathiesen ER, Hansen T, Pedersen O, Jensen DM, Lauenborg J, Damm P. High prevalence of type 2 diabetes and pre-diabetes in adult offspring of women with gestational diabetes mellitus or type 1 diabetes: the role of intrauterine hyperglycemia. Diabetes Care. 2008;31:340–6.
47. Wang H, Qiu X, Ni P, Qiu X, Lin X, Wu W, Xie L, Lin L, Min J, Lai X, Chen Y, Ho G, Ma L. Immunological characteristics of human umbilical cord mesenchymal stem cells and the therapeutic effects of their transplantion on hyperglycemia in diabetic rats. Int J Mol Med. 2014;33(2):263–70.
48. He H, Nagamura-Inoue T, Tsunoda H, Yuzawa M, Yamamoto Y, Yorozu P, Agata H, Tojo A. Stage-specific embryonic antigen 4 in Wharton's Jelly-derived mesenchymal stem cells is not a marker for proliferation and multipotency. Tissue Eng Part A. 2013;20:1314–24.
49. Sharma T, Kumari P, Pincha N, Mutukula N, Saha S, Jana SS, Ta M. Inhibition of non-muscle myosin II leads to G0/G1 arrest of Wharton's jelly-derived mesenchymal stromal cells. Cytotherapy. 2014;16:640–52.
50. Gatta V, D'Aurora M, Lanuti P, Pierdomenico L, Sperduti S, Palka G, Gesi M, Marchisio M, Miscia S, Stuppia L. Gene expression modifications in Wharton's Jelly mesenchymal stem cells promoted by prolonged in vitro culturing. BMC Genomics. 2013;14:635.
51. Tang W, Zeve D, Suh JM, Bosnakovski D, Kyba M, Hammer RE, Tallquist MD, Graff JM. White fat progenitor cells reside in the adipose vasculature. Science. 2008;322:583–6.
52. Tchoukalova Y, Koutsari C, Jensen M. Committed subcutaneous preadipocytes are reduced in human obesity. Diabetologia. 2007;50:151–7.
53. Pierdomenico L, Lanuti P, Lachmann R, Grifone G, Cianci E, Gialò L, Pacella S, Romano M, Vitacolonna E, Miscia S. Diabetes mellitus during pregnancy interferes with the biological characteristics of Wharton's jelly mesenchymal stem cells. Open Tissue Eng Regen Med J. 2011;4:103–11.
54. Ayuzawa R, Doi C, Rachakatla RS, Pyle MM, Maurya DK, Troyer D, Tamura M. Naive human umbilical cord matrix derived stem cells significantly attenuate growth of human breast cancer cells in vitro and in vivo. Cancer Lett. 2009;280:31–7.
55. Ganta C, Chiyo D, Ayuzawa R, Rachakatla R, Pyle M, Andrews G, Weiss M, Tamura M, Troyer D. Rat umbilical cord stem cells completely abolish rat mammary carcinomas with no evidence of metastasis or recurrence 100 days post-tumor cell inoculation. Cancer Res. 2009;69:1815–20.
56. Tamura M, Kawabata A, Ohta N, Uppalapati L, Becker KG, Troyer D. Wharton's jelly stem cells as agents for cancer therapy. Open Tissue Eng Regen Med J. 2011;4:39–47.
57. Nakamizo A, Marini F, Amano T, Khan A, Studeny M, Gumin J, Chen J, Hentschel S, Vecil G, Dembinski J, et al. Human bone marrow-derived mesenchymal stem cells in the treatment of gliomas. Cancer Res. 2005;65:3307–18.
58. Rachakatla RS, Marini F, Weiss ML, Tamura M, Troyer D. Development of human umbilical cord matrix stem cell-based gene therapy for experimental lung tumors. Cancer Gene Ther. 2007;14:828–35.
59. Matsuzuka T, Rachakatla RS, Doi C, Maurya DK, Ohta N, Kawabata A, Pyle MM, Pickel L, Reischman J, Marini F, et al. Human umbilical cord matrix-derived stem cells expressing interferon-beta gene significantly attenuate bronchioloalveolar carcinoma xenografts in scid mice. Lung Cancer. 2010;70:28–36.
60. Scheers I, Lombard C, Najimi M, Sokal E. Cell therapy for the treatment of metabolic liver disease: an update on the umbilical cord derived stem cells candidates. Open Tissue Eng Regen Med J. 2011;4:48–53.
61. Campard D, Lysy PA, Najimi M, Sokal EM. Native umbilical cord matrix stem cells express hepatic markers and differentiate into hepatocyte-like cells. Gastroenterology. 2008;134:833–48.
62. Anzalone R, Lo Iacono M, Corrao S, Magno F, Loria T, Cappello F, Zummo G, Farina F, La Rocca G. New emerging potentials for human Wharton's jelly mesenchymal stem cells:

immunological features and hepatocyte-like differentiative capacity. Stem Cells Dev. 2010;19:423–38.
63. Semenov O, Breymann C. Mesenchymal stem cells derived from Wharton's jelly and their potential for cardio-vascular tissue engineering. Open Tissue Eng Regen Med J. 2011;4: 64–71.
64. Mayer Jr JE. Uses of homograft conduits for right ventricle to pulmonary artery connections in the neonatal period. Semin Thorac Cardiovasc Surg. 1995;7:130–2.
65. Schoen FJ, Levy RJ. Tissue heart valves: current challenges and future research perspectives. J Biomed Mater Res. 1999;47:439–65.
66. Shinoka T, Ma PX, Shum-Tim D, Breuer CK, Cusick RA, Zund G, Langer R, Vacanti JP, Mayer Jr JE. Tissue-engineered heart valves. Autologous valve leaflet replacement study in a lamb model. Circulation. 1996;94:II164–168.
67. Kenar H, Kose GT, Toner M, Kaplan DL, Hasirci V. A 3d aligned microfibrous myocardial tissue construct cultured under transient perfusion. Biomaterials. 2011;32:5320–9.
68. Lo Iacono M, Anzalone R, Corrao S, Giuffrè M, Di Stefano A, Giannuzzi P, Cappello F, Farina F, La Rocca G. Perinatal and Wharton's jelly-derived mesenchymal stem cells in cartilage regenerative medicine and tissue engineering strategies. Open Tissue Eng Regen Med J. 2011;4:72–81.
69. Arufe MC, De la Fuente A, Mateos J, Fuentes I, De Toro FJ, Blanco FJ. Analysis of the chondrogenic potential and secretome of mesenchymal stem cells derived from human umbilical cord stroma. Stem Cells Dev. 2011;20:1199–212.
70. Fong CY, Subramanian A, Gauthaman K, Venugopal J, Biswas A, Ramakrishna S, Bongso A. Human umbilical cord Wharton's jelly stem cells undergo enhanced chondrogenic differentiation when grown on nanofibrous scaffolds and in a sequential two-stage culture medium environment. Stem Cell Rev. 2012;8:195–209.
71. Wang L, Seshareddy K, Weiss ML, Detamore MS. Effect of initial seeding density on human umbilical cord mesenchymal stromal cells for fibrocartilage tissue engineering. Tissue Eng Part A. 2009;15:1009–17.
72. Wang L, Zhao L, Detamore MS. Human umbilical cord mesenchymal stromal cells in a sandwich approach for osteochondral tissue engineering. J Tissue Eng Regen Med. 2011;5:712–21.
73. Chon BH, Lee EJ, Jing L, Setton LA, Chen J. Human umbilical cord mesenchymal stromal cells exhibit immature nucleus pulposus cell phenotype in a laminin-rich pseudo-three-dimensional culture system. Stem Cell Res Ther. 2013;4(5):120.
74. Hall S. Nerve repair: a neurobiologist's view. J Hand Surg. 2001;26:129–36.
75. Ishikawa N, Suzuki Y, Ohta M, Cho H, Suzuki S, Dezawa M, Ide C. Peripheral nerve regeneration through the space formed by a chitosan gel sponge. J Biomed Mater Res A. 2007;83:33–40.
76. Ohta M, Suzuki Y, Chou H, Ishikawa N, Suzuki S, Tanihara M, Suzuki Y, Mizushima Y, Dezawa M, Ide C. Novel heparin/alginate gel combined with basic fibroblast growth factor promotes nerve regeneration in rat sciatic nerve. J Biomed Mater Res A. 2004;71:661–8.
77. Dezawa M, Takahashi I, Esaki M, Takano M, Sawada H. Sciatic nerve regeneration in rats induced by transplantation of in vitro differentiated bone-marrow stromal cells. Eur J Neurosci. 2001;14:1771–6.
78. Matsuse D, Kitada M, Kohama M, Nishikawa K, Makinoshima H, Wakao S, Fujiyoshi Y, Heike T, Nakahata T, Akutsu H, et al. Human umbilical cord-derived mesenchymal stromal cells differentiate into functional schwann cells that sustain peripheral nerve regeneration. J Neuropathol Exp Neurol. 2010;69:973–85.
79. Kuroda Y, Kitada M, Wakao S, Dezawa M. Mesenchymal stem cells and umbilical cord as sources for schwann cell differentiation: their potential in peripheral nerve repair. Open Tissue Eng Regen Med J. 2011;4:54–63.
80. Peng J, Wang Y, Zhang L, Zhao B, Zhao Z, Chen J, Guo Q, Liu S, Sui X, Xu W, et al. Human umbilical cord Wharton's jelly-derived mesenchymal stem cells differentiate into a schwann-cell phenotype and promote neurite outgrowth in vitro. Brain Res Bull. 2011;84:235–43.

81. Xu Q, Zhang HT, Liu K, Rao JH, Liu XM, Wu L, Xu BN. In vitro and in vivo magnetic resonance tracking of sinerem-labeled human umbilical mesenchymal stromal cell-derived schwann cells. Cell Mol Neurobiol. 2011;31:365–75.
82. Zhang X, Zhang Q, Li W, Nie D, Chen W, Xu C, Yi X, Shi J, Tian M, Qin J, Jin G, Tu W. Therapeutic effect of human umbilical cord mesenchymal stem cells on neonatal rat hypoxic-ischemic encephalopathy. J Neurosci Res. 2014;92(1):35–45.
83. Hsieh JY, Wang HW, Chang SJ, Liao KH, Lee IH, Lin WS, Wu CH, Lin WY, Cheng SM. Mesenchymal stem cells from human umbilical cord express preferentially secreted -factors related to neuroprotection, neurogenesis, and angiogenesis. PLoS One. 2013;8(8), e72604.
84. Paldino E, Cenciarelli C, Giampaolo A, Milazzo L, Pescatori M, Hassan HJ, Casalbore P. Induction of dopaminergic neurons from human Wharton's jelly mesenchymal stem cell by forskolin. J Cell Physiol. 2014;229(2):232–44.
85. López Y, Seshareddy K, Trevino E, Cox J, Weiss ML. Evaluating the impact of oxygen concentration and plating density on human Wharton's jelly-derived mesenchymal stromal cells. Open Tissue Eng Regen Med J. 2011;4:82–94.
86. Breitbach M, Bostani T, Roell W, Xia Y, Dewald O, Nygren JM, Fries JW, Tiemann K, Bohlen H, Hescheler J, et al. Potential risks of bone marrow cell transplantation into infarcted hearts. Blood. 2007;110:1362–9.
87. Bittira B, Kuang JQ, Al-Khaldi A, Shum-Tim D, Chiu RC. In vitro preprogramming of marrow stromal cells for myocardial regeneration. Ann Thorac Surg. 2002;74:1154–9. discussion 1159–1160.
88. Tomita S, Mickle DA, Weisel RD, Jia ZQ, Tumiati LC, Allidina Y, Liu P, Li RK. Improved heart function with myogenesis and angiogenesis after autologous porcine bone marrow stromal cell transplantation. J Thorac Cardiovasc Surg. 2002;123:1132–40.
89. Wang HS, Hung SC, Peng ST, Huang CC, Wei HM, Guo YJ, Fu YS, Lai MC, Chen CC. Mesenchymal stem cells in the Wharton's jelly of the human umbilical cord. Stem Cells. 2004;22:1330–7.
90. Matsuura K, Nagai T, Nishigaki N, Oyama T, Nishi J, Wada H, Sano M, Toko H, Akazawa H, Sato T, et al. Adult cardiac sca-1-positive cells differentiate into beating cardiomyocytes. J Biol Chem. 2004;279:11384–91.
91. Maltsev VA, Rohwedel J, Hescheler J, Wobus AM. Embryonic stem cells differentiate in vitro into cardiomyocytes representing sinusnodal, atrial and ventricular cell types. Mech Dev. 1993;44:41–50.
92. Hollweck T, Hartmann I, Eblenkamp M, Wintermantel E, Reichart B, Überfuhr P, Eissner G. Cardiac differentiation of human wharton's jelly stem cells—experimental comparison of protocols. Open Tissue Eng Regen Med J. 2011;4:95–102.
93. Fathi F, Murasawa S, Hasegawa S, Asahara T, Kermani AJ, Mowla SJ. Cardiac differentiation of p19cl6 cells by oxytocin. Int J Cardiol. 2009;134:75–81.
94. Lopez Y, Lutjemeier B, Seshareddy K, Trevino EM, Hageman KS, Musch TI, Borgarelli M, Weiss ML. Wharton's jelly or bone marrow mesenchymal stromal cells improve cardiac function following myocardial infarction for more than 32 weeks in a rat model: a preliminary report. Curr Stem Cell Res Ther. 2013;8:46–59.
95. Schugar RC, Chirieleison SM, Wescoe KE, Schmidt BT, Askew Y, Nance JJ, Evron JM, Peault B, Deasy BM. High harvest yield, high expansion, and phenotype stability of cd146 mesenchymal stromal cells from whole primitive human umbilical cord tissue. J Biomed Biotechnol. 2009;2009:789526.
96. Farias VA, Linares-Fernandez JL, Penalver JL, Paya Colmenero JA, Ferron GO, Duran EL, Fernandez RM, Olivares EG, O'Valle F, Puertas A, et al. Human umbilical cord stromal stem cell express cd10 and exert contractile properties. Placenta. 2011;32:86–95.
97. Sarugaser R, Hanoun L, Keating A, Stanford WL, Davies JE. Human mesenchymal stem cells self-renew and differentiate according to a deterministic hierarchy. PLoS One. 2009;4, e6498.

98. Sarugaser R, Ennis J, Stanford WL, Davies JE. Isolation, propagation, and characterization of human umbilical cord perivascular cells (hucpvcs). Methods Mol Biol. 2009;482: 269–79.
99. Lange-Consiglio A, Corradetti B, Rutigliano L, Cremonesi F, Bizzarro D. In vitro studies of horse umbilical cord matrix-derived cells: from characterization to labeling for magnetic resonance imaging. Open Tissue Eng Regen Med J. 2011;4:120–33.
100. Li XX, Li KA, Qin JB, Ye KC, Yang XR, Li WM, Xie QS, Jiang ME, Zhang GX, Lu XW. In vivo MRI tracking of iron oxide nanoparticle-labeled human mesenchymal stem cells in limb ischemia. Int J Nanomedicine. 2013;8:1063–73.

Chapter 15
Placental Stem Cells and Culture Methods

Mofeedah Al Shammary and Felipe Mangoni Moretti

Introduction

Placental Compartments: Anatomical Features

Placenta is one of the most important organs in the uterine environment, which consists of both fetal- and maternal-derived cells. It enables the fetus to survive and develop during pregnancy by providing a wide range of hormones, growth factors, cytokines and transcription factors and protecting the fetus from various chemical, infectious and immune assaults by waste elimination and gas exchange via the mother's blood supply. As shown in (Figs. 15.1 and 15.2) the human placenta is discoid in shape with a diameter of 15–20 cm and a thickness of 2–3 cm. From the margins of the chorionic disc extend the fetal membranes, amnion and chorion, which enclose the fetus in the amniotic cavity, and the endometrial decidua. The chorionic plate is a multilayered structure that faces the amniotic cavity consisting of [1, 2]:

1- Fetal part (amniotic and chorionic structures)
   - i) The amniotic membrane (composed of epithelium, compact layer, amniotic mesoderm, and spongy layer)
   - ii) Chorionic structures (composed of mesenchyme and chorionic villi and extravillus developing from trophoblast layer)

M. Al Shammary, M.D. (✉) • F.M. Moretti, M.D.
Gynecology and Newborn Care, Division of Maternal Fetal Medicine, Department of Obstetrics and Gynecology, The Ottawa Hospital, University of Ottawa, 501 Smyth Road, Room 8472, Ottawa, ON K1H 8L6, Canada
e-mail: malshammary@toh.on.ca; Shammary02@yahoo.com; fmoretti@toh.on.ca

D.O. Fauza, M. Bani (eds.), *Fetal Stem Cells in Regenerative Medicine*, Stem Cell Biology and Regenerative Medicine,
DOI 10.1007/978-1-4939-3483-6_15

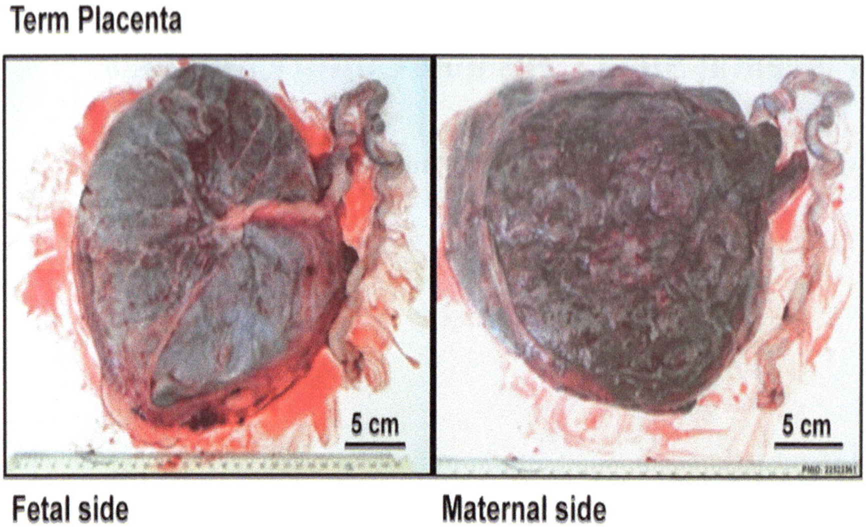

Fig. 15.1 The placenta structure viewed from both the fetal and maternal compartment at term (Adapted from Lynne K. Warrander, et al. PLoS One. 2012; 7(4): e34851. Published online 2012 Apr 16. Doi: 10.1371/journal.pone.0034851. *Maternal Perception of Reduced Fetal Movements Is Associated with Altered Placental Structure and Function*. PLos One PMID PMC3327709)

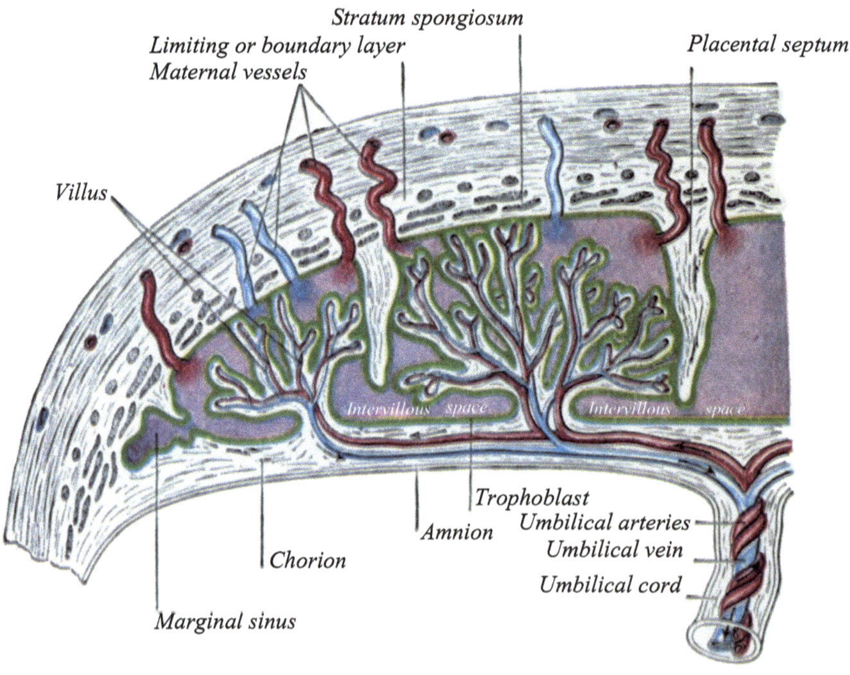

Fig. 15.2 Schematic structure of human placenta, showing both the fetal (amnion and chorion) and the maternal (decidua) components. Gray, Henry. *Anatomy of the Human Body*. Philadelphia: Lea & Febiger, 1918; Bartleby.com, 2000. www.bartleby.com/107/

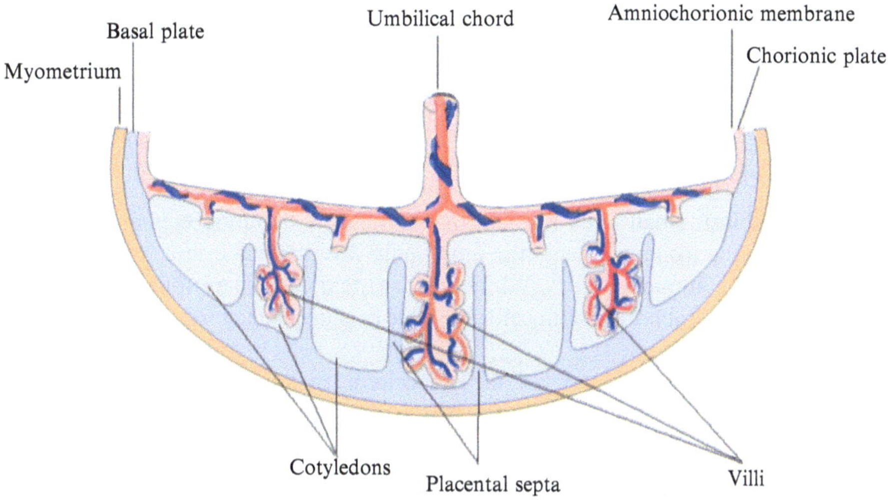

Fig. 15.3 Schematic view of placenta formation from the maternal side; protrusions of the basal plate within the chorionic villi produce the placental septa, which divide the parenchyma into irregular cotyledons. Fetal blood vessels are located within the branches of the villi (adapted from Parolini et al. 2008). Citation: Parolini, O., Strom, S. C. et al. (2008), Concise Review: Isolation and Characterization of Cells from Human Term Placenta: Outcome of the First International Workshop on Placenta Derived Stem Cells. STEM CELLS, 26: 300–311. doi: 10.1634/stemcells.2007-0594

2- Maternal part (decidua basalis)

i) Chorionic plate villus anchors the trophoblast of the basal plate and maternal endometrium, whereas others terminate freely in the intervillous space.
ii) Protrusions of the basal plate within the chorionic villi produce the placental septa, which divide subsequently the parenchyma into irregular cotyledons (Fig. 15.3).

In accordance with this classification, the placenta can be divided into four cellular components: amniotic epithelial cells, amniotic mesenchymal cells, chorionic mesenchymal cells and chorionic trophoblastic cells. Since one of the main issues about placental-derived stem cells is the contamination of fetal-derived with maternal-derived cells, we provide a brief overview of placental development before discussing the culturing methods used to differentiate placental mesenchymal stem cells.

Embryonic Development

In humans, 4–5 days after fertilization, the morula, which results from division of the zygote to a cluster of 16 cells (termed blastomeres), enters the uterine cavity. By day 6 and 7 post-fertilization, the blastocyst implants and placental development begins. During the implantation stage, the blastocyst is composed of an outer wall (trophoblast) that surrounds the blastocystic cavity. As the blastocyst adheres to the endometrium, it proliferates robustly, forming a bilayered trophoblast. The outer of the two layers becomes the syncytio-trophoblast (fusion of neighboring trophoblast cells) whereas the inner cells (cytotrophoblast) remain temporally unfused. At the second week after fertilization the inner cell mass differentiates into two layers:

a. The epiblast, small cells that later constitute the amniotic epithelium appear between the trophoblast and the embryonic disc and enclose a space that will become the amniotic cavity. While epiblast stem cells share some pluripotency factors with embryonic stem cells, the state of pluripotency for the former seems to be more progressed.
b. Hypoblast and Cytotrophoblast, the exocoelomic membrane and its cavity modify to form the yolk sac. Hypoblast serves as a transient structure for establishing axial patterning in the embryo.

To elaborate further, gastrulation has been depicted in Fig. 15.4. During this process the bilaminar disc differentiates into the three germ layers (ectoderm, mesoderm, and endoderm) and develops a defined form during the third week post-fertilization [3, 4].

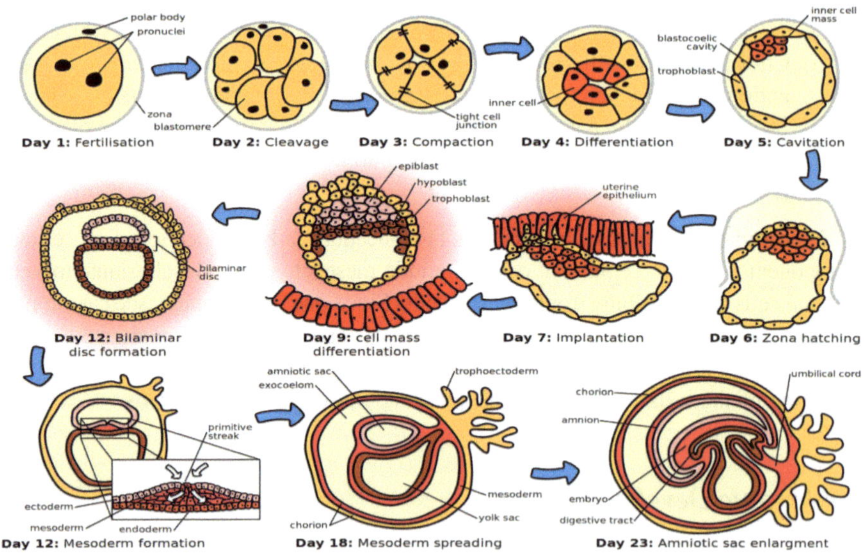

Fig. 15.4 Schematic section of structural stages in the gastrulation (Richard Wheeler.net). Citation: Richard Wheeler, RichardWheeler.net. *The Early Stage of Human Embryogenesis*. https://en.wikipedia.org/?title=User:Zephyris. Image: https://en.wikipedia.org/?title=User:Zephyris#/media/File:HumanEmbryogenesis.svg

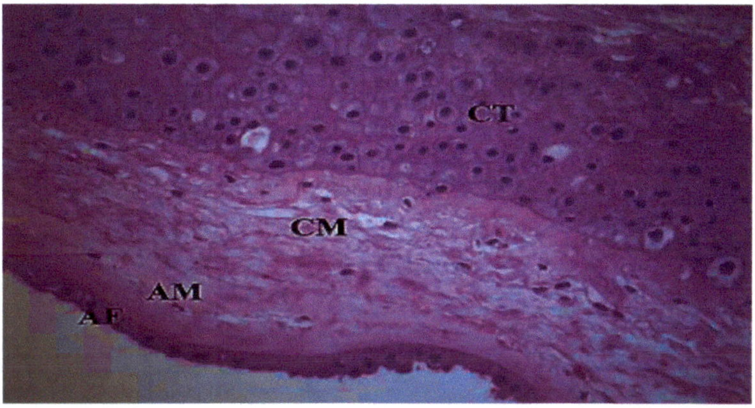

Fig. 15.5 Cross-sectional representation of human fetal membranes (amnion and chorion) stained with Hematoxylin and Eosin Staining. The amnion is composed of an epithelial layer of cuboidal and columnar cells, which lie on top of a mesodermal layer consisting of an upper a cellular compact layer and a deeper layer containing dispersed fibroblasts. The chorionic membrane consists of a mesodermal layer and a layer of extra villous trophoblast cells. *AE* amniotic epithelium, *AM* amniotic mesoderm, *CM* chorionic mesoderm, *CT* chorionic trophoblastic layer (M Evangelista. Cytotechnology 2008). Citation: Cytotechnology. 2008 Sep; 58(1): 33–42. Published online 2008 Sep 28. doi: 10.1007/s10616-008-9162-z. Springer Publishing Company. PMCID: PMC2593758. *Placenta-derived stem cells: new hope for cell therapy*? Marco Evangelista, Maddalena Soncini, and Ornella Parolini

Cellular and Molecular Markers

Using a wide range of cellular markers in the fetal membrane (Fig. 15.5), four distinct regions can be identified in the placenta: amniotic epithelial cells, amniotic mesenchymal cells, chorionic mesenchymal cells, and chorionic trophoblastic cells [1, 5, 6]. The amniotic epithelium (AE) is single layer of cuboidal and columnar epithelial cells that are in contact with the amniotic fluid. These cells are typically attached to a distinct basal lamina that is connected to the amniotic mesoderm (AM), a cellular compact layer composed of collagens I and III and fibronectin (Fig. 15.6). Deeper in the amniotic mesoderm, a network of dispersed fibroblast-like mesenchymal cells and few macrophages are observed. More recently, it has been reported that the mesenchymal layer of amnion contains two subdivisions, one having a mesenchymal phenotype referred as amniotic mesenchymal stromal cells (AMSCs), and the other composed of monocyte-like cells [7, 8]. Furthermore, there is a spongy layer, loosely arranged collagen fibers, extricating the amniotic and chorionic mesoderm. The chorionic membrane (chorion leave) consists of mesodermal and trophoblastic regions. A large and incomplete basal lamina separates the chorionic mesoderm from the extra villous trophoblast cells that are distributed within the fibrinoid layer and express various immunohistochemical markers of proliferation. The fibrinoid layer sequentially is composed of two diverse types:

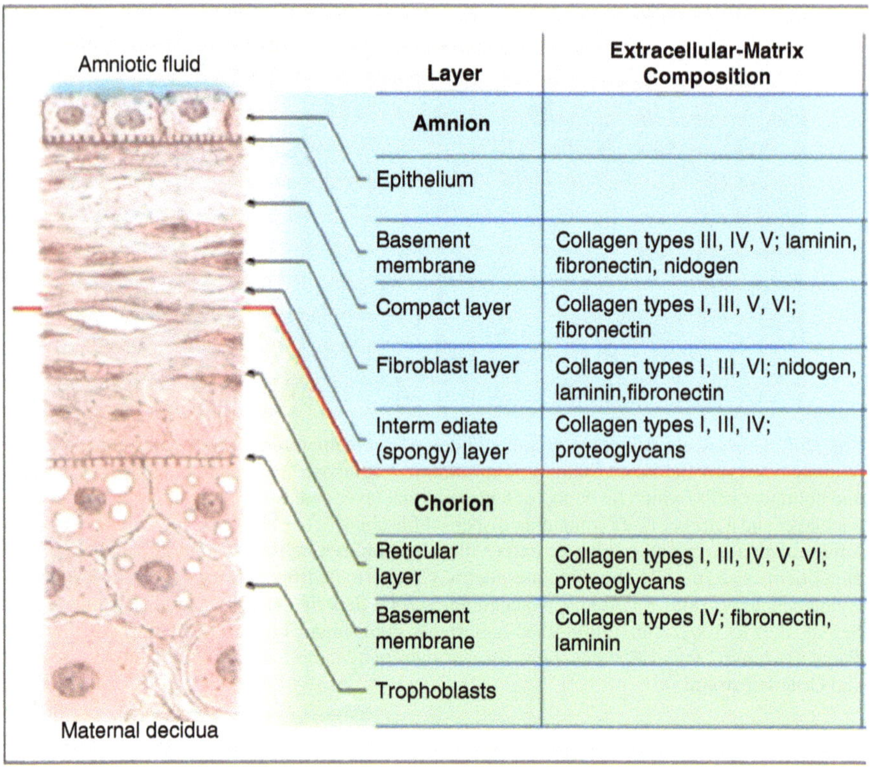

Fig. 15.6 Schematic representation of fetal membrane structure at term (Modified from Parry and Strauss, 1998). Citation: HTTP://WWW.NEJM.ORG/DOI/FULL/10.1056/NEJM199803053381006#T=ARTICLE. REVIEW ARTICLE. MECHANISMS OF DISEAS. Franklin H. Epstein, M.D., Editor. *Premature Rupture of the Fetal Membranes*. Samuel Parry, M.D., and Jerome F. Strauss, M.D., Ph.D. N Engl J Med 1998; 338:663–670. March 5, 1998. DOI: 10.1056/NEJM199803053381006

a matrix type on the inner side (more compact) and a fibrin type on the outer side (more reticulate) [9, 10].

Based on a number of cell culture techniques, the following cell populations have been sequestered from the placenta (Fig. 15.7):

a) Human amniotic epithelial cells (hAEC).
b) Human amniotic mesenchymal stromal cells (hAMSc)
c) Human chorionic mesenchymal stromal cells (hCMSc)

The cellular and molecular profiling of hAMSC and hCMSC has shown the expression of several pluripotent stem cell markers, including octamer-binding protein-4 (OCT-4), SRY-related HMG-box gene 2 (SOX-2), Nanog, CD117 (c-KIT), SSEA-3, SSEA-4, TRA-1-60 and TRA-1-81 [7, 11]. Furthermore, both hAMSC and hCMSC express the typical markers detected in bone marrow-derived MSC.

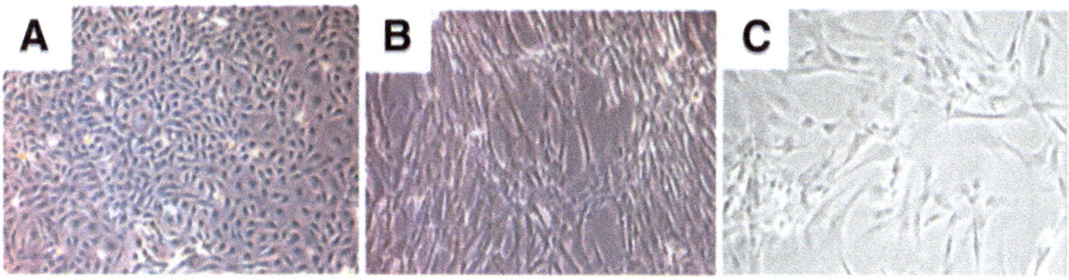

Fig. 15.7 The placenta derived cell populations (**a**): Human amniotic epithelial cells. (**b**): Human amniotic mesenchymal stromal cells. (**c**): Human chorionic mesenchymal stromal cells (Adapted from Parolini et al. 2008). Citation: Parolini, O., Strom, S. C. et al. (2008), Concise Review: *Isolation and Characterization of Cells from Human Term Placenta*: Outcome of the First International Workshop on Placenta Derived Stem Cells. STEM CELLS, 26: 300–311. doi: 10.1634/stemcells.2007-0594

Table 15.1 The expression profile at Passage 2–4 for human amniotic and chorionic mesenchymal stromal cells (Adapted from Dominici et al. 2006)

Positive (>95 %)	Negative (<2 %)
CD90	CD45
CD73	CD34
CD105	CD14
	HLA-DR

The immunophenotyping of placental MSCs are generally performed between Passages 2 and 15 (P2–P15). At early passages (P2–P4), nearly 95 % of the cells appear to be positive for mesenchymal markers (CD105 and CD73) and CD90, though lacking the manifestation of CD14, CD34, CD45 and HLA-DR (Table 15.1) [9].

Notably, a set of standards have been proposed by the International Society for Cellular Therapy (ISCT) to define and classify MSCs, including those isolated from hAMSC and hCMSC. The cells should demonstrate features such as adherence to plastic surfaces, formation of fibroblast colony-forming units, differentiation potential to one or more mesodermal lineages (osteogenic, adipogenic and chondrogenic) and the expression of CD90, CD73 and CD105, while lacking CD45, CD34, CD14 or HLA-DR (Table 15.2); [5, 7, 11–23].

Human Amniotic Epithelial Cells (hAEC)

Among the cell types in the placenta, human amniotic epithelial cells have been studied most extensively. From the cell biology and developmental perspectives, hAECs develop from the epiblast by 8 days post fertilization and they can be secluded from the amnion after delivery [24]. Several studies have shown the heterogeneous expression of a number of stem cell markers such as OCT-4, NANOG, SOX-2, TRA-1-60 and TRA-1-81 in hAECs [7, 11]. These cells have been shown to differentiate into the three germ layer cell lineages: ectoderm, mesoderm and endoderm.

Table 15.2 The differentiation potential and marker expression of different placental cells (Adapted from Parolini et al. 2008)

Type of stem cells	Differentiation potential	Markers	References
Haematopoietic	Haematopoietic	CD34, c-Kit, Sca-1, Gata-2, Gata-3 and Runx-1	Parolini et al. [5]
AE	Mesenchymal, haematopoietic, hepatic, cardiac, pancreatic and neural cells	OCT-4, Nanog, SOX-2, TRA-1-60,; TRA-1-81, EpCAM, E-cadherin, CD49d, CD49f, CK7	Miki et al. [12], Murphy et al. [13], Pratama et al. [14]
CM	Adipogenic, chondrogenic, osteogenic, skeletal myogenic and neurogenic	CD105, CD90, CD73, CD44, CD29, CD13, CD166, CD49e, CD10, HLA-ABC	Pasquinelli et al. [11], Sakuragawa et al. [15], Portmann-lanz et al. [16], Wolbank et al. [17], Zhang et al. [18], Soncini et al. [19]
AM	Adipose, chondrogenic, osteogenic, skeletal myogenic, angiogenic, neurogenic, pancreatic and myogenic	CD105, CD90, CD73, CD44, CD29, HLA-A,B,C, CD13, CD10, CD49c, CD49d, CD54, CD166	Pasquinelli et al. [11], Sakuragawa et al. [15], Portmann-lanz et al. [16], Wolbank et al. [17], Zhang et al. [18], Soncini et al. [19]

A number of key steps have been identified regarding the isolation of amniotic epithelial cells. For instance, rapid and efficient collection and refrigeration of the placenta, followed by the segregation of cells within 3 h seem to be critical for the efficient production of hAECs [25]. The serous layer on the maternal side of placenta is cut within the first 5 min after delivery and the placenta is turned over with the fetal side facing up. The organ is flushed with normal saline and rinsed with Citrate-Phosphate-Dextrose Adenine (CPDA-1), after which the amnion is stripped from the underlying chorion and digested with trypsin or other digestive enzymes. The resulting cells can be used directly or purified further by fluorescent-activated cell sorting (FACS). Under appropriate conditions, the expression of pluripotent stem cell markers becomes evident *in vitro* ([26, 27]; also see Table 15.3 for an exhaustive protocol). AE cells can be cultured in a serum-free medium with limitations in passage number [27]. Remarkably, hAECs express low levels of human leukocyte antigen (HLA)-A, B, C along with CD34, CD133, CD117 and CCR4 [15, 16, 24, 25, 27]. See Table 15.3 for the expression profile of various markers in hAECs.

Human Amniotic Mesenchymal Stromal Cells (hAMSc)

hAMSCs are derived from the extraembryonic mesoderm [5] and are originally disseminated in the collagenous stroma underlying the epithelial monolayer of the amniotic membrane [28, 29]. Human AMSCs have demonstrated multi-lineage

Table 15.3 Human amniotic epithelial cells: isolation protocols, phenotype, and *in vitro* differentiation (Adapted from Parolini et al. 2008)

Isolation protocols	Mechanical peeling of amnion membrane from the underlying chorion followed by digestion with the following: dispase II (2.4 U/ml) for 1 h or trypsin-EDTA (different concentrations and incubation times). Centrifugation (200 g for 10 min) of solution containing released cells, discarding the remaining membrane.
Phenotype at passages 2–4 [15, 16, 26]	*Mesenchymal and hematopoietic markers*: CD105+, CD90+, CD73+, CD44+, CD29+, HLA-A,B,C+, CD13+, CD10+, CD166+, CD49d–, CD49e+, CD117 (+/– very weak signal), CD14–, CD34–, CD45–, HLA-DR-
	Embryonic cell markers: SSEA-3+, SSEA-4+, TRA1-60+, TRA1-81+, SSEA-1–
	Others: CD324 (E-cadherin)+, POU5F1+, SOX2+, CFC1+, NANOG+, DPPA3+, PROM1+, PAX6+, FOXD3–, GDF3–, CD140b+, CD349–, GCTM2+
Adipogenic [16]	DMEM high glucose or DMEM/Ham's F-12 medium, 10 % FBS, 0.5 mM isobutyl-methylxanthine, 1 μM dexamethasone, 10 μM insulin, 200 μM indomethacin
Chondrogenic [26]	DMEM high glucose, 1 % fetal bovine serum (FBS) 6.25 μg/ml insulin, 10 ng/ml TGF-B1, 50 ng/ml fresh ascorbic acid
Osteogenic [16]	DMEM high glucose (or DMEM/Ham's F-12 medium), 10 % FBS, 10 μM dexamethasone, 10 nM 1-a,25-dihydroxyvitamin D3, 50 g/ml ascorbic acid, 10 mM B-glycerophosphate
Skeletal myogenic [16]	DMEM/Ham's F-12 medium (or DMEM high glucose), 10 % FBS, 5 % human serum (or horse serum), 50 μM hydrocortisone
Neurogenic [25]	DMEM high glucose, 10 % FBS, 30 μM all-trans retinoic acid
	DMEM, 10 % FBS, 55 μM 2-mercaptoethanol, 1 mM sodium pyruvate, 5×10^{-5} M all-trans retinoic acid, 10 ng/ml FGF-4
	DMEM/Ham's F-12 medium, 10 % fetal calf serum; FCS, 5×10^{-5} M alltrans retinoic acid, 10 ng/ml FGF4, N-2, B-27
Pancreatic [16, 25]	DMEM, 10 % FBS, 55 μM 2-mercaptoethanol, 1 mM sodium pyruvate, 10 mM nicotinamide on collagen I coated plate
	DMEM (or DMEM/Ham's F-12 medium) containing N2 supplement 10 mM nicotinamide
Hepatic [24, 25]	MEM, 10 % FBS, 55 M 2-mercaptoethanol, 1 mM sodium pyruvate, dexamethasone 10^{-7} M, 0.1 μM insulin for 3 weeks, addition of 1 mM phenobarbital for the final 3 days, on collagen I-coated plate
	DMEM, 10 % FBS, 20 ng/ml HGF, 10 ng/ml FGF-2, 10 ng/ml oncostatin M, 100 nM dexamethasone, 10 U/ml heparin sodium salt
	DMEM/Ham's F-12 medium, 10 % FCS + 0.1 μM insulin, 1×10^{-7} M dexamethasone

distinction potential. Besides differentiating into the typical mesenchymal lineages (osteogenic, chondrogenic, adipogenic), hAMSC are also capable to differentiate into other cell types, including neural and glial cells, skeletal muscle cells, cardiac myocytes, pancreatic and hepatocyte-like cells [7, 16].

Many protocols have been proposed for the isolation of hAMSCs. Most of these protocols contain key steps to digest the tissue with several enzymes (e.g., Collagenase I or II, Hyaluronidase and Trypsin), followed by filtration and or centrifugation [30, 31].

After complete removal of human amniotic epithelial cells (hAECs) by trypsin digestion, cells are obtained in subsequent enzymatic digestions, using collagenase (0.75–2 mg/mL) with or without adding DNase (20–75 μg/mL) [5, 7, 15, 19, 32].

Some groups have reported that expansion of hAMSCs is possible for at least five passages with no morphological or karyotypic alterations [24, 28]. The media used for expansion are typically composed of a basal medium, along with supplements such as fetal calf serum (FCS), epidermal growth factor (EGF), beta mercaptoethanol, non-essential amino acid (NEAA) and leukemia inhibitory factor (LIF). Slight variations such as addition of insulin, transferrin, tri-iodothyronine [33] or LIF [34, 35] have been recommended in some cases.

The expression profile of cell surface markers in hAMSCs is generally determined by RT-PCR, flow cytometry, and/or immunocytochemistry, using CD73, CD90, CD105, CD10, CD13, CD29, CD44, CD49c, CD49d, CD49e, CD54, CD140b, CD166, CD349, STRO-1 and HLA-ABC [5, 17, 36, 37]. Notably, low expression of CD271 [5, 38] and CD117 (c-kit) [39] has been also reported in these cells. In some cases, Transmission electron microscopy has been used to assess the ultrastructural characteristics of hAMSC [11].

Furthermore, immunostaining with SSEA-3 and SSEA-4 [5, 20, 39–41] along with RNA analysis for Oct-4 have been used to compare hAMSCs with bone marrow-derived MSCs [17, 20, 28, 40].

Human Chorionic Mesenchymal Stromal Cells (hCMSCs)

Chorionic mesenchymal stromal cells can be successfully isolated at different trimesters. For example, minimally invasive techniques such as chorionic villus sampling (CVS, Fig. 15.8) can be used to isolate these cells from chorionic villi between 9th and 12th weeks of gestation. However, these cells represent only 1 % of the population [42]. Generally, the chorion is digested with collagenase or

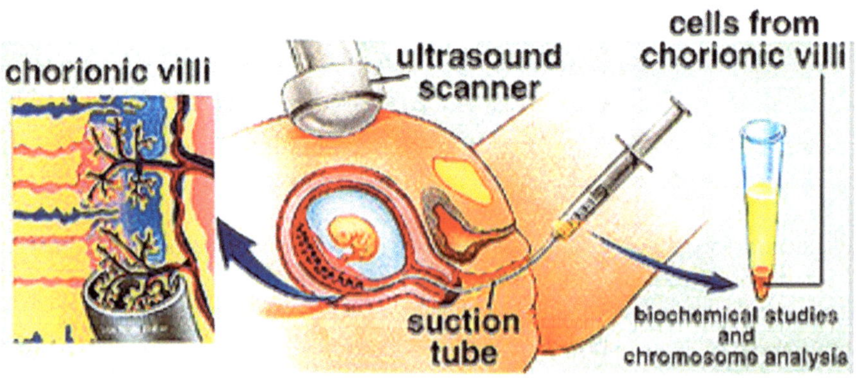

Fig. 15.8 Chorionic villus sampling (CVS, adapted from Mader, 1997). Citation: Sylvia S. Mader. Inquiry into Life, 8th edition, McGraw-Hill Higher Education, 1997. ISBN: 10: 0072418826

Table 15.4 Human chorionic mesenchymal stromal cells: isolation protocols, phenotype (Adapted from Parolini et al. 2008)

Procedure	References
Isolation Protocols	Mechanical removal of surrounding layers after treatment with 2.4 U/ml dispase II at 37 °C, followed by 1–3 h treatment with Worthington collagenase II (270 U/ml) or collagenase A (0.83 mg/ml) [16, 19, 46–48]
Phenotype at passages 2–4	*Mesenchymal and hematopoietic markers*: CD105+, CD90+, CD73+, CD44+, CD29+, HLA-A,B,C+, CD13+, CD166+, CD49e+, CD271low, CD10+, CD14–, CD34–, CD45–, CD117–, CD133–, HLA-DR– [16, 19, 48]
	Embryonic stem cell marker: SSEA-4–/+
	Others: CD349+, CD140b+, CD324– [49, 50]

DNase enzymes [43], and after a mechanical and enzymatic removal of the trophoblastic layers, mesenchymal stem cells can be further isolated with low contamination risk [44, 45]. On the other hand, to minimize maternal CMSC contamination, cells are isolated by dissection from the deflected part of the fetal amnion membranes [44].

A brief protocol implemented for CMSC isolation has been described in Table 15.4 along with the expression of mesenchymal and haematopoietic stem cell markers such as CD14, CD34, CD45 [46–50]. Interestingly, Chorionic villus cytotrophoblast cells expressing OCT4, NANOG, SOX2, c-MYC, KLF4, SSEA3, SSEA4, TRA-1-60 and TRA-1-81 are able to differentiate into the three germ layers [51].

Mesodermal Differentiation of Placental Mesenchymal Stem Cells

<u>Adipogenic differentiation</u>: Cells are cultured in complete medium with the addition of 0.5 mM isobutyl-methylxanthine IBMX, 1 μM dexamethasone, 10 μM insulin, and 60 μM indomethacin, 10 % FBS. Adipogenic differentiation can be confirmed within 3 weeks by observing the formation of intracellular lipid droplets under the microscope and via Oil Red O staining [52–54].

<u>Osteogenic differentiation</u>: Media containing 10 μM dexamethasone, 10 nM 1a,25-dihydroxyvitamin D3, 10 mM β-glycerophosphate and 50 μg/ml ascorbic acid-2-phosphate are used to differentiate towards osteogenic cell lineage. Differentiation can be confirmed by calcium precipitation and ALP production [55].

<u>Chondrogenic differentiation</u>: Media containing high glucose, 1 % FBS, 6.25 μg/ml insulin, proline and 10 ng/ml transforming growth factor-β1 (TGF-β1) and 50 ng/ml ascorbic acid induce differentiation of placental mesenchymal stem cells into chondrocytes. Cells are usually maintained for ~20 days and their differentiation can be confirmed by production of sulfated glycosaminoglycan (sGAG) and type II collagen [54, 56, 57].

Neurogenic Differentiation of Placental Mesenchymal Stem Cells

Differentiation into neurons has been achieved in media containing nerve-growth factor (NGF), DMSO, and 30 μM all-trans retinoic acid ([55, 56]). Neuronal morphology can be observed after 6–14 days and further confirmed with neuronal markers, including beta-III tubulin, NeuN, MAP2, neurofilament and neuron-specific enolase (NSE) [15, 16].

Prospective Clinical Applications of Placental Mesenchymal Stem Cells

Isolated from generally discarded, ethically non-concerning placental tissue, mesenchymal stem cells may serve as an attractive source of cells for different clinical applications. Since MSCs have the ability to down-regulate immune response and support tissue repair mechanisms, they can be used to treat a wide range of diseases, including neural tube defects, Alzheimer disease, multiple sclerosis, muscular dystrophy, graft vs. host disease (GvHD), hepatic cirrhosis, diabetes type 1 and cardiomyopathy [21, 58, 59]. In particular, spina bifida, a common and complex form of neural tube defect affecting approximately 1 in 1500 live births, leaves most survivors with a range of minor physical problems to severe physical and mental disabilities [60, 61]. Among different types of spina bifida (i.e., occulta, closed neural tube defects, meningocele and myelomeningocele), the latter is the most severe form, occurring when the spinal cord/neural elements are exposed through the opening in the spine and resulting in partial or complete paralysis. Fetal surgeries in selected cases of myelomeningocele may offer in utero closure of the defect to prevent further damage. However, surgical approaches alone may not be sufficient to improve motor and sensory functions. Thus, the use of stem cells from prenatal sources to repair the damaged region has been considered as a potential therapeutic avenue. In particular, Fauza's group has provided several reports, emphasizing that stem cell-based therapy may facilitate recovery in a sheep model of myelomeningocele [62–66]. In addition to MSCs, neural crest stem cells (NCSCs) have also gained attention for their therapeutic potential in spina bifida [67, 68]. Of relevance, the placenta has been considered as a source of NCSCs with applications in neural repair. Recently, California institute for regenerative medicine (CIRM) has organized translational studies that specifically focus on the applications of placental neural crest stem cells in spinal bifida [69, 70].

Notably, despite promising results in animal models, more work is required to make the use of placental MSCs or NCSCs in clinical setting a reality. Standard procedures yet to be developed for stem cell culture, banking/storage, quality control prior to transplantation in patients.

Taken together, the lessons learned from the research data and clinical trials on placental stem cells have proven that these cells provide a great potential for different therapeutic applications. Due to their high proliferative capacity and expansion capacity, their culturing period can be reduced with regards to time, passage number and contamination risks. Future studies will shed more light on the application of these cells in clinic.

References

1. Cunningham FG, MacDonald PC, Gant MF, et al. The placenta and fetal membranes. In: Williams obstetrics. 20th ed. Stamford, CT: Appleton and Lange; 1997. p. 95–125.
2. Benirschke K, Kaufmann P. Pathology of the human placenta. New York, NY: Springer; 2000. p. 42–6. 116, 281–297.
3. Sood R, Zehnder JL, Druzin ML, et al. Gene expression patterns in human placenta. Proc Natl Acad Sci U S A. 2006;103:5478–83.
4. Moore KL, Persaud TVN. The developing human. 6th ed. Philadelphia, PA: W.B. Saunders Company; 1998. p. 41–62.
5. Parolini O, Alviano F, Bagnara GP, Bilic G, Buhring HJ, Evangelista M, Hennerbichler S, Liu B, Magatti M, Mao N, Miki T, Marongiu F, Nakajima H, Nikaido T, Portmann-Lanz CB, Sankar V, Soncini M, Stadler G, Surbek D, Takahashi TA, Redl H, Sakuragawa N, Wolbank S, Zeisberger S, Zisch A, Strom SC. Concise review: isolation and characterization of cells from human term placenta: outcome of the first international workshop on placenta derived stem cells. Stem Cells. 2008;26:300–11.
6. Magatti M, De Munari S, Vertua E, et al. Human amnion mesenchyme harbors cells with allogeneic T cell suppression and stimulation capabilities. Stem Cells. 2008;26:182.
7. In't Anker PS, Scherjon SA, Kleijburg-van der Keur C, et al. Isolation of mesenchymal stem cells of fetal or maternal origin from human placenta. Stem Cells. 2004;22:1338–45.
8. Casey ML, MacDonald PC. Interstitial collagen synthesis and processing in human amnion: a property of the mesenchymal cells. Biol Reprod. 1996;55:1253–60.
9. Dominici M, Le Blanc K, Mueller I, et al. Minimal criteria for defining multipotent mesenchymal stromal cells. The International Society for Cellular Therapy position statement. Cytotherapy. 2006;8:315–7.
10. Pipino C, Shangaris P, Resca E, Zia S, Deprest J, Sebire NJ, David AL, Guillot PV, De Coppi P. Placenta as a reservoir of stem cells: an underutilized resource? Br Med Bull. 2013;105:43.
11. Pasquinelli G, Tazzari P, Ricci F, et al. Ultrastructural characteristics of human mesenchymal stromal (stem) cells derived from bone marrow and term placenta. Ultrastruct Pathol. 2007;31:23–31.
12. Miki T, Lehmann T, Cai H, et al. Stem cell characteristics of amniotic epithelial cells. Stem Cells. 2005;23:1549–59.
13. Murphy S, Rosli S, Acharya R, et al. Amnion epithelial cell isolation and characterization for clinical use. Curr Protoc Stem Cell Biol. 2010;13:1E.6.1–6.25.
14. Pratama G, Vaghjiani V, Tee JY, et al. Changes in culture expanded human amniotic epithelial cells: implications for potential therapeutic applications. PLoS One. 2011;6:e26136.
15. Sakuragawa N, Kakinuma K, Kikuchi A, et al. Human amnion mesenchyme cells express phenotypes of neuroglial progenitor cells. J Neurosci Res. 2004;78:208–14.
16. Portmann-lanz CB, Schoeberlein A, Huber A, et al. Placental mesenchymal stem cells as potential autologous graft for pre- and perinatal neuroregeneration. Am J Obstet Gynecol. 2006;194:664–73.
17. Wolbank S, Peterbauer A, Fahrner M, Hennerbichler S, van Griensven M, Stadler G, Redl H, Gabriel C. Dose-dependent immunomodulatory effect of human stem cells from amniotic membrane: a comparison with human mesenchymal stem cells from adipose tissue. Tissue Eng. 2007;13:1173–83.

18. Zhang X, Mitsuru A, Igura K, et al. Mesenchymal progenitor cells derived from chorionic villi of human placenta for cartilage tissue engineering. Biochem Biophys Res Commun. 2006;340:944–52.
19. Soncini M, Vertua E, Gibelli L, et al. Isolation and characterization of mesenchymal cells from human fetal membranes. J Tissue Eng Regen Med. 2007;1:296–305.
20. Evangelista M, Soncini M, Parolini O. Placenta-derived stem cells: new hope for cell therapy? Cytotechnology. 2008;58:33–42.
21. Baksh D, Song L, Tuan RS. Adult mesenchymal stem cells: characterization, differentiation, and application in cell and gene therapy. J Cell Mol Med. 2004;8:301–16.
22. Brooke G, Tong H, Levesque J-P, et al. Molecular trafficking mechanisms of multipotent mesenchymal stem cells derived from human bone marrow and placenta. Stem Cells Dev. 2008;17:929–40.
23. Chen K, Wang D, Du WT, Han ZB, Ren H, Chi Y, Yang SG, Zhu D, Bayard F, Han ZC. Human umbilical cord mesenchymal stem cells huc-mscs exert immunosuppressive activities through a pge2-dependent mechanism. Clin Immunol. 2010;135:448–58.
24. Alviano F, Fossati V, Marchionni C, et al. Term amniotic membrane is a high throughput source for multipotent mesenchymal stem cells with the ability to differentiate into endothelial cells in vitro. BMC Dev Biol. 2007;7:11.
25. Miki T, Marongiu F, Dorko K, et al. Isolation of amniotic epithelial stem cells. Curr Protoc Stem Cell Biol. 2010;12:1E.3.1–3.10.
26. Ilancheran S, Moodley Y, Manuelpillai U. Human fetal membranes: a source of stem cells for tissue regeneration and repair? Placenta. 2009;30:2–10.
27. Wolbank S, Stadler G, Peterbauer A, Gillich A, Karbiener M, Streubel B, Wieser M, Katinger H, van Griensven M, Redl H, et al. Telomerase immortalized human amnion- and adipose derived mesenchymal stem cells: maintenance of differentiation and immunomodulatory characteristics. Tissue Eng Part A. 2009;15:1843–54.
28. Bilic G, Zeisberger SM, Mallik AS, Zimmermann R, Zisch AH. Comparative characterization of cultured human term amnion epithelial and mesenchymal stromal cells for application in cell therapy. Cell Transplant. 2008;17:955–68.
29. Manuelpillai U, Moodley Y, Borlongan CV, Parolini O. Amniotic membrane and amniotic cells: potential therapeutic tools to combat tissue inflammation and fibrosis? Placenta. 2011;32 Suppl 4:S320–5.
30. Tamagawa T, Oi S, Ishiwata I, Ishikawa H, Nakamura Y. Differentiation of mesenchymal cells derived from human amniotic membranes into hepatocyte-like cells in vitro. Hum Cell. 2007;20:77–84.
31. Tsagias N, Koliakos I, Karagiannis V, Eleftheriadou M, Koliakos GG. Isolation of mesenchymal stem cells using the total length of umbilical cord for transplantation purposes. Transfus Med. 2011;21:253–61.
32. Manochantr S, Tantrawatpan C, Kheolamai P, Pratya Y, Supokawej A, Issaragrisil S. Isolation, characterization and neural differentiation potential of amnion derived mesenchymal stem cells. J Med Assoc Thai. 2010;93 Suppl 7:S183–91.
33. Bilic G, Ochsenbein-Kolble N, Hall H, Huch R, Zimmermann R. *In vitro* lesion repair by human amnion epithelial and mesenchymal cells. Am J Obstet Gynecol. 2004;190:87–92.
34. Whittle WL, Gibb W, Challis JR. The characterization of human amnion epithelial and mesenchymal cells: the cellular expression, activity and glucocorticoid regulation of prostaglandin output. Placenta. 2000;21:394–401.
35. Sudo K, Kanno M, Miharada K, Ogawa S, Hiroyama T, Saijo K, Nakamura Y. Mesenchymal progenitors able to differentiate into osteogenic, chondrogenic, and/or adipogenic cells in vitro are present in most primary fibroblast-like cell populations. Stem Cells. 2007;25:1610–7.
36. König J, Huppertz B, Desoye G, Parolini O, Frohlich JD, Weiss G, Dohr G, Sedlmayr P, Lang I. Amnion-derived mesenchymal stromal cells show angiogenic properties but resist differentiation into mature endothelial cells. Stem Cells Dev. 2012;21:1309–20.
37. Lisi A, Briganti E, Ledda M, Losi P, Grimaldi S, Marchese R, Soldani G. A combined synthetic-fibrin scaffold supports growth and cardiomyogenic commitment of human placental derived stem cells. PLoS One. 2012;7:e34284.

38. Roubelakis MG, Trohatou O, Anagnou NP. Amniotic fluid and amniotic membrane stem cells: marker discovery. Stem Cells Int. 2012;2012:107836.
39. Diaz-Prado S, Muinos-Lopez E, Hermida-Gomez T, Rendal-Vazquez ME, Fuentes-Boquete I, de Toro FJ, Blanco FJ. Multilineage differentiation potential of cells isolated from the human amniotic membrane. J Cell Biochem. 2010;111:846–57.
40. Tsuji H, Miyoshi S, Ikegami Y, Hida N, Asada H, Togashi I, Suzuki J, Satake M, Nakamizo H, Tanaka M, et al. Xenografted human amniotic membrane-derived mesenchymal stem cells are immunologically tolerated and transdifferentiated into cardiomyocytes. Circ Res. 2010;106:1613–23.
41. Bilic G, Hall H, Bittermann AG, Zammeretti P, Burkhart T, Ochsenbein-Kolble N, Zimmermann R. Human preterm amnion cells cultured in 3-dimensional collagen i and fibrin matrices for tissue engineering purposes. Am J Obstet Gynecol. 2005;193:1724–32.
42. Akle CA, Adinolfi M, Welsh KI, et al. Immunogenicity of human amniotic epithelial cells after transplantation into volunteers. Lancet. 1981;2:1003–5.
43. Terada S, Matsuura K, Enosawa S, et al. Inducing proliferation of human amniotic epithelial (HAE) cells for cell therapy. Cell Transplant. 2000;9:701–4.
44. Heazlewood CF, Sherrell H, Ryan J, Atkinson K, Wells CA, Fisk NM. High incidence of contaminating maternal cell overgrowth in human placental mesenchymal stem/stromal cell cultures: a systematic review. Stem Cells Transl Med. 2014;3:1305–11.
45. Bačenková D, Rosocha J, Tóthová T, et al. Isolation and basic characterization of human term amnion and chorion mesenchymal stromal cells. Cytotherapy. 2011;13:1047–56.
46. Fariha M-MN, Chua K-H, Tan G-C, et al. Human chorion-derived stem cells: changes in stem cell properties during serial passage. Cytotherapy. 2011;13:582–93.
47. Chang C-M, Kao C-L, Chang Y-L, et al. Placenta-derived multipotent stem cells induced to differentiate into insulin-positive cells. Biochem Biophys Res Commun. 2007;357:414–20.
48. Battula VL, Bareiss PM, Treml S, et al. Human placenta and bone marrow derived MSC cultured in serum-free, b-FGF-containing medium express cell surface frizzled-9 and SSEA-4 and give rise to multilineage differentiation. Differentiation. 2007;75:279–91.
49. Miao Z, Jin J, Chen L, Zhu J, Huang W, Zhao J, Qian H, Zhang X. Isolation of mesenchymal stem cells from human placenta: comparison with human bone marrow mesenchymal stem cells. Cell Biol Int. 2006;30:681–7.
50. Weissman IL, Shizuru JA. The origins of the identification and isolation of hematopoietic stem cells, and their capability to induce donor-specific transplantation tolerance and treat autoimmune diseases. Blood. 2008;112:3543–53.
51. Chien C-C, Yen BL, Lee F-K, et al. In vitro differentiation of human placenta-derived multipotent cells into hepatocyte-like cells. Stem Cells. 2006;24:1759–68.
52. Banas A, Teratani T, Yamamoto Y, et al. IFATS collection: in vivo therapeutic potential of human adipose tissue mesenchymal stem cells after transplantation into mice with liver injury. Stem Cells. 2008;26:2705–12.
53. Vellasamy S, Sandrasaigaran P, Vidyadaran S, George E, Ramasamy R. Isolation and characterisation of mesenchymal stem cells derived from human placenta tissue. World J Stem Cells. 2012;4(6):53–61.
54. Gonzalez R, Griparic L, Umana M, Burgee K, Vargas V, Nasrallah R, Silva F, Patel A. An efficient approach to isolation and characterization of pre- and postnatal umbilical cord lining stem cells for clinical applications. Cell Transplant. 2010;19:1439–49.
55. Jaiswal N, Haynesworth SE, Caplan AI, et al. Osteogenic differentiation of purified, culture-expanded human mesenchymal stem cells in vitro. J Cell Biochem. 1997;64:295–312.
56. Wei JP, Nawata M, Wakitani S, et al. Human amniotic mesenchymal cells differentiate into chondrocytes. Cloning Stem Cells. 2009;11:19–26.
57. Suzdal'tseva YG, Burunova VV, Vakhrushev IV, Yarygin VN, Yarygin KN. Capability of human mesenchymal cells isolated from different sources to differentiation into tissues of mesodermal origin. Bull Exp Biol Med. 2007;143:114121.
58. Yen BL, Huang HI, Chien CC, Jui HY, Ko BS, Yao M, Shun CT, Yen ML, Lee MC, Chen YC. Isolation of multipotent cells from human term placenta. Stem Cells. 2005;23:3–9.

59. Korbling M, Estrov A. Adult stem cells for tissue repair: a new therapeutic concept? N Engl J Med. 2003;349:570–82.
60. Hsu YC, Chen SL, Wang DY, Chiu IM. Stem cell-based therapy in neural repair. Biomed J. 2013;36:98–105.
61. Rintoul NE, Sutton LN, Hubbard AM, et al. A new look at myelomeningoceles: functional level, vertebral level, shunting, and the implications for fetal intervention. Pediatrics. 2002;109:409–13.
62. Turner CG, Klein JD, Wang J, Thakor D, Benedict D, Ahmed A, Teng YD, Fauza DO. The amniotic fluid as a source of neural stem cells in the setting of experimental neural tube defects. Stem Cells Dev. 2013;22(4):548–53.
63. Turner CG, Pennington EC, Gray FL, Ahmed A, Teng YD, Fauza DO. Intra-amniotic delivery of amniotic-derived neural stem cells in a syngeneic model of spina bifida. Fetal Diagn Ther. 2013;34(1):38–43.
64. Pennington EC, Rialon KL, Dionigi B, Ahmed A, Zurakowski D, Fauza DO. The impact of gestational age on targeted amniotic cell profiling in experimental neural tube defects. Fetal Diagn Ther. 2015;37(1):65–9.
65. Fauza DO, Jennings RW, Teng YD, Snyder EY. Neural stem cell delivery to the spinal cord in an ovine model of fetal surgery for spina bifida. Surgery. 2008;144(3):367–73.
66. Dionigi B, Ahmed A, Brazzo 3rd J, Connors JP, Zurakowski D, Fauza DO. Partial or complete coverage of experimental spina bifida by simple intra-amniotic injection of concentrated amniotic mesenchymal stem cells. J Pediatr Surg. 2015;50(1):69–73.
67. Saadai P, Wang A, Nout YS, Downing TL, Lofberg K, Beattie MS, Bresnahan JC, Li S, Farmer DL. Human induced pluripotent stem cell-derived neural crest stem cells integrate into the injured spinal cord in the fetal lamb model of myelomeningocele. J Pediatr Surg. 2013;48(1):158–63.
68. Sieber-Blum M. Human epidermal neural crest stem cells as candidates for cell-based therapies, disease modeling, and drug discovery. Birth Defects Res C Embryo Today. 2014;102(3):221–6.
69. https://www.cirm.ca.gov/our-progress/awards/developing-placental-neural-crest-stem-cells-utero-spina-bifida-repair
70. https://www.cirm.ca.gov/our-progress/awards/placental-neural-crest-stem-cells-nanofibrous-scaffold-fetal-repair-spina-bifida
71. Zephyris (Wheeler R). 2010. Human embryogenesis TheVisualMD.com
72. Mader SS. Inquiry into life. 8th ed. New York, NY: McGraw-Hill; 1997.

Part IV
Translational Developments

Chapter 16
Fetal Stem Cell Banking

Mohammad Z. Albanna and Erik J. Woods

Introduction

Fetal stem cells derived from various discarded fetal tissues and fluids, particularly mesenchymal stem cells (MSCs) derived from umbilical cord blood and placental tissues including chorionic plate and amniotic membrane, have been shown to be a safe and effective cell source for various cell-based therapeutics including hard-to-cure-diseases such as cardiac related diseases [1–3]. In comparison to bone marrow- and adipose-derived MSCs, placental tissue-derived stem cells have demonstrated comparable immunomodulatory properties with additional advantages of expressing higher levels of human leukocyte antigen HLA-ABC, HLA-G and other cytokines including interleukins 2, 4 and 13 [4]. Particularly, MSCs derived from chorionic villi or amniotic fluid were shown to have higher proliferation capacity compared to bone marrow-derived MSCs despite extensive *in-vitro* expansion with no signs of transformation, and maintained their karyotyping [5]. Hence, cells derived from various fetal tissues and fluids are emerging as a practical alternative to bone marrow-MSCs or other adult stem cells due to their immediate availability, abundance, and potential for extensive *in-vitro* expansion for clinical applications. However, the manufacturing of a clinical grade cellular product for therapeutics including a cryopreservation protocol specific for every cell type is still lacking. In recent years, more focus has been devoted to introduce efficient isolation, characterization and large-scale expansion protocols for specific cell types that are compliant with good tissue practice (GTP) and good manufacturing practice (GMP). On the other hand, less attention has been given to the development of safe, effective, and practical long-term cell banking protocols. The pace at which these promising cells will translate to the clinic as an off-the-shelf

M.Z. Albanna, Ph.D. • E.J. Woods, Ph.D. (✉)
Cook General Biotechnology, LLC, 1102 Indiana Ave, Indianapolis, IN 46202, USA
e-mail: Mohammad.Albanna@cookgbt.com; Erik.Woods@CookMedical.com

D.O. Fauza, M. Bani (eds.), *Fetal Stem Cells in Regenerative Medicine*,
Stem Cell Biology and Regenerative Medicine,
DOI 10.1007/978-1-4939-3483-6_16

therapy is extremely contingent on the advancements to be made on cell banking that will preserve their metabolic activities until needed.

Since the development of cryoprotectants (CPAs), which are additives provided to cells before freezing to preserve their metabolic, biological and structural properties post-thaw, cell and tissue banking have seen rapid development that led to tremendous advancements of cellular therapy applications. Broadly speaking, CPAs can be divided into two classes including permeating (*e.g.* polyols such as glycerol or ethylene glycol, propylene glycol, other hydroxylic compounds, and dimethyl

Cryoprotectant (abbreviation) Chemical Formula	**Structure**	**Putative Mechanism and Notes**
Dimethyl Sulfoxide (Me$_2$SO; DMSO) C_2H_6OS		Permeable- protects through colligative properties Very permeable to virtually all cells investigated (11); most widely used CPA; can be toxic at high concentrations (7)
Ethane 1,2 diol (ethylene glycol; EG) $C_2H_6O_2$	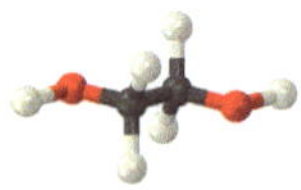	Permeable- protects through colligative properties Readily permeable to most cells (100, 113), good glass forming properties; often used in vitrification solutions in combination with other CPAs; can be hepatotoxic.
Propane 1,2 diol (propylene glycol; PG) $C_3H_8O_2$	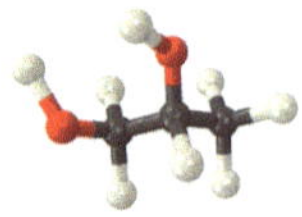	Permeable- protects through colligative properties Readily permeable to most cells; good glass forming properties; often used in vitrification solutions in combination with other CPAs (113); useful for embryo cryopreservation.
Glycerol (glycerine; Gly) $C_3H_8O_3$	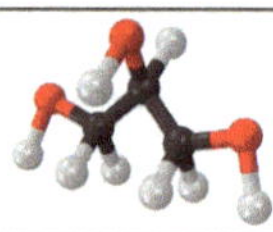	Permeable- protects through colligative properties Non-toxic; very slow cell permeation kinetics (124).
Sucrose $C_{12}H_{22}O_{11}$		Impermeable- protects through dehydration and stabilization Generally used in conjunction with permeable compounds in vitrification (14-16, 113).
D-glucose (dextrose) $C_6H_{12}O_6$	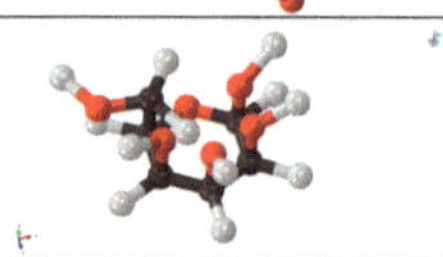	Impermeable- protects through dehydration and stabilization Generally used in conjunction with permeable compounds in vitrification.
Trehalose (mycose) $C_{12}H_{22}O_{11}$	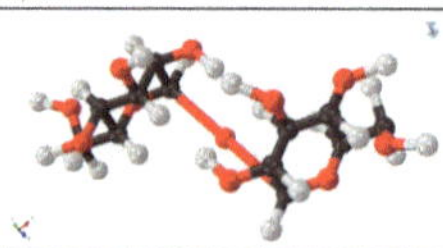	Impermeable- protects through dehydration and stabilization Demonstrated efficacy with or without additional compounds (14-16).
Hydroxyethyl Starch (HES) $C_{22}H_{44}O_{17}$	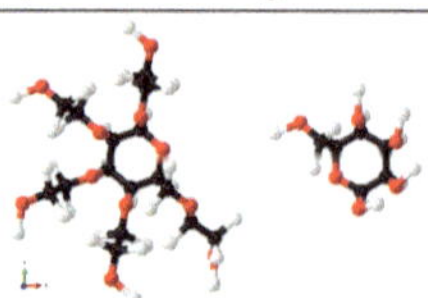	Impermeable- protects through dehydration and stabilization Demonstrated efficacy with or without additional compounds (22, 109).

Fig. 16.1 Some commonly used permeating and non-permeating cryoprotectant additives (CPAs)

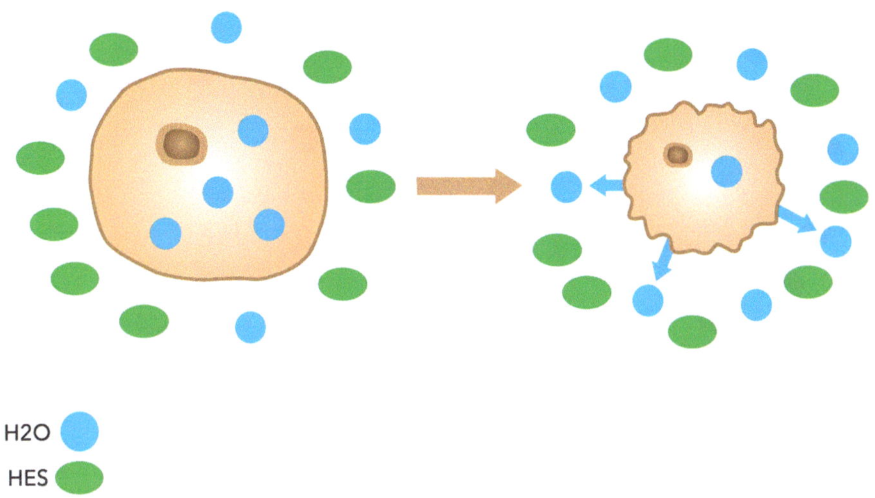

Fig. 16.2 The effects of the addition of a non-permeable cryoprotectant additive (CPA) on a cell. In the schematic, when hydroxyethyl starch (HES) is added to the extracellular solution, water moves from the intracellular to extracellular space to equalize the chemical potential gradient and the cell remains shrunken. The tolerance of a cell to shrinkage due to dehydration is generally cell type specific and cells can be damaged by this event prior to further processing or freezing

sulfoxide) and/or non-permeating (*e.g.* small carbohydrate sugars such as trehalose, sucrose and dextrose) (Fig. 16.1). Permeating CPAs are thought to protect via their colligative properties [6–8] while non-permeating CPAs are generally thought to protect by dehydration as well as membrane stabilization [9, 10]. While these CPAs are required for survival, their presence can cause direct and indirect damage as well. Some CPAs can be directly cytotoxic at the often molar concentrations required for use. Additionally, the dehydration produced by non-permeable CPAs can be damaging (Fig. 16.2) and even permeable CPAs cannot penetrate cells as efficiently as water, leading to transient osmotic events which can also be damaging (Fig. 16.3). The safety of various CPAs is one of the major emerging considerations in the field of cell therapy and regenerative medicine. While there are some inconsistent findings on the protective and toxic effects of permeable CPAs depending on cell type, their use is a common practice in most cell banking protocols.

Therapeutic applications of fetal stem cells have been considered widely for clinical applications for the treatment of several diseases as an off-the-shelf cell source. However, manufacturing of clinical grade cellular doses remains a major obstacle hindering the rapid advancement of these cells into the clinic due to the presence of animal derived products and perceived toxic CPAs such as dimethylsulfoxide (DMSO). Current cryopreservation practice often employs permeable CPAs such as DMSO supplemented to an animal or human protein-based solution. Several cryopreservation formulations have been studied and/or introduced to the market with varying results when compared to traditional cryopreservation media formulations.

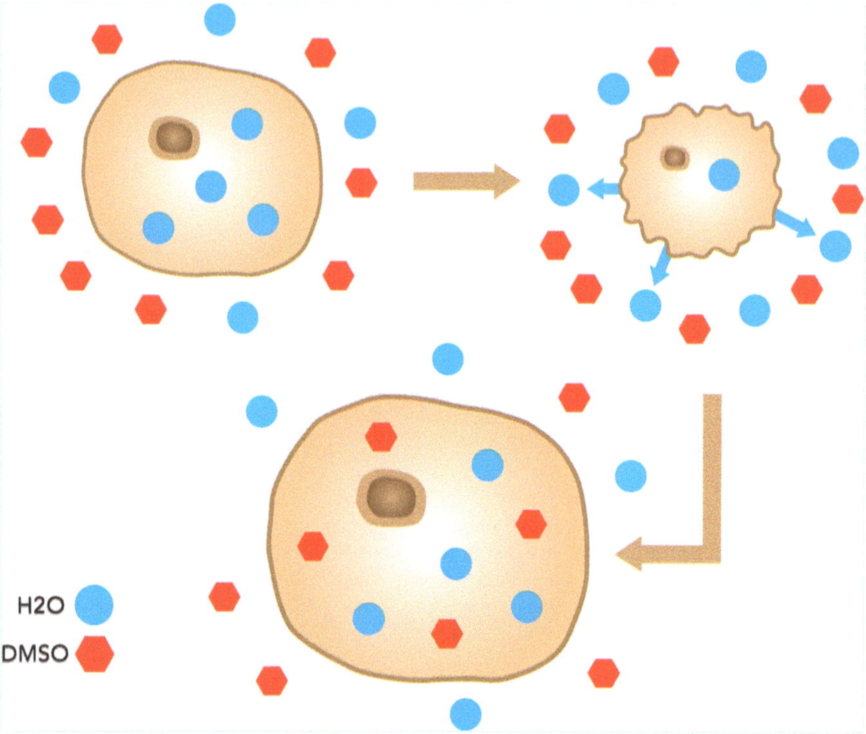

Fig. 16.3 The effects of the addition of a permeable cryoprotectant additive (CPA) on a cell. In the schematic, when dimethyl sulfoxide (DMSO) is added to the extracellular suspension, the cell initially shrinks as water can move more quickly across the cell membrane to equalize the chemical potential gradient. Over time, the permeable CPA moves into the cell (along with water) and the final volume of the cell will be slightly larger than it was previous to the addition of the CPA. When a cell is thawed and either re-suspended in CPA free media or transfused/transplanted, the opposite event occurs and the cell will transiently swell. Tolerance to such volume excursions can be damaging and are generally cell type specific

Several hundred adverse reactions have been reported from the infusion of stem cells cryopreserved in DMSO [11]. Clinical side effects of DMSO for patients infused with peripheral blood progenitor cells (PBPCs) have been linked directly to multiple complications including cardiac complications which were not seen in patients receiving allogeneic non-cryopreserved PBPCs [12]. Depletion of DMSO from cryopreserved peripheral blood-derived stem cells before administration to patients has been shown to have no negative effect on cell number of CD34+, viability, or colony forming unit capacity [13]. It actually demonstrated significant reduction in complications and side effects for patients within the first 12 h post injection including reduced rash, abdominal cramps, dysgeusia, dyspnea/cough, macrohematuria/proteinuria and cardiovascular problems. DMSO can also induce apoptotic degeneration in developing central nervous systems at different postnatal ages in mice which might have similar effect on children who undergo bone marrow transplantation [14]. It was also demonstrated that administration of DMSO to mice

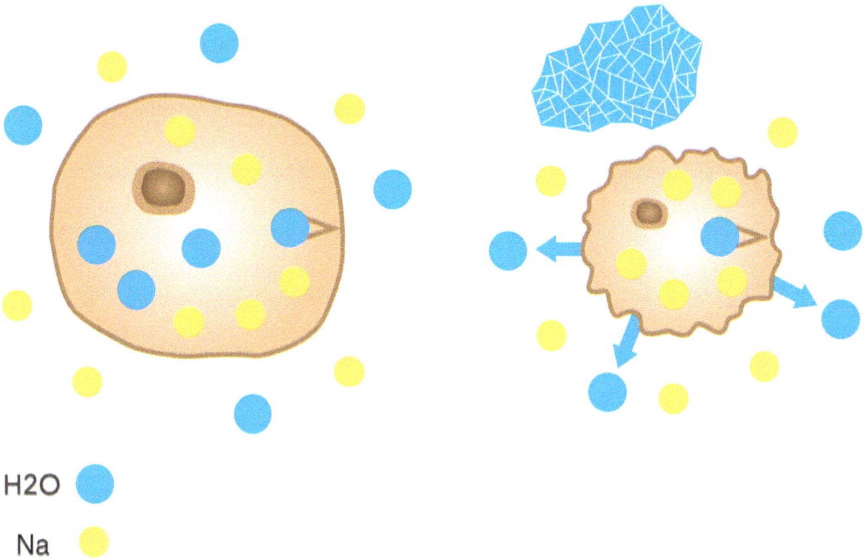

Fig. 16.4 The effects of extracellular ice on a cell during cryopreservation. As ice forms as pure water and concentrates the extracellular solutes (sodium is represented in the diagram, but all solutes are concentrated). As this occurs, water moves from the intra- to the extra-cellular space and the cell shrinks while solute concentration increases. If slow cooling continues long enough, only bound water will remain in the cell and solute concentration can become so high that damage to cell proteins can occur. Cryoprotectant additives can protect from these so called "solution effects" injury through their colligative properties

induce apoptosis in lymphoid organs such as thymus and spleen in a dose dependent manner [15]. Also, studies have shown the frequent addition and removal of DMSO can lead to osmotic injury to the cells due to the exiting and entering of water from and to the cells [16]. The damage typically associated with cells upon freezing is mainly due to the formation of ice crystals in the extracellular medium that surrounds the cells and in the intracellular compartment. Ice growth in the extracellular media concentrates solutes to levels which can be damaging ("solution effects" injury; Fig. 16.4) while intracellular ice can be directly damaging [10]. In addition, the formation of oxygen free radicals is another leading cause of reduced cell viability during or immediately post-thaw. Several studies have shown that the addition of bioantioxidants to the freezing solution has advantageous effect of cells recovery post-thaw [17]. One more reason for recued viability post-thaw is cell apoptosis due the activation of caspase-8 and caspase-9 through extrinsic and intrinsic pathways throughout the cryopreservation process, respectively [18]. Thus, the addition of caspase inhibitors can have a protective effect on cells for long term storage. So there is a pressing need to introduce a nontoxic CPA or reduce the concentration of these toxic CPAs.

Disaccharides including trehalose and sucrose have been investigated as nontoxic and non-permeable natural CPAs due to their ability to dehydrate and stabilize cells during the cryopreservation process [19–21]. Trehalose, a naturally occurring disaccharide, has found several uses in health and pharmaceutical industries in addition to food and cosmetics industries due to its safety and unique chemical and biological properties which distinguish it from other sugars [22]. It functions as an energy source or protectant during dehydration and freezing of cells. Trehalose is naturally synthesized by some cells as a response to environmental stresses like heat, cold, or oxidation [23]. Several studies have shown that the addition of small amounts of trehalose to DMSO improves the post-thaw viability of various mammalian cells and tissues [24, 25]. When compared to DMSO alone, a combination of trehalose and DMSO showed better cryoprotection properties for cord blood- and fetal liver-derived hematopoietic stem cells as evidenced by increased colony formation [26]. The capacity of other non-penetrating CPAs such as hydroxyethyl starch (HES) as a cryoprotectant to replace or decrease the concentration of DMSO in cryopreservation solutions has also been investigated by cryobiologists over the last few decades with promising preliminary results [27].

Clinical application of cryopreserved stem cell-based therapy is often hindered by three main drawbacks: reduced cell viability, decreased stemness and differentiation properties, and inclusion of animal-derived proteins in the freezing solution. Thus, most research efforts are focusing on eliminating or reducing DMSO concentration and introducing a xeno-free cryopreservation solution. The goal is to ensure high cell viability and recovery and maintenance of metabolic and biological activities post-thaw. The purpose of this chapter is to provide a brief overview on fetal stem cell types from different fetal tissues and fluids and discuss their current banking methods.

Fetal Stem Cell Types and Origins

Hematopoietic Stem Cells-Derived from Umbilical Cord Blood, Placental Blood, and Fetal Bone Marrow

Hematopoietic stem cells (HSCs) have been isolated from different sources including bone marrow, mobilized peripheral blood, umbilical cord blood, and placental blood [28]. HSCs represent CD34+ cell populations in these tissues and fluids. Cord blood is rising as an alternative rich source of HSCs to bone marrow in children and adults for the treatment of malignant and nonmalignant conditions due to the ease of access and abundance of cells. Studies that have shown the efficacy of these cells to restore hematopoiesis in patients are too numerous to list. The isolation of hematopoietic stem cells from fetal bone marrow is an invasive procedure and uncommon, therefore limited reports are available in the literature [29, 30]. Hematopoietic stem cells can be also isolated from fetal bone marrow between 20 and 24 weeks of gestation in practical numbers suitable for *in utero* or postnatal implantation [31].

Mesenchymal Stem Cells-Derived from Umbilical Cord Blood/Tissue and Fetal Bone Marrow

Mesenchymal stem cells can practically be isolated from umbilical cord Wharton's jelly postpartum and represent a primitive and potential autologous and allogeneic noncontroversial cell source for cell therapy applications [32, 33]. Tsuji *et al.* showed a total of 4×10^{15} cells per 1 g of umbilical cord tissue can be obtained from cultures 2 weeks post isolation [1]. Not only can human umbilical cord Wharton's Jelly-derived MSCs easily differentiate into chondrocytes, osteocytes, and adipocytes, but they have also demonstrated capacity for differentiation into special lineages such as nerve–like cells [34, 35], functional hepatocytes [36, 37], insulin-producing cells [38], endothelial cells [39], and steroidogenic cells [40]. In comparison with bone marrow-derived MSCs, umbilical cord Wharton's Jelly-derived MSC showed higher proliferative potential and differentiation capacity into certain lineages such as endothelial cells and pancreatic cells with higher expression of functional markers [38, 39]. Studies suggested MSCs derived from cord blood or umbilical cord Wharton's jelly have similar immunomodulatory properties to adipose-derived and bone marrow-derived MSCs and can substitute them in several allogeneic treatments including reconstruction of hematopoiesis [41]. Indeed, human umbilical cord Wharton's Jelly-derived MSCs showed comparable hematopoiesis supportive functions to bone marrow-derived MSCs making them excellent alternative for treatment of blood-related diseases [42].

Mesenchymal stem cells isolated from umbilical cord blood showed promise for various nonhematopoietic tissue regeneration and repair [43, 44]. They were successfully isolated from umbilical cord blood samples that had undergone several years of cryopreservation and demonstrated similar characteristics to bone marrow-derived MSCs [45–48] with ability to differentiate into functional-producing cells [49], neuron-like cells [50], Schwann-like cells [51], and have potential for cellular therapy of central neurological disorders [52]. Umbilical cord blood-derived MSCs were shown to increase the expansion capabilities of CD34+ cells and enhance the proliferative capacity of colony forming cells when cocultured with HSCs from cord blood [53]. While the isolation success rate of MSC from bone marrow- or adipose-derived is often 100 % and only 63 % from cord blood, however, cord blood MSCs had higher proliferation capacity and could be cultured longer to be expandable to a higher number than the other MSC types [54]. Umbilical cord blood-derived MSCs showed similar differentiation capacity to chondrogenic and osteogenic lineages compared to bone marrow- and adipose-derived MSCs but differ in the adipogenic capacity [55]. Kern *et al.* showed no adipogenic differentiation capacity of umbilical cord blood MSCs [54]. It is worth mentioning that MSCs isolated from different compartments of the umbilical cord including umbilical cord artery, vein and Wharton's jelly showed different proliferative and differentiation capacities based on the location [56].

Fetal bone marrow-derived MSCs are not a common cell source for therapeutic applications and very limited studies are available [57, 58]. MSCs isolated from human first-trimester fetal blood, liver and bone marrow expressed several pluripotent stem cell markers similar to embryonic stem cells, had longer telomeres, and could be expressed quicker and longer than their adult counterparts [59]. Moreover, MSCs isolated from adult or fetal bone marrow respond differently to transforming growth factor beta 3 ($TGF\beta_3$) and bone morphogenetic protein 2 (BMP2) signaling pathways during cartilage repair [60].

Mesenchymal/Stromal Stem Cells Derived from Amniotic Fluid

Amniotic fluid-derived mesenchymal stem cells (AFSCs) are often isolated from samples of amniotic fluid through a minimally invasive procedure called amniocentesis. AFSCs have the potential to be used as a source of autologous cells to the fetus in the future, fabricate tissue prior to birth or as an allogeneic cell source for related or unrelated individual to the fetus [61–64]. AFSCs were successfully isolated and expanded in culture for several weeks using human serum and maintained their multipotent markers and showed comparable proliferation rate to cells expanded in animal serum [65]. AFSCs seeded on biomaterials have been proposed as an autologous cell source for fetal neural tube defect through an intrauterine tissue engineering approach [66] or to augment fetal lung growth through their paracrine effects [67]. While AFSCs are widely considered multipotent with intermediary properties between embryonic and mesenchymal stem cells, these cells can attain pluripotent characteristics similar to embryonic stem cells under embryonic stem cell culturing conditions eliminating the need for viral vector reprogramming [68]. These cells share similar properties to bone marrow-derived MSCs, but also exhibit additional stem cell markers including the pluripotency marker Oct-4 and the stage-specific embryonic antigen-4 (SSEA-4), and proteins associated with proliferation and primitive phenotype which may explain the embryonic nature of these fetal cells [69]. Human amniotic fluid stem cells have been shown to potentially recover ovarian function in mice [70] and to differentiate into primordial Oocytes *in vitro* [71].

Mesenchymal Stem Cells Derived from Placental Tissue

The human placenta consists of placental villi and fetal membranes. The fetal membranes include the amniotic membrane and the chorionic membrane. These membranes provide coverage and bacterial protection for the fetus and also secrete a number of essential substances for the maintenance of the pregnancy. Placental tissues and fetal membranes are a rich source of mesenchymal stem/stromal cells [72–74]. Mesenchymal/stromal stem cells from both chorion and amnion were successfully isolated and showed characteristics of both stem cells and embryonic stem cells [75]. It is believed that the amniotic membrane retains different stem cells throughout the pregnancy as it arises from the embryonic epiblast prior to gastrulation.

Placental MSCs can be isolated from various locations within the human term placenta including placental tissue, chorionic villi, and decidua parietalis. The cells that are closer to the fetus are called fetal MSCs and the ones closer to the placenta (i.e. mother) are called maternal MSCs. Placental MSCs were successfully isolated from the fetal part of the chorionic villi or decidua parietalis and shown to secret several growth factors and cytokines that are responsible for cell migration and homing, angiogenesis, and immunomodulation, and have the capacity to differentiate into adipocytes, chondrocytes and osteocytes [76–78]. Amniotic membrane-derived MSCs are abundant and can generate nearly two million cells per gram of amnion immediately after isolation [79]. Other studies showed that a total of 20×10^{15} cells per 1 g of amniotic membrane and $\sim 3 \times 10^{15}$ cells per 1 g of chorionic plate (placenta) can be obtained 2 week post isolation [1]. MSCs derived from the amnion and chorion possess potent immunosuppressive properties evidenced by their ability to suppress lymphocytes response [80] and repair capacity for tissue damage associated with inflammatory and fibrotic degeneration [81].

Several cell populations from different particular parts of the placenta have been identified including mesenchymal, hematopoietic, trophoblastic and pluripotent stem cells [82]. MSCs derived from placenta were found to have similar stem cell markers and immune profile to bone marrow-derived MSCs [83]. While MSCs derived from the fetal part of the placenta (chorionic plate) showed a higher degree of immunomodulation than BM-MSC or ADMSCs [4], a study reported that MSCs derived from whole placental tissue are potentially less immunomodulatroy compared to bone marrow MSCs [84]. This observation might be due to the fact that MSCs isolated from whole placenta are often a mixed population of both maternal and fetal cells. Such observations warrant a need for developing standard protocols for good isolation of cells that only favor fetal cells over the maternal cells or eliminate the maternal cells completely from the cultures. Human placenta-derived MSCs were shown to be effective in suppressing the proliferation and function of umbilical cord blood T-lymphocytes [85, 86]. This may open new avenues for combinations of placental tissue-derived MSCs with hematopoietic stem cells derived from cord blood to suppress the graft-versus-host diseases in transplant patients. Zhang and colleagues showed that coculture of human placenta derived MSCs and umbilical cord blood cells caused a twofold increase of CD34+ cell proliferation and a fivefold increase in colony forming cells [87]. Lee and colleagues showed that MSC from placental tissue had higher proliferation capacity and expression of hepatogenic markers in differentiated cells compared to bone marrow-derived MSC and adipose-derived MSCs [88].

Epithelial Stem Cells Derived from Placental Tissue

Human epithelial cells are in direct contact with the amniotic fluid and provide support to the intrauterine milieu of the fetus [89]. Amniotic membrane-derived cells are believed to have multipotent characteristics similar to adult MSCs and even pluripotent characteristics similar to embryonic stem cells [90]. A good isolation protocol can generate around six million epithelial cells per gram of amnion from a term placenta [79], with a ~40–50 % reduction for preterm placenta [91].

Studies have shown that epithelial cells isolated from placental tissue showed expression of surface markers such as the stage-specific embryonic antigens (SSEA) and gene expression that is similar to embryonic stem cells [92]. Amniotic epithelial cells have the ability to differentiate into multiple derivatives including osteocytes, adipocytes, cardiomyocytes, myocytic, neuronal cells, astrocytes, pancreatic cells and hepatocytes under specific inducing protocols and are showed to be immune-privileged cells [92–94]. Amniotic epithelial cells did not form tumors 7 months post implantation in animals [94]. They have been shown to inhibit both the adaptive and innate immune system evidenced by their ability to significantly inhibit the chemotactic activities of neutrophils and macrophages and reduce the proliferation of T and B cells [95, 96]. Even more, epithelial cells derived from the amniotic membrane were shown to be reprogrammed into induced pluripotent stem cells (iPS) and were similar of human embryonic stem cells in morphology, proliferation, surface markers, gene expression [97]. Human amniotic epithelial cells were induced to differentiate into functional insulin-producing cells in a streptozotocin-induced diabetic C57 mouse model for 30 days post implantation [98] and granulosa cells and restored ovarian function [99]. While epithelial cells are poised to be a unique, easily accessible and off-the-shelf promising source for cell therapeutics, their success remains largely dependent upon on the ability of long term banking of these cells to have a ready-to-use source for clinicians when needed.

Banking of Fetal Stem Cells

Hematopoietic Stem Cells-Derived from Umbilical Cord Blood, Placental Blood, and Fetal Bone Marrow

Hematopoietic stem cells from umbilical cord blood are often cryopreserved in a 10 % v/v DMSO containing freezing solution which is based on the general practice of freezing the same cell population derived from bone marrow or peripheral blood derived stem cells. Due to concerns over increasing severity of adverse clinical events with the increase of DMSO concentration, the focus has been shifted toward reducing the concentration of DMSO or eliminating it from freezing solutions of hematopoietic stem cells. Slight reductions of DMSO concentration from 10 to 7.5 % has been shown to result in an increased number of colonies [100] and an increased reconstruction of hematopoiesis after autologous infusion as evident by a shorter recovery time of leukocytes [101]. Other protocols employ a much reduced level of DMSO at 5 % combined with 6 % hydroxyethyl starch (HES) [102]. Several studies have attempted to reduce the concentration of DMSO to cryopreserve CD34+ populations from peripheral blood progenitor cells [102–104]. These successful attempts triggered researchers to apply the same technique to HSCs from umbilical cord blood [105]. Despite that some studies have shown that hematopoietic stem cells can tolerate exposure to 10 % DMSO at 4 °C or even 37 °C incubation for up to 2 h [106, 107], cryopreservation practice in cell processing facilities tends to favor minimizing the exposure time to DMSO pre- and post-freezing. Studies have also shown that the addition of trehalose, a membrane stabilizer, or catalase, a bioantioxidant, or a combination of both to the freezing solution of hematopoietic stem cells resulted in increased colony formation at −196 and −80 °C storage temperatures compared to the conventional 10 % DMSO [26]. Another study showed that the addition of trehalose or sucrose to 2.5 % DMSO resulted in higher colony unit formation of hematopoietic stem cells from umbilical cord blood compared to 2.5 % DMSO alone [108]. The same study also showed that the addition of catalase to trehalose or sucrose at reduced concentration of 2.5 % DMSO increases the percentage of colony forming units compared to 2.5 % DMSO alone. It was also demonstrated from this study that the addition of disaccharides and catalase to 2.5 or 5 % DMSO improves cell viability. These studies confirmed that the addition of bioantioxidants is crucial to reduce the reactive oxygen species which lead to death in cryopreserved hematopoietic stem cells. Rodrigues and colleagues [20] screened 43 different cryopreservation solutions containing various combinations of DMSO, sucrose and trehalose at different concentrations to freeze HSCs derived from umbilical cord blood. They found that addition of 30 mmol/L trehalose to 2.5 % DMSO or 60 mmol/L to 5 % DMSO produces comparable results to conventional freezing solutions containing 10 % DMSO in terms of cell viability and clonogenic potential of umbilical cord blood cells 2 weeks post-thawing. Sasnoor and colleagues demonstrated that the addition of catalase and trehalose to the conventional 10 % DMSO freezing solution results in better protection, functionality and graft quality of hematopoietic stem cells from cord blood [109, 110]. Another study showed that the use of 10 % DMSO with 2 % human albumin at high cell concentration of mononuclear cells with fast addition and removal of DMSO result in high viability and recovery of hematopoietic stem cells [111].

It has also been demonstrated that cooling rate has a significant effect on HSCs derived from cord blood in terms of viability, recovery and clonogenic activities post-that. A cooling rate range between 1 and 2.5 °C/min maintains HSCs clonogenic capacity with 10 % DMSO [112]. Controlled-rate freezing has been shown to yield higher viability than non-controlled (e.g. "dump freezing") methods using isopropylalcohol (IPA) baths in static temperature chambers (e.g. −85 °C) freezers [105]. The combination of 10 % ethylene glycol (EG) and 2 % DMSO using controlled rate freezing methods resulted in improved recovery and viability of HSCs compared to the combination of 10 % DMSO and 2 % dextran-40 [105]. Different opinions remain debatable whether controlled rate freezing versus uncontrolled rate freezing, and slow versus fast cooling rates are optimal. A study showed that hematopoietic recovery of cord blood units frozen with controlled rate freezing and uncontrolled rate freezing with recommendation of using HES for volume reduction if URF is used [113]. Another animal study showed that straight freeze is comparable to controlled-rate freezing of DMSO and HES freezing solutions [114]. A study showed no difference in terms of viability and clonogenic recovery of cryopreserved, separated (MNCs), or fresh umbilical cord blood units [115]. It was also shown that the delay of cryopreservation of cord blood for a period of 3–4 days affects the viability of CD34+ and CD45+ cell populations [116] and decreases clonogenic activities [117], which may hinder their therapeutic potential after grafting. Long term freezing studies are still needed to demonstrate the safety and efficacy of reduced concentration of DMSO and disaccharide inclusion in cryopreservation solutions for umbilical blood stem cells before moving into clinical implimentation.

The cryopreservation protocols of hematopoietic stem cells derived from fetal bone marrow are limited and for the time being remain based on techniques developed for hematopoietic stem cells from cord blood or adult bone marrow.

Mesenchymal Stem Cells-Derived from Umbilical Cord Blood/Tissue and Fetal Bone Marrow

Mesenchymal stem cells isolated from umbilical cord blood are prepared and cryopreserved using the techniques that were originally developed for hematopoietic stem cells from cord blood or bone marrow. Most of the current literature on cryopreservation of umbilical cord blood-derived MSCs are derived from cord blood strategies and still employ 10 % DMSO as a CPA with or without human- or animal-based products in the cryopreservation solution for research purposes only. Wang *et al.* developed a DMSO free cryopreservation solution by including ethylene glycol (EG), 1,2-propylene glycol (PG) and sucrose as basic CPAs, supplemented with polyvinyl alcohol (PVA) as an additive [118]. Their results showed high post-thaw viability by vitrification and plunging into liquid nitrogen. Blaci and Can [119] reported a high post-that viability (>85 %) of cord blood-derived MSCs using a combination of 0.1 M sucrose and 10 % DMSO at 1 °C/min freezing rate in computer-controlled multistep slow freezing (MSSF). Vitrification on the other hand resulted in a very poor viability (<20 %). Most recently, cord blood-derived MSCs were frozen in 10 % DMSO and 90 % FBS with no adverse events noted in patients administrated with cells intravenously and/or intrathecally [120]. Optimized, cost-effective, and advanced cryopreservation techniques still remain to be developed to facilitate the clinical translation of this type of stem cell into the clinic [121–123]. Fetal bone marrow-derived MSCs are usually isolated for research purposes and often frozen based on the protocols that were originally developed for adult bone marrow MSCs. Due to the fact that MSCs derived from adult and fetal bone marrow showed different signaling pathways and biological properties during tissue regeneration, cryopreservation protocols specific for fetal bone marrow-derived MSCs should be developed to accommodate the nature of these fetal stem cells for therapeutic applications.

Mesenchymal/Stromal Stem Cells Derived from Amniotic Fluid

Conventional cryopreservation protocols employing 10 % DMSO at 1 °C/min and thawing rate >100 °C/min are currently being used to bank human amniotic fluid derived MSCs with no adverse effect of their biological properties [124]. A recent study showed an increase in post-thaw cell viability of AFSCs cryopreserved in freezing solutions containing trehalose, catalase and caspase enzyme benzyloxycarbonyl-Val-Ala-dl-Asp-fluoromethylketone (zVAD-fmk) with 5 % DMSO compared to traditional freezing solution containing 10 % DMSO and 30 % FBS [18]. Reduced concentrations of DMSO to 5 and 2.5 % produced comparable results in terms of cell growth and expression of surface markers. The same group in another study demonstrated that the addition of trehalose and catalase to 5 % DMSO supplemented with caspase enzyme zVAD-fmk showed similar post-thawing cell viability, proliferation, and apoptosis evidenced by the absence of Caspase-3, caspase-8, caspase-9 in the preserved cells after long term cryopreservation (1 year) compared to 10 % DMSO and 20 % FBS-based cryopreservation solution [125]. The enzyme zVAD-fmk was also shown to significantly enhance post-thaw survival of embryonic stem cells when added to the cryopreservation solution and post-thaw culture media [126].

While cell apoptosis can practically be induced at different stages including post isolation, pre-freezing, freezing or post thaw processing, caspase inhibitors may have an advantage if added at any of these stages or selective stages. Indeed, the incubation of cells post-thaw with pan-caspase inhibitor zVAD-fmk seemed to reduce apoptosis in transplanted cells and improved metabolic activities of the graft [127]. It was also shown that the addition of caspase inhibitor in the culture media after isolation enhanced cell yield, reduced apoptosis and contributed to insulin production of human islets cells [128]. Thus, there is a pressing research need to investigate whether the addition of an optimal concentration of pan caspase inhibitor zVAD-fmk or other caspase inhibitors at pre- and/or post-freezing stages is beneficial for maintaining metabolic and differentiation properties of fetal stem cells isolated from various sources. While several protocols have been introduced to freeze AFSCs and demonstrated ability to preserve the features of these cells including differentiation capacity, 20 % FBS containing media supplemented with either 5 or 10 % DMSO or glycerol showed workable cell viability while solutions containing sucrose or trehalose did not after 6 months of storage regardless of programmed or nonprogrammed freezing methods [129]. Other studies evaluating different combinations of FBS and DMSO have indicated 40 % FBS and 10 % DMSO provide the highest cell viability and maintenance of multipotencey of these cells 3 months post cryopreservation. Steigman *et al.* proposed a 3-stage isolation, expansion, and cryopreservation protocol for amniotic fluid-MSCs. In their cryopreservation protocol, they used 10 % DMSO supplemented with 2.5 % human serum albumin and maintained above 70 % post-thaw cell viability [130].

Mesenchymal Stem Cells and Epithelial Stem Cells Derived from Placental Tissue

Human amniotic fluid and serum have been used previously to cryopreserve embryos [131, 132]. Recently, amniotic fluid has been investigated as a potential source to cryoprotect amniotic epithelial cells and to replace FBS [133]. A combination of basal media, amniotic fluid and 10 % DMSO showed comparable post thaw viability to DMSO and FBS supplemented media. When DMSO was substituted with 50 % glycerol, viability was comparable to glycerol and FBS supplemented media, but 20 % less than DMSO-based freezing solution. Interestingly, when DMSO, glycerol, basal media and amniotic fluid were all combined, viability was comparable to glycerol based freezing solutions.

Murphy *et al.* compared a commonly used serum-based freezing solution supplemented with 10 % DMSO, Cryostor CS-5, a xeno-free cryoprotectant media containing 5 % DMSO, and Synth-a-freeze, a defined protein free, HEPES and bicarbonate buffered-based freezing solution containing 10 % DMSO [134]. They found comparable post-thaw viability between the three cryopreservation solutions. Cryostor was the preferred cryopreservation media for cell recovery and maintenance based on significantly higher post-thaw metabolic activity than the other two cryopreservation solutions.

MSCs derived from placental tissue are often frozen for short term uses in DMSO in the presence of human or animal serum for research purposes. A long-term evaluation of DMSO-free or at reduced concentrations on viability and post-thaw performance remains to be seen with these cells [122].

Commercialization and Regulation of Fetal Stem Cell Banking

Despite the few promising clinical outcomes of fetal stem cells interventions, many private entities started to offer services to bank fetal stem cells from newborns for future use by the donor or close relatives. In the United States private stem cell banks must comply with the food and drug administration (FDA) requirements such as establishment registration and listing, donor screening and testing for infectious diseases, reporting and labeling requirements, and compliance with good tissue manufacturing practice regulations. The FDA currently regulates the use of cord blood-derived stem cells without prior approval if used for the donor or first- or second-degree relatives only for the treatment of hematologic related diseases, and some inherited metabolic and immune system disorders. Other sources of stem cells are currently not approved by the FDA due to the lack of substantial clinical evidence of efficacy. The use of allogeneic stem cells often requires biological license application (BLA) under Section 351 of the Public Health Service Act or investigational of new drug under the Food, Drug and Cosmetic Act. Among the many private accrediting organizations are the American Association of Blood Banks (AABB) and the American Association of tissue banks (AATB) and the Foundation for the

Accreditation of Cell Therapy (FACT), all of which are international and non-profit organizations that work to develop and implement robust standards, accreditation and educational programs to govern all aspects of business operations for the individuals and institutions involved in stem cell banking.

The commercial side of stem cell banking has seen rapid increase in the last decade due to the increased knowledge of the possible treatments using stem cells. According to Bioinformant WorldWide, a market research company from Roseville, Minnesota, there were only 23 cord blood banks worldwide in 2006. As of today, there are nearly 500 banks worldwide from 97 countries that offer cord blood banking services. BCC Research, a research market company in Wellesley, Massachusetts, has estimated a market of $4.5 billion in 2010 for private banks based on fees paid by parents for banking cord blood stem cells. While stem cells have been successfully used to treat hematologic diseases, the thought that stem cells have great promise to cure other hard to treat diseases is hyped based on few clinical trials [135].

The challenges that face global regulators faced with new therapies have caused slow growth of cell therapy trials. Some countries have responded by developing legislation for faster approvals which has also lead to some reports of medical tourism. Patients who seek stem cell intervention for various diseases including multiple sclerosis and cerebral palsy often travel to China, Germany and Mexico with an average treatment cost of $21,500 excluding travel expenses [136].

Summary and Conclusions

Fetal stem cells are emerging as an alternative to autologous or allogeneic cell sources for many cellular therapeutic applications. The current freezing protocols for fetal stem cells are adopted from adult stem cells or hematopoietic stem cells and still use high concentrations of CPAs which can be problematic. The interval between the cell collection and clinical use may extend from days to weeks or even years. Thus, a good manufacturing freezing protocol should be in place to assure high cell recovery of frozen cells with full metabolic capacity to perform the intended function upon transplantation or infusion. Several cryopreservation formulations are currently under evaluation and have shown promise. Further animal and human studies are needed to confirm their safety and efficacy in fetal stem cell banking.

References

1. Tsuji H, Miyoshi S, Ikegami Y, Hida N, Asada H, et al. Xenografted human amniotic membrane-derived mesenchymal stem cells are immunologically tolerated and transdifferentiated into cardiomyocytes. Circ Res. 2010;106(10):1613–23.
2. Nishiyama N, Miyoshi S, Hida N, Uyama T, Okamoto K, et al. The significant cardiomyogenic potential of human umbilical cord blood-derived mesenchymal stem cells in vitro. Stem Cells. 2007;25(8):2017–24.

3. Okamoto K, Miyoshi S, Toyoda M, Hida N, Ikegami Y, et al. 'Working' cardiomyocytes exhibiting plateau action potentials from human placenta-derived extraembryonic mesodermal cells. Exp Cell Res. 2007;313(12):2550–62.
4. Lee JM, Jung J, Lee HJ, Jeong SJ, Cho KJ, et al. Comparison of immunomodulatory effects of placenta mesenchymal stem cells with bone marrow and adipose mesenchymal stem cells. Int Immunopharmacol. 2012;13(2):219–24.
5. Poloni A, Maurizi G, Babini L, Serrani F, Berardinelli E, et al. Human mesenchymal stem cells from chorionic villi and amniotic fluid are not susceptible to transformation after extensive in vitro expansion. Cell Transplant. 2011;20(5):643–54.
6. Lovelock JE. The mechanism of the protective action of glycerol against haemolysis by freezing and thawing. Biochim Biophys Acta. 1953;11(1):28–36.
7. Woods EJ, Liu J, Zieger MA, Lakey JR, Critser JK. Water and cryoprotectant permeability characteristics of isolated human and canine pancreatic islets. Cell Transplant. 1999;8(5):549–59.
8. Woods EJ, Liu J, Gilmore JA, Reid TJ, Gao DY, et al. Determination of human platelet membrane permeability coefficients using the Kedem-Katchalsky formalism: estimates from two- vs three-parameter fits. Cryobiology. 1999;38(3):200–8.
9. Woods EJ, Benson JD, Agca Y, Critser JK. Fundamental cryobiology of reproductive cells and tissues. Cryobiology. 2004;48(2):146–56.
10. Eroglu A, Russo MJ, Bieganski R, Fowler A, Cheley S, et al. Intracellular trehalose improves the survival of cryopreserved mammalian cells. Nat Biotechnol. 2000;18(2):163–7.
11. Cox MA, Kastrup J, Hrubiško M. Historical perspectives and the future of adverse reactions associated with haemopoietic stem cells cryopreserved with dimethyl sulfoxide. Cell Tissue Bank. 2012;13(2):203–15.
12. Donmez A, Tombuloglu M, Gungor A, Soyer N, Saydam G, et al. Clinical side effects during peripheral blood progenitor cell infusion. Transfus Apher Sci. 2007;36(1):95–101.
13. Syme R, Bewick M, Stewart D, Porter K, Chadderton T, et al. The role of depletion of dimethyl sulfoxide before autografting: on hematologic recovery, side effects, and toxicity. Biol Blood Marrow Transplant. 2004;10(2):135–41.
14. Hanslick JL, Lau K, Noguchi KK, Olney JW, Zorumski CF, et al. Dimethyl sulfoxide (DMSO) produces widespread apoptosis in the developing central nervous system. Neurobiol Dis. 2009;34(1):1–10.
15. Aita K, Irie H, Tanuma Y, Toida S, Okuma Y, et al. Apoptosis in murine lymphoid organs following intraperitoneal administration of dimethyl sulfoxide (DMSO). Exp Mol Pathol. 2005;79(3):265–71.
16. Woods EJ, Pollok KE, Byers MA, Perry BC, Purtteman J, et al. Cord blood stem cell cryopreservation. Transfus Med Hemother. 2007;34(4):276–85.
17. Limaye LS. Bone marrow cryopreservation: improved recovery due to bioantioxidant additives in the freezing solution. Stem Cells. 1997;15(5):353–8.
18. Seo JM, Sohn MY, Suh JS, Atala A, Yoo JJ, et al. Cryopreservation of amniotic fluid-derived stem cells using natural cryoprotectants and low concentrations of dimethylsulfoxide. Cryobiology. 2011;62(3):167–73.
19. Buchanan SS, Gross SA, Acker JP, Toner M, Carpenter JF, et al. Cryopreservation of stem cells using trehalose: evaluation of the method using a human hematopoietic cell line. Stem Cells Dev. 2004;13(3):295–305.
20. Rodrigues JP, Paraguassú-Braga FH, Carvalho L, Abdelhay E, Bouzas LF, et al. Evaluation of trehalose and sucrose as cryoprotectants for hematopoietic stem cells of umbilical cord blood. Cryobiology. 2008;56(2):144–51.
21. Leslie SB, Israeli E, Lighthart B, Crowe JH, Crowe LM. Trehalose and sucrose protect both membranes and proteins in intact bacteria during drying. Appl Environ Microbiol. 1995;61(10):3592–7.
22. Richards AB, Krakowka S, Dexter LB, Schmid H, Wolterbeek APM, et al. Trehalose: a review of properties, history of use and human tolerance, and results of multiple safety studies. Food Chem Toxicol. 2002;40(7):871–98.

23. Jain NK, Roy I. Effect of trehalose on protein structure. Protein Sci. 2009;18(1):24–36.
24. Erdag G, Eroglu A, Morgan J, Toner M. Cryopreservation of fetal skin is improved by extracellular trehalose. Cryobiology. 2002;44(3):218–28.
25. Beattie GM, Crowe JH, Lopez AD, Cirulli V, Ricordi C, et al. Trehalose: a cryoprotectant that enhances recovery and preserves function of human pancreatic islets after long-term storage. Diabetes. 1997;46(3):519–23.
26. Limaye LS, Kale VP. Cryopreservation of human hematopoietic cells with membrane stabilizers and bioantioxidants as additives in the conventional freezing medium. J Hematother Stem Cell Res. 2001;10(5):709–18.
27. Stolzing A, Naaldijk Y, Fedorova V, Sethe S. Hydroxyethylstarch in cryopreservation - mechanisms, benefits and problems. Transfus Apher Sci. 2012;46(2):137–47.
28. Kurtzberg J, Laughlin M, Graham ML, Smith C, Olson JF, et al. Placental blood as a source of hematopoietic stem cells for transplantation into unrelated recipients. N Engl J Med. 1996;335(3):157–66.
29. Jones DR, Bui TH, Anderson EM, Ek S, Liu D, et al. In utero haematopoietic stem cell transplantation: current perspectives and future potential. Bone Marrow Transplant. 1996;18(5):831–7.
30. Bui TH, Jones DR. Stem cell transplantation into the fetal recipient: challenges and prospects. Curr Opin Obstet Gynecol. 1998;10(2):105–8.
31. Golfier F, Barcena A, Harrison MR, Muench MO. Fetal bone marrow as a source of stem cells for in utero or postnatal transplantation. Br J Haematol. 2000;109(1):173–81.
32. Nagamura-Inoue T, He H. Umbilical cord-derived mesenchymal stem cells: their advantages and potential clinical utility. World J Stem Cells. 2014;6(2):195–202.
33. Bongso A, Fong CY. The therapeutic potential, challenges and future clinical directions of stem cells from the Wharton's jelly of the human umbilical cord. Stem Cell Rev. 2013;9(2):226–40.
34. Ma L, Feng XY, Cui BL, Law F, Jiang XW, et al. Human umbilical cord Wharton's Jelly-derived mesenchymal stem cells differentiation into nerve-like cells. Chin Med J (Engl). 2005;118(23):1987–93.
35. Mitchell KE, Weiss ML, Mitchell BM, Martin P, Davis D, et al. Matrix cells from Wharton's jelly form neurons and glia. Stem Cells. 2003;21(1):50–60.
36. Zhang YN, Lie PC, Wei X. Differentiation of mesenchymal stromal cells derived from umbilical cord Wharton's jelly into hepatocyte-like cells. Cytotherapy. 2009;11(5):548–58.
37. Zhao Q, Ren H, Li X, Chen Z, Zhang X, et al. Differentiation of human umbilical cord mesenchymal stromal cells into low immunogenic hepatocyte-like cells. Cytotherapy. 2009;11(4):414–26.
38. Wu LF, Wang NN, Liu YS, Wei X. Differentiation of Wharton's jelly primitive stromal cells into insulin-producing cells in comparison with bone marrow mesenchymal stem cells. Tissue Eng Part A. 2009;15(10):2865–73.
39. Chen MY, Lie PC, Li ZL, Wei X. Endothelial differentiation of Wharton's jelly-derived mesenchymal stem cells in comparison with bone marrow-derived mesenchymal stem cells. Exp Hematol. 2009;37(5):629–40.
40. Wei X, Peng G, Zheng S, Wu X. Differentiation of umbilical cord mesenchymal stem cells into steroidogenic cells in comparison to bone marrow mesenchymal stem cells. Cell Prolif. 2012;45(2):101–10.
41. Yoo KH, Jang IK, Lee MW, Kim HE, Yang MS, et al. Comparison of immunomodulatory properties of mesenchymal stem cells derived from adult human tissues. Cell Immunol. 2009;259(2):150–6.
42. Lu LL, Liu YJ, Yang SG, Zhao QJ, Wang X, et al. Isolation and characterization of human umbilical cord mesenchymal stem cells with hematopoiesis-supportive function and other potentials. Haematologica. 2006;91(8):1017–26.
43. van de Ven C, Collins D, Bradley MB, Morris E, Cairo MS. The potential of umbilical cord blood multipotent stem cells for nonhematopoietic tissue and cell regeneration. Exp Hematol. 2007;35(12):1753–65.

44. Korbling M, Robinson S, Estrov Z, Champlin R, Shpall E. Umbilical cord blood-derived cells for tissue repair. Cytotherapy. 2005;7(3):258–61.
45. Lee MW, Choi J, Yang MS, Moon YJ, Park JS, et al. Mesenchymal stem cells from cryopreserved human umbilical cord blood. Biochem Biophys Res Commun. 2004;320(1):273–8.
46. Lee MW, Yang MS, Park JS, Kim HC, Kim YJ, et al. Isolation of mesenchymal stem cells from cryopreserved human umbilical cord blood. Int J Hematol. 2005;81(2):126–30.
47. Barachini S, Trombi L, Danti S, D'Alessandro D, Battolla B, et al. Morpho-functional characterization of human mesenchymal stem cells from umbilical cord blood for potential uses in regenerative medicine. Stem Cells Dev. 2009;18(2):293–305.
48. Kedong S, Xiubo F, Tianqing L, Macedo HM, LiLi J, et al. Simultaneous expansion and harvest of hematopoietic stem cells and mesenchymal stem cells derived from umbilical cord blood. J Mater Sci Mater Med. 2010;21(12):3183–93.
49. Phuc PV, Nhung TH, Loan DT, Chung DC, Ngoc PK. Differentiating of banked human umbilical cord blood-derived mesenchymal stem cells into insulin-secreting cells. In Vitro Cell Dev Biol Anim. 2011;47(1):54–63.
50. Kang XQ, Zang WJ, Bao LJ, Li DL, Xu XL, et al. Differentiating characterization of human umbilical cord blood-derived mesenchymal stem cells in vitro. Cell Biol Int. 2006;30(7):569–75.
51. Zhang HT, Cheng HY, Zhang L, Fan J, Chen YZ, et al. Umbilical cord blood cell-derived neurospheres differentiate into Schwann-like cells. Neuroreport. 2009;20(4):354–9.
52. Luo YC, Zhang HT, Cheng HY, Yang ZJ, Dai YW, et al. Differentiation of cryopreserved human umbilical cord blood-derived stromal cells into cells with an oligodendrocyte phenotype. In Vitro Cell Dev Biol Anim. 2010;46(7):585–9.
53. Huang GP, Pan ZJ, Jia BB, Zheng Q, Xie CG, et al. Ex vivo expansion and transplantation of hematopoietic stem/progenitor cells supported by mesenchymal stem cells from human umbilical cord blood. Cell Transplant. 2007;16(6):579–85.
54. Kern S, Eichler H, Stoeve J, Kluter H, Bieback K. Comparative analysis of mesenchymal stem cells from bone marrow, umbilical cord blood, or adipose tissue. Stem Cells. 2006;24(5):1294–301.
55. Rebelatto CK, Aguiar AM, Moretao MP, Senegaglia AC, Hansen P, et al. Dissimilar differentiation of mesenchymal stem cells from bone marrow, umbilical cord blood, and adipose tissue. Exp Biol Med (Maywood). 2008;233(7):901–13.
56. Ishige I, Nagamura-Inoue T, Honda MJ, Harnprasopwat R, Kido M, et al. Comparison of mesenchymal stem cells derived from arterial, venous, and Wharton's jelly explants of human umbilical cord. Int J Hematol. 2009;90(2):261–9.
57. Campagnoli C, Roberts IA, Kumar S, Bennett PR, Bellantuono I, et al. Identification of mesenchymal stem/progenitor cells in human first-trimester fetal blood, liver, and bone marrow. Blood. 2001;98(8):2396–402.
58. In 't Anker PS, Noort WA, Scherjon SA, Kleijburg-van C, der Keur C, Kruisselbrink AB, et al. Mesenchymal stem cells in human second-trimester bone marrow, liver, lung, and spleen exhibit a similar immunophenotype but a heterogeneous multilineage differentiation potential. Haematologica. 2003;88(8):845–52.
59. Guillot PV, Gotherstrom C, Chan J, Kurata H, Fisk NM. Human first-trimester fetal MSC express pluripotency markers and grow faster and have longer telomeres than adult MSC. Stem Cells. 2007;25(3):646–54.
60. Brady K, Dickinson SC, Guillot PV, Polak J, Blom AW, et al. Human fetal and adult bone marrow-derived mesenchymal stem cells use different signaling pathways for the initiation of chondrogenesis. Stem Cells Dev. 2014;23(5):541–54.
61. Weber B, Zeisberger SM, Hoerstrup SP. Prenatally harvested cells for cardiovascular tissue engineering: fabrication of autologous implants prior to birth. Placenta. 2011;32(4):14.
62. Delo DM, De Coppi P, Bartsch Jr G, Atala A. Amniotic fluid and placental stem cells. Methods Enzymol. 2006;419:426–38.
63. Kaviani A, Guleserian K, Perry TE, Jennings RW, Ziegler MM, et al. Fetal tissue engineering from amniotic fluid. J Am Coll Surg. 2003;196(4):592–7.

64. Joo S, Ko IK, Atala A, Yoo JJ, Lee SJ. Amniotic fluid-derived stem cells in regenerative medicine research. Arch Pharm Res. 2012;35(2):271–80.
65. Kunisaki SM, Armant M, Kao GS, Stevenson K, Kim H, et al. Tissue engineering from human mesenchymal amniocytes: a prelude to clinical trials. J Pediatr Surg. 2007;42(6):974–9.
66. Hosper NA, Bank RA, van den Berg PP. Human Amniotic Fluid-Derived Mesenchymal Cells from Fetuses with a Neural Tube Defect Do Not Deposit Collagen Type I Protein After TGF-beta1 Stimulation In Vitro. Stem Cells Dev. 2014;23(5):555–62.
67. Di Bernardo J, Maiden MM, Hershenson MB, Kunisaki SM. Amniotic fluid derived mesenchymal stromal cells augment fetal lung growth in a nitrofen explant model. J Pediatr Surg. 2014;49(6):859–64.
68. Moschidou D, Mukherjee S, Blundell MP, Jones GN, Atala AJ, et al. Human mid-trimester amniotic fluid stem cells cultured under embryonic stem cell conditions with valproic acid acquire pluripotent characteristics. Stem Cells Dev. 2013;22(3):444–58.
69. Roubelakis MG, Pappa KI, Bitsika V, Zagoura D, Vlahou A, et al. Molecular and proteomic characterization of human mesenchymal stem cells derived from amniotic fluid: comparison to bone marrow mesenchymal stem cells. Stem Cells Dev. 2007;16(6):931–52.
70. Lai D, Wang F, Chen Y, Wang L, Wang Y, et al. Human amniotic fluid stem cells have a potential to recover ovarian function in mice with chemotherapy-induced sterility. BMC Dev Biol. 2013;13:34.
71. Yu X, Wang N, Qiang R, Wan Q, Qin M, et al. Human amniotic fluid stem cells possess the potential to differentiate into primordial follicle oocytes in vitro. Biol Reprod. 2014;90:73.
72. Kmiecik G, Niklinska W, Kuc P, Pancewicz-Wojtkiewicz J, Fil D, et al. Fetal membranes as a source of stem cells. Adv Med Sci. 2013;58(2):185–95.
73. Ilancheran S, Moodley Y, Manuelpillai U. Human fetal membranes: a source of stem cells for tissue regeneration and repair? Placenta. 2009;30(1):2–10.
74. Fauza D. Amniotic fluid and placental stem cells. Best Pract Res Clin Obstet Gynaecol. 2004;18(6):877–91.
75. Bacenkova D, Rosocha J, Tothova T, Rosocha L, Sarissky M. Isolation and basic characterization of human term amnion and chorion mesenchymal stromal cells. Cytotherapy. 2011;13(9):1047–56.
76. Abumaree MH, Al Jumah MA, Kalionis B, Jawdat D, Al Khaldi A, et al. Phenotypic and functional characterization of mesenchymal stem cells from chorionic villi of human term placenta. Stem Cell Rev. 2013;9(1):16–31.
77. Castrechini NM, Murthi P, Gude NM, Erwich JJ, Gronthos S, et al. Mesenchymal stem cells in human placental chorionic villi reside in a vascular Niche. Placenta. 2010;31(3):203–12.
78. Castrechini NM, Murthi P, Qin S, Kusuma GD, Wilton L, et al. Decidua parietalis-derived mesenchymal stromal cells reside in a vascular niche within the choriodecidua. Reprod Sci. 2012;19(12):1302–14.
79. Bilic G, Zeisberger SM, Mallik AS, Zimmermann R, Zisch AH. Comparative characterization of cultured human term amnion epithelial and mesenchymal stromal cells for application in cell therapy. Cell Transplant. 2008;17(8):955–68.
80. Bailo M, Soncini M, Vertua E, Signoroni PB, Sanzone S, et al. Engraftment potential of human amnion and chorion cells derived from term placenta. Transplantation. 2004;78(10):1439–48.
81. Cargnoni A, Gibelli L, Tosini A, Signoroni PB, Nassuato C, et al. Transplantation of allogeneic and xenogeneic placenta-derived cells reduces bleomycin-induced lung fibrosis. Cell Transplant. 2009;18(4):405–22.
82. Miki T, Mitamura K, Ross MA, Stolz DB, Strom SC. Identification of stem cell marker-positive cells by immunofluorescence in term human amnion. J Reprod Immunol. 2007;75(2):91–6.
83. Zhang Y, Li CD, Jiang XX, Li HL, Tang PH, et al. Comparison of mesenchymal stem cells from human placenta and bone marrow. Chin Med J (Engl). 2004;117(6):882–7.

84. Fazekasova H, Lechler R, Langford K, Lombardi G. Placenta-derived MSCs are partially immunogenic and less immunomodulatory than bone marrow-derived MSCs. J Tissue Eng Regen Med. 2011;5(9):684–94.
85. Li C, Zhang W, Jiang X, Mao N. Human-placenta-derived mesenchymal stem cells inhibit proliferation and function of allogeneic immune cells. Cell Tissue Res. 2007; 330(3):437–46.
86. Li CD, Zhang WY, Li HL, Jiang XX, Zhang Y, et al. Mesenchymal stem cells derived from human placenta suppress allogeneic umbilical cord blood lymphocyte proliferation. Cell Res. 2005;15(7):539–47.
87. Zhang Y, Li C, Jiang X, Zhang S, Wu Y, et al. Human placenta-derived mesenchymal progenitor cells support culture expansion of long-term culture-initiating cells from cord blood CD34+ cells. Exp Hematol. 2004;32(7):657–64.
88. Lee HJ, Jung J, Cho KJ, Lee CK, Hwang SG, et al. Comparison of in vitro hepatogenic differentiation potential between various placenta-derived stem cells and other adult stem cells as an alternative source of functional hepatocytes. Differentiation. 2012;84(3):223–31.
89. Iwasaki R, Matsubara S, Takizawa T, Takayama T, Yashiro T, et al. Human amniotic epithelial cells are morphologically homogeneous: enzyme histochemical, tracer, and freeze-substitution fixation study. Eur J Histochem. 2003;47(3):223–32.
90. Diaz-Prado S, Muinos-Lopez E, Hermida-Gomez T, Rendal-Vazquez ME, Fuentes-Boquete I, et al. Multilineage differentiation potential of cells isolated from the human amniotic membrane. J Cell Biochem. 2010;111(4):846–57.
91. Ochsenbein-Kolble N, Bilic G, Hall H, Huch R, Zimmermann R. Inducing proliferation of human amnion epithelial and mesenchymal cells for prospective engineering of membrane repair. J Perinat Med. 2003;31(4):287–94.
92. Miki T, Marongiu F, Ellis EC, Dorko K, Mitamura K, et al. Production of hepatocyte-like cells from human amnion. Methods Mol Biol. 2009;481:155–68.
93. Ilancheran S, Michalska A, Peh G, Wallace EM, Pera M, et al. Stem cells derived from human fetal membranes display multilineage differentiation potential. Biol Reprod. 2007;77(3):577–88.
94. Miki T, Lehmann T, Cai H, Stolz DB, Strom SC. Stem cell characteristics of amniotic epithelial cells. Stem Cells. 2005;23(10):1549–59.
95. Pozzobon M, Piccoli M, De Coppi P. Stem cells from fetal membranes and amniotic fluid: markers for cell isolation and therapy. Cell Tissue Bank. 2014;15(2):199–211.
96. Li H, Niederkorn JY, Neelam S, Mayhew E, Word RA, et al. Immunosuppressive factors secreted by human amniotic epithelial cells. Invest Ophthalmol Vis Sci. 2005;46(3):900–7.
97. Zhao H-X, Li Y, Jin H-F, Xie L, Liu C, et al. Rapid and efficient reprogramming of human amnion-derived cells into pluripotency by three factors OCT4/SOX2/NANOG. Differentiation. 2010;80(2–3):123–9.
98. Hou Y, Huang Q, Liu T, Guo L. Human amnion epithelial cells can be induced to differentiate into functional insulin-producing cells. Acta Biochim Biophys Sin (Shanghai). 2008;40(9):830–9.
99. Wang F, Wang L, Yao X, Lai D, Guo L. Human amniotic epithelial cells can differentiate into granulosa cells and restore folliculogenesis in a mouse model of chemotherapy-induced premature ovarian failure. Stem Cell Res Ther. 2013;4(5):124.
100. Smagur A, Mitrus I, Giebel S, Sadus-Wojciechowska M, Najda J, et al. Impact of different dimethyl sulphoxide concentrations on cell recovery, viability and clonogenic potential of cryopreserved peripheral blood hematopoietic stem and progenitor cells. Vox Sang. 2013;104(3):240–7.
101. Mitrus I, Smagur A, Giebel S, Gliwinska J, Prokop M, et al. A faster reconstitution of hematopoiesis after autologous transplantation of hematopoietic cells cryopreserved in 7.5 % dimethyl sulfoxide if compared to 10 % dimethyl sulfoxide containing medium. Cryobiology. 2013;67(3):327–31.
102. Abrahamsen JF, Bakken AM, Bruserud O. Cryopreserving human peripheral blood progenitor cells with 5-percent rather than 10-percent DMSO results in less apoptosis and necrosis in CD34+ cells. Transfusion. 2002;42(12):1573–80.

103. Galmes A, Besalduch J, Bargay J, Novo A, Morey M, et al. Long-term storage at -80 degrees C of hematopoietic progenitor cells with 5-percent dimethyl sulfoxide as the sole cryoprotectant. Transfusion. 1999;39(1):70–3.
104. Halle P, Tournilhac O, Knopinska-Posluszny W, Kanold J, Gembara P, et al. Uncontrolled-rate freezing and storage at -80 degrees C, with only 3.5-percent DMSO in cryoprotective solution for 109 autologous peripheral blood progenitor cell transplantations. Transfusion. 2001;41(5):667–73.
105. Son JH, Heo YJ, Park MY, Kim HH, Lee KS. Optimization of cryopreservation condition for hematopoietic stem cells from umbilical cord blood. Cryobiology. 2010;60(3):287–92.
106. Rowley SD, Anderson GL. Effect of DMSO exposure without cryopreservation on hematopoietic progenitor cells. Bone Marrow Transplant. 1993;11(5):389–93.
107. Branch DR, Calderwood S, Cecutti MA, Herst R, Solh H. Hematopoietic progenitor cells are resistant to dimethyl sulfoxide toxicity. Transfusion. 1994;34(10):887–90.
108. Motta JPR, Gomes BE, Bouzas LF, Paraguassú-Braga FH, Porto LC. Evaluations of bioantioxidants in cryopreservation of umbilical cord blood using natural cryoprotectants and low concentrations of dimethylsulfoxide. Cryobiology. 2010;60(3):301–7.
109. Sasnoor LM, Kale VP, Limaye LS. Supplementation of conventional freezing medium with a combination of catalase and trehalose results in better protection of surface molecules and functionality of hematopoietic cells. J Hematother Stem Cell Res. 2003;12(5):553–64.
110. Sasnoor LM, Kale VP, Limaye LS. A combination of catalase and trehalose as additives to conventional freezing medium results in improved cryoprotection of human hematopoietic cells with reference to in vitro migration and adhesion properties. Transfusion. 2005;45(4):622–33.
111. Meyer TP, Hofmann B, Zaisserer J, Jacobs VR, Fuchs B, et al. Analysis and cryopreservation of hematopoietic stem and progenitor cells from umbilical cord blood. Cytotherapy. 2006;8(3):265–76.
112. Hunt CJ, Armitage SE, Pegg DE. Cryopreservation of umbilical cord blood: 2. Tolerance of CD34+ cells to multimolar dimethyl sulphoxide and the effect of cooling rate on recovery after freezing and thawing. Cryobiology. 2003;46(1):76–87.
113. Solves P, Mirabet V, Planelles D, Carbonell-Uberos F, Roig R. Influence of volume reduction and cryopreservation methodologies on quality of thawed umbilical cord blood units for transplantation. Cryobiology. 2008;56(2):152–8.
114. Naaldijk Y, Staude M, Fedorova V, Stolzing A. Effect of different freezing rates during cryopreservation of rat mesenchymal stem cells using combinations of hydroxyethyl starch and dimethylsulfoxide. BMC Biotechnol. 2012;12:49.
115. Almici C, Carlo-Stella C, Wagner JE, Mangoni L, Garau D, et al. Clonogenic capacity and ex vivo expansion potential of umbilical cord blood progenitor cells are not impaired by cryopreservation. Bone Marrow Transplant. 1997;19(11):1079–84.
116. Guttridge MG, Soh TG, Belfield H, Sidders C, Watt SM. Storage time affects umbilical cord blood viability. Transfusion. 2014;54:1278.
117. Pereira-Cunha FG, Duarte AS, Costa FF, Saad ST, Lorand-Metze I, et al. Viability of umbilical cord blood mononuclear cell subsets until 96 hours after collection. Transfusion. 2013;53(9):2034–42.
118. Wang HY, Lun ZR, Lu SS. Cryopreservation of umbilical cord blood-derived mesenchymal stem cells without dimethyl sulfoxide. Cryo Letters. 2011;32(1):81–8.
119. Balci D, Can A. The assessment of cryopreservation conditions for human umbilical cord stroma-derived mesenchymal stem cells towards a potential use for stem cell banking. Curr Stem Cell Res Ther. 2013;8(1):60–72.
120. Gong W, Han Z, Zhao H, Wang Y, Wang J, et al. Banking human umbilical cord-derived mesenchymal stromal cells for clinical use. Cell Transplant. 2012;21(1):207–16.
121. Parolini O, Alviano F, Bagnara GP, Bilic G, Buhring HJ, et al. Concise review: isolation and characterization of cells from human term placenta: outcome of the first international Workshop on Placenta Derived Stem Cells. Stem Cells. 2008;26(2):300–11.
122. Thirumala S, Goebel WS, Woods EJ. Clinical grade adult stem cell banking. Organogenesis. 2009;5(3):143–54.

123. Thirumala S, Goebel WS, Woods EJ. Manufacturing and banking of mesenchymal stem cells. Expert Opin Biol Ther. 2013;13(5):673–91.
124. Miranda-Sayago JM, Fernandez-Arcas N, Benito C, Reyes-Engel A, Herrero JR, et al. Evaluation of a low cost cryopreservation system on the biology of human amniotic fluid-derived mesenchymal stromal cells. Cryobiology. 2012;64(3):160–6.
125. Cho HJ, Lee SH, Yoo JJ, Shon YH. Evaluation of cell viability and apoptosis in human amniotic fluid-derived stem cells with natural cryoprotectants. Cryobiology. 2014;68(2):244–50.
126. Heng BC, Clement MV, Cao T. Caspase inhibitor Z-VAD-FMK enhances the freeze-thaw survival rate of human embryonic stem cells. Biosci Rep. 2007;27(4-5):257–64.
127. Montolio M, Tellez N, Biarnes M, Soler J, Montanya E. Short-term culture with the caspase inhibitor z-VAD.fmk reduces beta cell apoptosis in transplanted islets and improves the metabolic outcome of the graft. Cell Transplant. 2005;14(1):59–65.
128. Nakano M, Matsumoto I, Sawada T, Ansite J, Oberbroeckling J, et al. Caspase-3 inhibitor prevents apoptosis of human islets immediately after isolation and improves islet graft function. Pancreas. 2004;29(2):104–9.
129. Janz Fde L, Debes Ade A, Cavaglieri Rde C, Duarte SA, Romao CM, et al. Evaluation of distinct freezing methods and cryoprotectants for human amniotic fluid stem cells cryopreservation. J Biomed Biotechnol. 2012;2012:649353.
130. Steigman SA, Armant M, Bayer-Zwirello L, Kao GS, Silberstein L, et al. Preclinical regulatory validation of a 3-stage amniotic mesenchymal stem cell manufacturing protocol. J Pediatr Surg. 2008;43(6):1164–9.
131. Ng SC, Sathananthan H, Bongso A, Lee MN, Mok H, et al. The use of amniotic fluid and serum with propanediol in freezing of murine 2-cell embryos. Fertil Steril. 1988;50(3):510–3.
132. Dorfmann AD, Bender SD, Robinson P, Fugger EF, Bustillo M, et al. Cell-free human amniotic fluid as culture medium for mouse and human embryos. Fertil Steril. 1989;51(4):671–4.
133. Niknejad H, Deihim T, Peirovi H, Abolghasemi H. Serum-free cryopreservation of human amniotic epithelial cells before and after isolation from their natural scaffold. Cryobiology. 2013;67(1):56–63.
134. Murphy S, Rosli S, Acharya R, Mathias L, Lim R et al. Amnion epithelial cell isolation and characterization for clinical use. Curr Protoc Stem Cell Biol 2010; Chapter 1: Unit 1E.6.
135. Zarzeczny A, Rachul C, Nisbet M, Caulfield T. Stem cell clinics in the news. Nat Biotechnol. 2010;28(12):1243–6. doi:10.1038/nbt1210-1243b.
136. Regenberg AC, Hutchinson LA, Schanker B, Mathews DJ. Medicine on the fringe: stem cell-based interventions in advance of evidence. Stem Cells. 2009;27(9):2312–9.

Chapter 17
In Utero Stem Cell Transplantation

Matthew M. Boelig and Alan W. Flake

Introduction

Over the past three decades, advances in fetal diagnosis and therapy have led to a paradigm shift in the management of congenital disease. The human fetus has become a candidate for preventive treatment, prior to the evolution of severe organ damage or death in utero. Fetal surgical interventions, however, are restricted to specific anatomic defects that benefit from early intervention, examples of which include myelomeningocele repair, thoracoamniotic shunting or resection of congenital lung lesions causing fetal hydrops, resection of sacrococcygeal teratomas causing high-output cardiac failure, and fetoscopic laser coagulation in twin-to-twin transfusion syndrome (TTTS) [101, 102].

In contrast to anatomic anomalies, there are many genetic disorders that do not place the fetus at risk before birth but have devastating manifestations during postnatal life. While fetal therapy is not usually considered for these diseases, it is our view that in many cases, there are distinct biological advantages that favor prenatal over postnatal cellular and genetic therapies. For instance, congenital hematopoietic disorders, such as the hemoglobinopathies and immunodeficiency disorders, exact a considerable burden on society and may be an ideal target for prenatal therapy. Currently, postnatal hematopoietic stem cell transplantation (HSCT) is the only curative therapy for many of these diseases; however, there are practical limitations

M.M. Boelig, M.D.
Department of Surgery, Hospital of the University of Pennsylvania,
4 Maloney, Philadelphia, PA 19104, USA
e-mail: boeligm@uphs.upenn.edu

A.W. Flake, M.D., F.A.C.S., F.A.A.P. (✉)
Division of Pediatric General, Thoracic and Fetal Surgery, The Children's Hospital of Philadelphia, 3401 Civic Center Boulevard, Philadelphia, PA 19104-4319, USA
e-mail: flake@email.chop.edu

D.O. Fauza, M. Bani (eds.), *Fetal Stem Cells in Regenerative Medicine*,
Stem Cell Biology and Regenerative Medicine,
DOI 10.1007/978-1-4939-3483-6_17

to HSCT, namely the toxicity associated with the requirement for myeloablation and immunosuppression resulting in significant morbidity and mortality. In contrast, in utero hematopoietic cell transplantation (IUHCT), offers the potential to avoid the need for toxic myeloablative treatment and immunosuppression by exploiting normal events in fetal immune and hematopoietic development. When performed prior to the maturation of the fetal immune system, IUHCT can achieve multilineage hematopoietic engraftment and donor specific tolerance (DST). With this in mind, potential targets for IUHCT include hematopoietic disorders diagnosed early in gestation, and any disorder that would benefit from prenatal allogeneic tolerance induction to facilitate postnatal cellular or solid organ transplantation.

In theory, any congenital disease that can be identified early in gestation and can be treated by stem cell therapy, is potentially better treated by prenatal cellular therapy, as long as the intervention is safe to both mother and fetus. The potential of IUHCT is further increased by recent advances in prenatal screening, as well as the advent of high-throughput molecular techniques applied to fetal DNA or fetal cells gathered from maternal blood, that may ultimately permit practical population screening for the known gamut of genetic disease [101, 102]. Thus the spectrum of fetal therapy in the near future will no longer be limited to anatomic anomalies that effect fetal development, but will include anticipated postnatal diseases for which there are biological advantages unique to fetal development that favor prenatal over postnatal treatment [85].

In this chapter, we will review the scientific rationale, experimental basis, and the remaining barriers to progress in the field prior to clinical application.

Rationale

Human immune and hematopoietic development is characterized by a well- defined series of events. Normal immunologic development provides an opportunity for the engraftment of allogeneic hematopoietic stem cells (HSCs) and achievement of donor specific tolerance (DST). Billingham and Medawar were the first to describe the concept of "actively acquired" tolerance to a specific antigen as a consequence of the presence and processing of the antigen during fetal development [9]. This process occurs in the fetal thymus and can be understood as a form of self-education with two major components: (1) positive selection of prelymphocytes for recognition of self-major histocompatibility complex (MHC), and (2) negative selection (removal) of potentially autoreactive prelymphocytes, i.e. those that demonstrate high affinity interactions with self-antigen in association with self-MHC [74, 93]. This complex process—the foundation of central tolerance—generates a broad array of lymphocytes that recognize foreign antigen in association with self-MHC. Self-reactive T cells that elude negative selection are suppressed by other mechanisms in a process known as peripheral tolerance, the principle mediator of which is the regulatory T cell (Treg). T regulatory cells play a critical role in the prevention

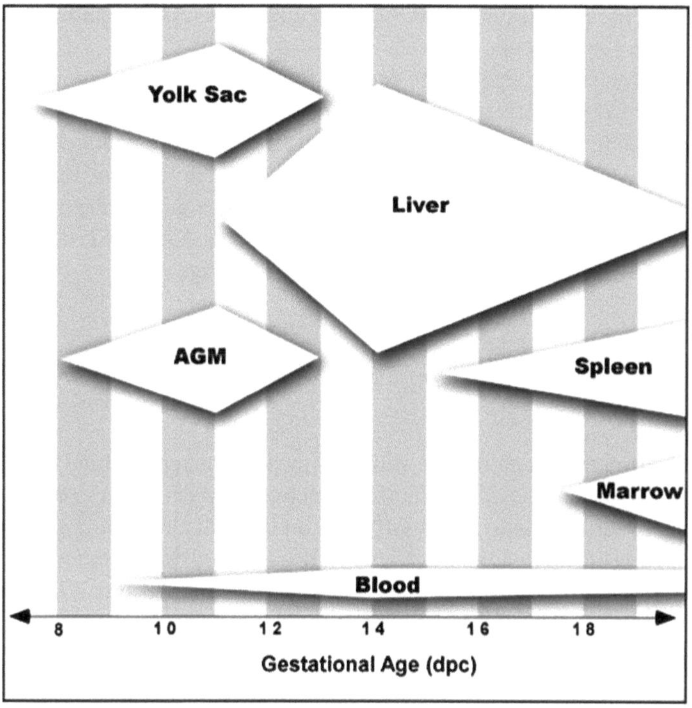

Fig. 17.1 Model depicting the location and relative frequencies of fetal murine hematopoietic stem cells (HSCs) in utero. HSCs undergo distinct spatiotemporal migrations and expansions in the murine embryo. *AGM* aorta-gonad-mesonephros region, *dpc* days post-conception. Christensen JL, Wright DE, Wagers AJ, Weissman IL. Circulation and chemotaxis of fetal hematopoietic stem cells. PLoS Biol. 2004 Mar; 2 (3): E75

of autoimmune disease [51, 92], and are implicated in the prevention of rejection of solid organ grafts and limiting GVHD after HSCT [59]. The therapeutic success of IUHCT is predicated on the idea that the introduction of allogeneic donor cells prior to thymic self-education can lead to the establishment of complete central and peripheral tolerance to the donor. Achievement of DST in the fetus would also facilitate additional postnatal transplantation (of cells, solid organs, etc.) with a minimal-conditioning regimen and without the need for traditional myeloablation and its associated toxicities.

Understanding normal events in hematopoietic development is critical for implementation of IUHCT. Fetal murine hematopoiesis, for example, undergoes a specific spatiotemporal migration and expansion, moving from yolk sac, to the aorta-gonad-mesonephros region and placenta, to the fetal liver and, finally, to the bone marrow [19, 62, 66, 94] (Fig. 17.1). The predictable nature of this process may offer opportunities for manipulations to favor the introduction and engraftment of donor HSCs, although host cell competition represents a formidable barrier. In sum, there is a rapidly expanding fetal hematopoietic microenvironment that, under specific circumstances, is amenable to receiving and supporting donor cells.

The relatively small size of the human fetus early in gestation allows for the introduction of comparatively large doses of donor cells. These conditions help to generate a stoichiometric advantage for donor cells, allowing donor cells to home to and competitively occupy available host hematopoietic niches. Finally, IUHCT offers the opportunity to mitigate or cure certain diseases prior to the onset of serious end-organ damage, a compelling rationale applicable to metabolic storage diseases that cause early and irreversible neurologic damage (Fig. 17.1).

Scientific Basis

Several experiments of nature have validated the key concepts behind IUHCT. Some of the most convincing evidence is the hematopoietic chimerism that is naturally observed between dizygotic twins with shared placental circulation. Ray Owen first reported on this phenomenon in 1945, when he observed that male/female dizygotic twins fathered by different sires often share the same blood type [73]. Owen was familiar with the existence of naturally-occurring placental vascular connections, which were first described by the American zoologist Frank Lillie in 1916 while working on a theory of freemartin cattle and placental hormonal exchange [61]. Owen hypothesized that placental vascular anastomoses permit the exchange of "embryonal cells ancestral to the erythrocyte of the adult animal," precursors that are then capable of long-term engraftment in both twins. Peter Medawar, a British biologist, subsequently hypothesized that skin grafting between bovine twins would be a foolproof way to distinguish between identical and fraternal twins. Surprisingly, Medawar and colleagues found that both identical and fraternal twin calves readily accepted skin grafts [4]. Medawar allegedly learned about Owen's work while reading a newly published text on antigen recognition [14] and made the connection between skin grafting and actively acquired immune tolerance [9]. Further experimental efforts in cattle twins expanded the understanding of acquired immune tolerance. Routine acceptance of renal allografts between bovine twins promoted the principle of donor specific tolerance as a consequence of the continuous exchange of hematopoietic cells early in gestation [23, 24].

Of clinical importance, hematopoietic chimeras have also been observed in human [35, 38, 95] and nonhuman primate twins [81, 82]. The frequency of chimerism in human dizygotic twins has been demonstrated to be fairly high—8 % for monozygotic twins and 21 % for triplets [99]. Levels of chimerism are often high enough to cure most hematologic diseases [38]. One must note that placental vascular anastomoses permit the development of long-term hematopoietic chimerism via the early and continuous admixture of circulating HSC/progenitors and their antigenic repertoire. This is clearly a difficult set of circumstances to replicate experimentally; however, IUHCT has had some notable successes in certain animal models, as will be discussed in the next section.

Experimental Progress in Animal Models

Over the last 30 years, small and large animal models of IUHCT have been developed and investigated, which have heightened our current understanding of the possibilities and pitfalls of IUHCT. This progress has been essential for laying the groundwork for future human clinical application.

The most informative animal model for analysis of the requirements for engraftment and tolerance has been the murine model. Stage for stage, mice and humans share similar hematologic and immunologic ontogenies, making the mouse a relevant model for IUHCT. In addition, the short gestation, large litter size, breadth of available inbred and transgenic strains, and resistance to engraftment, make the murine model attractive for investigating IUHCT. Roger Fleischman and Beatrice Mintz conducted the first studies on IUHCT and chose a c-kit-deficient mouse as their model in a study focused on stem cell biology rather than therapeutic applications [34]. This model featured an underlying stem cell survival and proliferation deficiency that granted a substantial advantage to donor HSCs. Gestational day 11 (E11) fetal mice were injected with allogeneic donor fetal liver cells via transplacental injection. The authors observed that the degree of erythroid replacement correlated with the magnitude of the host's anemia. Lethally anemic homozygous (W/W) mice experienced early and near total replacement of its erythroid lineage by donor-derived red blood cells. In later work, Mintz was able to demonstrate that the hematopoietic compartment could be reconstituted in this model from a single donor HSC [67]. Subsequent studies confirmed that IUHCT achieved multilineage chimerism in stem cell deficient anemic mice [12, 13]. In a different set of experiments by Blazar and colleagues using a murine model of severe combined immunodeficiency (SCID), in which mice have severely impaired T- and B-cell proliferation, hematopoietic reconstitution after IUHCT was limited to the lymphoid compartment [12, 13]. Thus, only split chimerism was achieved with the SCID model. IUHCT achieved multilineage engraftment in the peripheral blood and spleens in the non-obese diabetic (NOD)/SCID murine model, but this strain has additional defects in macrophage function (including antigen presentation) and in the natural killer lineage [5]. The early success in these hematopoietically defective murine models contrast with the failure to achieve significant engraftment in strains with an intact competitive barrier [16, 27, 52] and highlight the difficulty of donor cell engraftment in a normal fetal host.

Until recently, experience with IUHCT in animal models without stem cell or lineage deficiencies has met with limited success in achieving long-term allogeneic chimerism due to the barriers to engraftment discussed below. Relevant to the early clinical experience, initial studies of IUHCT in the intact murine model were disappointing in terms of engraftment [16, 52, 53]. As discussed below, recent improvements in the technique of donor cell administration, *ex vivo* donor cell treatment, and recognition of a maternal adaptive immune barrier have led to the ability to engraft almost 100 % of injected murine fetuses and achieve consistent levels of mixed hematopoietic chimerism across full major histocompatibility (MHC) barriers.

IUHCT in non-deficient large animal models have confirmed the difficulty of multilineage engraftment in the presence of a competitive, intact host hematopoietic compartment. The sheep model is the primary exception and has proven to be more permissive for successful IUHCT. Flake and colleagues were able to achieve allogeneic multilineage engraftment in the ovine model after a single intraperitoneal injection of fetal liver-derived cells [30]. Peripheral blood erythroid chimerism at 6 months was durable and as high as 23 % in this model. The sheep model also proved to be permissive of xenogeneic engraftment after the introduction of human fetal liver-derived HSCs into the preimmune ovine fetus [90, 109, 110]. Other large animal models of significance to IUHCT include the nonhuman primate [22, 41, 88], the pig [56, 57], and the dog [11]. Harrison observed long-term but very low-level multilineage engraftment and no GVHD after IUHCT in fetal rhesus monkeys using fetal liver cells [41]. In the same model, IUHCT with T cell depleted parental bone marrow led to only microchimerism with inconsistent donor specific tolerance [22]. In neither experiment did peripheral blood chimerism achieve levels of therapeutic significance. Experience in the pig model with IUHCT using T cell depleted bone marrow (1.5 %) has led to stable multilineage, low level engraftment without GVHD across full swine leukocyte antigen barriers, permitting tolerance to donor-matched renal allotransplantation [56, 57]. More recently, the canine model has proven to be a promising pre-clinical model for IUHCT and our results will be summarized below.

Barriers to Engraftment

Despite the theoretical advantages of IUHCT, the reality of experimental and clinical failures to achieve significant engraftment after IUHCT suggested that there are barriers to engraftment that must be elucidated before IUHCT can become a plausible therapeutic modality. The logical barriers are: (1) competition from an intact and robust host hematopoietic compartment, (2) lack of available space within the host hematopoietic niche, and (3) unsuspected innate or adaptive immune barriers of the developing fetus [33].

Due to the ethical issues surrounding the use of embryonic stem cells and fetal tissue, and the limited number of HSCs in a unit of cord blood, adult-derived cells will likely be the source of donor cells in human application of IUHCT. Adult-derived HSCs are known to be at a profound competitive disadvantage when compared to the various fetal progenitor populations, including fetal liver HSCs [39, 50] and cord blood HSCs [39, 58, 84]. The fetal host, replete with fetal HSCs that cycle and expand more rapidly than their adult equivalents, is an unfriendly milieu for adult-derived donor cells. Moreover, initial predictions that the rapid expansion of the fetal hematopoietic compartment would yield extra space for donor cells have proven to be unfounded [86]. Fetal growth is accompanied by known migrations in hematopoiesis as well as an expansion in the number of niches. The stromal environment in these burgeoning niches (liver, spleen, marrow) must mature prior to engraftment; however, once ready, it appears that host hematopoietic progenitors rapidly engraft these sites [107]. Once the host has populated a niche, donor cell engraftment depends on cycling of host cells in the niche and the availability of host cells to occupy the niche. While little is known about the kinetics of host cell cycling in fetal hematopoietic niches, there is a well-documented excess of circulating host HSCs that can compete with donor cells [39, 40].

Despite differences between the fetal and postnatal microenvironments, one could argue that the most analogous system to consider is the postnatal non-myeloablated syngeneic model of HSCT in which dose escalation of cells can lead to dose-dependent augmentation of engraftment [33]. When compared to the myeloablated postnatal model, enormous doses of donor cells by weight are required to obtain even modest degrees of mixed hematopoietic chimerism in the mouse. Even in the congenic murine model of IUHCT using the intravenous technique, in which an immune barrier is absent, host competition limits peripheral blood chimerism to <10 % at 6 months, despite the administration of enormous doses of adult bone marrow mononuclear cells (2×10^{11} BM MNCs/kg) [78]. Cell tracking experiments have confirmed that donor cells do participate in the normal migrations of host hematopoietic cells. Using GFP-transgenic donor bone marrow MNCs and fluorescence microscopy, it has been confirmed that, in allogeneic murine experiments with E14 IUHCT, donor cells home to the liver, engraft, and then migrate to and engraft definitive hematopoietic niches (spleen, marrow, thymus) (Fig. 17.2). Because of the competitive barrier present in IUHCT, strategies to selectively improve donor cell competition, i.e. improve the relative ability of donor cells to home, engraft and proliferate, are an important future goal of IUHCT directed research. The converse of inhibition of host cell hematopoiesis also has potential but carries more risk due to the lack of specificity of most host myeloablative strategies. These concepts will be further explored below (Fig. 17.2).

Although the fetal recipient was predicted to be pre-immune, experimental evidence suggested the presence of an immune barrier to engraftment. Prior murine studies comparing engraftment after congenic versus allogeneic IUHCT failed to demonstrate a significant advantage in the congenic setting [16, 46]; however, these studies were limited by low engraftment (microchimerism). Additionally, it was noted that the clinical success of IUHCT was limited to disorders featuring a host immunologic impairment, such as bare lymphocyte syndrome and X-linked severe combined immunodeficiency [31, 96, 105]. Finally, studies have uncovered alloreactive T cells in the human fetal liver early in gestation (16 weeks), raising the possibility that the "pre-immune" host may have some ability to mount an immune response prior to targeted gestational dates for IUHCT [83]. However, in the murine model, partial deletion of donor reactive lymphocytes has been clearly documented in several studies of IUHCT, suggesting that IUHCT may be able to replicate certain aspects of immunologic ontogeny [42, 44, 53, 76].

Improvements in the murine model in our laboratory, and the delivery of higher doses of donor cells, allowed the achievement of macrochimerism in the murine model (>1–2 %) with the intraperitoneal technique for IUHCT. These mice had associated DST across full MHC barriers as confirmed by mixed lymphocyte reaction

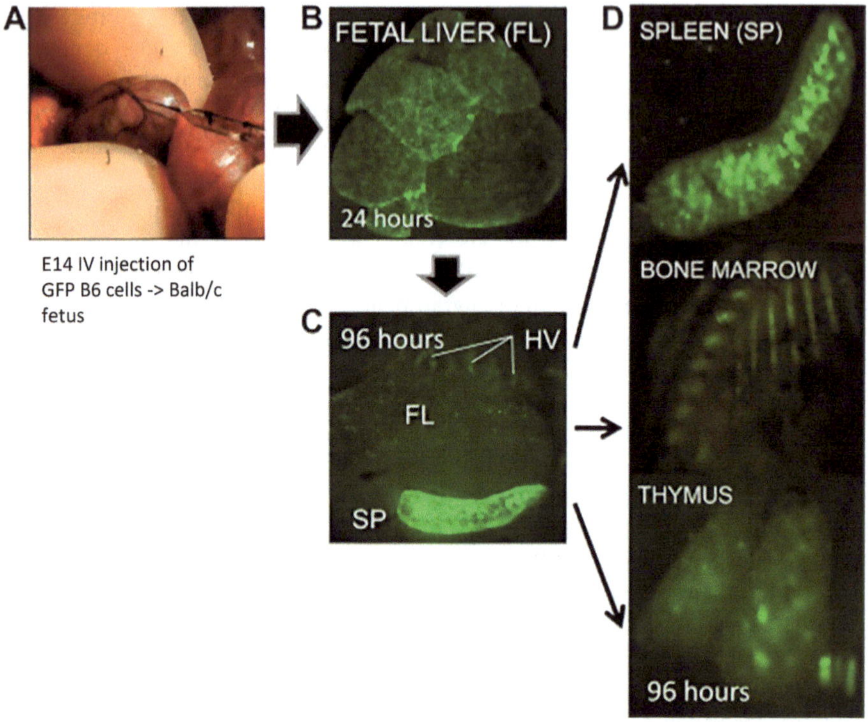

Fig. 17.2 Migration of donor hematopoiesis after IUHCT. Stereoscopic fluorescence microscopy can be used to track the homing, engraftment, and migrations of green fluorescent protein (GFP)-positive allogeneic bone marrow (BM) mononuclear cells (MNCs) following intravenous injection at embryonic day 14 (E14). (**a**) Technique of E14 intravenous injection of donor cells via the vitelline vein, using India ink for illustration. (**b**) A representative fetal liver (FL) harvested 24 h post-injection (E15), demonstrating clear evidence of homing and engraftment of GFP-positive donor cells. (**c**) At 96-h post-injection (E18), GFP-positive donor cells can be seen streaming out of the hepatic veins (HV) and localizing to definitive sites of hematopoiesis (e.g. spleen). (**d**) Representative images of the spleen, BM (vertebrae and ribs), and thymus engrafted by GFP-positive donor cells. Reprinted from Cytotherapy, Volume 15(5), Vrecenak JD and Flake AW, In utero hematopoietic cell transplantation—recent progress and potential for clinical application, 525–35, Copyright 2013, with permission from Elsevier

and acceptance of donor skin grafts for more than 8 weeks. This allowed testing of the strategy of in utero tolerance induction to facilitate postnatal boosting with minimally ablative HSCT to obtain engraftment levels that would be therapeutic for human disease. Three such strategies were tested and proved to be fruitful in the murine model. Robust replacement of host hematopoiesis was achieved across major MHC barriers (without major morbidity or GVHD) after (1) donor-specific lymphocyte infusion (DLI) [42], (2) low-dose total body irradiation (TBI) followed by T-cell depleted BMT [76], and (3) low-dose busulfan followed by BMT [6]. DLI was dose-dependent and led to complete donor peripheral blood chimerism in all of the mice receiving the highest dose (30 million donor congenic splenocytes) without significant incidence of GVHD (1/56 animals). In the TBI experiment, enhancement of engraftment was radiation dose-dependent and based on a competitive advantage of donor cells over irradiated host hematopoietic cells. Low-dose busulfan was found to be an effective single myelosuppressive agent and achieved dose-dependent high-level mixed macrochimerism by blunting the proliferative potential of host hematopoietic progenitor cells and conferring a competitive advantage on donor cells.

The development of an intravenous technique for IUHCT marked a major improvement in the murine model [49, 78]. Injection of donor cells via the vitelline vein allows the delivery of larger volumes (and thereby larger doses) of donor cells and permits visual confirmation of injection via clearance of the vein by the injectate, removing the uncertainty of cell delivery with intraperitoneal and intrahepatic injection. The intravenous technique led to substantial improvements in levels of engraftment in the congenic and allogeneic murine models.

The concept of IUHCT hinges on the presumption that the fetus is in a preimmune state at the time of transplantation. However, little is known regarding the potential of the innate immune system and specifically natural killer (NK) cells, as a barrier to IUHCT. Shabaan and colleagues found that a threshold level of initial chimerism (>1.8 %) predicted long-term engraftment and induction of host NK cell tolerance to donor alloantigen via reduced expression of surface inhibitory Ly49A receptor [28]. However, dose dependent modification of NK activation and inhibitory receptors negates this barrier at higher levels of chimerism, making this barrier inconsequential if chimerism higher than 2 % can be achieved. A more disturbing observation was uncovered by our ability to achieve higher levels of chimerism with the intravascular technique. Peranteau et al. revisited the question of an adaptive barrier by using the intravenous technique of donor cell injection in the murine model [78]. Despite the ability to consistently deliver cells to the murine fetus, only approximately 30 % of animals maintained engraftment in the allogeneic setting, whereas congenic IUHCT led to mixed hematopoietic chimerism 100 % of the time. Allogeneic chimerism disappeared between birth and 4 weeks of age, suggesting the presence of an adaptive immune barrier.

In a subsequent pivotal study, Merianos et al. demonstrated the presence of an alloresponse in all pups that lost chimerism after allogeneic IUHCT [64]. In this study, non-chimeric pups exhibited an increased frequency of alloreactive T cells comparable to immunized controls whereas chimeric pups demonstrated reduced frequencies of alloreactive lymphocytes comparable to naïve controls. This important observation was consistent with a mechanism of partial clonal deletion of alloreactive cells in pups that sustained their chimerism with achievement of donor specific tolerance. Moreover, chimeric pups were found to have more frequent and more suppressive T regulatory cells, suggesting that peripheral mechanisms of tolerance are also critical. There were two critical observations in this study that implicated the maternal immune system as the cause of an adaptive immune response in pups: (1) assessment of maternal allospecific immune response demonstrated clear immunization of the mothers that underwent IUHCT; and (2) fostering of pups by a naïve mother resulted in a 100 % rate of mixed hematopoietic chimerism in allogeneic recipients. The loss of chimerism was not dependent on maternal lymphocytes (cell-free maternal serum transferred orally to pups led to loss of chimerism). By fostering pups after birth, it became possible to consistently achieve multilineage engraftment across full MHC barriers. Most importantly, in the absence of a maternal immune response, immunologic ontogeny was clearly recapitulated in that central and peripheral mechanisms of tolerance were observed.

A subsequent study by Nijagal et al. [70] investigated maternal-fetal cellular trafficking across the placenta after allogeneic IUHCT and concluded that maternal T cells—not the maternal humoral response—constitute the main immunologic barrier to consistent engraftment. The authors observed maternal leukocytes in the fetal blood at levels of nearly 10 % of the fetal CD45$^+$ population at the time of IUHCT (E14). Interestingly, IUHCT appeared to increase levels of maternal T cells in the fetal circulation. This study implicated maternal-fetal T-cell trafficking as the principle cause of engraftment loss by demonstrating better engraftment in T cell-deficient maternal strains. The discrepancies in findings the Nijagal [71] and Merianos [63] studies are likely due in part to differences in their respective murine models (intrahepatic versus intravenous, 5 million fetal liver-derived cells versus 20 million adult bone marrow-derived cells). The cellular composition of these grafts is quite different—the graft used in the Nijagal study contains fewer HSCs, antigen-presenting cells (APCs), and T cells resulting in lower levels of chimerism and minimal maternal immunization. It would be expected that non-immunized mothers and immune-deficient maternal strains would fail to demonstrate a role for the maternal humoral response [101, 102]. Flake and colleagues have not observed any limitations in allogeneic chimerism after intravenous transplantation and fostering. These observations suggest that maternal-fetal lymphocyte trafficking plays a minimal role as a barrier to engraftment after allogeneic IUHCT in the mouse. Until the role of the maternal immune response is defined in a relevant large animal model, the use of maternal cells would be practical and safer in human trials of IUHCT.

Pre-clinical Studies in the Canine Model

We have chosen the canine model to pursue pre-clinical studies because: (1) immunologic ontogeny in the dog is similar to humans [29]; (2) dogs have been used extensively as a model for postnatal BMT regimens and have proven to be a highly predictive pre-clinical model for GVHD [25, 91]; (3) the canine model until recently has been difficult to engraft by IUHCT, similar to the human; (4) canines are outbred and have similar genetic diversity to humans, and (5) a number of canine models exist that are relevant to human disorders potentially treatable by IUHCT. In prior studies, short term microchimerism (<1 %) had been observed after canine IUHCT via intraperitoneal injection of enriched paternal CD34+ cells [11]. In our initial studies we utilized the canine leukocyte adhesion deficiency (CLAD) model, a canine version of human leukocyte adhesion deficiency (LAD-1) [79]. CLAD-affected dogs have a severe immunodeficiency and the disease is lethal by 6 months of age, whereas the carrier has a normal phenotype. By achieving low level engraftment (<2 %) by intraperitoneal injection of paternal cells, we were able to improve or cure the CLAD phenotype. In the same study, we were able to generate low level chimerism (<2 %) in CLAD carrier dogs. While these levels were too low for therapeutic relevance, they were associated with inconsistent DST. When DST was present, this facilitated enhancement of engraftment by minimal conditioning, same donor, postnatal HSCT in two of seven dogs. In these dogs, the levels of chimerism were 20–50 %, well within the range considered therapeutic for numerous congenital hematopoietic/immunologic diseases. Importantly, graft-versus-host disease (GVHD) was not observed and we have continued to follow the animals for over 5 years with observation of stable levels of chimerism.

The CLAD carrier state does not convey a proliferation or survival deficit on host HSCs or downstream lineages. Therefore, this model is representative of the degree of host hematopoietic competition that one would expect to encounter for most target diseases. However, given the very low levels of initial chimerism and the inconsistent ability to enhance engraftment observed in this initial dog study, the outcomes were not adequate for consideration of a clinical trial. We have since optimized the model by defining the ideal ontologic timing of IUHCT, and by using the intracardiac (intravascular) mode of injection (Vrecenak et al. Blood [in press]). In the optimized model, ultrasound-guided intracardiac injection of maternal donor cells has resulted in sustained macrochimerism in approximately 90 % of transplanted animals with an average initial level of chimerism >10 %. This represents a major improvement over the results of IUHCT in previous large animal studies. Stable multi-lineage chimerism has been sustained in all animals for as long as 2 years without evidence of GVHD. When tested, animals from this cohort have also demonstrated associated DST for renal transplants. These studies demonstrate that the host competitive barrier to engraftment of adult derived HSC after IUHCT can be overcome, and that immune barriers are not prohibitive to engraftment of maternal cells. We feel that these studies support consideration of a clinical trial of IUHCT for hemoglobinopathies and other disorders.

Strategies to Improve Engraftment

Now that consistent and sustained mixed hematopoietic chimerism can be achieved in the allogeneic murine model of IUHCT, the model can be used to study various means for improving engraftment. Ideally, engraftment can be maximized after the administration of a single transplant in utero, obviating the need for postnatal (albeit minimally ablative) booster protocols in humans. Engraftment can be optimized by increasing the delivered cell dose, providing a competitive edge to donor cells, increasing homing of donor cells to hematopoietic stromal environments, augmenting host receptiveness to the graft, and avoiding the immune barrier.

The combination of the intravenous technique with fostering has allowed for consistent relatively high level engraftment after IUHCT in the murine model and set the stage for testing of a variety of approaches for further improvement of engraftment. Efforts to augment the competitiveness of donor cells have focused on *ex vivo* manipulation of the donor graft prior to IUHCT. *Ex vivo* treatments are safer—mother and fetus aren't exposed to the agent—and allow for specific titration of drugs/reagents. A successful example of this strategy demonstrated by our laboratory is that Diprotin, an inhibitor of the cell surface dipeptidylpeptidase CD26, can be used for successful *ex vivo* manipulation of donor cells immediately prior to IUHCT [77]. Diprotin improves homing of allogeneic hematopoietic progenitors to the fetal liver at E14, increases the competitive capacity of donor cells, and results in higher short- and long-term engraftment. Diprotin acts by inhibiting the cleavage of SDF-1α by CD26, thereby enhancing signaling through CXCR4 on HSCs and improving homing to the fetal hematopoietic niche. There is abundant evidence in the literature that numerous small molecules, many of which are natural pro-inflammatory agents, may increase HSC homing. It remains to be seen if there is a "cocktail" of agents that will maximize homing of donor cells to the niche.

Mobilization of host HSCs from the niche is a second promising technique for improving engraftment and may have synergistic effects if paired with agents that improve donor cell homing. Several of these agents are currently being investigated for safety and efficacy in the murine model of IUHCT. The niche is a dynamic microenvironment with rapid cycling of cells in and out, cell proliferation, and complex interactions between hematopoietic cells and the supporting stroma. Mobilization of host cells may tilt the competitive balance towards donor cells that have been manipulated to home more efficiently to the niche.

Perhaps the most exciting and intriguing strategy to improve engraftment involves selective myeloablation of the fetal hematopoietic compartment, i.e. the niche. The maternal administration of myeloablative agents, such as busulfan, has been shown to increase engraftment in both mice and sheep [1, 108]; however, administration of a non-specific myeloablative agent to the mother and fetus is prohibitively dangerous. Agents that cause selective, non-toxic fetal myeloablation prior to IUHCT are currently being investigated in the murine model and may have clinical promise. Potential strategies include RNA interference, antibodies targeting fetal hematopoietic progenitors, and targeted delivery of pro-apoptotic peptides to host HSCs. Our laboratory is currently investigating these strategies for safety and efficacy.

Other proposed methods for increasing engraftment after IUHCT include co-transplantation with various support cell populations, including plasmacytoid-precursor dendritic cells [48] and stromal cells [2, 3], and *ex vivo* treatment with growth factors [87], although the mechanisms for improving engraftment have not been delineated. Optimizing donor cell content may lead to improvements in engraftment. Immunologically active cells may have a graft-versus-hematopoietic effect (GVH) and increase engraftment, and T cell depleted grafts have demonstrated impairments in engraftment that are reversible with addition of donor T cells [20, 80, 88]. T cell addition, however, is known to be associated with an increased risk of GVHD. For a given animal model of IUHCT, there may be a donor T cell content in the graft that enhances engraftment without causing clinical GVHD. For example, the authors have been able to achieve clinically significant chimerism in a preclinical canine model of IUHCT using low doses of T cells (Vrecenak et al. Blood 2014).

No evidence of GVHD was observed in this model after numerous trials of IUHCT. Another strategy to make space for donor cells is to enhance GVH by using donor cells primed to target host hematopoietic cells. Two prior studies involving immunization of donor T cells with host antigen followed by *ex vivo* treatment to reduce the risk of GVHD (LLME or photochemical therapy) resulted in complete donor chimerism; however, some animals did suffer from GVHD [8, 44].

In Utero Transplantation with Non-hematopoietic Cells

While the HSC has been the classic cellular candidate for in utero cell transplantation (IUCT), other stem cell populations may be utilized for in utero therapy targeting non-hematopoietic congenital diseases. The feasibility of mesenchymal stem cell (MSC) use for in utero cell transplantation (IUCT) has been most extensively investigated. The MSC is a multipotent cell and can differentiate to form bone, cartilage, tendon, muscle, adipose, and bone marrow stroma. IUCT with human MSCs in the sheep model has led to persistence of detectable donor MSCs in multiple tissues for over a year beyond transplantation [60]. Intriguingly, these findings were noted when IUCT was performed both before and after the establishment of fetal immune competence. IUCT with MSCs has also been used in murine models of mesenchymal disorders. Fetal MSCs in particular have been used in IUCT because they have superior expansion and self-renewal properties compared to their adult counterparts and express unique cell adhesion molecules that facilitate transmigration across endothelial barriers into target tissue compartments [15, 17]. Intraperitoneal delivery of human fetal MSCs at E14-16 in a mouse model of Duchenne muscular dystrophy led to preferential long-term engraftment of muscle, although levels of chimerism were sub-therapeutic [18]. Guillot and colleagues demonstrated that IUCT of human first trimester fetal blood MSCs into a mouse model of intermediate severity type III osteogenesis imperfecta (OI) mice resulted in significantly fewer fractures and improved bone strength, length, and thickness [37]. Donor-derived cells differentiated into mature osteoblasts and produced normal COL1a2 protein [100]. Donor-derived osteoblasts also clustered at healed fracture sites, participating in active bone formation and remodeling.

Two human cases of allogeneic fetal MSC transplantation for OI have been reported with incomplete follow up. Le Blanc and colleagues reported on their injection of a female fetus with severe OI with HLA-mismatched male fetal MSCs in the 32nd week of gestation, at which point the fetus is immunocompetent [55]. They observed engraftment in the recipient bone marrow at 9 months of age (median of 7.4 % by whole Y genome fluorescent *in situ* hybridization staining), normal bone histology, and tolerance to donor MSC as proven by a negative mixed lymphocyte reaction. Serial dual-energy x-ray absorptiometry (DEXA) scans demonstrated improving bone mineralization compared to age-matched controls out to 22-months (76 % normal), although some of the effect may have been attributable to early and aggressive bisphosphonate therapy. A subsequent report by Götherström et al.

described follow up at 8 years on this same patient and reported an additional patient with OI type III who they transplanted with fetal MSCs at 31 weeks gestation [36]. Although there appeared to be reduced fractures and evidence of clinical benefit in both patients, the effects plateaued and both patients were treated by reinfusion of MSCs at intervals postnatally to maintain beneficial effect. Although promising, these results leave room for much improvement and raise many questions regarding the applied stem cell biology. There is controversy regarding what cell is capable of reconstitution of the osteoblastic niche and about the mechanism of the transient improvement in the two patients treated. It may be that MSCs act by a paracrine mechanism rather than directly by engraftment and differentiation into osteoblasts and that another progenitor cell is required for osteoblast replacement. If these questions can be experimentally addressed, there are several congenital mesenchymal diseases that may be amenable to IUCT with appropriate stem cell populations including the muscular dystrophies, hypophosphatasia, and osteopetrosis.

Clinical Applications and Future Directives

Although much progress has been made in the mouse and large animal models, the historical clinical experience with IUHCT has been disappointing. It is important to note that most of these attempts occurred after promising results were observed in the sheep model between the years of 1986 and 2000. During that interval, the 50 or so reported attempts of IUHCT have targeted numerous congenital diseases and utilized a variety of transplantation protocols and donor cell populations. A conclusion that can be clearly drawn is that IUHCT has failed in disorders in which the fetal hematopoietic compartment is highly competitive. In fact, the only successes have been for treatment of XSCID [31, 96–98, 103, 105], a disease in which donor T cells have a clear survival and proliferative advantage. Predictably, donor engraftment was limited to the T cell compartment. At this point there is no clearly documented advantage of IUHCT over non-myeloablated postnatal BMT in SCID patients because both achieve split chimerism. Other forms of SCID (adenosine deaminase deficiency, mutations in Jak 3 and ZAP-70) and Bruton disease (X-linked agammaglobulinemia) should theoretically experience similar donor cell amplification in the affected lineage after IUHCT [32]. Attempts to cure other immunodeficiency disorders, namely chronic granulomatous disease [69] and Chediak-Higashi syndrome [21], have not resulted in detectable engraftment after birth.

Hemoglobinopathies are an appealing target for IUHCT, and several attempts have been made to treat them by IUHCT. Unfortunately, attempts to cure beta- or alpha thalassemia in humans by IUHCT have resulted in no engraftment with the exception of microchimerism in a few patients. No evidence of DST has been observed in any of these studies [26, 45, 72, 89, 97, 104].

Metabolic storage diseases are another category of diseases that IUHCT may ameliorate. These inborn errors of metabolism constitute a heterogeneous class of disorders characterized by deficiencies in specific lysosomal hydrolases required for the metabolism of lipids, glycoproteins, or mucopolysaccharides. The intracellular accumulation of intermediate metabolic substrates causes specific patterns of tissue damage and organ failure [68]. IUHCT, thus, could potentially prevent early and irreversible insults from the disease, giving it a distinct advantage over postnatal BMT. There have been several reported attempts at IUHCT for metabolic storage diseases, including Hurler Disease (Flake 1995, unpublished) and the demyelinating diseases metachromatic leukodystrophy (MCDL) [89] and globoid cell leukodystrophy [7]. All have failed to produce significant engraftment with the exception of Bambach et al. who reported a peculiar and unexplained observation of overwhelming donor myelopoiesis in their patient with globoid cell leukodystrophy. The fetus died at 7 weeks after IUHCT. Even if high level engraftment can be achieved, the blood–brain barrier may prevent donor HSC-derived glial cells from crossing into the CNS and providing the enzymatic reinforcement needed to halt disease progression [32]. Thus, in diseases that cause CNS substrate accumulation, a prenatal approach may require a combined strategy of IUHCT and CNS directed gene therapy.

Given this track record, there have been few recent attempts at IUHCT in humans. The rationale for IUHCT, however, remains compelling and historical failures are better understood now due to the progress that has been made in mice and large animal models. Lack of clinical success can be largely attributed to deficiencies in protocol design, including the timing of IUHCT, inappropriate donor cell content, and inadequate delivery of donor cells. Additional investigation into the optimization of donor cell homing, optimal graft content, and maternal-fetal immune mechanisms in relevant preclinical large animal models should continue and may lead to further improvements in engraftment that can be applied clinically. Some diseases that can be clinically ameliorated with low levels of chimerism, such as chronic granulomatous disease (CGD), hyper IgM syndrome, and leukocyte adhesion deficiency are particularly attractive targets. CGD, for example, can be cured with 5 % normal neutrophils [10]. Diseases that require high levels of chimerism for clinical improvement, such as hemoglobinopathies, are more likely to require a combined protocol of IUHCT and minimal conditioning, same donor, postnatal HSCT.

For the time being, a compelling argument can be made for the initiation of a clinical trial for IUHCT using maternal donor cells to avoid maternal immunization. The ideal timing for IUHCT would in our opinion be prior to 15 weeks gestation, when the fetal liver is the site of active hematopoiesis, thymic selection is initiating, and the fetus is small. We consider Sickle Cell Disease (SCD) to be an appropriate target because of its high prevalence and the natural history that results in the gradual acquisition of end-organ damage and subsequent morbidities. While curative, postnatal BMT is rarely employed due to the low frequency of HLA-matched sibling donors and the prohibitive side effects of mismatched BMT [65]. Of the approximately 80,000 people with SCD in the United States, only a few hundred have undergone myeloablative allo-HSCT, and GVHD remains a concern [47]. Mixed hematopoietic chimerism levels as low as 20 % can cure SCD, due to the dramatically longer half-life of donor-derived erythrocytes [43]. Early, preemptive intervention (before 14 weeks gestation) via IUHCT would prevent the manifestations of disease from occurring. Recipients with sustained chimerism <20 % would be candidates for minimal conditioning, same donor postnatal HSCT, to achieve therapeutic levels of chimerism. As a new therapeutic modality for the treatment of SCD, IUHCT has the potential to yield significant improvements in quality of life, life expectancy, and health care savings, converting a chronic, debilitating disease into clinical quiescence.

Summary

The compelling rationale for IUHCT is that donor cells may partake in normal immunologic and hematopoietic ontogeny if introduced in early gestation, resulting in DST and mixed hematopoietic chimerism. Current evidence in animal models validates the rationale. Host competition and limited space remain potent barriers to engraftment in all models but may be addressed by increasing cell doses, *ex vivo* manipulation of cells to improve their competitiveness, and potentially selective myeloablation of the fetal hematopoietic niche. While immune barriers in the murine model have been mitigated, larger animals and humans present a more sophisticated immunologic challenge due to early maturation of the fetal immune system in utero and the question of maternal immunization and maternal fetal cell trafficking after IUHCT. Use of maternal cells as a donor graft would provide a clear, immediate way to bypass an immunologic barrier. With improvements in technique and early diagnosis, as well as an excellent safety profile, initiation of human trials for congenital hematopoietic disorders can be justified.

References

1. Abe T, Masuda S, Tanaka Y, Nitta S, Kitano Y, Hayashi S, et al. Maternal administration of busulfan before in utero transplantation of human hematopoietic stem cells enhances engraftments in sheep. Exp Hematol. 2012;40(6):436–44.
2. Almeida-Porada G, Flake AW, Glimp HA, Zanjani ED. Cotransplantation of stroma results in enhancement of engraftment and early expression of donor hematopoietic stem cells in utero. Exp Hematol. 1999;27(10):1569–75.
3. Almeida-Porada G, Porada CD, Tran N, Zanjani ED. Cotransplantation of human stromal cell progenitors into preimmune fetal sheep results in early appearance of human donor cells in circulation and boosts cell levels in bone marrow at later time points after transplantation. Blood. 2000;95(11):3620–7.
4. Anderson D, Billingham R, Lampkin G, Medawar P. The use of skin grafting to distinguish between monozygotic and dizygotic twins in cattle. Heredity. 1951;5:379–97.
5. Archer DR, Turner CW, Yeager AM, Fleming WH. Sustained multilineage engraftment of allogeneic hematopoietic stem cells in NOD/SCID mice after in utero transplantation. Blood. 1997;90(8):3222–9.
6. Ashizuka S, Peranteau WH, Hayashi S, Flake AW. Busulfan-conditioned bone marrow transplantation results in high-level allogeneic chimerism in mice made tolerant by in utero hematopoietic cell transplantation. Exp Hematol. 2006;34(3):359–68.

7. Bambach BJ, Moser HW, Blakemore K, Corson VL, Griffin CA, Noga SJ, et al. Engraftment following in utero bone marrow transplantation for globoid cell leukodystrophy. Bone Marrow Transplant. 1997;19(4):399–402.
8. Bhattacharyya S, Chawla A, Smith K, Zhou Y, Talib S, Wardwell B, et al. Multilineage engraftment with minimal graft-versus-host disease following in utero transplantation of S-59 psoralen/ultraviolet a light-treated, sensitized T cells and adult T cell-depleted bone marrow in fetal mice. J Immunol. 2002;169(11):6133–40.
9. Billingham RE, Brent L, Medawar PB. 'Actively acquired tolerance' of foreign cells. Nature. 1953;172(4379):603–6.
10. Björgvinsdóttir H, Ding C, Pech N, Gifford M, Li L, Dinauer M. Retroviral-mediated gene transfer of gp91phox into bone marrow cells rescues defect in host defense against Aspergillus fumigatus in murine X-linked chronic granulomatous disease. Blood. 1997;89(1):41–8.
11. Blakemore K, Hattenburg C, Stetten G, Berg K, South S, Murphy K, et al. In utero hematopoietic stem cell transplantation with haploidentical donor adult bone marrow in a canine model. Am J Obstet Gynecol. 2004;190(4):960–73.
12. Blazar BR, Taylor PA, Vallera DA. Adult bone marrow-derived pluripotent hematopoietic stem cells are engraftable when transferred in utero into moderately anemic fetal recipients. Blood. 1995;85(3):833–41.
13. Blazar BR, Taylor PA, Vallera DA. In utero transfer of adult bone marrows cells into recipients with severe combined immunodeficiency disorder yields lymphoid progeny with T- and B-cell functional capabilities. Blood. 1995;86(11):4353–66.
14. Burnet FM, Fenner F. The production of antibodies. London: Macmillan; 1949.
15. Campagnoli C, Roberts IA, Kumar S, Bennett PR, Bellantuono I, Fisk NM. Identification of mesenchymal stem/progenitor cells in human first-trimester fetal blood, liver, and bone marrow. Blood. 2001;98(8):2396–402.
16. Carrier E, Lee TH, Busch MP, Cowan MJ. Induction of tolerance in nondefective mice after in utero transplantation of major histocompatibility complex-mismatched fetal hematopoietic stem cells. Blood. 1995;86(12):4681–90.
17. Chan J, O'Donoghue K, de la Fuente J, Roberts IA, Kumar S, Morgan JE, et al. Human fetal mesenchymal stem cells as vehicles for gene delivery. Stem Cells. 2005;23(1):93–102.
18. Chan J, Waddington SN, O'Donoghue K, Kurata H, Guillot PV, Gotherstrom C, et al. Widespread distribution and muscle differentiation of human fetal mesenchymal stem cells after intrauterine transplantation in dystrophic mdx mouse. Stem Cells. 2007;25(4):875–84.
19. Christensen JL, Wright DE, Wagers AJ, Weissman IL. Circulation and chemotaxis of fetal hematopoietic stem cells. PLoS Biol. 2004;2(3), E75.
20. Crombleholme TM, Harrison MR, Zanjani ED. In utero transplantation of hematopoietic stem cells in sheep: the role of T cells in engraftment and graft-versus-host disease. J Pediatr Surg. 1990;25(8):885–92.
21. Cowan MJ, Golbus M. In utero hematopoietic stem cell transplants for inherited diseases. Am J Pediatr Hematol Oncol. 1994;16(1):35–42.
22. Cowan MJ, Tarantal AF, Capper J, Harrison M, Garovoy M. Long-term engraftment following in utero T cell-depleted parental marrow transplantation into fetal rhesus monkeys. Bone Marrow Transplant. 1996;17(6):1157–65.
23. Cragle R, Stone WH. Preliminary results of kidney grafts between cattle chimeric twins. Transplantation. 1967;5:328–35.
24. Crow JF. A golden anniversary: cattle twins and immune tolerance. Genetics. 1996;144(3): 855–9.
25. Diaconescu R, Storb R. Allogeneic hematopoietic cell transplantation: from experimental biology to clinical care. J Cancer Res Clin Oncol. 2005;131(1):1–13.
26. Diukman R, Golbus MS. In utero stem cell therapy. J Reprod Med. 1992;37(6):515–20.
27. Donahue J, Gilpin E, Lee TH, Busch MP, Croft M, Carrier E. Microchimerism does not induce tolerance and sustains immunity after in utero transplantation. Transplantation. 2001;71(3):359–68.

28. Durkin ET, Jones KA, Rajesh D, Shaaban AF. Early chimerism threshold predicts sustained engraftment and NK-cell tolerance in prenatal allogeneic chimeras. Blood. 2008;112(13): 5245–53.
29. Felsburg PJ. Overview of immune system development in the dog: comparison with humans. Hum Exp Toxicol. 2002;21(9-10):487–92.
30. Flake AW, Harrison MR, Adzick NS, Zanjani ED. Transplantation of fetal hematopoietic stem cells in utero: the creation of hematopoietic chimeras. Science. 1986;233(4765):776–8.
31. Flake AW, Roncarolo MG, Puck JM, Almeida-Porada G, Evans MI, Johnson MP, et al. Treatment of X-linked severe combined immunodeficiency by in utero transplantation of paternal bone marrow. N Engl J Med. 1996;335(24):1806–10.
32. Flake AW, Zanjani ED. In utero hematopoietic stem cell transplantation: a status report. JAMA. 1997;278(11):932–7.
33. Flake AW, Zanjani ED. In utero hematopoietic stem cell transplantation: ontogenic opportunities and biologic barriers. Blood. 1999;94(7):2179–91.
34. Fleischman RA, Mintz B. Prevention of genetic anemias in mice by microinjection of normal hematopoietic stem cells into the fetal placenta. Proc Natl Acad Sci U S A. 1979;76(11): 5736–40.
35. Gill T. Chimerism in humans. Transplant Proc. 1977;9(2):1423–31.
36. Götherström C, Westgren M, Shaw SW, Aström E, Biswas A, Byers PH, et al. Pre- and postnatal transplantation of fetal mesenchymal stem cells in osteogenesis imperfecta: a two-center experience. Stem Cells Transl Med. 2014;3(2):255–64.
37. Guillot PV, Abass O, Bassett JH, Shefelbine SJ, Bou-Gharios G, Chan J, et al. Intrauterine transplantation of human fetal mesenchymal stem cells from first-trimester blood repairs bone and reduces fractures in osteogenesis imperfecta mice. Blood. 2008;111(3):1717–25.
38. Hansen HE, Niebuhr E, Lomas C. Chimeric twins. T.S. and M.R. reexamined. Hum Hered. 1984;34(2):127–30.
39. Harrison DE, Zhong RK, Jordan CT, Lemischka IR, Astle CM. Relative to adult marrow, fetal liver repopulates nearly five times more effectively long-term than short-term. Exp Hematol. 1997;25(4):293–7.
40. Harrison DE, Astle CM. Short- and long-term multilineage repopulating hematopoietic stem cells in late fetal and newborn mice: models for human umbilical cord blood. Blood. 1997;90(1):174–81.
41. Harrison MR, Slotnick RN, Crombleholme TM, Golbus MS, Tarantal AF, Zanjani ED. In-utero transplantation of fetal liver haemopoietic stem cells in monkeys. Lancet. 1989;2(8677):1425–7.
42. Hayashi S, Peranteau WH, Shaaban AF, Flake AW. Complete allogeneic hematopoietic chimerism achieved by a combined strategy of in utero hematopoietic stem cell transplantation and postnatal donor lymphocyte infusion. Blood. 2002;100(3):804–12.
43. Hayashi S, Abdulmalik O, Peranteau WH, Ashizuka S, Campagnoli C, Chen Q, et al. Mixed chimerism following in utero hematopoietic stem cell transplantation in murine models of hemoglobinopathy. Exp Hematol. 2003;31(2):176–84.
44. Hayashi S, Hsieh M, Peranteau WH, Ashizuka S, Flake AW. Complete allogeneic hematopoietic chimerism achieved by in utero hematopoietic cell transplantation and cotransplantation of LLME-treated, MHC-sensitized donor lymphocytes. Exp Hematol. 2004;32(3):290–9.
45. Hayward A, Ambruso D, Battaglia F, Donlon T, Eddelman K, Giller R, et al. Microchimerism and tolerance following intrauterine transplantation and transfusion for alpha-thalassemia-1. Fetal Diagn Ther. 1998;13(1):8–14.
46. Howson-Jan K, Matloub YH, Vallera DA, Blazar BR. In utero engraftment of fully H-2-incompatible versus congenic adult bone marrow transferred into nonanemic or anemic murine fetal recipients. Transplantation. 1993;56(3):709–16.
47. Hsieh MM, Fitzhugh CD, Tisdale JF. Allogeneic hematopoietic stem cell transplantation for sickle cell disease: the time is now. Blood. 2011;118(5):1197–207.
48. Huang Y, Bozulic LD, Miller T, Xu H, Hussain LR, Ildstad ST. CD8alpha+plasmacytoid precursor DCs induce antigen-specific regulatory T cells that enhance HSC engraftment in vivo. Blood. 2011;117(8):2494–505.

49. Javazon EH, Merchant AM, Danzer E, Flake AW. Reconstitution of hematopoiesis following intrauterine transplantation of stem cells. Methods Mol Med. 2005;105:81–94.
50. Jordan CT, Astle CM, Zawadzki J, Mackarehtschian K, Lemischka IR, Harrison DE. Long-term repopulating abilities of enriched fetal liver stem cells measured by competitive repopulation. Exp Hematol. 1995;23(9):1011–5.
51. Josefowicz SZ, Lu LF, Rudensky AY. Regulatory T cells: mechanisms of differentiation and function. Annu Rev Immunol. 2012;30:531–64.
52. Kim HB, Shaaban AF, Yang EY, Liechty KW, Flake AW. Michrochimerism and tolerance after in utero bone marrow transplantation in mice. J Surg Res. 1998;77(1):1–5.
53. Kim HB, Shaaban AF, Milner R, Fichter C, Flake AW. In utero bone marrow transplantation induces donor-specific tolerance by a combination of clonal deletion and clonal anergy. J Pediatr Surg. 1999;34(5):726–9.
54. Vrecenak JD, Pearson EG, Santore MT, Todorow CA, Li H, Radu A, Johnson MP, Flake AW. Stable longterm mixed hematopoietic chimerism achieved in a canine model of allogeneic in utero hematopoietic cell transplantation. Blood. 2014;124(12):1987–1995.
55. Le Blanc K, Götherström C, Ringdén O, Hassan M, McMahon R, Horwitz E, et al. Fetal mesenchymal stem-cell engraftment in bone after in utero transplantation in a patient with severe osteogenesis imperfecta. Transplantation. 2005;79(11):1607–14.
56. Lee PW, Cina RA, Randolph MA, Goodrich J, Rowland H, Arellano R, et al. Stable multilineage chimerism across full MHC barriers without graft-versus-host disease following in utero bone marrow transplantation in pigs. Exp Hematol. 2005;33(3):371–9.
57. Lee PW, Cina RA, Randolph MA, Arellano R, Goodrich J, Rowland H, et al. In utero bone marrow transplantation induces kidney allograft tolerance across a full major histocompatibility complex barrier in Swine. Transplantation. 2005;79(9):1084–90.
58. Leung W, Ramirez M, Civin CI. Quantity and quality of engrafting cells in cord blood and autologous mobilized peripheral blood. Biol Blood Marrow Transplant. 1999;5(2):69–76.
59. Leventhal J, Huang Y, Xu H, Goode I, Ildstad ST. Novel regulatory therapies for prevention of Graft-versus-host disease. BMC Med. 2012;10:48.
60. Liechty KW, MacKenzie TC, Shaaban AF, Radu A, Moseley AM, Deans R, et al. Human mesenchymal stem cells engraft and demonstrate site-specific differentiation after in utero transplantation in sheep. Nat Med. 2000;6(11):1282–6.
61. Lillie FR. The theory of the free-martin. Science. 1916;43(1113):611–3.
62. Medvisnky A, Dzierzak E. Definitive hematopoiesis is autonomously initiated by the AGM region. Cell. 1996;86(6):897–906.
63. Merianos D, Heaton T, Flake AW. In utero hematopoietic stem cell transplantation: progress towards clinical application. Biol Blood Marrow Transplant. 2008;14(7):729–40.
64. Merianos DJ, Tiblad E, Santore MT, Todorow CA, Laje P, Endo M, et al. Maternal alloantibodies induce a postnatal immune response that limits engraftment following in utero hematopoietic cell transplantation in mice. J Clin Invest. 2009;119(9):2590–600.
65. Michlitsch JG, Walters MC. Recent advances in bone marrow transplantation in hemoglobinopathies. Curr Mol Med. 2008;8(7):675–89.
66. Mikkola HK, Gekas C, Orkin SH, Dieterlen-Lievre F. Placenta as a site for hematopoietic stem cell development. Exp Hematol. 2005;33(9):1048–54.
67. Mintz B, Anthony K, Litwin S. Monoclonal derivation of mouse myeloid and lymphoid lineages from totipotent hematopoietic stem cells experimentally engrafted in fetal hosts. Proc Natl Acad Sci U S A. 1984;81(24):7835–9.
68. Moses S. Pathophysiology and dietary treatment of the glycogen storage diseases. J Pediatr Gastroenerol Nutr. 1990;11(2):155–74.
69. Muench MO, Rae J, Barcena A, Leemhuis T, Farrell J, Humeau L, et al. Transplantation of a fetus with paternal Thy-1(+)CD34(+) cells for chronic granulomatous disease. Bone Marrow Transplant. 2001;27(4):355–64.
70. Nijagal A, Wegorzewska M, Jarvis E, Le T, Tanq Q, MacKenzie TC. Maternal T cells limit engraftment after in utero hematopoietic cell transplantation in mice. J Clin Invest. 2011;121(2):582–92.

71. Nijagal A, Flake AW, MacKenzie TC. In utero hematopoietic cell transplantation for the treatment of congenital anomalies. Clin Perinatol. 2012;39(2):301–10.
72. Orlandi F, Giambona A, Messana F, Marino M, Abate I, Calzolari R, et al. Evidence of induced non-tolerance in HLA-identical twins with hemoglobinopathy after in utero fetal transplantation. Bone Marrow Transplant. 1996;18(3):637–9.
73. Owen RD. Immunogenetic consequences of vascular anastomoses between bovine twins. Science. 1945;102(2651):400–1.
74. Palmer E. Negative selection—clearing out the bad apples from the T-cell repertoire. Nat Rev Immunol. 2003;3(5):383–91.
75. Peranteau WH, Hayashi S, Hsieh M, Shaaban AF, Flake AW. High-level allogeneic chimerism achieved by prenatal tolerance induction and postnatal nonmyeloablative bone marrow transplantation. Blood. 2002;100(6):2225–34.
76. Peranteau WH, Endo M, Adibe OO, Merchant A, Zoltick PW, Flake AW. CD26 inhibition enhances allogeneic donor-cell homing and engraftment after in utero hematopoietic-cell transplantation. Blood. 2006;108(13):4268–74.
77. Peranteau WH, Endo M, Adibe OO, Flake AW. Evidence for an immune barrier after in utero hematopoietic-cell transplantation. Blood. 2007;109(3):1331–3.
78. Peranteau WH, Heaton TE, Gu YC, Volk SW, Bauer TR, Alcorn K, et al. Haploidentical in utero hematopoietic cell transplantation improves phenotype and can induce tolerance for postnatal same-donor transplants in the canine leukocyte adhesion deficiency model. Biol Blood Marrow Transplant. 2009;15(3):293–305.
79. Petersen SM, Gendelman M, Murphy KM, Torbenson M, Jones RJ, Althaus JE, et al. Use of T-cell antibodies for donor dosaging in a canine model of in utero hematopoietic stem cell transplantation. Fetal Diagn Ther. 2007;22(3):175–9.
80. Picus J, Aldrich WR, Letvin NL. A naturally occurring bone marrow-chimeric primate. I. Integrity of its immune system. Transplantation. 1985;39(3):297–303.
81. Picus J, Holley K, Aldrich WR, Griffin JD, Letvin NL. A naturally occurring bone marrow-chimeric primate. II. Environment dictates restriction on cytolytic T lymphocyte-target cell interactions. J Exp Med. 1985;162(6):2035–52.
82. Renda MC, Fecarotta E, Dieli F, Markling L, Westgren M, Damiani G, et al. Evidence of alloreactive T lymphocytes in fetal liver: implications for fetal hematopoietic stem cell transplantation. Bone Marrow Transplant. 2000;25(2):135–41.
83. Rosler ES, Brandt JE, Chute J, Hoffman R. An in vivo competitive repopulation assay for various sources of human hematopoietic stem cells. Blood. 2000;96(10):3414–21.
84. Santore MT, Roybal JL, Flake AW. Prenatal stem cell transplantation and gene therapy. Clin Perinatol. 2009;36(2):451–71.
85. Shaaban AF, Kim HB, Milner R, Flake AW. A kinetic model for the homing and migration of prenatally transplanted marrow. Blood. 1999;94(9):3251–7.
86. Shaaban AF, Kim HB, Gaur L, Liechty KW, Flake AW. Prenatal transplantation of cytokine-stimulated marrow improves early chimerism in a resistant strain combination but results in poor long-term engraftment. Exp Hematol. 2006;34(9):1278–87.
87. Shields LE, Gaur LK, Gough M, Potter J, Sieverkropp A, Andrews RG. In utero hematopoietic stem cell transplantation in nonhuman primates: the role of T cells. Stem Cells. 2003;21(3):304–14.
88. Slavin S, Naparstek E, Ziegler M, Lewin A. Clinical application of intrauterine bone marrow transplantation for treatment of genetic diseases—feasibility studies. Bone Marrow Transplant. 1992;9 Suppl 1:189–90.
89. Srour EF, Zanjani ED, Brandt JE, Leemhuis T, Briddell RA, Heerema NA, et al. Sustained human hematopoiesis in sheep transplanted in utero during early gestation with fractionated adult human bone marrow cells. Blood. 1992;79(6):1404–12.
90. Storb R. Allogeneic hematopoietic stem cell transplantation—yesterday, today, and tomorrow. Exp Hematol. 2003;31(1):1–10.

91. Takahashi T, Tagami T, Yamazaki S, Uede T, Shimizu J, Sakaguchi N, et al. Immunologic self-tolerance maintained by CD25(+)CD4(+) regulatory T cells constitutively expressing cytotoxic T lymphocyte-associated antigen 4. J Exp Med. 2000;192(2):303–10.
92. Takahama Y. Journey through the thymus: stromal guides for T-cell development and selection. Nat Rev Immunol. 2006;6(2):127–35.
93. Tavian M, Péault B. Embryonic development of the human hematopoietic system. Int J Dev Biol. 2005;49(2-3):243–50.
94. Thomsen M, Hansen HE, Dickmeiss E. MLC and CML studies in the family of a pair of HLA haploidentical chimeric twins. Scand J Immunol. 1977;6(5):523–8.
95. Touraine JL, Raudrant D, Royo C, Rebaud A, Roncarolo MG, Souillet G, et al. In-utero transplantation of stem cells in bare lymphocyte syndrome. Lancet. 1989;1(8651):1382.
96. Touraine JL, Raudrant D, Rebaud A, Roncarolo MG, Laplace S, Gebuhrer L, et al. In utero transplantation of stem cells in humans: immunological aspects and clinical follow-up of patients. Bone Marrow Transplant. 1992;9 Suppl 1:121–6.
97. Touraine JL, Raudrant D, Golfier F, Rebaud A, Sembeil R, Roncarolo MG, et al. Reappraisal of in utero stem cell transplantation based on long-term results. Fetal Diagn Ther. 2004;19(4):305–12.
98. Van Dijk BA, Boomsma DI, de Man AJ. Blood group chimerism in human multiple births is not rare. Am J Med Genet. 1996;61(3):264–8.
99. Vanleene M, Saldanha Z, Cloyd KL, Jell G, Bou-Gharios G, Bassett JH, et al. Transplantation of human fetal blood stem cells in the osteogenesis imperfect mouse leads to improvement in multiscale tissue properties. Blood. 2011;117(3):1053–60.
100. Vrecenak JD, Flake AW. In utero hematopoietic cell transplantation – recent progress and potential for clinical application. Cytotherapy. 2013;15(5):525–35.
101. Vrecenak JD, Flake AW. Fetal surgical intervention: progress and perspectives. Pediatr Surg Int. 2013;29(5):407–17.
102. Wengler GS, Lanfranchi A, Frusca T, Verardi R, Neva A, Burgnoni D, et al. In-utero transplantation of parental CD34 hematopoietic progenitor cells in a patient with X-linked severe combined immunodeficiency (SCIDX1). Lancet. 1996;348(9040):1484–7.
103. Westgren M, Ringden O, Eik-Nes S, Ek S, Anvret M, Brubakk AM, et al. Lack of evidence of permanent engraftment after in utero fetal stem cell transplantation in congenital hemoglobinopathies. Transplantation. 1996;61(8):1176–9.
104. Westgren M, Ringden O, Bartmann P, Bui TH, Lindton B, Mattsson J, et al. Prenatal T-cell reconstitution after in utero transplantation with fetal liver cells in a patient with X-linked severe combined immunodeficiency. Am J Obstet Gynecol. 2002;187(2):475–82.
105. Wolf NS, Bertoncello I, Jiang D, Priestley G. Developmental hematopoiesis from prenatal to young-adult life in the mouse model. Exp Hematol. 1995;23(2):142–6.
106. Yoder MC, Cumming JG, Hiatt K, Mukherjee P, Williams DA. A novel method of myeloablation to enhance engraftment of adult bone marrow cells in newborn mice. Biol Blood Marrow Transplant. 1996;2(2):59–67.
107. Zanjani ED, Pallavicini MG, Ascensao JL, Flake AW, Langlois RG, Reitsma M, et al. Engraftment and long-term expression of human fetal hemopoietic stem cells in sheep following transplantation in utero. J Clin Invest. 1992;89(4):1178–88.
108. Zanjani ED, Flake AW, Rice H, Hedrick M, Tavassoli M. Long-term repopulating ability of xenogeneic transplanted human fetal liver hematopoietic stem cells in sheep. J Clin Invest. 1994;93(3):1051–5.

Chapter 18
Fetal Tissue Engineering

Christina Feng and Dario O. Fauza

Introduction

The fetus is an ideal tissue engineering subject, both as host and donor. The developmental and long-term impacts of tissue implantations into a fetus, along with the unique characteristics of fetal cells, combine to significantly expand the reach of tissue engineering. Inklings of such a perspective have led to attempts at harnessing these potential benefits since long before the modern eras of transplantation and regenerative medicine. The first reported transplantation of human fetal tissue took place as early as 1922, when a fetal adrenal graft was transplanted into a patient with Addison's disease [1]. This and all the other initial experiments that followed involving human fetal tissue transplantation failed [2, 3]. Only in the late twentieth century did fetal tissue transplantation in humans start to yield favorable, though still quite variable, outcomes. Although the vast majority of studies to date have involved simply fetal cell, tissue, or organ transplantation, various engineered open systems using fetal cells have been tested in animal models, with their first clinical applications expected for the not too distant future.

Fetal tissue has been used as an investigational tool in biomedical science since the 1930s [4, 5]. *In vitro* applications of fetal tissue have long been established and are fairly common. Cultures of different fetal cell lines and commercial

C. Feng, M.D.
Department of Surgery, Beth Israel Deaconess Medical Center,
110 Francis St, Boston, MA 02215, USA
e-mail: cfeng@bidmc.harvard.edu

D.O. Fauza, M.D., Ph.D. (✉)
Department of Surgery, Boston Children's Hospital, Harvard Medical School,
300 Longwod Avenue, Fegan 3, Boston, MA 02115, USA
e-mail: Dario.Fauza@childrens.harvard.edu

D.O. Fauza, M. Bani (eds.), *Fetal Stem Cells in Regenerative Medicine*,
Stem Cell Biology and Regenerative Medicine,
DOI 10.1007/978-1-4939-3483-6_18

preparations of human fetal tissue have been routinely used in a wide assortment of studies, as well as in viral isolation and culture, and to produce vaccines. Fetal cells and extra-embryonic structures such as placenta, amnion and the umbilical cord have been used to develop new products and to screen them for toxicity, teratogenicity and carcinogenicity. Fetal tissue banks have been operating in the United States and abroad for many years supplying fetal cell lines for many of these activities.

Compared with such a large body of data having come out of research and other endeavors involving fetal cells or tissues, much less has been done on "true" engineering of fetal tissue, through culture and placement of fetal cells into matrices or membranes—so-called open system tissue engineering—or through other *in vitro* manipulations prior to implantation. Controlled human trials of actual fetal tissue engineering have yet to be performed and a still relatively modest number of animal experiments have been reported to date. Fetal cells were first used experimentally in engineered constructs by Vacanti *et al.* in the mid to late 1980s, interestingly as part of the introduction of the notion of selective cell transplantation using bioabsorbable, synthetic polymers as matrices [6]. They used fetal cells from the liver, intestine and pancreas, which were cultured, seeded on bioabsorbable matrices and later implanted in rats in heterologous fashion and heterotopically, namely in the interscapular fat, omentum, and mesentery, with no structural replacement. Specimens were removed for histological analysis no later than 2 weeks after implantation, with successful engraftment observed in some animals that received hepatic and intestinal constructs, but in none that received pancreatic ones. Only in 1995 was a second study performed, by the same group, that time involving only fetal liver constructs, also implanted in heterologous and heterotopic fashion in rats [7]. Fetal hepatocytes were shown to proliferate to a greater extent than adult ones in culture and to yield higher cross-sectional cell area at the implant. As in the first experiment, neither structural replacement, nor functional studies were included. The use of fetal constructs as a means of structural and functional replacement was first reported experimentally only in 1997 [8, 9].

This chapter offers an overview of various facets of the still young field of fetal tissue engineering, along with a few insights on fetal cell and tissue transplantation.

Peculiarities of Fetal Cells

Perennial complications of tissue engineering, such as immunological rejection (in non-autologous applications), growth limitations, differentiation and function restraints, incorporation barriers and cell/tissue delivery difficulties and others can all be minimized, if not totally prevented, when fetal cells are used. Due to their properties both *in vitro* and *in vivo*, fetal cells are among the best "building blocks" for tissue engineering.

In Vitro

Most fetal cells multiply more rapidly and more often in culture than postnatal counterparts. Depending on the cell line considered, however, such enhanced proliferation is more or less pronounced, or, in a few cases, not evident at all. Due, at least in part, to their proliferation patterns and generally more immature state, fetal cells have long been recognized as ideal targets for gene transfers. Because they are very plastic in differentiation potential, fetal cells respond better than mature cells to environmental signals. Just as a more commonly explored example, younger mesenchymal stem cells (MSCs) from fetal tissues, blood, or annexes are significantly more plastic and grow faster than adult, bone marrow-derived MSCs. Fetal MSCs are biologically closer to embryonic stem cells and have unique markers and characteristics not found in adult bone marrow MSCs, which are advantageous for cell therapy.

Fetal cells, including from annexes, can survive at lower oxygen tensions than those tolerated by phenotypically comparable mature cells and respond better to hypoxia during *in vitro* manipulations [10]. They also tend to lack long extensions and strong intercellular adhesions. In large part because of those characteristics, fetal cells display better survival after refrigeration and cryopreservation protocols when compared with adult cells. This enhanced endurance during cryopreservation, however, seems to be tissue-specific. For instance, data from primates and humans have shown that fetal hematopoietic stem cells, as well as fetal lung, kidney, intestine, thyroid and brain tissues can be well preserved at low temperatures, whereas non-hematopoietic liver and spleen tissues can also be cryopreserved, but not so easily.

It has long been known that fetal cells are "immunologically privileged". For instance, fetal MSCs may express human leucocyte antigens (HLA) class I but not HLA class II. The presence of interferon gamma (IFN-gamma) in the growth medium may initiate the intracellular synthesis and cell surface expression of HLA class II, but neither undifferentiated nor differentiated fetal MSCs tend to induce proliferation of allogenic lymphocytes in mixed cultures. In fact, fetal MSCs treated with IFN-gamma typically suppress alloreactive lymphocytes in that setting. These and other characteristics contribute to explain why both undifferentiated and differentiated fetal MSCs may not elicit much alloreactive lymphocyte proliferation, thus potentially rendering these cells particularly suitable for heterologous transplantation.

In Vivo

Engineered constructs made with fetal cells tend to be less susceptible to rejection in allologous applications. Certain xenograft implantations are also viable, as studies suggest that fetal cells are also better tolerated in cross-species transplantations, including in humans [11–13]. Fetal allograft survival, growth, maturation and function in immunocompetent recipients is age and tissue specific, as is the expression of major histocompatibility complex (H-2) antigens. At least in fetal mice, the precise gestational time of first detection of H-2 antigen expression and the proportion of cells expressing these determinants depend on inbred strain, specific haplotype, tissue of origin and antiserum batch employed. Nonetheless, the precise factors governing the timing and tissue-specificity of H-2 antigen expression have yet to be determined in most species, including humans.

Mechanisms other than H-2 antigen expression also seem to govern fetal immunogenicity. For example, at least in certain mammalian species, the conceptus suppresses T cell activity by catabolizing tryptophan and defends itself against rejection by the mother. In humans, fetal cells are found in the maternal circulation in most pregnancies and fetal progenitor cells have been found to persist in the circulation of women decades after child birth [14, 15]. Pregnancy results in the acquisition of cells with stem-cell-like properties that are thought to influence maternal health post-partum, by triggering disease and/or avoiding/combating it, for example by to differentiating into select phenotypes in diseased/injured maternal tissue.

Fetal cells can produce high levels of angiogenic and trophic factors, which enhance their ability to grow once grafted. Those factors may also facilitate regeneration of surrounding host tissues. Interestingly, significant clinical and hematological improvement has been described after fetal liver stem cell transplantation in humans, even when there is no evidence of sustained engraftment. These improvements have been attributed to the restoration of autologous hematopoiesis and inhibition of tumor cell growth promoted by the infused cells, through mechanisms yet to be fully understood. The more immature state of fetal cells also optimize engraftment, by allowing them to grow, elongate, migrate and establish functional connections with other cells [12].

Due to many of the general benefits derived from the use of fetal cells, along with other properties specific to each cell line, several types of fetal cellular transplantation have been investigated experimentally or employed in humans for decades. Clinically, fetal cells have been (mostly anecdotally) proven useful in a number of different conditions, including, but not limited to: Parkinson's and Huntington's disease; *Diabetes mellitus*; aplastic anemia; Wiskott-Aldrich syndrome; thymic aplasia (DiGeorge syndrome) and thymic hypoplasia with abnormal immunoglobulin syndrome (Nezelof syndrome); thalassemia; Fanconi anemia; acute myelogenous and lymphoblastic leukemia; Philadelphia chromosome-positive chronic myeloid leukemia; X-linked lymphoproliferative syndrome; neuroblastoma; severe combined immunodeficiency disease; hemophilia; osteogenesis imperfecta; skin reconstruction; acute fatty liver of pregnancy; neurosensory hypoacusis; malaria; and HIV. They have also been applied in attempts to repair inborn errors of metabolism, including Gaucher's disease, Fabry's disease, fucosidosis, Hurler's syndrome, metachromatic leucodystrophy, Hunter's syndrome, glycogenosis, Sanfilippo's syndrome, Morquio syndrome type B and Niemann-Pick disease. Experimentally, fetal cell and organ transplantation continue to be studied in an ever-expanding array of diseases. *In utero* haematopoietic stem cell transplantation is a promising, entirely nonmyeloablative approach to achieve mixed hematopoietic chimerism and associated donor-specific tolerance for the treatment of a variety of genetic disorders, as discussed in detail elsewhere in this textbook.

Alternative Sources of Fetal Cells

Fetal cells amenable to processing for tissue engineering can be obtained from a variety of sources, besides the fetus, including the amniotic fluid and membrane, placenta, Wharton's jelly and umbilical cord blood. Although peripheral maternal blood can also be a source of fetal cells, their consistent isolation in numbers and phenotypes compatible with tissue engineering remains to be shown. So far, of all these sources, the amniotic fluid and the placenta have been the most appealing clinically. They are the least invasive ones for both the mother and the fetus (until peripheral maternal blood proves consistently viable, if ever) and both amniocentesis and chorionic villus sampling (CVS) are widely used forms of prenatal diagnostic screening. The fact that a diagnostic amniocentesis and/or CVS are routinely offered when a fetus is diagnosed with a structural anomaly in prenatal imaging eliminates an additional risk from the therapeutic processing of the sample procured; rendering moot eventual ethical concerns (Fig. 18.1). Further, these sources actually expand the ethically sensible realm of fetal tissue engineering as a form of perinatal therapy beyond life-threatening structural anomalies.

The full spectrum of cell types that can be obtained from sources other than the fetus for tissue engineering purposes is discussed in dedicated chapters elsewhere in this textbook. This section will discuss only certain aspects not covered in these other chapters.

Amniotic Fluid

The cellular profile of the amniotic fluid varies with gestational age [16]. In addition to a common origin with the mesenchymal portion of the placenta, the amniotic cavity/fluid receives cells shed from the fetus and, quite possibly, from the placenta as well (although the latter has yet to be definitely confirmed). The mechanisms responsible for the production and turnover of the amniotic fluid are thought to also influence the cell types present in the amniotic cavity. In the first half of gestation, most of the amniotic fluid derives from active sodium and chloride transport across the amniotic membrane and fetal skin, with simultaneous passive movement of water. In the second half, most of the fluid derives from fetal micturition. An additional significant source of amniotic fluid is active secretion from the respiratory tract. Fetal swallowing and gastrointestinal tract excretions, while not voluminous, of course also play a role in the composition of the amniotic fluid. As a result of such complex fluid dynamics, cells present in the urinary, respiratory, and gastrointestinal tracts are shed into the amniotic cavity.

Amniotic fluid composition changes predictably throughout gestation. In humans, it is isotonic with fetal plasma during early pregnancy, due to the transudation of fetal plasma through the maternal deciduas, or through the fetal skin prior to keratinization, which occurs at approximately 24 weeks. Subsequently and until term, it becomes increasingly hypotonic relative to maternal or fetal plasma. All these variables that

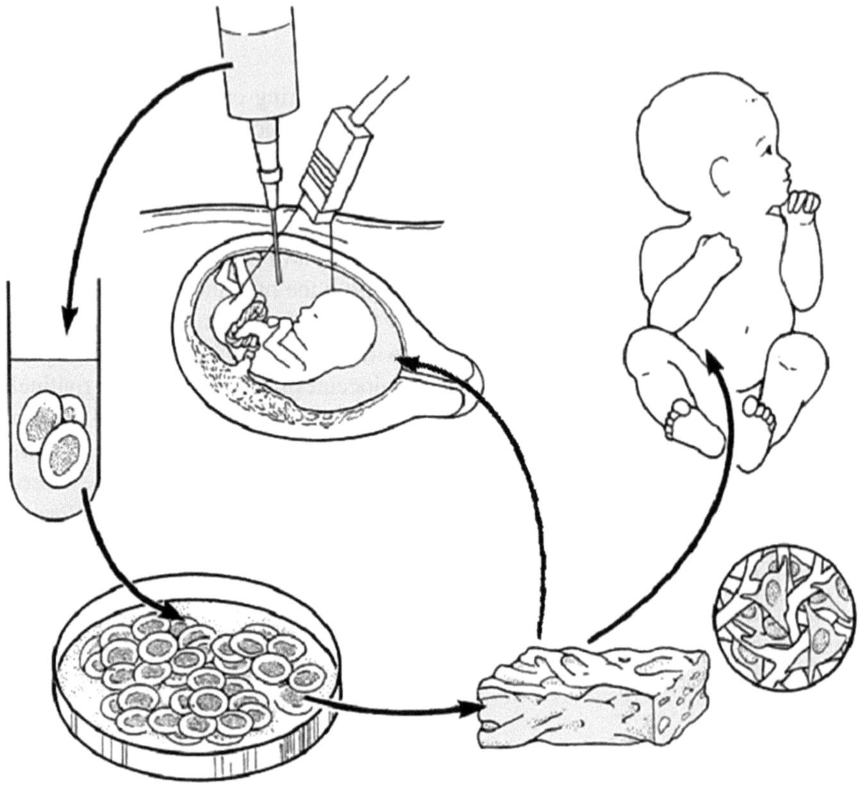

Fig. 18.1 Diagram representing the concept of fetal tissue engineering from amniotic fluid cells for the treatment of birth defects. A small aliquot of amniotic fluid is obtained from a routine amniocentesis, typically performed when a structural anomaly is diagnosed by routine prenatal imaging screening. Fetal tissue is then engineered *in vitro* from amniotic progenitor cells while pregnancy continues, so that the newborn, or a fetus, can benefit from having autologous, expanded tissue promptly available for surgical reconstruction at birth or in utero. Reproduced, with permission, from "Fauza DO. Tissue engineering and transplantation in the fetus. In: Lanza R, Langer R, Vacanti JP, editors. Principles of tissue engineering. San Diego: Academic Press; 2014. p. 511–30"

play a role in amniotic fluid composition seem to contribute to the changeable profile of the cellular component present therein [16]. Still, much remains to be clarified about the ontogeny of many subsets of amniocytes at any gestational age, as discussed in other chapters.

As also discussed in more detail elsewhere, MSCs can be isolated from the amniotic fluid throughout gestation. Our group has described a fairly simple and easily reproducible protocol for isolation of amniotic fluid MSCs (afMSCs), now widely employed [17–19]. A number of other protocols have also been reported and used by many. Ever since our initial demonstrations that afMSCs can be used for tissue engineering purposes, a plethora of experimental reports have described amniotic fluid cell-based constructs at a variety of anatomical sites and health/disease states—too many to list here [17, 20]. The translational relevance of afMSCs in fetal tissue engineering has been underscored by the viability of scaled-up manufacturing of human afMSCs in compliance with regulatory guidelines and by the establishment of amniotic cell banks [21, 22].

Placenta

Different cell types are found in the placenta at different gestational ages, as a result of the mechanisms driving placental development. Mesenchymal placental villi are the first structures providing the morphological requisites for maternal-fetal exchange of gases, nutrients, and waste. They are also the precursors of all other villous types. The differentiation of the mesenchymal villi into immature or mature intermediate villi is a determining factor in the balance between growth and maturation of the placenta, which, in turn, has a direct impact on the cell types that can be isolated from it at different gestational ages. Placental mesenchymal stem cells (pMSCs) are part of a large mesenchymal component. Recruitment of these cells supports the so-called vasculogenesis that occur during vascularization of the villous sprouts, in addition to the angiogenesis based on the proliferation of endothelial precursors. Mesenchymal cells also play other roles in placental development, such as the paracrine control of the stability of the cytotrophoblast column, which in turn determines the degree of trophoblast invasiveness. Thus, perhaps not surprisingly, pMSCs can be valuable in tissue engineering applications.

Also here, a number of different protocols for isolation and expansion of pMSCs have been reported, including one analogous to that described for the separation of afMSCs, which can be employed in both "full thickness" and CVS placental specimens [19, 23].

Maternal Blood

Fetal cells can be documented in the maternal circulation in the majority of human pregnancies and have increasingly been used for diagnostic purposes, often substituting amniocentesis, albeit mostly for the detection of fetal aneuploidy, with relatively low fetal cell yields [24]. Fetal progenitor cells have been found to persist in the circulation of women for as much as decades after child birth [14]. Among the fetal cells that have been identified therein are trophoblasts, lymphocytes, and nucleated erythroid cells. A certain population of fetal cells, so-called pregnancy-associated, or pregnancy-acquired fetal progenitor cells, seems to differentiate in diseased or injured maternal tissue. They are found in higher frequency in maternal tissue injury sites when compared to healthy areas, and exhibit plasticity and site-appropriate phenotypes. Indeed, their exact phenotype is still unknown, though recent research has established that they contain cells of ectodermal, endodermal, and mesodermal lineages [25]. At this time, peripheral maternal blood has yet to be proven as a viable source of fetal cells of predictable phenotype and in consistently high enough numbers for tissue engineering applications. This enticing perspective, however, surely deserves continued scrutiny.

The Fetus as a Transplantation Host

A number of advantages of implanting an engineered construct *in utero* can be envisioned, not only from a theoretical perspective, but also from clinical and experimental evidence derived from intra-uterine cellular transplantation studies already reported. Those (potential or documented) advantages include: induction of graft tolerance in the fetus, due to its immunologic immaturity; induction of donor-specific tolerance in the fetus by concurrent or previous intra-uterine transplantation of hematopoietic progenitor cells; a completely sterile environment; the presence of hormones, cytokines and other inter-cellular signaling factors that may enhance graft survival and development; the unique wound healing properties of the fetus; and minimization, or early prevention of clinical manifestations of disease, before they can cause irreversible damage. Most of those advantages are closely dependent on gestational time of transplantation.

Fetal Immune Development

The uniqueness of the fetal immune system deserves special attention among the potential benefits of *in utero* transplantation. Studies involving pre and postnatal transplantation of lymphohematopoietic fetal cells have led to a better understanding of the fetal immune response.

Fetal tolerance with permanent chimerism has long been shown to occur in nature in non-identical twins with shared placental circulation [26, 27]. Still relatively little is known, however, about precisely when and by what mechanism this tolerance is lost. The precursors of the hematopoietic stem cells arise in the yolk sac, migrate to the fetal liver and then to the thymus, spleen and bone marrow. The fetal liver has its highest concentration of hematopoietic stem cells between the 4th and the 20th week of gestation. Because of their cellular immunologic "immaturity", the fetal liver, umbilical cord blood and, to a lesser extent, the fetal thymus have been studied as potential sources of hematopoietic stem cells for major histocompatibility complex—incompatible bone marrow transplantation for more than half a century now, with other chapters devoted to discussing this in detail.

During gestation, lymphocytes capable of eliciting graft-versus-host disease (GVHD) are found in the thymus by the 14th week of gestation, but not detectable in the liver until the 18th week. Thus, despite considerable numbers of granulocyte-macrophage colony-forming cells, there is an almost complete absence of mature T cells up to the 14th week in human fetal livers. While B cell development takes place mostly in the liver, T cell development occurs predominantly in the thymus. This is probably why fetal liver cells are immuno-incompetent for cell-mediated and T cell-supported humoral reactions, such as graft rejections and GVHD. However, by 18 weeks of gestation, the spleen is considered fully immunocompetent. In principle, tissue matching is not necessary in fetal liver transplantation, if procurement happens up to a certain point in gestation. In a number of animal models and small clinical series, fetal liver cells have induced no or only moderate GVHD in histoincompatible donor/recipient pairs. Recently, unique sub-populations of innate lymphoid cells (ILCs), a family of effectors and regulators of innate immunity and tissue remodeling, have been isolated from fetal lung and intestine [28]. The significance of this finding and its potential translational implications has only started to be investigated [29, 30].

Although fetal liver stem cells should not cause GVHD, they can still be subject to rejection when transplanted. Because of that, fetal liver stem cell transplantation has been attempted in the clinical setting in patients with depressed immune function, such as in immunodeficiencies, bone marrow insufficiency and during fetal life (*in utero* transplantation). The same principle applies to the use of fetal thymus. Fatal cases of GVHD are much less likely in patients who receive fetal liver stem cells harvested before the 14th week of gestation. Umbilical cord blood stem cell transplantation has been associated with a minimal incidence of GVHD [31, 32].

Another central aspect of intrauterine transplantation is the fact that maternal cells trafficking into the fetus may pose as the chief barrier to effective engraftment of allogeneic cells or tissues delivered prenatally, as also discussed in more detail in another chapter. It has been recently shown that there may be macrochimerism of maternal leukocytes in the fetal blood, with substantial increases in T cell trafficking after intrauterine transplantation [33, 34]. This suggests that clinical viability of intrauterine transplantation, at least of hematopoietic stem cells, may be enhanced by transplanting cells matched also to the mother.

In Utero Transplantation

Over the past few decades, cellular intrauterine transplantation has been employed clinically to treat a variety of diseases, including lymphohematopoietic diseases, beta-thalassemia, inborn errors of metabolism, and genetic disorders, with some success. *In utero* hematopoietic stem cell transplantation is a nonmyeloablative approach to achieve mixed hematopoietic chimerism and resulting donor-specific tolerance, improving survival of other grafts later in life. Through prenatal transplantation of hematopoietic progenitor cells, both allogeneic and xenogeneic chimerisms have been induced in animal models and allogeneic chimerism has been achieved in humans [35–38]. A recent development in prenatal cell transplantation is neural stem cell delivery to the fetal spinal cord as a means to reverse at least some of the local damage associated with experimental spina bifida [12, 39].

Tolerance of allogeneic intrauterine implantation of an engineered construct has been first demonstrated in an ovine model of fetal tracheal reconstruction with heterologous cartilage engineered from afMSCs [40]. In like manner, tolerance to an airway construct made with autologous afMSCs and a xenologous decellularized matrix has also been shown in a fetal ovine model [41]. Clinically, the prenatal repair of structural airway anomalies could be justified in select patients with defects so severe that breathing would otherwise be simply impossible at birth.

Fetal Tissue Engineering

As mentioned above, fetal constructs as a means of structural and functional replacement, in autologous fashion, in large animal models, were first reported experimentally in 1997 [8, 9]. Those studies introduced the concept of minimally invasive procurement of fetal cells, which could then used to engineer tissue *in vitro* in parallel to the remainder of gestation, so that an infant, or a fetus, with a prenatally diagnosed birth defect could benefit from having autologous, expanded tissue readily available for surgical implantation, either in the neonatal period or before birth. Fetal tissue engineering as a therapeutic strategy has only begun to be explored, with relatively few studies undertaken thus far [8, 9, 17, 18, 20, 23, 40–62].

Current Applications

Major congenital anomalies are present in approximately 3 % of all newborns, responding for nearly 20 % of deaths occurring in the neonatal period and virtually immeasurable morbidity rates throughout postnatal life. Fetal tissue engineering has emerged experimentally as a viable new alternative for the treatment of some of these anomalies. The following are brief but illustrative examples of some of the current research in preclinical applications of fetal tissue engineering to diseases such as congenital anomalies and genetic disorders. It is not meant to be a comprehensive, all-inclusive list.

Diaphragmatic Hernia

According to the Congenital Diaphragmatic Hernia (CDH) Study Group, the majority of infants with CDH cannot have their diaphragmatic defect closed primarily and most of these patients end up receiving a prosthetic diaphragmatic patch. Prosthetic diaphragmatic repair is a well-known risk factor for recurrence of the hernia, which can occur in a sizeable proportion of these patients [63–66]. Hernia recurrence is believed to stem from normal growth, which leads to traction and eventual detachment of the prosthesis. The use of a patch made of living autologous tissue able to remodel with growth could conceivably overcome much of the morbidity associated with prosthetic diaphragmatic repair. Given the timing of CDH repair in the neonatal period, an autologous construct would have to be made with fetal cells.

The first experimental report of engineered diaphragmatic replacement dates to 2000, when our group used the concept of fetal tissue engineering with myoblast-based constructs to that end, in an ovine model [42]. In subsequent efforts we focused on the engineering of a diaphragmatic tendon, rather than a muscle patch, for a variety of reasons [67]. A substantial portion of the normal diaphragm is comprised of a tendon. Except for the rare cases of complete diaphragmatic agenesis, the native residual diaphragmatic muscle seems to develop and function normally in the vast majority of children with CDH. In a comparison between tendon-based and muscle-based constructs, we actually noticed improved structural and biomechanical outcomes *in vivo* in the former, with eventual loss of myogenic identity of the donor cells in the latter [50].

afMSCs have proven particularly effective and practical for diaphragmatic tendon engineering, in large part because, by default, these cells tend to assume a fibroblastic/myofibroblastic phenotype, which is the expected phenotype in tendons (Fig. 18.2) [20, 50, 57]. These previous studies have also pointed to scaffold composition and architecture as a concurrent determining factor of outcome.

Although much can still be optimized in diaphragmatic graft engineering, initial clinical experience with fetal tissue engineering as a means to repair CDH is expected for the not too distant future.

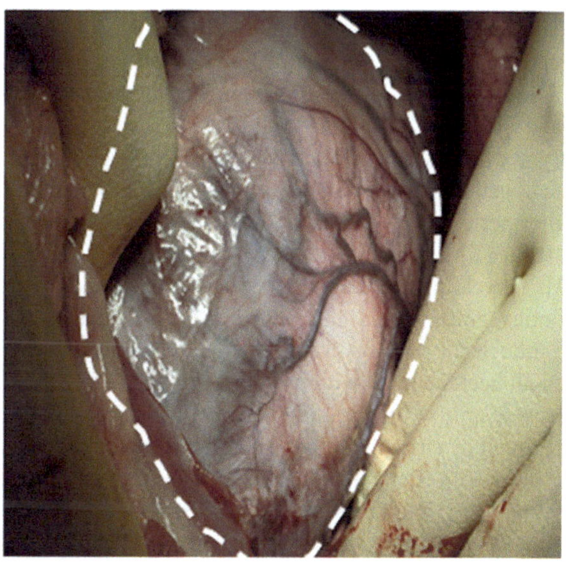

Fig. 18.2 An intact ovine diaphragmatic tendon seen from the chest, 12 months after autologous repair with an engineered, mesenchymal amniocyte-based construct. The *dotted line* encircles the area of the graft. Reproduced, with permission, from "Kunisaki, S. M., J. R. Fuchs, A. Kaviani, J. T. Oh, D. A. LaVan, J. P. Vacanti, J. M. Wilson and D. O. Fauza (2006). "Diaphragmatic repair through fetal tissue engineering: a comparison between mesenchymal amniocyte- and myoblast-based constructs." *J Pediatr Surg* **41**(1): 34–9; discussion 34–9"

Airway Anomalies

Severe forms of congenital tracheal anomalies include long-segment stenosis, atresia, and agenesis. The ideal treatment for these diseases still remains fundamentally unsolved. A variety of surgical techniques have been attempted, usually entailing the use of autologous or synthetic grafts, which are prone to infection, re-stenosis, implant extrusion, inconsistent functional outcomes and inability to remodel and grow with the child [68–73]. Cadaveric tracheal transplantation has not improved this scenario and overall mortality remains high [72–75].

Unfortunately, over one-half of the cases involve more than 50 % of the trachea, rendering primary reconstruction impossible without excessive anastomotic tension. A suitable repair of long-length tracheal deformities would require a conduit that is rigid enough to prevent collapse on inspiration while remaining flexible enough to bend with the neck. In principle, a tissue engineered conduit could offer these features, as well as have the potential for growth with the child. To date, however, anecdotal clinical experience with engineered airway constructs based on bone marrow MSCs has shown that the methodology is not yet mature enough for widespread use, at least using postnatal cells [76]. Our group has applied the principle of fetal tissue engineering in different experimental models of airway repair. We have reported on cartilage engineered from either fetal auricular chondrocytes or bone marrow-derived MSCs as a means to repair tracheal defects *in utero* [43, 44]. We have shown that cartilage can be engineered from fetal umbilical cord blood MSCs [47]. A substantial limitation of these early studies, however, was the fact all these cell sources are not easily accessible, are not without risk to the mother and fetus, and are associated with relatively prolonged intervals between cell procurement and airway repair.

Subsequently, we have shown that cartilaginous grafts engineered from afMSCs have a unique extracellular matrix composition when compared with cartilage engineered from other perinatal MSCs, as well as when compared with native hyaline and elastic cartilage, proving particularly conducive to surgical implantation (Fig. 18.3) [77]. Thereafter, we have shown, in a large animal model, that afMSC-based cartilaginous grafts could be a means for tracheal reconstruction [40]. Interestingly, although the grafts were all engineered in the absence of respiratory epithelium, they all became lined with pseudostratified columnar epithelium *in vivo* and the animals were able to breathe spontaneously post-operatively. However, stridor eventually ensued in virtually all subjects, likely because portions of the grafts remodeled into fibrous cartilage, leading to variable degrees of stenosis over time. Comparable results were obtained when the afMSCs were seeded onto decellularized airway, although epithelialization seemed somewhat enhanced [41] (Fig. 18.4). Still, given the potential impact of fetal tissue engineering on the currently dismal prognosis of fetuses with major airway disease, further development of this approach is certainly warranted.

Another application of fetal airway engineering is in the treatment of congenital cervical tumors. The prenatal diagnosis of a cervical mass raises the prospect of clinically relevant airway compromise and respiratory distress at birth, often associated with significant airway damage, either from the disease process itself,

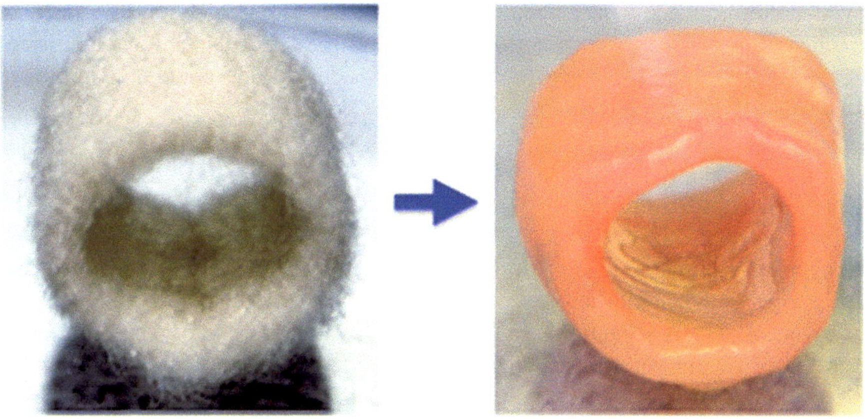

Fig. 18.3 Representative, cross-sectional view of a 3-D cartilaginous tube engineered from afMSCs (*right*) seeded onto a poly(lactic-co-glycolic acid) scaffold (*left*), previously maintained in a bioreactor under chondrogenic conditions. Reproduced, with permission, from "Fauza DO. Tissue engineering and transplantation in the fetus. In: Lanza R, Langer R, Vacanti JP, editors. Principles of tissue engineering. San Diego: Academic Press; 2014. p. 511–30"

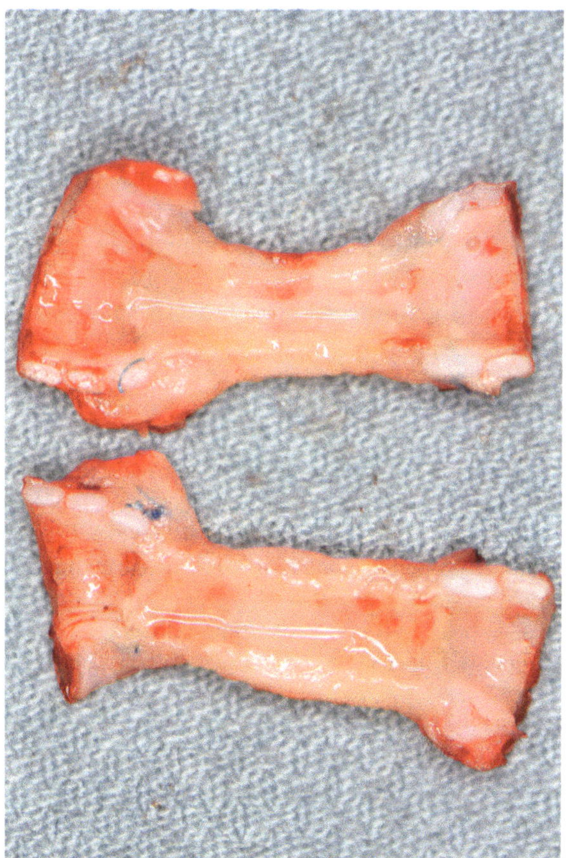

Fig. 18.4 Representative gross view of a longitudinally opened afMSC-engineered tracheal implant at necropsy showing the intraluminal glistening typical of epithelialization, confirmed histologically. Reproduced, with permission, from "Fauza DO. Tissue engineering and transplantation in the fetus. In: Lanza R, Langer R, Vacanti JP, editors. Principles of tissue engineering. San Diego: Academic Press; 2014. p. 511–30"

or as a consequence of therapeutic measures such as emergency intubations or tracheostomies and/or the resection of the mass [78–80]. Planned availability of autologous, suitably sized engineered fetal airway constructs could help minimize some of these complications.

Cardiovascular Anomalies

Congenital heart anomalies constitute the most common types of birth defects and still carry significant mortality and morbidity. Most of them involve variable degrees of myocardial, valvar, and/or vascular deformities. While primary repairs are often possible, different types of prosthesis may need to be implanted in more complex cases. Complications of prosthetic cardiac repair include thrombogenesis, absent contractility, lack of remodeling/growth, material-related failures, and suture line ruptures [81]. Myocardial, valvar, and vascular tissue engineering are potentially improved alternatives to current methods of cardiac reconstruction, all of which have been studied for a number of years now.

A contractile engineered patch could be valuable in the repair of severe septal defects, or in certain cases of cardiac hypoplasia. Several groups are working on this concept using two basic approaches. One of them is cellular cardiomyoplasty, which in essence is the simple direct injection of select cell suspensions, such as MSCs, into the myocardium, so as to possibly overcome the heart's inability to regenerate [82]. However, the notion of differentiating MSCs, including some of fetal origin, into cardiomyocytes has met with conflicting results [83–85]. The other approach is to actually create three-dimensional implantable grafts by seeding cells onto a scaffold, so as to form constructs with a defined structure which could lead to more meaningful myocardial augmentation when transplanted. Our group has shown experimentally that an autologous fetal myoblast-based engineered muscle patch implanted onto the myocardium can display prolonged donor cell survival and engraftment, with eventual expression of proteins typical of a cardiomyocyte-like lineage on the donor cells [48]. Conclusive documentation of myocardial transdifferentiation of these cells, however, along with functional analyses, remain to be described.

The engineering of heart valves normally involves two different layers, mimicking native valve architecture. An inner myofibroblast/fibroblast-based layer would produce the extracellular matrix profile typically responsible for the unique biomechanical properties found in heart valves, while an endothelial cell-based layer would produce an anti-thrombogenic and blood-compatible surface. The fabrication of such heart valves from both umbilical-cord derived and amniotic fluid-derived progenitor cells has been described [52, 53]. The ability of these structures to function as valve replacements *in vivo* in the long term remains to be conclusively described.

Deformities of the large vessels of the base of the heart are yet another major component of congenital cardiac disease. A Japanese group was the first to accumulate considerable clinical experience with the use of conduits engineered from endothelial cells obtained from a peripheral vein as vascular replacements in low-pressure systems, in children with varying forms of complex congenital cardiovascular anomalies [86, 87]. Current methods are focusing on the use of autologous bone marrow progenitor cells for seeding scaffold tubes and on enhanced scaffold fabrication [88–90]. Thus far, the mid- to long-term outcome of engineered vascular grafts in children have shown continued patency, with balloon angioplasty required in some cases to treat tissue overgrowth at anastomotic sites. The use of fetal cells in this setting could possibly lead to improved results and is currently under investigation [61].

Musculoskeletal Defects

Various congenital anomalies involve some degree of bone loss, such as craniofacial, chest wall, and limb defects. Tissue engineering has proven viable for bone replacement both experimentally and clinically since the 1990s [91]. Within the realm of perinatal medicine, prenatal delivery of MSCs, for example from first trimester peripheral fetal blood, has also been shown to ameliorate genetic bone disorders, such as osteogenesis imperfect [92–96]. Fetal chondrocytes seeded onto synthetic scaffolds have been implanted into chest wall defects in large animal models [45]. Our group has demonstrated that bone grafts can be engineered from afMSCs and electrospun nanofibers (Fig. 18.5). These grafts have been used in leporine models of postnatal repair of full thickness sternal and craniofacial defects [54, 56, 60]. Research utilizing afMSCs in various 3-dimensional constructs and materials to create bone and/or cartilage *in vitro* and *in vivo* remains ongoing [62, 97–100].

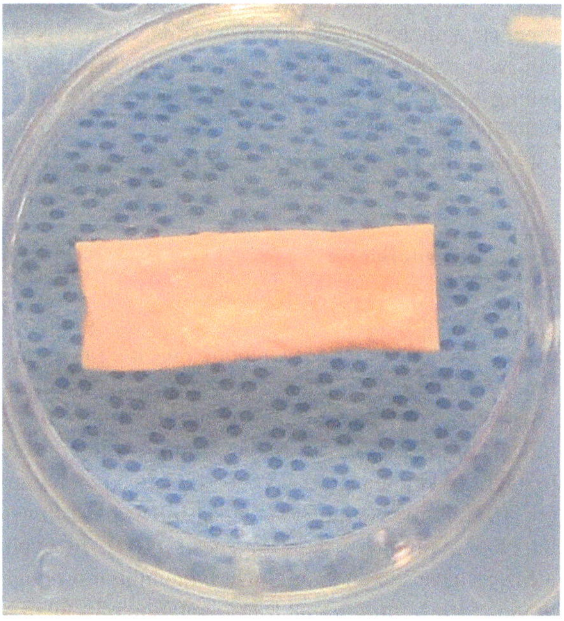

Fig. 18.5 Representativea gross view of a 3-D osseous construct engineered from afMSCs seeded onto poly-L-lactic acid electrospun nanofibrous scaffolds, previously maintained in a bioreactor under osteogenic conditions. Reproduced, with permission, from "Fauza DO. Tissue engineering and transplantation in the fetus. In: Lanza R, Langer R, Vacanti JP, editors. Principles of tissue engineering. San Diego: Academic Press; 2014. p. 511–30"

Neural Tube Defects

Failure of the neural tube to close by the end of the fourth week of gestation leads to different forms of midline vertebral defects (spina bifida) and/or of cranial defects (exencephaly). Spina bifida is the common major congenital anomaly of the central nervous system compatible with life. It leads to injury/loss of spinal cord tissue at and below the lesion, with common manifestations including paraplegia, urinary and fecal incontinence, sexual dysfunction and secondary musculoskeletal deformities. Overall mortality and morbidity rates remain high [101].

Classical treatment of spina bifida consists of surgical closure of the spinal canal soon after birth. Lifelong support and rehabilitation are typically necessary. Given that the neural damage associated with this disease is, in large part, secondary to the exposure of the spinal cord to the amniotic fluid and local trauma, prenatal surgical closure of the defect has been performed at a few centers in an attempt to improve outcome. Although a large multicenter clinical trial of fetal repair (the Management of Myelomeningocele Study) has shown that prenatal repair does offer some benefits, it is viable only in small subset of maternal-fetal units, only relatively late into the pathophysiological process, and not without complications, most notable preterm labor [101].

Neural stem cells (NSCs) have been shown to mediate repair in a variety of settings of postnatal central nervous system damage, including in the spinal cord. Our group proposed the notion of prenatal delivery of (fetal) NSCs as a potential means of promoting spinal cord repair and enhancing prenatal surgical coverage of spina bifida [12]. That initial study has shown that donor NSCs selectively engrafted within the most damaged areas of the spinal cord and retained an undifferentiated state *in vivo*, producing neurotrophic factors locally. These early findings, taken together with the large body of data on the use of NSCs in other forms of spinal cord injury, support further investigation into this multi-faceted prenatal therapy, combining local NSC delivery to the cord with mechanical/surgical repair aimed at inducing local protective and/or regenerative processes. This notion has been corroborated by results from other groups [39].

Other Disorders

Fetal stem cells have also been increasingly studied within the domain of other disorders, such as genetic, metabolic and lung diseases. Human MSCs derived from fetal pancreatic tissue have been shown to functionally engraft in the pancreas of fetal lambs, differentiate, and retain the ability to secrete insulin [102]. In mouse models of lung injury, systemic administration of select amniotic stem cells has led to engraftment and expression of specific alveolar and bronchiolar epithelial markers [103]. In mouse lung explants, pMSCs have shown to be potent paracrine stimulators of pulmonary morphogenesis [104]. In hypoplastic mouse lung explants, afMSCs have shown to augment branching morphogenesis and lung epithelial maturation [105]. Intrauterine transplantation of human MSCs from first trimester fetal blood can improve renal glomerulopathy postnatally in a collagen type I-deficient mouse model, which carries implications for the prenatal treatment of renal diseases such as Alport syndrome and polycystic kidney disease [106].

Ethical Considerations

The use of fetal tissue or cells has always been and will continue to be object of intense ethical debate. Much of the ethical controversies come from the fact that the primary source of fetal tissue is induced abortion. Spontaneous abortion generally does not raise moral issues. However, spontaneous abortion generally does not yields suitable fetal samples, in that it is frequently compromised by pathology such as chromosomal abnormalities, infections and/or anoxia. The National Institutes of Health, the American Obstetrical and Gynecological Society and the American Fertility Society have been regulating the use of fetal specimens for decades in accordance with the provisions that control the use of adult human tissue. Perhaps not surprisingly, despite their efforts and that of other national and international ethical committees and governmental bodies, a consensus has not yet been reached.

This polemic notwithstanding, tissue engineering, as a relatively novel development in fetal tissue/cell processing, adds a new dimension to that discussion. If specimens from a live, diseased fetus, or cells from a routine prenatal diagnostic procedure such as an amniocentesis or CVS, are to be used for the engineering of tissue, which in turn is to be implanted in autologous fashion for therapeutic purposes, no ethical objections should be anticipated, as long as the procedure is a valid treatment choice for a given perinatal condition. In that case scenario, ethical considerations are the very same that apply in any form of fetal intervention. On the other hand, if fetal engineered tissue is to be implanted in heterologous fashion, ethical issues are analogous to those involving fetal tissue/organ transplantation, regardless of whether the original specimen comes from a live or deceased fetus, or from banked fetal cells obtained from the amniotic fluid, placenta, or umbilical cord blood.

The distinction between autologous and heterologous implantation of engineered fetal tissue is germane to this debate, in that no condemnation of autologous use could be ethically justified. At the same time, regardless of whether an autologous or heterologous application is being considered, the reality of the amniotic fluid and other fetal annexes being sources of unique stem cells should contribute to ease the ethical objections surrounding the use of fetal cells therapeutically, including for non-life threatening diseases.

References

1. Hurst AF, Tanner WE, Osman AA. Addison's disease with severe anemia treated by suprarenal grafting. Proc R Soc Med. 1922;15:19.
2. Fichera G. Implanti omoplastici feto-umani nei cancro e nel diabete. Tumori. 1928;14:434.

3. Thomas ED, Lochte HL, Lu WC, et al. Intravenous infusion of bone marrow in patients receiving radiation and chemotherapy. N Engl J Med. 1957;247:491.
4. Klopper A, Diczfalusy E. Foetus and placenta. Oxford: Blackwell; 1969.
5. Diczfalusy E. Reproductive endocrinology in 1974. In: Crosignani PG, James VE, editors. Recent progress in reproductive endocrinology. London: Academic; 1974. p. 833–40.
6. Vacanti JP, Morse MA, Saltzman WM, Domb AJ, Perez-Atayde A, Langer R. Selective cell transplantation using bioabsorbable artificial polymers as matrices. J Pediatr Surg. 1988;23(1 Pt 2):3–9.
7. Cusick RA, Sano K, Lee H, Pollok J, Mooney D, Langer R, et al. Heterotopic fetal rat hepatocyte transplantation on biodegradable polymers. Surg Forum. 1995;XLVI:658–61.
8. Fauza DO, Fishman SJ, Mehegan K, Atala A. Videofetoscopically assisted fetal tissue engineering: bladder augmentation. J Pediatr Surg. 1998;33(1):7–12.
9. Fauza DO, Fishman SJ, Mehegan K, Atala A. Videofetoscopically assisted fetal tissue engineering: skin replacement. J Pediatr Surg. 1998;33(2):357–61.
10. Dionigi B, Ahmed A, Pennington EC, Zurakowski D, Fauza DO. A comparative analysis of human mesenchymal stem cell response to hypoxia in vitro: implications to translational strategies. J Pediatr Surg. 2014;49(6):915–8.
11. Liechty KW, MacKenzie TC, Shaaban AF, Radu A, Moseley AM, Deans R, et al. Human mesenchymal stem cells engraft and demonstrate site-specific differentiation after in utero transplantation in sheep. Nat Med. 2000;6(11):1282–6.
12. Fauza DO, Jennings RW, Teng YD, Snyder EY. Neural stem cell delivery to the spinal cord in an ovine model of fetal surgery for spina bifida. Surgery. 2008;144(3):367–73.
13. Pereira MC, Secco M, Suzuki DE, Janjoppi L, Rodini CO, Torres LB, et al. Contamination of mesenchymal stem-cells with fibroblasts accelerates neurodegeneration in an experimental model of Parkinson's disease. Stem Cell Rev. 2011;7(4):1006–17.
14. Bianchi DW, Johnson KL, Salem D. Chimerism of the transplanted heart. N Engl J Med. 2002;346(18):1410–2. author reply 2.
15. Khosrotehrani K, Bianchi DW. Multi-lineage potential of fetal cells in maternal tissue: a legacy in reverse. J Cell Sci. 2005;118(Pt 8):1559–63.
16. Bili C, Divane A, Apessos A, Konstantinos T, Apostolos A, Ioannis B, et al. Prenatal diagnosis of common aneuploidies using quantitative fluorescent PCR. Prenat Diagn. 2002; 22(5):360–5.
17. Kaviani A, Perry TE, Dzakovic A, Jennings RW, Ziegler MM, Fauza DO. The amniotic fluid as a source of cells for fetal tissue engineering. J Pediatr Surg. 2001;36(11):1662–5.
18. Kaviani A, Guleserian K, Perry TE, Jennings RW, Ziegler MM, Fauza DO. Fetal tissue engineering from amniotic fluid. J Am Coll Surg. 2003;196(4):592–7.
19. Klein JD, Fauza DO. Amniotic and placental mesenchymal stem cell isolation and culture. Methods Mol Biol. 2011;698:75–88.
20. Fuchs JR, Kaviani A, Oh JT, LaVan D, Udagawa T, Jennings RW, et al. Diaphragmatic reconstruction with autologous tendon engineered from mesenchymal amniocytes. J Pediatr Surg. 2004;39(6):834–8. discussion 8.
21. Kunisaki SM, Armant M, Kao GS, Stevenson K, Kim H, Fauza DO. Tissue engineering from human mesenchymal amniocytes: a prelude to clinical trials. J Pediatr Surg. 2007;42(6):974–9. discussion 9–80.
22. Steigman SA, Armant M, Bayer-Zwirello L, Kao GS, Silberstein L, Ritz J, et al. Preclinical regulatory validation of a 3-stage amniotic mesenchymal stem cell manufacturing protocol. J Pediatr Surg. 2008;43(6):1164–9.
23. Kaviani A, Perry TE, Barnes CM, Oh JT, Ziegler MM, Fishman SJ, et al. The placenta as a cell source in fetal tissue engineering. J Pediatr Surg. 2002;37(7):995–9.
24. Norwitz ER, Levy B. Noninvasive prenatal testing: the future is now. Rev Obstet Gynecol. 2013;6(2):48–62.
25. Seppanen E, Fisk NM, Khosrotehrani K. Pregnancy-acquired fetal progenitor cells. J Reprod Immunol. 2013;97(1):27–35.

26. Lillie FR. The theory of the free-martin. Science. 1916;43:611–3.
27. Owen RD. Immunogenic consequences of vascular anastomosis between bovine twins. Science. 1945;102:400–1.
28. Mjosberg JM, Trifari S, Crellin NK, Peters CP, van Drunen CM, Piet B, et al. Human IL-25- and IL-33-responsive type 2 innate lymphoid cells are defined by expression of CRTH2 and CD161. Nat Immunol. 2011;12(11):1055–62.
29. Constantinides MG, McDonald BD, Verhoef PA, Bendelac A. A committed precursor to innate lymphoid cells. Nature. 2014;508(7496):397–401.
30. Spits H, Di Santo JP. The expanding family of innate lymphoid cells: regulators and effectors of immunity and tissue remodeling. Nat Immunol. 2011;12(1):21–7.
31. Benito AI, Diaz MA, Gonzalez-Vicent M, Sevilla J, Madero L. Hematopoietic stem cell transplantation using umbilical cord blood progenitors: review of current clinical results. Bone Marrow Transplant. 2004;33(7):675–90.
32. Rocha V, Garnier F, Ionescu I, Gluckman E. Hematopoietic stem-cell transplantation using umbilical-cord blood cells. Rev Invest Clin. 2005;57(2):314–23.
33. Nijagal A, Wegorzewska M, Jarvis E, Le T, Tang Q, MacKenzie TC. Maternal T cells limit engraftment after in utero hematopoietic cell transplantation in mice. J Clin Invest. 2011;121(2):582–92.
34. Nijagal A, Wegorzewska M, Le T, Tang Q, Mackenzie TC. The maternal immune response inhibits the success of in utero hematopoietic cell transplantation. Chimerism. 2011;2(2):55–7.
35. Moustafa ME, Srivastava AS, Nedelcu E, Donahue J, Gueorguieva I, Shenouda SS, et al. Chimerism and tolerance post-in utero transplantation with embryonic stem cells. Transplantation. 2004;78(9):1274–82.
36. Lee PW, Cina RA, Randolph MA, Goodrich J, Rowland H, Arellano R, et al. Stable multilineage chimerism across full MHC barriers without graft-versus-host disease following in utero bone marrow transplantation in pigs. Exp Hematol. 2005;33(3):371–9.
37. Javazon EH, Merchant AM, Danzer E, Flake AW. Reconstitution of hematopoiesis following intrauterine transplantation of stem cells. Methods Mol Med. 2005;105:81–94.
38. Ashizuka S, Peranteau WH, Hayashi S, Flake AW. Busulfan-conditioned bone marrow transplantation results in high-level allogeneic chimerism in mice made tolerant by in utero hematopoietic cell transplantation. Exp Hematol. 2006;34(3):359–68.
39. Saadai P, Wang A, Nout YS, Downing TL, Lofberg K, Beattie MS, et al. Human induced pluripotent stem cell-derived neural crest stem cells integrate into the injured spinal cord in the fetal lamb model of myelomeningocele. J Pediatr Surg. 2013;48(1):158–63.
40. Kunisaki SM, Freedman DA, Fauza DO. Fetal tracheal reconstruction with cartilaginous grafts engineered from mesenchymal amniocytes. J Pediatr Surg. 2006;41(4):675–82.
41. Gray FL, Turner CG, Ahmed A, Calvert CE, Zurakowski D, Fauza DO. Prenatal tracheal reconstruction with a hybrid amniotic mesenchymal stem cells-engineered construct derived from decellularized airway. J Pediatr Surg. 2012;47(6):1072–9.
42. Fauza DO, Marler JJ, Koka R, Forse RA, Mayer JE, Vacanti JP. Fetal tissue engineering: diaphragmatic replacement. J Pediatr Surg. 2001;36(1):146–51.
43. Fuchs JR, Terada S, Ochoa ER, Vacanti JP, Fauza DO. Fetal tissue engineering: in utero tracheal augmentation in an ovine model. J Pediatr Surg. 2002;37(7):1000–6.
44. Fuchs JR, Hannouche D, Terada S, Vacanti JP, Fauza DO. Fetal tracheal augmentation with cartilage engineered from bone marrow-derived mesenchymal progenitor cells. J Pediatr Surg. 2003;38(6):984–7.
45. Fuchs JR, Terada S, Hannouche D, Ochoa ER, Vacanti JP, Fauza DO. Fetal tissue engineering: chest wall reconstruction. J Pediatr Surg. 2003;38(8):1188–93.
46. Wu X, Rabkin-Aikawa E, Guleserian KJ, Perry TE, Masuda Y, Sutherland FW, et al. Tissue-engineered microvessels on three-dimensional biodegradable scaffolds using human endothelial progenitor cells. Am J Physiol Heart Circ Physiol. 2004;287(2):H480–7.
47. Fuchs JR, Hannouche D, Terada S, Zand S, Vacanti JP, Fauza DO. Cartilage engineering from ovine umbilical cord blood mesenchymal progenitor cells. Stem Cells. 2005;23(7):958–64.

48. Fuchs JR, Nasseri BA, Vacanti JP, Fauza DO. Postnatal myocardial augmentation with skeletal myoblast-based fetal tissue engineering. Surgery. 2006;140(1):100–7.
49. Krupnick AS, Balsara KR, Kreisel D, Riha M, Gelman AE, Estives MS, et al. Fetal liver as a source of autologous progenitor cells for perinatal tissue engineering. Tissue Eng. 2004;10(5-6):723–35.
50. Kunisaki M, Fuchs JR, Kaviani A, Oh JT, LaVan DA, Vacanti JP, et al. Diaphragmatic repair through fetal tissue engineering: a comparison between mesenchymal amniocyte- and myoblast-based constructs. J Pediatr Surg. 2006;41(1):34–9. discussion 9.
51. Kunisaki SM, Jennings RW, Fauza DO. Fetal cartilage engineering from amniotic mesenchymal progenitor cells. Stem Cells Dev. 2006;15(2):245–53.
52. Schmidt D, Mol A, Breymann C, Achermann J, Odermatt B, Gossi M, et al. Living autologous heart valves engineered from human prenatally harvested progenitors. Circulation. 2006;114(1 Suppl):1125–31.
53. Schmidt D, Achermann J, Odermatt B, Breymann C, Mol A, Genoni M, et al. Prenatally fabricated autologous human living heart valves based on amniotic fluid derived progenitor cells as single cell source. Circulation. 2007;116(11 Suppl):164–70.
54. Steigman SA, Ahmed A, Shanti RM, Tuan RS, Valim C, Fauza DO. Sternal repair with bone grafts engineered from amniotic mesenchymal stem cells. J Pediatr Surg. 2009;44(6):1120–6. discussion 6.
55. Turner CG, Fauza DO. Fetal tissue engineering. Clin Perinatol. 2009;36(2):473–88. xii.
56. Klein JD, Turner CG, Ahmed A, Steigman SA, Zurakowski D, Fauza DO. Chest wall repair with engineered fetal bone grafts: an efficacy analysis in an autologous leporine model. J Pediatr Surg. 2010;45(6):1354–60.
57. Turner CG, Klein JD, Steigman SA, Armant M, Nicksa GA, Zurakowski D, et al. Preclinical regulatory validation of an engineered diaphragmatic tendon made with amniotic mesenchymal stem cells. J Pediatr Surg. 2011;46(1):57–61.
58. Weber B, Schoenauer R, Papadopulos F, Modregger P, Peter S, Stampanoni M, et al. Engineering of living autologous human umbilical cord cell-based septal occluder membranes using composite PGA-P4HB matrices. Biomaterials. 2011;32(36):9630–41.
59. Weber B, Zeisberger SM, Hoerstrup SP. Prenatally harvested cells for cardiovascular tissue engineering: fabrication of autologous implants prior to birth. Placenta. 2011;32 Suppl 4:S316–9.
60. Turner CG, Klein JD, Gray FL, Ahmed A, Zurakowski D, Fauza DO. Craniofacial repair with fetal bone grafts engineered from amniotic mesenchymal stem cells. J Surg Res. 2012;178(2):785–90.
61. Weber B, Kehl D, Bleul U, Behr L, Sammut S, Frese L, et al. In vitro fabrication of autologous living tissue-engineered vascular grafts based on prenatally harvested ovine amniotic fluid-derived stem cells. J Tissue Eng Regen Med. 2013. doi: 10.1002/term.1781
62. Pennington EC, Dionigi B, Gray FL, Ahmed A, Brazzo J, Dolinko A, et al. Limb reconstruction with decellularized, non-demineralized bone in a young leporine model. Biomed Mater. 2015;10(1):015021.
63. Moss RL, Chen CM, Harrison MR. Prosthetic patch durability in congenital diaphragmatic hernia: a long- term follow-up study. J Pediatr Surg. 2001;36(1):152–4.
64. Lund DP, Mitchell J, Kharasch V, Quigley S, Kuehn M, Wilson JM. Congenital diaphragmatic hernia: the hidden morbidity. J Pediatr Surg. 1994;29(2):258–62. discussion 62–4.
65. Kimber CP, Dunkley MP, Haddock G, Robertson L, Carey FA, Cuschieri A. Patch incorporation in diaphragmatic hernia. J Pediatr Surg. 2000;35(1):120–3.
66. Laituri CA, Garey CL, Valusek PA, Fike FB, Kaye AJ, Ostlie DJ, et al. Outcome of congenital diaphragmatic hernia repair depending on patch type. Eur J Pediatr Surg. 2010;20(6):363–5.
67. Fauza DO. Tissue engineering in congenital diaphragmatic hernia. Semin Pediatr Surg. 2014;23(3):135–40.
68. Hill SA, Milam M, Manaligod JM. Tracheal agenesis: diagnosis and management. Int J Pediatr Otorhinolaryngol. 2001;59(1):63–8.

69. Backer CL, Mavroudis C, Dunham ME, Holinger L. Intermediate-term results of the free tracheal autograft for long segment congenital tracheal stenosis. J Pediatr Surg. 2000;35(6):813–8. discussion 8–9.
70. Bando K, Turrentine MW, Sun K, Sharp TG, Matt B, Karmazyn B, et al. Anterior pericardial tracheoplasty for congenital tracheal stenosis: intermediate to long-term outcomes. Ann Thorac Surg. 1996;62(4):981–9.
71. Jacobs JR. Investigations into tracheal prosthetic reconstruction. Laryngoscope. 1988;98(11):1239–45.
72. Wright CD, Graham BB, Grillo HC, Wain JC, Mathisen DJ. Pediatric tracheal surgery. Ann Thorac Surg. 2002;74(2):308–13. discussion 14.
73. Backer CL, Mavroudis C, Gerber ME, Holinger LD. Tracheal surgery in children: an 18-year review of four techniques. Eur J Cardiothorac Surg. 2001;19(6):777–84.
74. Jacobs JP, Quintessenza JA, Andrews T, Burke RP, Spektor Z, Delius RE, et al. Tracheal allograft reconstruction: the total North American and worldwide pediatric experiences. Ann Thorac Surg. 1999;68(3):1043–51. discussion 52.
75. Chiu PP, Kim PC. Prognostic factors in the surgical treatment of congenital tracheal stenosis: a multicenter analysis of the literature. J Pediatr Surg. 2006;41(1):221–5. discussion 5.
76. Elliott MJ, De Coppi P, Speggiorin S, Roebuck D, Butler CR, Samuel E, et al. Stem-cell-based, tissue engineered tracheal replacement in a child: a 2-year follow-up study. Lancet. 2012;380(9846):994–1000.
77. Kunisaki SM, Fuchs JR, Steigman SA, Fauza DO. A comparative analysis of cartilage engineered from different perinatal mesenchymal progenitor cells. Tissue Eng. 2007;13(11):2633–44.
78. Hirose S, Farmer DL, Lee H, Nobuhara KK, Harrison MR. The ex utero intrapartum treatment procedure: looking back at the EXIT. J Pediatr Surg. 2004;39(3):375–80. discussion 80.
79. Hirose S, Sydorak RM, Tsao K, Cauldwell CB, Newman KD, Mychaliska GB, et al. Spectrum of intrapartum management strategies for giant fetal cervical teratoma. J Pediatr Surg. 2003;38(3):446–50. discussion 50.
80. Steigman SA, Nemes L, Barnewolt CE, Estroff JA, Valim C, Jennings RW, et al. Differential risk for neonatal surgical airway intervention in prenatally diagnosed neck masses. J Pediatr Surg. 2009;44(1):76–9.
81. Li RK, Yau TM, Weisel RD, Mickle DA, Sakai T, Choi A, et al. Construction of a bioengineered cardiac graft. J Thorac Cardiovasc Surg. 2000;119(2):368–75.
82. Taylor DA, Atkins BZ, Hungspreugs P, Jones TR, Reedy MC, Hutcheson KA, et al. Regenerating functional myocardium: improved performance after skeletal myoblast transplantation. Nat Med. 1998;4(8):929–33.
83. Sartore S, Lenzi M, Angelini A, Chiavegato A, Gasparotto L, De Coppi P, et al. Amniotic mesenchymal cells autotransplanted in a porcine model of cardiac ischemia do not differentiate to cardiogenic phenotypes. Eur J Cardiothorac Surg. 2005;28(5):677–84.
84. Zhao P, Ise H, Hongo M, Ota M, Konishi I, Nikaido T. Human amniotic mesenchymal cells have some characteristics of cardiomyocytes. Transplantation. 2005;79(5):528–35.
85. Iop L, Chiavegato A, Callegari A, Bollini S, Piccoli M, Pozzobon M, et al. Different cardiovascular potential of adult- and fetal-type mesenchymal stem cells in a rat model of heart cryoinjury. Cell Transplant. 2008;17(6):679–94.
86. Shin'oka T, Imai Y, Ikada Y. Transplantation of a tissue-engineered pulmonary artery. N Engl J Med. 2001;344(7):532–3.
87. Matsumura G, Hibino N, Ikada Y, Kurosawa H, Shin'oka T. Successful application of tissue engineered vascular autografts: clinical experience. Biomaterials. 2003;24(13):2303–8.
88. Hibino N, Shin'oka T, Matsumura G, Ikada Y, Kurosawa H. The tissue-engineered vascular graft using bone marrow without culture. J Thorac Cardiovasc Surg. 2005;129(5):1064–70.
89. Shin'oka T, Matsumura G, Hibino N, Naito Y, Watanabe M, Konuma T, et al. Midterm clinical result of tissue-engineered vascular autografts seeded with autologous bone marrow cells. J Thorac Cardiovasc Surg. 2005;129(6):1330–8.

90. Naito Y, Lee YU, Yi T, Church SN, Solomon D, Humphrey JD, et al. Beyond burst pressure: initial evaluation of the natural history of the biaxial mechanical properties of tissue-engineered vascular grafts in the venous circulation using a murine model. Tissue Eng Part A. 2014;20(1-2):346–55.
91. Vacanti CA, Bonassar LJ, Vacanti MP, Shufflebarger J. Replacement of an avulsed phalanx with tissue-engineered bone. N Engl J Med. 2001;344(20):1511–4.
92. Horwitz EM, Prockop DJ, Fitzpatrick LA, Koo WW, Gordon PL, Neel M, et al. Transplantability and therapeutic effects of bone marrow-derived mesenchymal cells in children with osteogenesis imperfecta. Nat Med. 1999;5(3):309–13.
93. Horwitz EM, Prockop DJ, Gordon PL, Koo WW, Fitzpatrick LA, Neel MD, et al. Clinical responses to bone marrow transplantation in children with severe osteogenesis imperfecta. Blood. 2001;97(5):1227–31.
94. Le Blanc K, Gotherstrom C, Ringden O, Hassan M, McMahon R, Horwitz E, et al. Fetal mesenchymal stem-cell engraftment in bone after in utero transplantation in a patient with severe osteogenesis imperfecta. Transplantation. 2005;79(11):1607–14.
95. Guillot PV, Abass O, Bassett JH, Shefelbine SJ, Bou-Gharios G, Chan J, et al. Intrauterine transplantation of human fetal mesenchymal stem cells from first-trimester blood repairs bone and reduces fractures in osteogenesis imperfecta mice. Blood. 2008;111(3):1717–25.
96. Gotherstrom C, Westgren M, Shaw SW, Astrom E, Biswas A, Byers PH, et al. Pre- and postnatal transplantation of fetal mesenchymal stem cells in osteogenesis imperfecta: a two-center experience. Stem Cells Transl Med. 2014;3(2):255–64.
97. Maraldi T, Riccio M, Resca E, Pisciotta A, La Sala GB, Ferrari A, et al. Human amniotic fluid stem cells seeded in fibroin scaffold produce in vivo mineralized matrix. Tissue Eng Part A. 2011;17(21-22):2833–43.
98. Peister A, Deutsch ER, Kolambkar Y, Hutmacher DW, Guldberg RE. Amniotic fluid stem cells produce robust mineral deposits on biodegradable scaffolds. Tissue Eng Part A. 2009;15(10):3129–38.
99. Riccio M, Maraldi T, Pisciotta A, La Sala GB, Ferrari A, Bruzzesi G, et al. Fibroin scaffold repairs critical-size bone defects in vivo supported by human amniotic fluid and dental pulp stem cells. Tissue Eng Part A. 2012;18(9-10):1006–13.
100. Sun H, Feng K, Hu J, Soker S, Atala A, Ma PX. Osteogenic differentiation of human amniotic fluid-derived stem cells induced by bone morphogenetic protein-7 and enhanced by nanofibrous scaffolds. Biomaterials. 2010;31(6):1133–9.
101. Adzick NS, Thom EA, Spong CY, Brock 3rd JW, Burrows PK, Johnson MP, et al. A randomized trial of prenatal versus postnatal repair of myelomeningocele. N Engl J Med. 2011;364(11):993–1004.
102. Ersek A, Pixley JS, Goodrich AD, Porada CD, Almeida-Porada G, Thain DS, et al. Persistent circulating human insulin in sheep transplanted in utero with human mesenchymal stem cells. Exp Hematol. 2010;38(4):311–20.
103. Carraro G, Perin L, Sedrakyan S, Giuliani S, Tiozzo C, Lee J, et al. Human amniotic fluid stem cells can integrate and differentiate into epithelial lung lineages. Stem Cells. 2008;26(11):2902–11.
104. Di Bernardo J, Maiden MM, Jiang G, Hershenson MB, Kunisaki SM. Paracrine regulation of fetal lung morphogenesis using human placenta-derived mesenchymal stromal cells. J Surg Res. 2014;190(1):255–63.
105. Di Bernardo J, Maiden MM, Hershenson MB, Kunisaki SM. Amniotic fluid derived mesenchymal stromal cells augment fetal lung growth in a nitrofen explant model. J Pediatr Surg. 2014;49(6):859–64.
106. Guillot PV, Cook HT, Pusey CD, Fisk NM, Harten S, Moss J, et al. Transplantation of human fetal mesenchymal stem cells improves glomerulopathy in a collagen type I alpha 2-deficient mouse. J Pathol. 2008;214(5):627–36.

Chapter 19
Perinatal Gene Therapy

Panicos Shangaris and Anna L. David

Introduction

Non invasive prenatal diagnosis using circulating fetal DNA in the maternal blood now allows clinicians to detect a congenital disease in the fetus as early as 10 weeks of gestation [1, 2]. This allows time for the couple to consider their options early in gestation and to act accordingly. The options available for the parents of an affected fetus are usually termination of pregnancy or monitoring until delivery and subsequently postnatal treatment. What if a third option was available, an in utero treatment to correct the genetic disorder and to provide the corrected protein to a therapeutic level? Preclinical studies in animal models in the last 15 years have shown that prenatal application of gene therapy to the fetus can cure severe genetic disease. More recently, structural anomalies in the fetus have been shown to be preventable using a gene therapy approach. In addition, for some non-genetic conditions, timely expression of a particular protein, for example during the last third of pregnancy, may alleviate pathology. Direct gene delivery using vectors carrying the correct gene, or autologous transplantation of gene manipulated/corrected stem cells derived from the fetus may achieve these objectives. In this chapter we describe progress in the field of perinatal gene therapy over the last 20 years, the hurdles to clinical translation and the first moves of therapy into the clinic.

P. Shangaris, M.B.B.Ch., B.A.O., M.Sc. • A.L. David, Ph.D., M.R.C.O.G. (✉)
Institute for Women's Health, University College London, 86-96 Chenies Mews, London WC1E 6HX, UK
e-mail: p.shangaris@ucl.ac.uk; a.david@ucl.ac.uk

D.O. Fauza, M. Bani (eds.), *Fetal Stem Cells in Regenerative Medicine*,
Stem Cell Biology and Regenerative Medicine,
DOI 10.1007/978-1-4939-3483-6_19

The Concept of Perinatal Gene Therapy

Gene therapy uses a vector to deliver a gene to its required site where expression of the protein can produce a therapeutic effect. Perinatal application of gene therapy can be directly to the fetus, to fetal stem cells for autologous transplantation or even to the mother where maternal pathology affects the fetus *in utero*, for example in the case of uteroplacental insufficiency. An advantage of the perinatal approach is the ability to deliver a therapeutic gene to an individual before the onset of organ damage, an important issue for many congenital metabolic diseases, such as some types of mucopolysaccharidoses, for example, where irreversible brain damage can occur before birth. Application of a therapy to the fetus targets a rapidly dividing population of stem cells, providing a large population of transduced cells to produce a better therapeutic effect [3]. The fetus also presents a size advantage, allowing a higher vector-to-target cell ratio. Organs that are difficult to target after birth may be more easily accessible during fetal life because of their developmental stages or relative immaturity [4]. For example, the fetal epidermis undergoes remodeling by programmed cell death to be replaced by mature keratinocytes which form a thick barrier to gene transfer postnatally [5] but which may be targeted *in utero* [6].

A major obstacle to postnatal gene therapy has been the development of an immune response against the transgenic (therapeutic) protein or the vector itself [7]. This complication is particularly important when gene therapy is aiming to correct a genetic disease in which complete absence of a gene product is observed. Some individuals may even have pre-existing antibodies to the viral vector that will prevent long-term expression of the transgenic protein, limiting therapeutic efficacy and preventing repeated vector administration. For example, pre-existing neutralizing antibodies against adeno-associated virus (AAV) serotype 2 have been shown to interfere with AAV2 vector-mediated factor IX (FIX) gene transfer to the liver [7–9]. Delivering foreign protein to the fetus can take advantage of immune tolerance which is induced during fetal life, a concept that was first proposed nearly 60 years ago [10, 11]. Induction of tolerance depends on the foreign protein being expressed sufficiently early in gestation before the immune system is fully developed and the protein being maintained at a detectable level within the fetus and being presented to the thymus at the correct time. For human gestation, transgenic protein expression will need to last at least 6 months if the vector is given early in pregnancy, which limits the types of viral vectors that can be applied. Proof of principle studies have shown long-term expression of proteins at therapeutic levels and induction of immune tolerance [12] in both small [13] and large animals [14, 15] and cured congenital disease in some animal models.

Selecting the Right Disease for Perinatal Gene Therapy

As with any new therapeutic modality, the risks of perinatal gene therapy are not well characterized and careful thought must be given to decide on the right disease to select for a first-in-woman trial. Where vectors are given directly to the fetus for correction of genetic disease there has been guidance given by the National Institutes for Health Recombinant DNA Advisory Committee that reported on a pre-proposal for the initial application of fetal gene therapy [16]. The recommendations were that application should be limited only to those diseases that:

- are associated with serious morbidity and mortality risks for the fetus either *in utero* or postnatally
- do not have an effective postnatal therapy, or have a poor outcome using available postnatal therapies
- are not associated with serious abnormalities that are not corrected by the transferred gene
- can be definitively diagnosed *in utero* and have a well defined genotype/phenotype relationship, and
- have an animal model for *in utero* gene transfer that recapitulates the human disease or disorder.

Achieving all this with a preferably single direct fetal vector injection approach is challenging. An alternative strategy that has been used most effectively in neonatal life is to combine stem cell transplantation (SCT) with gene therapy. The UK Gene Therapy Advisory Committee (GTAC) considered such an *in utero* stem cell gene therapy (IUSCGT) approach in their broader judgments about fetal gene therapy. The New and Emerging Technologies subgroup of GTAC found that the use of genetically modified stem cells in stem cell transplantation to the fetus was a possibility stating “such *ex vivo* modification would be unlikely to carry with it any higher risk to the germ line than the trials of postnatal somatic gene therapy which have already been approved”.

Some of the diseases that may be suitable for fetal treatment are listed in Table 19.1. Preclinical studies of direct fetal gene transfer are encouraging. Fetal application of gene therapy in mouse models of congenital disease such as haemophilia A [17] and B [18], congenital blindness [19], Crigler-Najjar type 1 syndrome [20] and Pompe disease (glycogen storage disease type II) [13, 17–21] have shown phenotypic correction of the condition. More recently the application of perinatal gene therapy has broadened with positive results in previously untreatable conditions. For structural anomalies, transient transduction of the periderm via intra-amniotic delivery of adenoviral vector encoding TGFβ3 prevents cleft palate in a mouse model of disease [22]. For obstetric conditions that affect the fetus, maternal uterine artery injection of adenovirus containing the vascular endothelial growth factor gene improves fetal growth in growth restricted sheep pregnancy [23].

Table 19.1 Examples of candidate diseases for perinatal gene therapy

Disease	Therapeutic gene product	Target cells/organ	Age at onset	Incidence	Life expectancy
Cystic fibrosis	Cystic fibrosis transmembrane conductance regulator	Airway and intestinal epithelial cells	In utero	1:4000	Mid-30s
Duchenne muscular dystrophy	Dystrophin	Myocytes	2 years	1:4500	25 years
Lysosomal storage disease: Gaucher	Glucocerebrosidase	Hepatocytes	9–11 years	1:9000 overall	<2 years
Lysosomal storage diseases in general	Lysosomes	Protein deficiency	At birth	1:5000	Very severely affected patients may die within the first 5 or 10-years of life, but disease severity and life expectancy are highly variable.
Leukocyte adhesion deficiency	Leucocyte integrin CD18 on chromosome 21	Deficiency of the β-2 integrin subunit	Neonatal period	1:100,000	<1 year
Osteogenesis imperfect	Col1a1	Collagen	In utero	1:20,000	variable
Sickle cell disease	α-Globin	Erythrocyte precursors	In utero	4:100 (in Africa)	Homozygous patients survive on average to 53 years (men), 58 years (women)
Spinal muscular atrophy	Survival motor neuron protein	Motor neurons	In utero type 0, 6 months type 1	1:10,000	2 years
Tay-Sachs disease	Hexosaminidase A	Neuronal cells	At birth	1:30 in Ashkenazi Jewish population	<5 years
Urea cycle defects: ornithine transcarbamalase deficiency	Ornithine transcarbamylase	Hepatocytes	2 days	1:30,000 overall	2 days (severe neonatal onset)
Haemophilias: A	Human VIII clotting factor	Hepatocytes	1 year	1:6000	Adulthood with treatment
Haemophilias: B	Human factor IX clotting factor	Hepatocytes	1 year	1:20,000	Adulthood with treatment

Haemophilias: Factor VII deficiency	Human factor VII clotting factor	Hepatocytes	In utero	1:500,000	Adulthood with treatment
Thalassaemias: α-thalassaemia	α-Globin	Erythrocyte precursors	In utero	1:2700	Lethal
Thalassaemias: β-thalassaemia	β-Globin	Erythrocyte precursors	First 2 years of life	1:100,000	For thalassemia major affected patients die before the age of 20 if untreated
SCID: X linked	γ*c* Cytokine receptor	Haematopoietic precursor cells	birth	1:1,000,000	<6 months if no bone marrow transplant
SCID: Adenosine deaminase deficiency	Adenosine deaminase	Haematopoietic precursor cells	birth	1:1,000,000	<6 months if no bone marrow transplant
Epidermolysis bullosa	Type VII collagen	Keratinocytes	birth	1:40,000	Adulthood
Severe fetal growth restriction	Vascular endothelial growth factor	Uterine arteries	In utero	1:100	Adulthood if individual survives the neonatal period
Congenital diaphragmatic hernia	Lung growth factors	Alveoli	In utero	1:2200	Adulthood if individual survives neonatal surgery
Cleft lip and palate	Transforming growth factor β3	Palatal shelf medial edge epithelium	In utero	1:500	Adulthood

SCID severe combined immunodeficiency

Selecting the Right Vector for Perinatal Gene Therapy

Vectors are vehicles that are used to carry the therapeutic gene into the cell. An ideal vector for perinatal gene therapy is one that can produce long-term regulated and therapeutic expression of the transferred gene through the use of a single and efficient gene delivery method, is safe to the mother and fetus, thus allowing incorporation into clinical practice. For example a vector carrying the beta globin gene should deliver and express the gene only to erythroid specific cells and lineages. These and other essential characteristics are described in Table 19.2. The most commonly tested vectors in gene therapy pre-clinical studies in the fetus have been adenovirus and adeno-associated virus, lentivirus and retrovirus vectors. Less is known about the effect of other viral vectors and non-viral vectors prenatally.

A summary of vectors is provided in Table 19.3.

Non viral Vectors

Non-viral vectors are an attractive option because of their perceived better safety profile and their ability to transfer very large fragments of genetic material. Some fetal studies have shown encouraging results. Intrahepatic injection of the cationic polymer polyethylenimine (PEI) in late-gestation fetal mice enhanced

Table 19.2 Characteristics of the ideal vector for prenatal gene therapy

Characteristic	Reason
Highly efficient, regulated transgene expression	Provide therapeutic protein expression
Length of time of transgene expression to suit disease	Example (1) Long term transgene expression for a monogenic disorder requires protein expression to last the lifetime of the individual eg haemophilias
	Example (2) Transient transgene expression for a developmental or obstetric disorder requires protein expression at a critical window of fetal development or gestation eg fetal growth restriction
Specific tropism to target organ	Avoid systemic gene transfer
Large carrying capacity	Accommodate therapeutic gene and any required regulatory elements
No toxicity	Safe for mother, fetus and future progeny
No immunogenicity	Avoid generating a fetal immune response
No mutagenic properties	Safe for fetus and future progeny
High Concentration	To allow as many cells to be infected as possible
Reproducibility of production	Able to be reproduced in various laboratories and under GMP/GLP conditions
Ability to Transduce dividing and non dividing cells	Homogenous transduction of specific stem cell population

Table 19.3 Types of vector and considerations in relation to perinatal gene therapy application

Vector	DNA	Efficiency	Tropism	Advantages	Disadvantages	Prenatal considerations
Non-viral DNA	No limit	+	Limited	Low toxicity Low immunogenicity	Low transduction efficiency	Expression may not last through gestation
Adenovirus	7.5 kb	+++	Depends on serotype	Can grow to high titre. Highly efficient gene transfer	Short term expression and immunogenic	Expression may not last through gestation
Helper-dependent Adenovirus	35 kb	+++	Broad	Low immunogenicity, high capacity, long-term expression in quiescent cells	Inefficient production	Insufficient vector concentration
Adeno-associated virus	4.7 kb generally	++	Depends on sub-type	Long term expression Low immunogenicity Very high titre	Liver toxicity in adult trials due to anti-capsid T cells	Some subtypes associated with miscarriage. Low level integration into active genes so theoretical mutagenesis risk.
Gamma retrovirus	10 kb	+	Depends on pseudotyping	Long term gene transfer	Potential for insertional mutagenesis. Infect dividing cells only.	Risk of germ-line transmission and insertional mutagenesis. Virus inactivated by amniotic fluid
Lentivirus	10 kb	++	Depends on pseudotyping	Long term gene transfer Infects dividing and non-dividing cells	Potential for insertional mutagenesis	Risk of germ-line transmission and insertional mutagenesis
Non-integrating lentivirus	10 kb	++	Depends on pseudotyping	Insertional mutagenesis unlikely	Short term expression	Rapidly dividing fetal cells may result in long term low transgenic protein expression
Foamy virus (type of retrovirus)	10 kb	++	Broad	Long term expression. Not associated with disease in humans.	Potential for insertional mutagenesis	Less likely to integrate into transcriptionally active regions when compared to classical retroviral vectors, such as lentiviruses
Adeno-retro hybrids	7.5 kb	++	As per adenovirus	Highly efficient gene transfer with long term expression	Potential for insertional mutagenesis	Risk of insertional mutagenesis
Herpes simplex	40–50 kb	++	Broad: CNS	Retro-axonal transduction Infects non-dividing cells	Mainly infect sensory neurons.	Expression can be altered from short term to latent by alteration of viral promoter
Vaccinia	25 kb	+	Broad	Infects dividing and non-dividing cells	Elicits immune responses when used as a vaccine	Not used in prenatal therapy to date

gene transfer to the liver as compared with administration of naked DNA. Encouragingly, the marker gene expression was 40-fold higher marker per milligram of protein in fetuses compared with adults [24]. Low-level gene transfer to the fetal sheep airway epithelium was also achieved using guanidium-cholesterol cationic liposomes delivered into the trachea in mid-gestation fetal sheep [25].

Unfortunately, current non-viral systems are hindered by their low transduction efficiency and short expression time. Manipulation of non-viral vector particles can overcome some of their problems. For example, altering the chemical structure of carbon bonds within cationic liposome-DNA complexes improves their transfection efficiency and reduces their toxicity *in vivo* [26]. Other developments include artificial chromosomes and Epstein—Barr-virus-based plasmids. DNA introduced as plasmid molecules remains episomal and will be lost with cell division, which is rapid in the fetus and could be a particular disadvantage. However transient gene transfer might be useful in the management of a developmental condition in which therapy is only required for a relatively short time. For instance, short-term transgene expression via injection of a liposome that inhibited fibronectin synthesis into the ductus arteriosis of mid-trimester fetal sheep maintained a patent ductus arteriosus prior before for congenital heart defects in neonatal sheep [27].

Adenovirus

These are useful vectors for proof-of-principle studies in gene therapy because they achieve highly efficient gene transfer in a wide range of fetal tissues depending on the route of administration [28]. Although they do not specifically have a tropism for the liver, these vectors strongly infect liver tissue after intravenous delivery. Gene expression is usually transient because the vector does not integrate into the host genome and is rapidly diluted by the active cellular proliferation that takes place in the fetus. Although the vector is highly immunogenic in adults, fetal administration can produce extended gene expression and induction of immune tolerance to the transgene and—in some cases—also to the vector, although immune responses to adenovirus are reported after fetal application, even in early gestation [28–30]. To reduce the immunogenicity and toxicity of the vector, all adenoviral coding sequences can be eliminated to generate so called 'gutless vectors'. Novel hybrid vectors that take advantage of adenovirus infectivity and the permanent nature of integrative vectors such as retroviruses and lentiviruses might also prove useful in the fetus [31, 32].

Adenoviral vectors have also been shown to be useful for the treatment of fetal growth restriction (FGR) in a sheep model of FGR [23].

Adeno-Associated Virus Vectors (AAV)

AAV vectors are considered to be less toxic and immunogenic than early-generation adenovirus vectors, although an immune response to transgenic protein has been observed after fetal intramuscular injection of AAV. Long-term transgene expression can be achieved after muscular, peritoneal or amniotic injection into the fetal mouse and rat [30, 33–36]. AAV vectors integrate into the genome only at low frequency and they are therefore likely to be diluted rapidly by the increasing tissue mass that occurs in the fetus. Integration of the wild-type virus is predominantly at a specific functionally unimportant location on human chromosome, reducing the theoretical risk of insertional mutagenesis [37].

Replication deficient adenoviral vectors (rAAV) seem to genetically modify cells in the retina which can be of therapeutic benefit in the treatment of certain inherited degenerative conditions that compromise photoreceptor, and hence visual, function. Promising outcomes have been seen in recent clinical trial with rAAV vectors encoding RPE65 in patients with Leber's congenital amaurosis (LCA) [38].

AAVs have the capability of crossing the adult BBB and achieving widespread CNS transduction after systemic gene delivery to the CNS in marmosets. Systemically delivered rAAVrh.10 can transduce the CNS efficiently and its transgene expression can be limited in the periphery by endogenous miRNAs in adult marmosets [39].

Retrovirus and Lentivirus

Retroviruses and the closely related lentiviruses can integrate permanently into the genome, thus offering the possibility of permanent gene delivery. Moloney leukaemia retrovirus (MLV) was the first vector to be applied fetally to investigate the dispersion of neuronal clones across the developing cerebral cortex of fetal mice. Since then, MLV has been used in a number of fetal studies of gene therapy, giving long-term expression in rats, sheep and non-human primates after intraperitoneal and intrahepatic delivery. Retroviruses require dividing cells for gene transfer, which suggests that they might be better suited for use in fetal rather than adult tissues where cells are rapidly dividing. Human serum can almost completely inactivate some retroviral particles, which limits their use *in vivo*, although increased resistance to serum inactivation can be achieved by pseudotyping, which replaces the natural envelope of the retrovirus with an envelope from another virus. A particular problem with in-utero application is that amniotic fluid has a mild inhibitory effect on retrovirus infection *in vitro*. This was probably responsible for the poor gene transfer observed after intra-amniotic application of retroviruses in fetal sheep and non-human primates [40–42].

Lentiviruses such as those based on HIV can also infect non-dividing cells, although gene transfer to the liver is improved by cell cycling in some lentiviruses.

Pseudotyping improves lentivirus stability and allows vector titres to be improved by ultracentrifugation. Different viral envelopes allow gene transfer to be targeted to specific tissues. For example, intramuscular and intrahepatic injection into fetal mice of an HIV vector pseudotyped with vesicular stomatitis virus protein G (VSVG) envelope preferentially transduced the fetal liver, whereas pseudotyping with mokola or ebola envelope proteins gave the most efficient transduction of the myocytes. Lentivirus vectors integrate into the genome randomly and—theoretically—are therefore able to cause insertional mutagenesis [32, 43, 44].

Gene Editing Approaches

The design of zinc finger nuclease (ZFN)–mediated repair was the first approach used to edit genes. ZFNs are artificial restriction enzymes generated by fusing a zinc finger DNA-binding domain to a DNA-cleavage domain. Zinc finger domains are engineered to target desired DNA sequences within complex genomes. ZFNs can be used in a variety of ways. They can disable dominant mutations in heterozygous individuals by producing double-strand breaks (DSBs) in the DNA. Alternatively the sequence of an allele can be rewritten by invoking the homologous recombination machinery to repair the DSB using the supplied DNA fragment as a template. ZFN-encoding plasmids can also be used to transiently express ZFNs to target a DSB to a specific gene locus in human cells providing a targeted gene editing approach [45].

This approach has been recently improved on by the use of transcription activator–like effector nucleases (TALENs) Chandrakasan and Malik. TALENs use DNA-recognition modules that recognize single base pairs, linked to the same FokI-derived cleavage domain that is used in ZFNs [46, 47]. Natural TALE proteins have several different modules for each of the four base pairs, but a code has been developed based on the most common modules, and this allows simple and effective assembly of new binding domains [48–51]. Reports of successful applications to genomic targets are appearing at an accelerating rate [52–54].

RNA-guided engineered nucleases (RGENs) derived from the bacterial clustered regularly interspaced short palindromic repeat (CRISPR)-Cas (CRISPR-associated) system are now available. In bacteria, the CRISPR system provides acquired immunity against invading foreign DNA via RNA-guided DNA cleavage. 'Spacers', which are small segments of foreign DNA are integrated within the CRISPR genomic loci and transcribed and processed into short CRISPR RNA (crRNA). These crRNAs anneal to transactivating crRNAs (tracrRNAs) and direct sequence-specific cleavage and silencing of pathogenic DNA by Cas proteins. CRISPR/Cas-mediated genome editing has been successfully demonstrated in zebrafish and bacterial cells [46].

It is likely that developments in these gene editing approaches may be used in human disease to induce site-specific DNA cleavage in the genome and repair, through endogenous mechanisms, giving high-precision genome editing.

Targeting of Vectors

Targeting vectors to organs or specific tissues is the ultimate goal, and will most likely require the use of several combined approaches. Choosing an appropriate route of delivery and relying on vector tropism alone is unlikely to achieve efficient yet highly-targeted tissue delivery. At one extreme, local injection could usefully restrict expression to the site of delivery yet this would probably be insufficient for systemic diseases. At the other extreme, systemic delivery would be subject to less restriction, mainly conferred by vector tropism.

Vector receptors in the developing fetus and the physical availability of cells to vector infection are likely to be different to those in adults. Differences in receptor profiles were probably responsible for alternative targeting in newborn and adult mice following adenoviral vector delivery [55]. Differential cell availability through development was elegantly demonstrated by Endo and colleagues when comparing gene expression from lentiviral vectors following intra-amniotic injection, ranging from 8 to 18 days post-coitus. At 8 days GFP expression was observed in tissues derived from mesoderm and neural ectoderm, whereas beyond 11 days, expression was limited to epithelial cells. This expression profile correlated closely with the cell types exposed to the amniotic fluid at these different developmental stages [56].

Additional steps therefore may be required to target vectors. Transcriptional targeting by use of appropriate promoters will probably translate to fetal application with few hurdles. However, expression patterns are changing rapidly during development as tissues mature, and the epigenetic environment is likely to be very different from that found in the terminally-differentiated tissues of the adult. This was observed in mouse embryonic stem cells (ESCs), where the type of promoter in a lentivirus vector determined the time of transgenic protein expression during ESC differentiation [57].

Micro-RNA (miRNA) technology has recently been used in gene transfer vectors to downregulate gene expression in certain cell types. Incorporation of two miRNA sites into a lentivirus vector restricted transgenic GFP expression solely to hepatocytes, where the base vector directed expression in hepatocytes, liver endothelium and Kupffer cells [58]. *In vivo*, this resulted in specific expression of human factor IX from hepatocytes, avoiding immune responses resulting from unwanted expression of the transgenic protein from Kupffer cells [59].

The rate of mitosis in the target cell is of considerable importance in the developing fetus when choosing a method of gene targeting. Gamma-retroviral and lentiviral vectors integrate at a high frequency into the host genome whereas integration defective lentivirus vectors and those based upon adenovirus and adeno-associated virus remain predominantly episomal. Genomes of vectors which are episomally maintained will become diluted when they infect vigorously-dividing cells such as exist in the fetus, whereas genomes will persist in those tissues which undergo little or no mitosis. Where the disease target may be neurons reaching a terminal stage of differentiation, vectors that are episomally maintained may be most appropriate. In contrast integrating vectors may be more appropriate where, for example, hematopoietic stem cells are the target for application to blood diseases such as sickle cell anaemia or thalassemia.

Administering Perinatal Gene Therapy

If gene therapy to the fetus is to be clinically applicable, developments in vector technology must be accompanied by improvements in minimally invasive methods of delivering vectors to the fetus. Traditionally, invasive surgical techniques such as maternal laparotomy or hysterotomy have been performed to access the fetus in small- and even large-animal models. However, in clinical practice, minimally invasive techniques such as ultrasound-guided injection, or even fetoscopy, are much more likely to be used in clinical practice to deliver gene therapy to the fetus with less morbidity and mortality. Delivery to the placenta or amniotic fluid can be achieved using ultrasound guided techniques, and delivery to the uterine vessels is feasible using interventional radiology.

The route to the clinic is likely to follow a step-wise progression from small animal disease models, usually the mouse, into larger animals such as the sheep or rabbit to test out feasibility and safety of the delivery method, and finally into non-human primates for safety studies in the immediate preparation for a clinical trial of fetal gene therapy.

There are numerous advantages to rodent models including a short gestation time of around 20–22 days, the large litter size, the ease of colony maintenance due to the small physical size and their relatively low cost. The mouse genome is well defined with many transgenic models particularly of human monogenetic disorders. Recent advances in high frequency ultrasound and improvements in technical expertise have made it feasible to perform procedures on the fetal mouse with suitable accuracy such as intracardiac injection for example.

Sheep are much easier to breed and maintain and are a well-established animal model of human fetal physiology. Sheep have a consistent gestation period of 145 days; the development of the fetus and of the immune system is very similar to humans. Using the pregnant sheep, we have adapted ultrasound-guided injection techniques from fetal medicine practice and developed new methods to deliver gene therapy to the fetal sheep. Maternal mortality in the pregnant sheep was negligible and fetal mortality was between 3 and 15 %, depending on the route of injection. Over 90 % of the fetal mortality was due to iatrogenic infection, usually with known fleece commensals. Invasive procedures such as tracheal injection had a complication rate of 6 %, which was related to blood vessel damage within the thorax. Intracardiac and umbilical vein injection had an unacceptably high procedure-related fetal mortality in first-trimester fetal sheep and umbilical vein injection was only reliably achieved from 70 days of gestation, equivalent to 20 weeks of gestation in humans.

The high maintenance costs and breeding conditions of non-human primates prohibit their use in the routine development of novel injection techniques. Using techniques developed in the sheep, ultrasound guidance has been used in non human primates to deliver gene therapy into the amniotic cavity or for direct injection of the lung and liver parenchyma by teams in the US. The relevant time windows for the different application routes in humans still need to be established with respect to technical feasibility, fetal physiology and the development of the fetal immune system. In the human, fetus the immune system develops from 12 to 14 weeks of gestation, when profound increases in circulating T lymphocytes can be observed. Thus it might be necessary to deliver gene therapy before this gestational age, which will limit the routes of application that can be safely used. Experiments in the non-human primate are likely to be useful prior to clinical application.

In clinical fetal medicine practice, complications following cordocentesis do occur although they are rare (approximately 1 % of cases). Fetal distress may occur during or after an intrauterine transfusion procedure and may result in emergency preterm delivery. Other complications include cord rupture, spasm, tamponade from a hematoma or excessive bleeding, volume overload, chorioamnionitis, preterm premature rupture of the membranes or preterm labour [60]. An intraperitoneal approach was initially described by Liley in 1963 [61] and is currently used occasionally in clinical practice when the clinical situation requires early gestation blood transfusion when intra-umbilical vein transfusion is considered more risky for the fetus.

Pre-clinical Success with Perinatal Gene Therapy

Congenital Blood Disorders

Inherited blood disorders would be a relatively simple target for fetal application of gene therapy as the fetal circulation can be reached safely during its circulation through the umbilical vein (UV) at the placental cord insertion or the intrahepatic UV from approximately mid-gestation. If earlier gestation access is required, for example to avoid the immune system, the peritoneal cavity is a route used successfully to transfuse anaemic fetuses in fetal medicine clinical practice.

Congenital blood disorders are relatively common in some populations, and prenatal screening and diagnostic services are available in many countries. For many of these conditions, the disease could be corrected by the availability of the correct therapeutic proteins in the blood and they can be secreted functionally from a variety of tissues, thus the actual site of production is not so important as long as therapeutic plasma levels are realized. This means that ectopic sites of protein production such as the liver and muscle could be targeted.

The Haemophilias

Application of gene therapy to the fetus is probably most advanced when considering congenital blood disorders such as the haemophilias. Deficiency in factor VIII (FVIII) and FIX proteins of the blood coagulation cascade, result in hemophilias A and B, respectively, and have a combined incidence of around 1 in 8000 people [62]. Other deficiencies can also occur, for example Factor VII deficiency, which results in life-threatening hemorrhage at birth. Current treatment of haemophilia A or B uses replacement therapy with human FVIII or FIX which is expensive but effective [63]. Beneficial effects occur after achieving only 1 % of the normal levels of clotting factor. Unfortunately, a proportion of patients develop antibodies to therapy leading to ineffective treatment and occasional anaphylaxis [64]. The complications of haemophilia treatment which include the major risk of HIV and hepatitis B infections, although less of an issue now that blood donors are screened effectively, have in some cases been far worse than the diseases themselves, increasing their morbidity and mortality [65]. The clotting proteins are required in the blood and can be secreted functionally from a variety of tissues, thus the actual site of production is not so important as long as therapeutic plasma levels are realized.

Adult gene therapy strategies have concentrated on application to the muscle or the liver, achieving sustained FIX expression in adult haemophiliac dogs or mice after intramuscular or intravascular injection of adeno-associated virus (AAV) vectors [66–68]. AAV serotype appears to be important. In mice injection of AAV serotype 1 resulted in tenfold higher levels of canineFIX when compared with serotype 2 [66]. In clinical trials using AAV2hFIX vectors in haemophilia B patients, only short-term and low level FIX expression was observed, however, which was probably caused by a cell-mediated immune response to transduced hepatocytes [7, 69]. Recent clinical success with self-complimentary AAV8 vectors has made adult gene therapy for haemophilias a clinical reality [70]. Unfortunately however, prior exposure to the wild-type virus from which the vector is engineered is common and population screening of individuals reveals that the worldwide prevalence of AAV antibodies approaches 40 % [71]. Non-human primates also carry pre-existing antibodies formed in response to infection with the wild-type virus, and studies suggest that even modest titres completely inhibit transduction when vector is delivered through the circulation [72]. Strategies to circumvent this obstacle in the adult include excluding individuals that have pre-existing neutralizing antibodies to AAV detectable, meaning that currently approximately 40 % of affected men would be excluded from treatment [73].

Research in fetal application of gene therapy has progressed from demonstrating proof of principle in mouse models, into larger animal models such as the sheep and non-human primate, where delivery techniques, long term transgene expression and safety can be better addressed (Table 19.3). Waddington et al. demonstrated permanent phenotypic correction of immune-competent haemophiliac mice by intravascular injection of a lentivirus vector encoding the human Factor IX (hFIX) protein at 16 days of gestation (term = 22 days) [13, 18]. Plasma factor levels remained at 10–15 % of normal in treated animals for their lifetime. Sabatino et al. subsequently demonstrated induction of tolerance after AAV-1-hFIX administration in Factor IX-deficient fetal mice [74].

Translation to large animals has been slower because of the need for more long-term gene transfer and a higher vector dose when compared to small animals, but recent studies demonstrates the potential for this route of delivery. Long-term transduction of hematopoietic stem cells in the bone marrow and blood could be demonstrated 5 years after delivery of retroviral vectors into the peritoneal cavity of early gestation fetal sheep at laparotomy [41]. In early gestation, delivery of adenovirus vectors into the umbilical vein of fetal sheep via hysterotomy gives widespread transient transduction of fetal tissues [75]. Using ultrasound guided injection, systemic vector spread and widespread tissue transduction was demonstrated after first trimester intraperitoneal injection of adenovirus vectors into fetal sheep, although direct injection of the umbilical vein was limited by the procedure-related high mortality in late first trimester [28].

More recently, using the same self-complementary AAV8 vector expressing the human factor IX (hFIX) gene used for the clinical trials, long term hFIX expression was observed after ultrasound guided intraperitoneal injection of fetal sheep in early and late gestation [14]. No functional antibodies could be detected against the vector or transgene product and there was no liver toxicity observed. Antibodies to the therapeutic gene were detectable when the animals were challenged at 6 months of age postnatally with the hFIX recombinant protein, showing that induction of immune tolerance was not achieved. This is probably due to the fall in hFIX expression, that was undetectable by 1 year after birth. Umbilical vein delivery in fetal non-human primates of a tenfold higher dose of the same self-complementary AAV system in late gestation produced clinically-relevant levels of hFIX sustained for over a year, with liver-specific expression and a non-neutralizing immune response [15].

Although haemophilia A and B do not usually manifest until after birth, deficiency of some clotting factors, however, can lead to life threatening neonatal central nervous system (CNS) haemorrhage. Congenital Factor VII deficiency, the most common autosomal bleeding disorder, is typified by severe or lethal bleeding in around 20 % of patients with a homozygous or compounds heterozygous genotype. Affected babies often experience bleeding in the CNS hours or days after birth, but increasing expression even above 1 % would be sufficient to improve the risk and incidence of spontaneous haemorrhage [76, 77]. A therapy delivered during the fetal period could avoid long term pathology and provide therapeutic transgene expression for life [76, 77]. In a mouse model of FVII deficiency a single tail vein administration of AAV to neonatal mice, equivalent to late gestation injection in humans, resulted in at least 10 weeks of murine FVII expression which mediated protection against fatal hemorrhage and significantly improved survival. Third trimester ultrasound guided intraperitoneal injection to fetal monkeys conferred high level expression of hFVII at birth with a gradual decline to >1 % by 7 weeks. Readministration of an alternative serotype at 12 months postnatal age, gave therapeutic expression for at least 6 months [78].

The Thalassaemias and Sickle Cell Disorders

Inherited abnormalities of haemoglobin (Hb), a tetramer of two α-like and two β-like globin chains, are a common global problem. Over 330,000 affected infants are born annually worldwide, 83 % with sickle cell disorders and 17 % with thalassaemias [79]. Screening strategies can be premarital and/or antenatal depending on socio-cultural and religious customs in different populations and countries. In many countries worldwide, prenatal diagnosis is available from 11 weeks of gestation using chorionic villus sampling or amniocentesis from 15 weeks, and increasingly non-invasive prenatal diagnosis is becoming available [80].

The β-thalassemias, including the hemoglobin E disorders, are increasingly common in Europe, the Americas and Australia owing to migration of people from endemic in the Mediterranean and South-East Asia. Approximately 1.5 % of the global population are heterozygotes or carriers of the β-thalassemias. Profound anaemia which if untreated leads to death in the first year of life is the most severe form of β-thalassaemia, Cooley's anaemia or β-thalassaemia major. Even a mild correction of the globin chain imbalance in a fraction of maturing erythroblasts reduces the morbidity caused by ineffective erythropoiesis, and improves outcome [81]. Postnatal allogeneic haematopoietic stem cell transplantation (HSCT) can cure the condition with recent results of 90 % survival and 80 % thalassaemia-free survival [82] but it is only available in approximately 30 % of cases because of the lack of a suitable matched donor.

In alpha-thalassemia there is a deficit in the production of the Hb α globin chains which gives rise to excess β-like globin chains that form tetramers, called Hb Bart's in fetal life and HbH in adult life. Compound heterozygotes and some homozygotes for α thalassaemia have a moderately severe anaemia with HbH in the peripheral blood. Individuals who make very little or no α globin chains, have a severe anaemia, termed Hb Bart's hydrops fetalis syndrome which is commonly diagnosed prenatally and which, if untreated causes death in the neonatal period [83].

Sickle cell disorders are caused by Hb gene variants that, similar to thalassaemia, reduce mortality from falciparum malaria in carriers, and leads to high carrier levels in malaria endemic countries. The abnormal HbS cells in the circulation leads to recurrent painful sickle cell crises. Current treatment relies on a number of strategies to prevent crises from occurring using for example prophylactic antibiotics, pneumococcal vaccination and good hydration, and effective crisis management using oxygen and pain-relief [84].

Treatment of hemoglobinopathies holds challenges for gene therapy since the vector is required to carry a large cargo of the globin gene and its regulatory elements, so as to ensure high levels of expression of β-/γ- globin genes for therapeutic correction. In β-thalassaemia, the tissue and developmental-specific expression of the individual globin genes is governed by interactions between the upstream β-globin locus control region (β-LCR) and the globin promoters [85]. Amelioration or even cure of mouse models of human sickle cell disease [86] and β-thalassemia major [87–89] has been achieved using lentivirus vectors containing complex regulatory sequences from the LCR region. Yolk sac vessel injection of this optimized lentiviral vector into mid-gestation fetal mice resulted in human alpha-globin gene expression in the liver, spleen, and peripheral blood in newborn mice with expression that peaked at 3–4 months and reached 20 % in some recipients [90]. Expression declined at 7 months of age (normal life-span 2–3 years) possibly due to insufficient HSC transduction or the late stage of mouse gestation at which the vector was introduced.

Recent advances in vector design have improved gene transfer for these diseases. The ubiquitous chromatin opening element (UCOE) augmented Spleen Focus Forming Virus (SFFV) promoter/enhancer provides lentivirus vectors with a natural tropism for the haematopoietic system [91, 92] resulting in reproducible and stable function in bone marrow stem and all differentiated, peripheral haematopoietic cell lineages [93]. Encouraging results from the first successful gene therapy for a patient with hemoglobin E-β-thalassemia in a French trial has opened up gene therapy as a potential definitive treatment option for patients with β-hemoglobinopathies [43]. Trials with different versions of the β-globin and γ-globin genes in self-inactivating lentivirus vectors are beginning in Italy and four centers in the United States to treat thalassemia or sickle cell disease.

Recently human β-thalassemia induced pluripotent stem cells were generated from amniotic fluid cells using a single excisable lentiviral stem cell cassette vector. Amniotic fluid cells from the prenatal diagnosis of a β-thalassemia patient were reprogrammed by expression of the four human reprogramming factors Oct4, KLF4, SOX2 and c-MYC using a doxycycline lentiviral system and demonstrated teratoma formation [94]. This type of cell manipulation may provide clinicians with corrected autologous patient-specific iPS cells to use in a combination IUSCGT approach for the treatment of thalassaemia [94].

Lung Diseases

The fetal lung is an ideal target for gene therapy because transduction of the fluid filled fetal lungs may be achieved more easily than in post-natal life, where there is an air-tissue interface. Post-natal gene transfer is reduced by the inherent lung damage that occurs with lung pathology [95] Candidate diseases for lung-directed fetal gene transfer include cystic fibrosis (CF), alpha-1 antitrypsin deficiency, surfactant protein B deficiency and pulmonary hypoplasia.

Cystic fibrosis (CF) is a common lethal autosomal recessive disease in which tissue injury begins in the prenatal period [96]. The potential targets for CF lung manifestations are the ciliated epithelial cells and ducts of the submucosal glands, where the wild-type CF transmembrane conductance regulator (CFTR) protein is expressed. *In vitro* studies suggest that correction of as few as 6 % of the defective cells may be sufficient to correct the chloride transport defect [95] [97]. Although CF is a multisystem disease, much of the morbidity and mortality derives from the diseased lung. Here the classical gene therapy target is the ciliated epithelial cells and ducts of the submucosal glands in the lungs where the wild-type CFTR protein is normally expressed [98]. Gastrointestinal manifestations of CF are now increasingly recognized as an important contributor to morbidity in those patients who reach adulthood [99], as well as affecting 15 % of neonates with the life-threatening condition of meconium ileus [100]. With the advent of prenatal screening for CF, the possibility of offering treatment to couples whose fetus is affected becomes more real [101].

Cystic fibrosis was one of the first diseases in which gene therapy was applied with around 400 CF patients receiving treatment postnatally using viral and nonviral gene transfer agents through mainly nebulized systems [102]. Early trials established the safety of adenovirus and nonviral vectors but CFTR expression was hindered by the low transduction efficiency of both vector classes on the respiratory epithelium, partly due to the location of the adenovirus receptor in the basolateral membrane of the respiratory epithelium which is isolated from the lumen by tight junctions [103]. In addition, a robust immune response caused a dose-dependent inflammation and pneumonia related to the immunogenicity of the viral proteins that prevented repeat administration [104]. AAV2 have shown in clinical trials to have reduced toxicity and immunogenicity [105, 106] but these phase I/II trials were in general unsuccessful due to neutralizing antibodies that prevented reliable repeat vector administration. Non-viral approaches are being investigated by the UK CF Gene Therapy Consortium that has developed a translation programme with two products. Proof-of-principle studies on the nasal epithelium show a 25 % correction of the molecular defect [107] and expression of hCFTR is seen in sheep transfected with a human CFTR plasmid, complexed with GL67 [108]. For Wave 1, the key milestone has been to undertake a multidose trial to assess whether repeated administration of the non-viral vector over a lengthy period (1 year) can improve CF lung disease. Wave 2 is working towards preparing a modified lentivirus vector for clinical trials. Studies so far have demonstrated lifetime gene expression and efficient repeat administration in mouse lung, lack of chronic toxicity and persistent gene expression in human *ex vivo* models [109].

One of the barriers to effective gene transfer to the airways in the adult or neonate with CF is that inflammation and damage of the lung precludes effective gene delivery. This could be overcome if gene therapy is applied at a stage of prenatal life where no or minimal lung damage has occurred and particularly because transduction of the fluid filled fetal airways may be more easily achieved. Fluorocarbon liquids such as perflubron have been used to push vector into the distal fetal airways from injection at the trachea [4] and enhance adenovirus mediated gene expression in normal and diseased rat lungs [95]. The proliferating cell population in CF airways are mainly basal cells [110] and these would be the best target in any gene therapy approach.

Initial studies appeared promising, with a report that CFTR-knockout mice could be cured by prenatal adenovirus administration into the amniotic fluid [111]. Since the fetus draws amniotic fluid into the lungs during fetal breathing movements, intra-amniotic delivery could provide an efficient route of gene transfer to the airways. Two further studies using the same vector, delivery method and mouse strain as well as a different CFTR-knockout mouse strain have, however, have been unable to replicate these findings [112, 113]. The high spontaneous survival rate of the CF mouse strain used in the original experiments may explain the initial enthusiasm for the results observed [111]. In addition, the inability to cure CF in this model might be due to the strain of mice used, the vector construct which only gives short term gene transfer [114], or because of insufficient fetal breathing movements. Several studies have applied AAV vectors using the amniotic route in an attempt to transduce the lung epithelium. Injection of AAV2 into rabbit amniotic fluid transduced the trachea and pulmonary epithelium of the fetus [115] and prolonged gene expression was seen in mice, rats and macaques [37] but probably insufficient for phenotypic cure of CF.

Targeting transgenic protein expression to the fetal lung can be better achieved in the mouse by increasing fetal breathing movements using a combination of intra-amniotic theophylline administration and exposure of the dam to elevated CO_2 levels [112]. Theophylline has a similar effect on breathing movements in fetal sheep [116]. Much of the vector was diluted in the amniotic fluid volume and not concentrated in the organ(s) required for CF therapy as strong gene transfer to the skin occurred after intra-amniotic delivery [112]. In large animals such as the fetal sheep we were unable to produce significant airways gene transfer after intra-amniotic adenovirus vector injection in the first trimester although the nasal passages were transduced [28]. Fetal breathing movements are not present in the first trimester human or sheep fetus, and the large amniotic fluid volume even at this gestation dilutes the vector meaning that a more targeted approach to the lung may be required in clinical practice.

Local injection of the lung parenchyma is an alternative to the amniotic route but gives only local gene transfer in fetal rats [117, 118] and non-human primates [119]. The stage of gestation is important, with transgene expression more local to the lung after vector injection in early second trimester (pseudoglandular stage) when compared to the late first trimester (embryonic stage) [120]. Gene delivery to the lung parenchyma can also been achieved by indirect means using AAV, by intraperitoneal injection for example [34]. Similarly injection of AAV1 and AAV2 into mouse muscle, peritoneal cavity and intravenously gave lung expression of the transgenic protein [121].

In larger animals, injection of the fetal trachea by transthoracic ultrasound-guided injection [28] targets gene transfer to the medium to small airways [4]. Increased transgene expression in the fetal trachea and bronchial tree was seen after complexation of the virus with DEAE-dextran, which confers a positive charge to the virus, and pretreatment of the airways with sodium caprate, which opens tight junctions in the airway epithelia thereby improving vector access to the coxsackie-adenovirus receptors [4, 122]. A novel surgical approach has also been developed in the mouse to deliver vectors directly into the fetal trachea [123] resulting in strong gene transfer up to 1 month after birth [124]. Tracheal readministration of AAV2/5 3 and 6 months after neonatal tracheal instillation resulted in efficient and stable gene transfer up to 7 months after birth without any evidence of a strong immune response [125]. This suggests that a late pregnancy gene transfer, equivalent to a neonatal mouse in terms of fetal development, with postnatal readministration might be suitable to achieve long term pulmonary gene transfer. Gene transfer to human fetal lungs has been achieved in a xenograft model in SCID mice with long-term expression in the surface epithelial and submucosal gland cells observed up to 4 weeks and 9 months after administration of AAV and lentiviral vectors respectively [126, 127].

For gastrointestinal CF pathology widespread intestinal transduction was achieved using ultrasound guided gastric injection in the early gestation fetal sheep [128] that had an associated low morbidity and mortality. Transgene expression was enhanced after pretreatment of the fetal gut with sodium caprate after adenovirus complexation with DEAE-dextran. In addition, instillation of the fluorocarbon perflubron after virus delivery resulted in tissue transduction from the fetal stomach to the colon.

Diseases of the Nervous System

Early lethal genetic diseases of the nervous system are individually rare, yet collectively lead to a large disease burden, and in some populations, have a high prevalence [129]. Conditions can directly affect the nerves themselves, such as spinal muscular atrophy (SMA), a disease primarily of the peripheral nervous system. Alternatively enzyme deficiencies can lead to a damaging build up of lysosomal substrates that damage neurons, as well as other organs in the body. Examples include the lysosomal storage diseases such as acute neuronopathic (Type II) Gaucher disease, neuronal ceroid lipofuscinoses and Niemann–Pick disease type C. In some cases, these conditions are not recognized during fetal life on prenatal ultrasound examination. For example, there are a few case reports that some fetuses with SMA have increased nuchal translucency, however a recent study in 12 women with affected fetuses did not find any association [130]. In Niemann-Pick disease type C however, *in utero* splenomegaly, hepatomegaly, ascites, fetal growth restriction and oligohydramnios (reduced liquor volume) are common [131]. Screening programs are available in populations with high prevalence such as the Ashkenazi Jews, where triple disease screening for Tay-Sachs disease, type 1 Gaucher disease and cystic fibrosis (CF) are commonly performed together [132]. Prenatal diagnosis is available for these conditions via chorionic villus sampling (CVS) assuming the gene defect is known. Creation of iPS from CVS might be a promising route for personalized autologous gene/cell therapy approach for the treatment of congenital CNS disorders [133, 134].

The Lysosomal Storage Diseases

The lysosomal storage diseases are inherited deficiencies of lysosomal enzymes that lead to intracellular substrate accumulation. In mucopolysaccharidosis type VII (MPS type VII) for example, a deficiency of β-glucuronidase activity leads to accumulation of glycosaminoglycans in lysosomes [135] leading to enlarged liver and spleen, growth and mental retardation and death from cardiac failure. Lysosomal storage diseases can manifest during intrauterine life as non-immune hydrops. Although rare, MPS VII has been a disease of choice to investigate gene therapy because of the availability of a mouse and dog model. Correction of the MPS phenotype theoretically requires only low levels of the therapeutic gene product [136]. Neonatal injection of a retrovirus vector in MPS VII dogs and mice resulted in hepatocyte transduction, with uptake of the enzyme from the circulation by other organs. The treated animals did not develop cardiac disease or corneal clouding and skeletal, cartilage and synovial disease was ameliorated [137]. Non-viral mediated gene transfer to the liver of MPS I and VII mice also improved the phenotype [138]. Still, the major challenge remains to target the brain which currently requires multiple brain injections with accompanying risks [139, 140] and immunosuppression to prevent pan-encephalitis that develops secondary to an immune response to the transgene [140]. Widespread correction of the pathological lesions in an MPS VII mouse was recently observed with adeno-associated virus gene transfer [141], a vector which elicits less of an immune response. Prenatal gene delivery is an alternative strategy. Injection of adenovirus into the cerebral ventricles of fetal mice led to widespread and long term gene expression throughout the brain and the spinal cord [142]. In the same study, delivery of a therapeutic gene to the cerebral ventricles of fetal MPS type VII mice prevented damage in most of the brain cells before and until 4 months after birth. A similar study using an AAV vector had comparable results but with longer expression [143].

From a translational perspective, direct vector administration into the fetal brain or ventricles for prenatal gene transfer is unappealing. There are technical difficulties in injecting the fetal brain through the skull using minimally invasive injection techniques, although this has been achieved in non-human primate [144] and sheep (A.L. David, unpublished work) under ultrasound guidance. In contrast, ultrasound-guided access to the human fetal circulation is commonly used for fetal blood sampling and transfusions in clinical practice, with minimal fetal loss rate or complications [145]. This triggered the hunt for vectors that could cross the blood/brain barrier.

Recently AAV vectors of serotypes 2/9 have been shown to have an astonishing ability to transduce cells of the nervous system, achieved not by intracranial but via intravenous injection in neonatal mice [146, 147], cats [146] and non-human primates [148]. The ability of the vector to cross the blood-brain barrier may depend on specific populations of receptors within the brain that facilitate transfer for particular AAV serotypes [149]. A recent study describing fetal intravenous injection of AAV 2/9 in either single-stranded or self-complementary format showed comprehensive transduction of the central nervous system, including all areas of the brain and retina, and the peripheral nervous system including the myenteric plexus. Interestingly the single stranded version, containing a woodchuck hepatitis virus post-transcriptional regulatory element (WPRE) achieved far higher and more comprehensive levels of expression than the self-complementary vector lacking WPRE [150].

Prenatal gene transfer has also been applied with some success in glycogen storage disease type II (GSDII), which is caused by a deficiency in acid α-glucosidase (GAA). This leads to lysosomal accumulation of glycogen in all cell types and abnormal myofibrillogenesis in striated muscle with death from respiratory failure. Delivery of the AAV-*GAA* vector by intraperitoneal injection to the mouse embryo in knockout models gave high-level transduction of the diaphragm and restoration of its normal contractile function [21].

Neuronal ceroid lipofuscinoses (NCLs), known collectively as Batten disease, are autosomal recessive lysosomal storage diseases which have lead to significant central nervous system pathology. Infantile NCL, caused by mutations in the CLN1 gene, results in deficiency in palmitoyl protein thioesterase 1 (PPT1). Patients with the disease are born with no pathological manifestations but by 12 months of age they show signs of mental retardation, motor dysfunction and visual problems [151] and survive only to 6 years of age, on average. A mouse model of this disease shows many of the same pathological symptoms and premature death occurs by 8.5 months [152]. Although there have been no attempts as yet at treating this model by fetal gene therapy, there is a strong case to be made for this in a preclinical and clinical setting [153, 154].

Spinal Muscular Atrophy

Spinal muscular atrophy (SMA) is characterized by degeneration of the lower motor neurons in the anterior horn of the spinal cord and the brainstem. Although rare (incidence 1:10,000) [155], SMA is invariably fatal after a course of progressive muscle weakness and atrophy. It is caused by homozygous loss or mutation in the telomeric survival motor neuron gene 1 (SMN 1) with subsequent neuronal cell death through apoptosis. Affected individuals can be partially protected by the presence of an increased copy number of the SMN2 gene [155], a nearly identical copy gene of SMN1, that produces only 10 % of full-length SMN RNA/protein. This suggests that SMN2 may play a disease-modifying role and could be a target for gene therapy of the disease.

The childhood forms which are all autosomal recessive can be divided into three types depending on their severity [156]. The fetal form of the disease, type 0, presents in utero with diminished fetal movements and arthrogryposis [157]. Neuronal degeneration and loss in SMA type I begins during intra-uterine life which makes prenatal gene therapy an attractive option [158]. Vectors derived from adenovirus, herpes simplex virus (HSV), adeno-associated virus (AAV), and lentivirus are capable of transducing neurons *in vitro* and *in vivo* [156]. Neuroprotective factors such as cardiotrophin1 [159] or anti-apoptotic proteins such as Bcl-xL can be used, for example adenovirus Bcl-xL has been shown to inhibit neuronal cell death in rat cell cultures [160]. SMN gene replacement is also possible. Multiple intramuscular injections of a RabG–EIAV lentivirus vector containing the human SMN gene increased SMN protein levels in SMA type 1 fibroblasts and in SMA mice and reduced motor neuron death [161].

Lower motor neurons can be targeted by direct injection of the spinal cord which although successfully achieved in the rat and mouse post-natally [162, 163], is technically risky in the adult human. Injection of an AAV8 containing the human SMN gene into the CNS of SMA mice improved mortality [164] although they still died prematurely despite continual, high-level expression from the viral vector, which may have been due to a failure to correct the autonomic system that regulates cardiac function.

An alternative is remote delivery, and motor neuron gene expression has been achieved by intra-muscular or intra-axonal injection with subsequent retrograde axonal transport in small animals [165, 166]. A therapeutic effect was documented after intramuscular injection of adenovirus-cardiotrophin 1 in a mouse model of SMA [159]. It is not clear, however, if remote delivery will be effective in larger animals or an affected human where the peripheral nerves are much longer and retrograde transport is impaired secondary to the disease. Fetal application in this context may provide the advantage of a shorter and healthy axon and recent results in mice suggest that a fetal approach is feasible. Using lentivirus vectors, that are efficient at infecting non-dividing cells Rahim et al. observed transduction of multiple dorsal root ganglia and efferent nerves following intrathecal injection of an EIAV (equine infectious anaemia virus) lentivirus vector into fetal mice at 16 dpc [167].

Systemic delivery using AAV vectors is probably the way forward and might also correct the cardiac dysfunction that occurs. Foust et al. incorporated SMN cDNA into the 2/9 vector serotype and showed that neonatal intravenous injection of this vector into the corresponding mouse model of SMA resulted in an unprecedented improvement in survival and motor function [168]. Using a self-complimentary AAV9 vector containing a codon-optimized SMN1 sequence injected intravenously on day 1 postnatal, Dominguez et al. achieved 100 % rescue of a mouse model of severe SMA, completely corrected motor function and reduced the weight loss associated with this model [169].

The Urea Cycle Defects

The inherited inborn errors of metabolism result from enzyme deficiencies in different metabolic pathways. One of the first metabolic disorders targeted by gene therapy is the defect in the urea cycle, ornithine transcarbamylase deficiency (OTC), an X-linked condition which results in accumulation of ammonia with resultant repeated episodes of hyperammonemia within 1 week of life, damaging the central nervous system and jeopardizing life [170, 171]. A phase I trial targeting the liver through intra-arterial adenovirus injection ended with low level gene transfer and a fatal immune reaction in one of the 17 patients [172, 173]. Subsequent investigation in small animals focused on less immunogenic vectors and showed that long-term correction of the metabolic defect in OTC deficiency could be achieved using a helper-dependant adenovirus vector [174] and AAV [175]. Because of the difficulties with post-natal OTC deficiency gene therapy and the severity and very early onset of the complete form, fetal gene therapy may be a good approach.

A notable success in small animal models is in the long-term correction of bilirubin UDP-glucuronyltransferase deficiency in fetal rats using a lentivirus vector [20]. Humans who suffer from this defect are classified as having Crigler-Najjar type 1 syndrome and suffer severe brain damage early in childhood due to the inability to conjugate and excrete bilirubin. A rat model of Criggler-Najjar was injected with a lentivirus vector carrying the gene for bilirubin UDP-glucuronyltransferase. The treated rats sustained a 45 % decrease in serum bilirubin levels for more than a year, a level that would be considered therapeutic in the human [20]. Despite the long-term expression, these rats developed antibodies against bilirubin UDP-glucuronyltransferase [176] which may be related to the fact that the fetal injection was done late in fetal life and to the unusual immunogenicity of the transgenic protein.

Intravascular vector delivery in small animals can give excellent liver transduction but in larger animals such as the fetal sheep, the intra-peritoneal route seems to be the best route to target gene transfer to the fetal liver [28], and no immune response to the transgenic protein was detected after injection of adenovirus vectors in early gestation. Direct intrahepatic injection resulted in low level gene transfer with necrosis of the liver around the injection site, which was thought to be due to a direct toxic effect of adenovirus vector on hepatocytes [28]. Studies of intrahepatic injection using other vectors such as lentivirus or AAV show better results. In the non-human primate, intrahepatic or intraperitoneal (IP) vector injection resulted in widespread gene transfer and particularly to the liver, with no trans-placental transfer to the mother [177–180]. In one of these studies however, IP injection of lentivirus vector at the end of the first trimester showed that a subset of female but not male germ cells were transduced [178]. In the ovary, meiosis begins in the innermost areas of the cortex during the 12th and 13th weeks of gestation, while proliferating primordial germ cells forming the oocytes are found in the most superficial areas of the cortical region of the developing ovary, and these may be vulnerable to lentiviral gene transfer when delivered early in gestation via the IP route. Since IP injection is a relatively safe and well-studied ultrasound guided fetal injection method [181], this is likely to be the route of choice when compared to liver injection, which is used rarely in fetal medicine for diagnosis of congenital liver disease. The risk of germline gene transfer in female fetuses will need to be evaluated carefully.

Metabolic diseases other than ornithine transcarbamylase deficiency and bilirubin UDP-glucuronyltransferase deficiency that could benefit from fetal gene therapy to the liver are phenylketonuria, galactosemia, and long-chain acyl-CoA dehydrogenase deficiency [182].

Muscular Dystrophy

Targeting the muscle for gene delivery could be a successful strategy for treatment of muscular dystrophies. Duchenne muscular dystrophy (DMD) is the commonest form, and is X-linked. Abnormal or absent dystrophin leads to progressive muscle weakness in early childhood, culminating in death secondary to respiratory or cardiac failure during the third decade of life. For a one step prenatal gene therapy, the striated muscles in the limbs and chest, and the cardiac muscle would need to be transduced. Prenatal diagnosis is available, and carriers can be screened for the presence of a male fetus using non-invasive prenatal diagnosis, avoiding the need for invasive tests and the associated miscarriage risk in 50 % [181].

In adult clinical trials, dystrophin gene transfer to striated muscle using viral [183] and non-viral vectors [184] has been hampered by low efficacy because of the development of cellular and humoral immunity to the transgenic dystrophin gene [185, 186]. This could be avoided by prenatal application, which would also target a rapidly proliferating population of myocytes that are present in the fetus. Satellite cells that are capable of regenerating muscle fibers are transduced after intramuscular lentivirus vector delivery to fetal mice [187]. Importantly inducible dystrophin expression begun during fetal life corrected the phenotype in a DMD mouse model, where postnatal expression did not [188], supporting a fetal approach.

Intramuscular fetal injection of an adenovirus containing the full-length murine dystrophin gene in the mdx mouse model of DMD conferred effective protection from cycles of degeneration and regeneration normally seen in affected muscle fibers [189], but gene transfer level was low. More efficient gene transfer to all necessary muscle groups was seen after delivery of lentivirus vectors to fetal mice using multiple routes of injection. Systemic delivery targeted the heart, direct injection transduced the limb musculature and intraperitoneal injection reached the diaphragm and innermost costal musculature. Expression lasted for over 15 months and did not stimulate any immune response [190].

Large animal muscle gene transfer has been investigated. Gene delivery to the hindlimb musculature of the early gestation fetal sheep using ultrasound-guided injection of adenovirus vectors resulted in highly efficient gene transfer with a low procedure complication rate [28]. A clinically relevant method for respiratory muscle gene transfer has also been evaluated in early gestation fetal sheep and showed that transduction of intercostal muscles occurred after ultrasound guided creation of a hydrothorax into which adenovirus vectors were introduced [191].

There has been considerable recent success using AAV vectors to transduce fetal musculature. Early studies on AAV showed long term local transgenic protein expression following direct injection into fetal mouse muscle [33] and transduction of the diaphragm after IP injection [33]. Using this route to administer AAV1, Rucker and colleagues restored acid α-glucosidase activity to the diaphragm in a mouse model of Pompe disease [21]. This was the first demonstration that fetal gene transfer could correct a model of congenital muscle pathology. More recently studies on intraperitoneal delivery of AAV8 into normal fetal mice show high levels of marker gene expression in all the muscle groups affected by congenital muscular dystrophies [192], and in the mdx mouse model of DMD, delivery of an AAV containing dystrophin significantly improved the dystrophic phenotype [193]. Postnatal application of AAV6 containing full length and micro-dystrophins in neonatal mice can almost entirely prevent and partially reverse the pathology associated with DMD, but only near the site of injection [194]. In late-gestation macaques umbilical vein injection of AAV9 results in very high levels of transgenic protein expression in many tissues including skeletal and cardiac muscle [15]. Systemic delivery of AAV vectors in the fetus may finally provide a solution to target the necessary muscle groups for muscular dystrophy therapy. However the packaging capacity of AAV is restricted to delivery of truncated dystrophin minigenes which may negatively counteract the efficiency of this vector system.

Exon-skipping is another strategy that is proving quite successful. An antisense oligonucleotide is used to modify splicing, particularly of exons, such as skipping of the mutated exon 51, which allows a partly functional dystrophin protein to be produced from the muscle. This therapy has been successful in the *mdx* mouse and a dog model of DMD, and there are currently three phase III trials internationally [195].

The Genodermatoses

The genodermatoses are a group of genetic skin diseases that may be associated with significant morbidity and morality. Examples include the epidermolysis bullosa (EB) disorders, the ichthyotic disorders, and disorders of pigmentation such as oculocutaneous albinism. Methods of prenatal diagnosis are varied. Where the molecular defect is known, amniocentesis or chorionic villus sampling is commonly used. X-linked ichthyosis is associated with low levels of unconjugated estriol, one of the markers used for Down's syndrome screening, and this can prompt prenatal diagnosis in the mother. When the gene mutation is unknown however, fetal skin biopsy is necessary, that unfortunately carries a slightly higher fetal loss than other invasive tests (1–3 %), results in scarring, and may need to be performed at quite late gestations [196]. Ultrasonography can be used in the diagnosis of a few of these disorders. In harlequin ichthyosis for example, typical sonographic features include echogenic amniotic fluid, large joint and digital contractures and facial dysmorphism, including flat face and wide mouth with thickened lips.

The genodermatoses may be good candidates for prenatal gene therapy, where gene transfer to the skin via the amniotic fluid may provide obvious advantage to cumbersome post-natal therapy. Transgenic protein expression is seen in the skin after intraamniotic delivery of adenoviral vectors to mice (12 days post conception (dpc)) [112], and sheep in the early first trimester (day 33 of 145 days of gestation) using ultrasound-guided injection [28]. In a mouse model of Herlitz junctional epidermolysis bullosa, a lethal skin disease, a combination of adenovirus and AAV vectors injected into the amniotic cavity of fetal mice (14 dpc) led to expression of the laminin-5 transgenic protein although only a minor increase in the lifespan of treated mice was seen [36]. In all these studies, only the most superficial layers of the skin, the periderm and epidermis were transduced. Several strategies have been used in small animals to target the deeper layers, such as intra-amniotic injection with subsequent electroporation [197] or application of microbubble-enhanced ultrasound (shot-gun method) [198, 199]. Translation to clinical practice will be challenging. Earlier in gestation, epidermal stem cell populations are accessible for gene transfer using the intraamniotic delivery route. Injection of lentivirus vectors between day 8 and 12 dpc in fetal mice resulted in long term transgenic protein expression in basal epidermal stem cells into adulthood [200]. Using a skin specific keratin 5 promoter instead of the cytomegalovirus promoter also improved epidermal gene transfer.

Primary Immune Deficiencies

The primary immune deficiencies result from inherited mutations in genes required for the production, function or survival of specific leukocytes such as T, B or NK lymphocytes, neutrophils and antigen-presenting cells, or are caused by cytotoxic metabolites. The leukocytes are produced from the pluripotent hematopoietic stem cells (HSC) in the bone marrow, and therefore allogeneic bone marrow transplantation (BMT) from a healthy donor into an affected patient can restore the immune system. Successful BMT has been achieved in deficiencies such as Wiskott-Aldrich Syndrome (WAS), Chronic Granulomatous Disease (CGD) and Adenosine Deaminase Deficiency where there is toxicity rather than a defect of the proliferation gene and in X-linked Severe Combined Immunodeficiency (SCID). With the exception of X-SCID all of the other primary immune deficiencies require extra therapeutic steps such as pre-transplant conditioning, marrow cytoreduction to "make space" in the marrow for the transplanted HSC and immune ablation to prevent rejection of the donor HSC. These strategies carry risks for the patient and in some cases, a haploidentical donor is unavailable. Gene therapy has therefore been developed for treatment of some patients.

Post-natal gene therapy using *in-vitro* transduced autologous HSCs with subsequent transplantation into the same patient has been used successful in adenosine deaminase deficient SCID [201, 202], X-linked SCID [203] and CGD [204]. Despite the encouraging results, 4 out of 26 subjects subsequently developed a T cell leukemia-like condition which may have been related to integration of the retroviral vector near a suspected proto-oncogene [205]. Newer approaches to decrease this risk have used lentivirus vectors that have been studied non-human primates [206]. Also, semi-viral systems have been developed with the aim to offer stable gene transfer along with a favorable pattern of integration [138, 207]. These semi-viral systems are still limited by their low transduction efficiency as in the case of the sleeping beauty transposon [138, 207].

Prenatal diagnosis of the primary immune deficiencies is available where the gene mutation is known. For example, it has been applied in families that have been identified to be at risk of these conditions, such as those harboring mutations in both of the recombination activating genes RAG1 and RAG2 that are involved SCID syndromes [208, 209].

Diseases of the Sensory Organs

Prenatal gene delivery looks promising in eye and ear diseases but translation to man will be challenging. In animal models of Leber congenital amaurosis, a severe retinal dystrophy, fetal gene therapy using AAV or lentivirus vectors resulted in an efficient transduction of retinal pigment epithelium and restoration of visual function [210]. Similarly AAV was able to efficiently transfect the developing cochlea in fetal mice [211]. All these previous studies relied on injection into the developing sensory organ itself, which will be difficult to achieve in clinical practice. Vector delivery via the amniotic cavity early during embryonic development depends critically on the stage of gestation. For example, intraamniotic delivery of lentivirus specifically at day 8 dpc resulted in gene transfer to the mouse retina [212] but later delivery time points were only able to target the lens and cornea. Greater tissue specificity and safety can probably be accomplished by the use of tissue-specific promoters, or regulated transgene expression, but there will still be the need for accurate prenatal diagnosis at a time of gestation equivalent to 3–5 weeks in human pregnancy, something that will be difficult to achieve with current diagnostic techniques.

In the guinea pig, neurotrophin gene therapy integrated into the cochlear implant and improved its performance by stimulating spiral ganglion neurite regeneration [213].

Fetal Growth Restriction

Severe fetal growth restriction (FGR) affects 1:500 pregnancies and is a major cause of neonatal morbidity and mortality. The underlying abnormality in many cases is uteroplacental insufficiency, whereby the normal physiology process of trophoblast invasion that converts the uterine spiral arteries into a high-flow large conduit for blood, fails to occur. Currently there is no therapy available that can improve fetal growth or delay delivery to allow fetal maturity. FGR is commonly diagnosed on routine fetal ultrasound when the fetal growth velocity falls below the expected gestational age charts. Abnormally low uterine artery Doppler blood flow and increased vascular resistance is also classically seen in mid-gestation.

A targeted approach to the uteroplacental circulation is needed, since intravascular infusion of sildenafil citrate, a nitric oxide donor, causes a drop in systemic blood pressure and had detrimental effects on growth restricted sheep fetuses [214]. In the pregnant sheep transient local over-expression of vascular endothelial growth factor (VEGF) mediated via adenovirus vector injection into the uterine arteries increases uterine artery blood flow and significantly reduced vascular contractility [215]. VEGF expression was confined to the perivascular adventitia of the uterine arteries, together with new vessel formation, supporting the local effect of gene transfer. These effects are long term, lasting from mid-gestation (80 days) through to term (145 days) [216] with reduced intima to media ratio suggesting vessel remodeling, and adventitial angiogenesis demonstrated. Recent work in an FGR sheep model in which uterine blood flow is reduced by 35 % in mid-gestation, demonstrates that uterine artery injection of the same dose of Ad.VEGF significantly improved fetal growth in late gestation [23].

In the clinical context vector delivery into the uterine artery could be achieved through interventional radiology, which is supported by the RCOG as a prophylactic measure before delivery in women at high risk of postpartum haemorrhage [217]. While this is more invasive than administering oral medication it has the potential advantage of targeting vasoactive changes to the maternal uteroplacental circulation. The EVERREST Project (EVERREST [218]) aims to carry out a phase I/IIa clinical trial to assess the safety and efficacy of maternal uterine artery Ad.VEGF gene therapy for severe early onset FGR. The project, funded by the European Union, involves a multinational, multidisciplinary consortium, including experts in bioethics, fetal medicine, fetal therapy, obstetrics, and neonatology.

Fetal Structural Malformations

Fetal structural malformations are another potentially important application of perinatal gene therapy. Although individually rare, collectively up to 1 % of all fetuses are affected by a structural malformation, that for some are lethal or are associated with significant morbidity. Congenital diaphragmatic hernia (CDH) for example, is a condition where there is a defect of the diaphragm resulting in herniation of some or all of the intra-abdominal organs into the fetal chest. This compresses the fetal lungs preventing adequate growth, which results in poor lung function at birth. With surgical correction of the diaphragmatic defect, many neonates do well. Current management of severe CDH involves fetoscopic placement of an inflatable balloon in the fetal trachea to block outflow of the tracheal fluid, which encourages lung growth [219]. There is however, an underlying lung defect which may contribute to the lung pathology, and gene transfer may play a part in correcting this problem.

Short term expression of growth factors at a critical stage of lung growth may be useful for this serious condition. In a rat model of CDH, adenovirus mediated prenatal CFTR expression enhanced saccular density and air space in the lungs [220]. After surgical creation of CDH in fetal sheep, non-viral vector expression of keratinocyte growth factor in the trachea lead to increased surfactant protein B synthesis in the lungs suggesting better maturation of the regrowing lung [221].

Use of Manipulated Stem Cells for Perinatal Therapy

Gene therapy in early gestation before the maturation of the immune system could, theoretically, eliminate the need for marrow conditioning or the restriction to an HLA-matched donor. Prenatal treatment with hematopoetic stem cell transplantation has been attempted for a variety of immunodeficiencies and hemoglobinopathies using IP transfer of paternal or maternal hematopoietic cells or fetal liver [222] however the clinical successes were mainly in cases of X-linked SCID [223, 224] where no immune response to the transplanted cells could be mounted.

An autologous stem cell gene transfer approach [225] using fetal stem cells from a number of sources within the fetus including the blood, liver, amniotic fluid (AF) and placenta could be adopted. Fetal liver or blood sampling at an early gestational age carries a significant risk of miscarriage [226, 227]. It is now apparent that pluripotent stem cells can be readily derived from fetal samples collected at amniocentesis [228] or chorionic villus sampling [229–232], procedures that have a low fetal mortality. Human AFS cells have the potential to differentiate into a variety of cell types and can be transduced easily without altering their characteristics [228, 233–235]. Recent work in sheep described good fetal survival after autologous AF mesenchymal stem cell (MSC) transplantation using ultrasound-guided amniocentesis, and subsequent IP injection of selected, expanded and transduced AFMSCs into the donor fetus. Widespread cell migration and engraftment, particularly in the liver, heart, muscle, placenta, umbilical cord and adrenal gland was seen [236].

Recently our group studied the functional haematopoietic potential of transduced GFP+ sheep AF-derived stem cells, before and after autologous IUSCT. First trimester sheep AF was collected by ultrasound-guided amniocentesis or at post mortem examination. Sheep CD34+AF or adult bone marrow cells were selected and transduced overnight with an HIV lentivirus vector containing eGFP. Transduced fresh or frozen CD34+AF, or bone marrow cells, were injected intravenously into NOD-SCID-gamma (NSG) mice. GFP+ cells were detected in the haematopoietic organs and peripheral blood of NSG mice primary and secondary recipients 3 months later. Autologous IUSCT was performed in fetal sheep using ultrasound-guided intraperitoneal injection of fresh transduced GFP+CD34+AF cells. GFP+ cells were detected in the peripheral blood of injected lambs up to 6 months postnatally and 3 months after secondary transplantation of bone marrow from autologous IUSCT lambs into NSG mice, GFP+ cells were detected in haematopoietic organs. This demonstration of autologous IUSCT of CD34+AF cells in a large animal model supports the concept for clinical translation to treat congenital haematopoietic diseases *in utero* (Fig. 19.1) (unpublished data, under review).

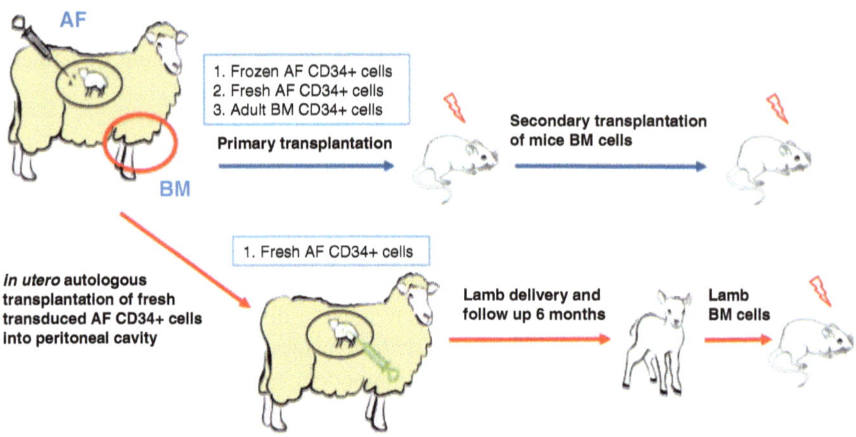

Fig. 19.1 Transduced sheep eGFP+ CD34+ selected fresh or frozen AF and adult BM cells were transplanted into immunocompromised NSG mice (primary and secondary xenogeneic transplantation). Transduced sheep eGFP+CD34+ fresh AF were also injected into donor sheep fetuses (*in utero* autologous transplantation) that were subsequently delivered and followed for up to 3 months of age. Bone marrow from these primary sheep recipients was then used to perform xenogeneic secondary transplantation into NSG mice. *AF* amniotic fluid, *BM* bone marrow

Considerations for Translation of Perinatal Gene Therapy to Human Application

Preclinical testing in animal models of disease will be an important step before clinical translation is realized. There is no ideal animal model and a balance is needed, taking into consideration the gestational development of the organ to be targeted and how that relates to the human, the type of placentation, fetal size, number and lifespan, parturition, and the fetal and maternal immune response.

Toxicology studies will be needed using animals such as the pregnant rabbit, in which reproductive toxicology is commonly performed, with good historical datasets, and which is a model that is understood by the regulators. A variety of guidelines and regulations such as those described by the Committee for Medicinal Products for Human Use (CHMP) of the European Medicines Agency, will need to be taken into consideration when planning preclinical study protocols. These could include for example, the guidelines on the non-clinical testing for inadvertent germline transmission of gene transfer vectors [237] or on the non-clinical studies required before first clinical use of gene therapy medicinal products [238].

In addition to animal studies, the effect of gene therapy vectors on the human placenta can be assessed *in vitro*. Two models are available, cultured villous explants or perfused whole placental cotyledons. Villi isolated from different lobules of the placenta can be cultured in net-wells and submerged in growth medium. In this model, the syncytiotrophoblast routinely undergoes shedding *in vitro* after about 1 day of culture, but this barrier consistently regenerates through the differentiation of underlying cytotrophoblast cells 2 days later [239]. Cellular integrity and apoptosis can be assessed using specific markers such as lactate dehydrogenase levels, released into the culture medium. The method of placental perfusion has been adapted so as to preserve cellular and tissue architecture whilst allowing a dual fetal- and maternal-side haemodynamic compartment to be maintained [240]. Movement of substances applied to the maternal or fetal side of the placenta can be studied in the opposite side of the placenta using this model, over a 5–9 h time period after delivery of the placenta. This model has provided a wealth of data on the physiology of normal and pathological human placentae [241] and may be useful in measuring spread of vector from the fetus to the mother or vice versa.

Phase I Trials

Phase I human trials are likely to face hurdles because of difficulties in testing pregnant women where toxicological studies are usually contraindicated. Thus, when human application becomes possible, extensive un-biased parental counselling and informed consent is paramount because of the uncertainties about the efficacy and long term safety of prenatal gene therapy which may not become evident until much later in the individual's life. This can be difficult when the decision to participate in a fetal gene therapy trial will occur close to the time of prenatal diagnosis of the condition. Because the risks involve the mother, fetus and possibly future progeny, parents will also be required to consent their offspring and themselves to lifelong follow up.

One criticism levelled at fetal gene therapy is a belief that couples pregnant with an affected child would be unlikely to proceed with prenatal therapy and would opt for a termination instead. This concern is not solely applicable to perinatal gene therapy however, but also can be raised for any fetal treatment such as fetal surgery and *in utero* stem cell transplantation. The general public have been concerned that ethical discussion about issues such as gene therapy, cloning and the Human Genome Project are falling behind the technology. There is almost no research in this area, and the views of the general public and patient groups will need to be solicited as the technology is being translated into the clinic. In this regard, the EVERREST consortium that is translating maternal perinatal gene therapy into the clinic for treatment of fetal growth restriction is conducting a detailed bioethics investigation with stakeholders, patients and the public to gauge their opinion.

Conclusions

Perinatal gene therapy offers the potential for clinicians not only to diagnose but also to treat inherited genetic disease. Structural disease in the fetus and maternal obstetric conditions that affect the fetus may also be amenable to treatment using gene transfer technology. Fetal application may prove better than application in the adult for treatment, or even prevention of early onset genetic disorders such as CNS and liver disorders. Gene transfer to the developing fetus targets rapidly expanding stem cell populations that are inaccessible after birth. Integrating vector systems give permanent gene transfer. In animal models of congenital disease the functionally immature fetal immune system does not respond to the product of the introduced gene, and therefore immune tolerance can be induced. For the treatment to be acceptable, it must be safe for both mother and fetus, and preferably avoid germline transmission. Recent developments in the understanding of genetic disease, vector design, and minimally invasive delivery techniques have brought fetal gene therapy closer to clinical practice. However more research needs to be done to answer the questions below before it can be introduced as a therapy. Which vectors can provide long-term regulated gene expression preferably for the lifetime of the individual? What is the best route of administration and the optimum gestational age to target gene therapy for specific diseases? How can informed consent best be obtained from the couple who are embarking on gene therapy treatment for their fetus?

Currently, perinatal gene therapy remains an experimental procedure but it is rapidly moving into the clinic with the potential of a first-in-woman study within the next 5 years, to treat fetal growth restriction.

References

1. Lewis C, Hill M, Chitty LS. Non-invasive prenatal diagnosis for single gene disorders: experience of patients. Clin Genet. 2014;85(4):336–42.
2. Twiss P, et al. Non-invasive prenatal testing for Down syndrome. Semin Fetal Neonatal Med. 2014;19(1):9–14.
3. Flake AW. In utero stem cell transplantation. Best Pract Res Clin Obstet Gynaecol. 2004;18(6):941–58.
4. Peebles D, et al. Widespread and efficient marker gene expression in the airway epithelia of fetal sheep after minimally invasive tracheal application of recombinant adenovirus in utero. Gene Ther. 2004;11(1):70–8.
5. Polakowska RR, et al. Apoptosis in human skin development: morphogenesis, periderm, and stem cells. Dev Dyn. 1994;199(3):176–88.
6. Endo M, et al. Early intra-amniotic gene transfer using lentiviral vector improves skin blistering phenotype in a murine model of Herlitz junctional epidermolysis bullosa. Gene Ther. 2012;19(5):561–9.
7. Manno CS, et al. Successful transduction of liver in hemophilia by AAV-Factor IX and limitations imposed by the host immune response. Nat Med. 2006;12(3):342–7.
8. Nathwani AC, et al. Adenovirus-associated virus vector–mediated gene transfer in hemophilia B. N Engl J Med. 2011;365(25):2357–65.
9. Nathwani AC, et al. Self-complementary adeno-associated virus vectors containing a novel liver-specific human factor IX expression cassette enable highly efficient transduction of murine and nonhuman primate liver. Blood. 2006;107(7):2653–61.
10. Billingham RE, Brent L. Acquired tolerance of foreign cells in newborn animals. Proc R Soc Lond B Biol Sci. 1956;146(922):78–90.
11. Billingham RE, Brent L, Medawar PB. Actively acquired tolerance of foreign cells. Nature. 1953;172(4379):603–6.
12. Waddington SN, et al. Fetal gene transfer. Curr Opin Mol Ther. 2007;9(5):432–8.
13. Waddington SN, et al. Fetal and neonatal gene therapy: benefits and pitfalls. Gene Ther. 2004a;11 Suppl 1:S92–7.
14. David AL, et al. Recombinant adeno-associated virus-mediated in utero gene transfer gives therapeutic transgene expression in the sheep. Hum Gene Ther. 2011;22(4):419–26.
15. Mattar CNZ, et al. Stable human FIX expression after 0.9G intrauterine gene transfer of self-complementary adeno-associated viral vector 5 and 8 in macaques. Mol Ther. 2011; 19(11):1950–60.
16. Anon. Prenatal gene transfer: scientific, medical, and ethical issues: a report of the Recombinant DNA Advisory Committee. Hum Gene Ther. 2000;11(8):1211–29.
17. Lipshutz GS, et al. Short-term correction of factor VIII deficiency in a murine model of hemophilia A after delivery of adenovirus murine factor VIII in utero. Proc Natl Acad Sci U S A. 1999;96(23):13324–9.
18. Waddington SN, et al. Permanent phenotypic correction of hemophilia B in immunocompetent mice by prenatal gene therapy. Blood. 2004b;104(9):2714–21.
19. Dejneka NS, et al. In utero gene therapy rescues vision in a murine model of congenital blindness. Mol Ther. 2004;9(2):182–8.
20. Seppen J, et al. Long-term correction of bilirubin UDPglucuronyltransferase deficiency in rats by in utero lentiviral gene transfer. Mol Ther. 2003;8(4):593–9.
21. Rucker M, et al. Rescue of enzyme deficiency in embryonic diaphragm in a mouse model of metabolic myopathy: pompe disease. Development. 2004;131(12):3007–19.
22. Wu C, et al. Intra-amniotic transient transduction of the periderm with a viral vector encoding TGFβ3 prevents cleft palate in Tgfβ3(-/-) mouse embryos. Mol Ther. 2013;21(1):8–17.
23. Carr DJ, et al. Uteroplacental adenovirus vascular endothelial growth factor gene therapy increases fetal growth velocity in growth-restricted sheep pregnancies. Hum Gene Ther. 2014;25(4):375–84.

24. Gharwan H, et al. Nonviral gene transfer into fetal mouse livers (a comparison between the cationic polymer PEI and naked DNA). Gene Ther. 2003;10(9):810–7.
25. Luton D, et al. Gene transfection into fetal sheep airways in utero using guanidinium-cholesterol cationic lipids. J Gene Med. 2004;6(3):328–36.
26. Fletcher S, et al. In vivo studies of dialkynoyl analogues of DOTAP demonstrate improved gene transfer efficiency of cationic liposomes in mouse lung. J Med Chem. 2006;49(1):349–57.
27. Mason CA, et al. Gene transfer in utero biologically engineers a patent ductus arteriosus in lambs by arresting fibronectin-dependent neointimal formation. Nat Med. 1999;5(2):176–82.
28. David A, et al. Ultrasound-guided percutaneous delivery of adenoviral vectors encoding the beta-galactosidase and human factor IX genes to early gestation fetal sheep in utero. Hum Gene Ther. 2003;14(4):353–64.
29. Lipshutz GS, Flebbe-Rehwaldt L, Gaensler KM. Reexpression following readministration of an adenoviral vector in adult mice after initial in utero adenoviral administration. Mol Ther. 2000;2(4):374–80.
30. Waddington SN, et al. In utero gene transfer of human factor IX to fetal mice can induce postnatal tolerance of the exogenous clotting factor. Blood. 2003;101(4):1359–66.
31. Crystal RG. Adenovirus: the first effective in vivo gene delivery vector. Hum Gene Ther. 2014;25(1):3–11.
32. Joyeux L, et al. In utero lung gene transfer using adeno-associated viral and lentiviral vectors in mice. Hum Gene Ther Methods. 2014;25(3):197–205.
33. Bouchard S, et al. Long-term transgene expression in cardiac and skeletal muscle following fetal administration of adenoviral or adeno-associated viral vectors in mice. J Gene Med. 2003;5(11):941–50.
34. Lipshutz GS, et al. In utero delivery of adeno-associated viral vectors: intraperitoneal gene transfer produces long-term expression. Mol Ther. 2001;3(3):284–92.
35. Mitchell M, et al. Long-term gene transfer to mouse fetuses with recombinant adenovirus and adeno-associated virus (AAV) vectors. Gene Ther. 2000;7(23):1986–92.
36. Mühle C, et al. Evaluation of prenatal intra-amniotic LAMB3 gene delivery in a mouse model of Herlitz disease. Gene Ther. 2006;13(23):1665–76.
37. Garrett DJ, et al. In utero recombinant adeno-associated virus gene transfer in mice, rats, and primates. BMC Biotechnol. 2003;3:16.
38. Hellström M, Harvey AR. Retinal ganglion cell gene therapy and visual system repair. Curr Gene Ther. 2011;11(2):116–31.
39. Yang B, et al. Global CNS transduction of adult mice by intravenously delivered rAAVrh.8 and rAAVrh.10 and nonhuman primates by rAAVrh.10. Mol Ther. 2014;22(7):1299–309.
40. Burguete T. Evidence for infection of the human embryo with adeno-associated virus in pregnancy. Hum Reprod. 1999;14(9):2396–401.
41. Porada CD, et al. In utero gene therapy: transfer and long-term expression of the bacterial neo(r) gene in sheep after direct injection of retroviral vectors into preimmune fetuses. Hum Gene Ther. 1998;9(11):1571–85.
42. Schoenhard JA, Hatzopoulos AK. Stem cell therapy: pieces of the puzzle. J Cardiovasc Transl Res. 2010;3(1):49–60.
43. Cavazzana-Calvo M, et al. Transfusion independence and HMGA2 activation after gene therapy of human β-thalassaemia. Nature. 2010;467(7313):318–22.
44. Levasseur DN, et al. Correction of a mouse model of sickle cell disease: lentiviral/antisickling beta-globin gene transduction of unmobilized, purified hematopoietic stem cells. Blood. 2003;102(13):4312–9.
45. Kim H, Kim J-S. A guide to genome engineering with programmable nucleases. Nat Rev Genet. 2014;15(5):321–34.
46. Gaj T, Gersbach CA, Barbas CF. ZFN, TALEN, and CRISPR/Cas-based methods for genome engineering. Trends Biotechnol. 2013;31(7):397–405.
47. Hwang WY, et al. Efficient genome editing in zebrafish using a CRISPR-Cas system. Nat Biotechnol. 2013;31(3):227–9.

48. Boch J, et al. Breaking the code of DNA binding specificity of TAL-type III effectors. Science. 2009;326(5959):1509–12.
49. Cermak T, et al. Efficient design and assembly of custom TALEN and other TAL effector-based constructs for DNA targeting. Nucleic Acids Res. 2011;39(12):e82.
50. Christian M, et al. Targeting DNA double-strand breaks with TAL effector nucleases. Genetics. 2010;186(2):757–61.
51. Moscou M. A simple cipher governs DNA recognition by TAL effectors. Science. 2009;326:1501.
52. Bedell VM, et al. In vivo genome editing using a high-efficiency TALEN system. Nature. 2012;491(7422):114–8.
53. Beumer KJ, et al. Comparing zinc finger nucleases and transcription activator-like effector nucleases for gene targeting in Drosophila. G3 (Bethesda). 2013;3(10):1717–25.
54. Xiao A, et al. Chromosomal deletions and inversions mediated by TALENs and CRISPR/Cas in zebrafish. Nucleic Acids Res. 2013;41(14):e141.
55. Huard J, et al. Differential short-term transduction efficiency of adult versus newborn mouse tissues by adenoviral recombinants. Exp Mol Pathol. 1995;62(2):131–43.
56. Endo M, et al. The developmental stage determines the distribution and duration of gene expression after early intra-amniotic gene transfer using lentiviral vectors. Gene Ther. 2010;17(1):61–71.
57. Hong S, et al. Functional analysis of various promoters in lentiviral vectors at different stages of in vitro differentiation of mouse embryonic stem cells. Mol Ther. 2007;15(9):1630–9.
58. Brown BD, Gentner B, et al. Endogenous microRNA can be broadly exploited to regulate transgene expression according to tissue, lineage and differentiation state. Nat Biotechnol. 2007;25(12):1457–67.
59. Brown BD, Cantore A, et al. A microRNA-regulated lentiviral vector mediates stable correction of hemophilia B mice. Blood. 2007;110(13):4144–52.
60. Lindenburg ITM, van Kamp IL, Oepkes D. Intrauterine blood transfusion: current indications and associated risks. Fetal Diagn Ther. 2014;36:263.
61. Liley AW. Errors in the assessment of hemolytic disease from amniotic fluid. Am J Obstet Gynecol. 1963;86:485–94.
62. Furie B, Limentani SA, Rosenfield CG. A practical guide to the evaluation and treatment of hemophilia. Blood. 1994;84(1):3–9.
63. Di Minno MND, et al. Cost of care of haemophilia with inhibitors. Haemophilia. 2010;16(1):e190–201.
64. Lusher JM. Inhibitors in young boys with haemophilia. Baillieres Best Pract Res Clin Haematol. 2000;13(3):457–68.
65. Von Drygalski A, et al. Prevalence and risk factors for hypertension in hemophilia. Hypertension. 2013;62(1):209–15.
66. Chao H, et al. Sustained and complete phenotype correction of hemophilia B mice following intramuscular injection of AAV1 serotype vectors. Mol Ther. 2001;4(3):217–22.
67. Herzog RW, et al. Long-term correction of canine hemophilia B by gene transfer of blood coagulation factor IX mediated by adeno-associated viral vector. Nat Med. 1999;5(1):56–63.
68. Snyder RO, et al. Correction of hemophilia B in canine and murine models using recombinant adeno-associated viral vectors. Nat Med. 1999;5(1):64–70.
69. Manno CS, et al. AAV-mediated factor IX gene transfer to skeletal muscle in patients with severe hemophilia B. Blood. 2003;101(8):2963–72.
70. Allay JA, et al. Good manufacturing practice production of self-complementary serotype 8 adeno-associated viral vector for a hemophilia B clinical trial. Hum Gene Ther. 2011;22(5):595–604.
71. Calcedo R, et al. Worldwide epidemiology of neutralizing antibodies to adeno-associated viruses. J Infect Dis. 2009;199(3):381–90.
72. Jiang H, et al. Effects of transient immunosuppression on adenoassociated, virus-mediated, liver-directed gene transfer in rhesus macaques and implications for human gene therapy. Blood. 2006;108(10):3321–8.

73. High KH, et al. Current status of haemophilia gene therapy. Haemophilia. 2014;20 Suppl 4:43–9.
74. Sabatino DE, et al. Persistent expression of hF.IX After tolerance induction by in utero or neonatal administration of AAV-1-F.IX in hemophilia B mice. Mol Ther. 2007;15(9):1677–85.
75. Yang EY, et al. BAPS Prize--1997. Fetal gene therapy: efficacy, toxicity, and immunologic effects of early gestation recombinant adenovirus. British Association of Paediatric Surgeons. J Pediatr Surg. 1999;34(2):235–41.
76. McVey JH, et al. Factor VII deficiency and the FVII mutation database. Hum Mutat. 2001;17(1):3–17.
77. Peyvandi F, et al. Rare bleeding disorders. Haemophilia. 2006;12 Suppl 3:137–42.
78. Binny C, et al. AAV-mediated gene transfer in the perinatal period results in expression of FVII at levels that protect against fatal spontaneous hemorrhage. Blood. 2012;119(4):957–66.
79. Modell B, Khan M, Darlison M. Survival in β-thalassaemia major in the UK: data from the UK Thalassaemia Register Meningococcal disease among children of Indian subcontinent ethnic origin For personal use only. Not to be reproduced without permission of The Lancet. Lancet. 2000;355:2051–2.
80. Ge H, et al. Noninvasive prenatal detection for pathogenic CNVs: the application in α-thalassemia. PLoS One. 2013;8(6):e67464.
81. Lucarelli G, Andreani M, Angelucci E. The cure of thalassemia by bone marrow transplantation. Blood Rev. 2002;16(2):81–5.
82. Angelucci E. Hematopoietic stem cell transplantation in thalassemia. Hematology Am Soc Hematol Educ Program. 2010;2010:456–62.
83. Harteveld CL, Higgs DR. Alpha-thalassaemia. Orphanet J Rare Dis. 2010;5:13.
84. Meremikwu MM, Okomo U. 2011. Sickle cell disease. Clinical evidence, 2011.
85. Quek L, Thein SL. Molecular therapies in beta-thalassaemia. Br J Haematol. 2007;136(3):353–65.
86. Pawliuk R, et al. Correction of sickle cell disease in transgenic mouse models by gene therapy. Science. 2001;294(5550):2368–71.
87. Persons DA, et al. Successful treatment of murine beta-thalassemia using in vivo selection of genetically modified, drug-resistant hematopoietic stem cells. Blood. 2003;102(2):506–13.
88. Puthenveetil G, et al. Successful correction of the human beta-thalassemia major phenotype using a lentiviral vector. Blood. 2004;104(12):3445–53.
89. Rivella S, et al. A novel murine model of Cooley anemia and its rescue by lentiviral-mediated human beta-globin gene transfer. Blood. 2003;101(8):2932–9.
90. Han X, et al. Fetal gene therapy of α-thalassemia in a mouse model. Proc Natl Acad Sci U S A. 2007;104(21):9007–11.
91. Antoniou M, et al. Transgenes encompassing dual-promoter CpG islands from the human TBP and HNRPA2B1 loci are resistant to heterochromatin-mediated silencing. Genomics. 2003;82(3):269–79.
92. Williams S, et al. CpG-island fragments from the HNRPA2B1/CBX3 genomic locus reduce silencing and enhance transgene expression from the hCMV promoter/enhancer in mammalian cells. BMC Biotechnol. 2005;5:17.
93. Zhang F, et al. Lentiviral vectors containing an enhancer-less ubiquitously acting chromatin opening element (UCOE) provide highly reproducible and stable transgene expression in hematopoietic cells. Blood. 2007;110(5):1448–57.
94. Fan Y, et al. Generation of human β-thalassemia induced pluripotent stem cells from amniotic fluid cells using a single excisable lentiviral stem cell cassette. J Reprod Dev. 2012;58(4): 404–9.
95. Weiss DJ, Bonneau L, Liggitt D. Use of perfluorochemical liquid allows earlier detection of gene expression and use of less vector in normal lung and enhances gene expression in acutely injured lung. Mol Ther. 2001;3(5 Pt 1):734–45.
96. Ornoy A, et al. Pathological confirmation of cystic fibrosis in the fetus following prenatal diagnosis. Am J Med Genet. 1987;28(4):935–47.

97. Johnson LG, et al. Efficiency of gene transfer for restoration of normal airway epithelial function in cystic fibrosis. Nat Genet. 1992;2(1):21–5.
98. Kreda SM, et al. Characterization of wild-type and deltaF508 cystic fibrosis transmembrane regulator in human respiratory epithelia. Mol Biol Cell. 2005;16(5):2154–67.
99. Dray X, et al. Distal intestinal obstruction syndrome in adults with cystic fibrosis. Clin Gastroenterol Hepatol. 2004;2(6):498–503.
100. Boué A, et al. Prenatal diagnosis in 200 pregnancies with a 1-in-4 risk of cystic fibrosis. Hum Genet. 1986;74(3):288–97.
101. Scotet V, et al. Prenatal diagnosis of cystic fibrosis: the 18-year experience of Brittany (western France). Prenat Diagn. 2008;28(3):197–202.
102. Boyd AC. Gene and stem cell therapy. In: Bush A, Alton EWFW, Davies JC, Griesenbach U, Jaffe A, editors. Cystic fibrosis in the 21st century. Karge: Basel; 2006. p. 221–9.
103. Pickles RJ, et al. Retargeting the coxsackievirus and adenovirus receptor to the apical surface of polarized epithelial cells reveals the glycocalyx as a barrier to adenovirus-mediated gene transfer. J Virol. 2000;74(13):6050–7.
104. Harvey B-G, et al. Safety of local delivery of low- and intermediate-dose adenovirus gene transfer vectors to individuals with a spectrum of morbid conditions. Hum Gene Ther. 2002;13(1):15–63.
105. Flotte TR, et al. Phase I trial of intranasal and endobronchial administration of a recombinant adeno-associated virus serotype 2 (rAAV2)-CFTR vector in adult cystic fibrosis patients: a two-part clinical study. Hum Gene Ther. 2003;14(11):1079–88.
106. Moss RB, et al. Repeated adeno-associated virus serotype 2 aerosol-mediated cystic fibrosis transmembrane regulator gene transfer to the lungs of patients with cystic fibrosis: a multicenter, double-blind, placebo-controlled trial. Chest. 2004;125(2):509.
107. Alexander BL, et al. Progress and prospects: gene therapy clinical trials (part 1). Gene Ther. 2007;14(20):1439–47.
108. Griesenbach U, Geddes DM, Alton EWFW. Gene therapy progress and prospects: cystic fibrosis. Gene Ther. 2006;13(14):1061–7.
109. Armstrong DK, et al. Gene therapy in cystic fibrosis. Arch Dis Child. 2014;99(5):465–8.
110. Voynow JA, et al. Basal-like cells constitute the proliferating cell population in cystic fibrosis airways. Am J Respir Crit Care Med. 2005;172(8):1013–8.
111. Larson JE, et al. Reversal of cystic fibrosis phenotype in mice by gene therapy in utero. Lancet. 1997;349(9052):619–20.
112. Buckley SMK, et al. Factors influencing adenovirus-mediated airway transduction in fetal mice. Mol Ther. 2005;12(3):484–92.
113. Davies LA, et al. Adenovirus-mediated in utero expression of CFTR does not improve survival of CFTR knockout mice. Mol Ther. 2008;16(5):812–8.
114. Egan M, et al. Defective regulation of outwardly rectifying Cl- channels by protein kinase A corrected by insertion of CFTR. Nature. 1992;358(6387):581–4.
115. Boyle MP, et al. In utero AAV-mediated gene transfer to rabbit pulmonary epithelium. Mol Ther. 2001;4(2):115–21.
116. Moss IR, Scarpelli EM. Stimulatory effect of theophylline on regulation of fetal breathing movements. Pediatr Res. 1981;15(5):870–3.
117. Henriques-Coelho T, et al. Targeted gene transfer to fetal rat lung interstitium by ultrasound-guided intrapulmonary injection. Mol Ther. 2007;15(2):340–7.
118. Toelen J, et al. Long term pulmonary gene therapy with a lentiviral vector in a fetal rat model. Am J Obstet Gynecol. 2006;195(6):S22.
119. Tarantal AF, et al. Intrapulmonary and intramyocardial gene transfer in rhesus monkeys (Macaca mulatta): safety and efficiency of HIV-1-derived lentiviral vectors for fetal gene delivery. Mol Ther. 2005;12(1):87–98.
120. Tarantal AF, Lee CI, et al. Lentiviral vector gene transfer into fetal rhesus monkeys (Macaca mulatta): lung-targeting approaches. Mol Ther. 2001;4(6):614–21.
121. Bilbao R, et al. Patterns of gene expression from in utero delivery of adenoviral-associated vector serotype 1. Hum Gene Ther. 2005;16(6):678–84.

122. Gregory LG, et al. Enhancement of adenovirus-mediated gene transfer to the airways by DEAE dextran and sodium caprate in vivo. Mol Ther. 2003;7(1):19–26.
123. Carlon MS et al. A novel surgical approach for intratracheal administration of bioactive agents in a fetal mouse model. Journal of visualized experiments. J Vis Exp 2012; (68): e4219.
124. Carlon M, et al. Efficient gene transfer into the mouse lung by fetal intratracheal injection of rAAV2/6.2. Mol Ther. 2010;18(12):2130–8.
125. Carlon MS, et al. Immunological ignorance allows long-term gene expression after perinatal recombinant adeno-associated virus-mediated gene transfer to murine airways. Hum Gene Ther. 2014;25(6):517–28.
126. Lim F-Y, et al. Adeno-associated virus (AAV)-mediated gene transfer in respiratory epithelium and submucosal gland cells in human fetal tracheal organ culture. J Pediatr Surg. 2002;37(7):1051–7. discussion 1051–7.
127. Lim F-Y, et al. Human fetal trachea-SCID mouse xenografts: efficacy of vesicular stomatitis virus-G pseudotyped lentiviral-mediated gene transfer. J Pediatr Surg. 2003;38(6):834–9.
128. David AL, et al. Clinically applicable procedure for gene delivery to fetal gut by ultrasound-guided gastric injection: toward prenatal prevention of early-onset intestinal diseases. Hum Gene Ther. 2006;17(7):767–79.
129. Meikle PJ, et al. Prevalence of lysosomal storage disorders. JAMA. 1999;281(3):249–54.
130. Zadeh N, Hudgins L, Norton ME. Nuchal translucency measurement in fetuses with spinal muscular atrophy. Prenat Diagn. 2011;31(4):327–30.
131. Spiegel R, et al. The clinical spectrum of fetal Niemann-Pick type C. Am J Med Genet A. 2009;149A(3):446–50.
132. Eng CM, et al. Prenatal genetic carrier testing using triple disease screening. JAMA. 1997;278(15):1268–72.
133. Mulcahy PJ, et al. Gene therapy: a promising approach to treating spinal muscular atrophy. Hum Gene Ther. 2014;25(7):575–86.
134. Okano H, Yamanaka S. iPS cell technologies: significance and applications to CNS regeneration and disease. Mol Brain. 2014;7:22.
135. Sly WS, et al. Beta glucuronidase deficiency: report of clinical, radiologic, and biochemical features of a new mucopolysaccharidosis. J Pediatr. 1973;82(2):249–57.
136. Sands MS, Davidson BL. Gene therapy for lysosomal storage diseases. Mol Ther. 2006;13(5):839–49.
137. Mango RL, et al. Neonatal retroviral vector-mediated hepatic gene therapy reduces bone, joint, and cartilage disease in mucopolysaccharidosis VII mice and dogs. Mol Genet Metab. 2004;82(1):4–19.
138. Aronovich EL, et al. Prolonged expression of a lysosomal enzyme in mouse liver after Sleeping Beauty transposon-mediated gene delivery: implications for non-viral gene therapy of mucopolysaccharidoses. J Gene Med. 2007;9(5):403–15.
139. Berges BK, et al. Widespread correction of lysosomal storage in the mucopolysaccharidosis type VII mouse brain with a herpes simplex virus type 1 vector expressing beta-glucuronidase. Mol Ther. 2006;13(5):859–69.
140. Ciron C, et al. Gene therapy of the brain in the dog model of Hurler's syndrome. Ann Neurol. 2006;60(2):204–13.
141. Cearley CN, Wolfe JH. A single injection of an adeno-associated virus vector into nuclei with divergent connections results in widespread vector distribution in the brain and global correction of a neurogenetic disease. J Neurosci. 2007;27(37):9928–40.
142. Shen J-S, et al. Widespread and highly persistent gene transfer to the CNS by retrovirus vector in utero: implication for gene therapy to Krabbe disease. J Gene Med. 2005;7(5):540–51.
143. Karolewski BA, Wolfe JH. Genetic correction of the fetal brain increases the lifespan of mice with the severe multisystemic disease mucopolysaccharidosis type VII. Mol Ther. 2006;14(1):14–24.
144. Tarantal AF, et al. Sonographic heat generation in vivo in the gravid long-tailed macaque (Macaca fascicularis). J Ultrasound Med. 1993;12(5):285–95.
145. Daffos F, Capella-Pavlovsky M, Forestier F. A new procedure for fetal blood sampling in utero: preliminary results of fifty-three cases. Am J Obstet Gynecol. 1983;146(8):985–7.

146. Duque S, et al. Intravenous administration of self-complementary AAV9 enables transgene delivery to adult motor neurons. Mol Ther. 2009;17(7):1187–96.
147. Foust KD, et al. Intravascular AAV9 preferentially targets neonatal neurons and adult astrocytes. Nat Biotechnol. 2009;27(1):59–65.
148. Bevan AK, et al. Systemic gene delivery in large species for targeting spinal cord, brain, and peripheral tissues for pediatric disorders. Mol Ther. 2011;19(11):1971–80.
149. Manfredsson FP, Rising AC, Mandel RJ. AAV9: a potential blood-brain barrier buster. Mol Ther. 2009;17(3):403–5.
150. Rahim AA, et al. Intravenous administration of AAV2/9 to the fetal and neonatal mouse leads to differential targeting of CNS cell types and extensive transduction of the nervous system. FASEB J. 2011;25(10):3505–18.
151. Vanhanen SL, et al. EEG and evoked potentials in infantile neuronal ceroid-lipofuscinosis. Dev Med Child Neurol. 1997;39(7):456–63.
152. Macauley SL, et al. Cerebellar pathology and motor deficits in the palmitoyl protein thioesterase 1-deficient mouse. Exp Neurol. 2009;217(1):124–35.
153. Ferla R, et al. Similar therapeutic efficacy between a single administration of gene therapy and multiple administrations of recombinant enzyme in a mouse model of lysosomal storage disease. Hum Gene Ther. 2014;25(7):609–18.
154. Shayman JA. Thematic review series: Recent advances in the treatment of lysosomal storage diseases. J Lipid Res. 2014;55(6):993–4.
155. Wirth B, et al. Mildly affected patients with spinal muscular atrophy are partially protected by an increased SMN2 copy number. Hum Genet. 2006;119(4):422–8.
156. Federici T, Boulis NM. Gene-based treatment of motor neuron diseases. Muscle Nerve. 2006;33(3):302–23.
157. Dubowitz V. Very severe spinal muscular atrophy (SMA type 0): an expanding clinical phenotype. Eur J Paediatr Neurol. 1999;3(2):49–51.
158. Fidziańska A, Rafalowska J. Motoneuron death in normal and spinal muscular atrophy-affected human fetuses. Acta Neuropathol. 2002;104(4):363–8.
159. Lesbordes J-C, et al. Therapeutic benefits of cardiotrophin-1 gene transfer in a mouse model of spinal muscular atrophy. Hum Mol Genet. 2003;12(11):1233–9.
160. Matsuoka N, et al. Adenovirus-mediated gene transfer of Bcl-xL prevents cell death in primary neuronal culture of the rat. Neurosci Lett. 1999;270(3):177–80.
161. Azzouz M, et al. Lentivector-mediated SMN replacement in a mouse model of spinal muscular atrophy. J Clin Invest. 2004;114(12):1726–31.
162. Boulis NM, et al. Adenoviral nerve growth factor and beta-galactosidase transfer to spinal cord: a behavioral and histological analysis. J Neurosurg. 1999;90(1 Suppl):99–108.
163. Tenenbaum L, et al. Recombinant AAV-mediated gene delivery to the central nervous system. J Gene Med. 2004;6 Suppl 1:S212–22.
164. Passini MA, et al. CNS-targeted gene therapy improves survival and motor function in a mouse model of spinal muscular atrophy. J Clin Invest. 2010;120(4):1253–64.
165. Kahn A, et al. Gene therapy of neurological diseases. C R Seances Soc Biol Fil. 1996;190(1):9–11.
166. Yamamura J, et al. Long-term gene expression in the anterior horn motor neurons after intramuscular inoculation of a live herpes simplex virus vector. Gene Ther. 2000;7(11):934–41.
167. Rahim AA, et al. In utero gene transfer to the mouse nervous system. Biochem Soc Trans. 2010;38(6):1489–93.
168. Foust KD, et al. Rescue of the spinal muscular atrophy phenotype in a mouse model by early postnatal delivery of SMN. Nat Biotechnol. 2010;28(3):271–4.
169. Dominguez E, et al. Intravenous scAAV9 delivery of a codon-optimized SMN1 sequence rescues SMA mice. Hum Mol Genet. 2011;20(4):681–93.
170. Maestri NE, Clissold D, Brusilow SW. Neonatal onset ornithine transcarbamylase deficiency: a retrospective analysis. J Pediatr. 1999;134(3):268–72.
171. Stratford-Perricaudet LD, et al. Evaluation of the transfer and expression in mice of an enzyme-encoding gene using a human adenovirus vector. Hum Gene Ther. 1990;1(3):241–56.

172. Raper SE, et al. A pilot study of in vivo liver-directed gene transfer with an adenoviral vector in partial ornithine transcarbamylase deficiency. Hum Gene Ther. 2002;13(1):163–75.
173. Raper SE, et al. Fatal systemic inflammatory response syndrome in a ornithine transcarbamylase deficient patient following adenoviral gene transfer. Mol Genet Metab. 2003;80(1-2):148–58.
174. Mian A, et al. Long-term correction of ornithine transcarbamylase deficiency by WPRE-mediated overexpression using a helper-dependent adenovirus. Mol Ther. 2004;10(3):492–9.
175. Moscioni D, et al. Long-term correction of ammonia metabolism and prolonged survival in ornithine transcarbamylase-deficient mice following liver-directed treatment with adeno-associated viral vectors. Mol Ther. 2006;14(1):25–33.
176. Seppen J, et al. Immune response to lentiviral bilirubin UDP-glucuronosyltransferase gene transfer in fetal and neonatal rats. Gene Ther. 2006;13(8):672–7.
177. Lai L, et al. A preliminary evaluation of recombinant adeno-associated virus biodistribution in rhesus monkeys after intrahepatic inoculation in utero. Hum Gene Ther. 2002;13(17):2027–39.
178. Lee CCI, et al. Fetal gene transfer using lentiviral vectors and the potential for germ cell transduction in rhesus monkeys (Macaca mulatta). Hum Gene Ther. 2005;16(4):417–25.
179. Tarantal AF, et al. Fetal gene transfer using lentiviral vectors: in vivo detection of gene expression by microPET and optical imaging in fetal and infant monkeys. Hum Gene Ther. 2006;17(12):1254–61.
180. Tarantal AF, O'Rourke JP, et al. Rhesus monkey model for fetal gene transfer: studies with retroviral- based vector systems. Mol Ther. 2001;3(2):128–38.
181. Raymond FL, et al. Molecular prenatal diagnosis: the impact of modern technologies. Prenat Diagn. 2010;30(7):674–81.
182. Häberle J. Urea cycle disorders revisited–clinical, biochemical and therapeutical aspects. J Pediat Biochem. 2014;4:1–3.
183. Van Deutekom JCT, van Ommen G-JB. Advances in Duchenne muscular dystrophy gene therapy. *Nature reviews*. Genetics. 2003;4(10):774–83.
184. Romero NB, et al. Phase I study of dystrophin plasmid-based gene therapy in Duchenne/ Becker muscular dystrophy. Hum Gene Ther. 2004;15(11):1065–76.
185. Tang Y, et al. AAV-directed muscular dystrophy gene therapy. Expert Opin Biol Ther. 2010;10(3):395–408.
186. Wells DJ, Ferrer A, Wells KE. Immunological hurdles in the path to gene therapy for Duchenne muscular dystrophy. Expert Rev Mol Med. 2004;4(23):1.
187. MacKenzie TC, et al. Transduction of satellite cells after prenatal intramuscular administration of lentiviral vectors. J Gene Med. 2005;7(1):50–8.
188. Ahmad A. Mdx mice inducibly expressing dystrophin provide insights into the potential of gene therapy for Duchenne muscular dystrophy. Hum Mol Genet. 2000;9(17):2507–15.
189. Reay DP, et al. Full-length dystrophin gene transfer to the mdx mouse in utero. Gene Ther. 2008;15(7):531–6.
190. Gregory LG, et al. Highly efficient EIAV-mediated in utero gene transfer and expression in the major muscle groups affected by Duchenne muscular dystrophy. Gene Ther. 2004;11(14):1117–25.
191. Weisz B, et al. Targeting the respiratory muscles of fetal sheep for prenatal gene therapy for Duchenne muscular dystrophy. Am J Obstet Gynecol. 2005;193(3 Pt 2):1105–9.
192. Koppanati BM, et al. Systemic delivery of AAV8 in utero results in gene expression in diaphragm and limb muscle: treatment implications for muscle disorders. Gene Ther. 2009;16(9):1130–7.
193. Koppanati BM, et al. Improvement of the mdx mouse dystrophic phenotype by systemic in utero AAV8 delivery of a minidystrophin gene. Gene Ther. 2010;17(11):1355–62.
194. Wang B, Li J, Xiao X. Adeno-associated virus vector carrying human minidystrophin genes effectively ameliorates muscular dystrophy in mdx mouse model. Proc Natl Acad Sci U S A. 2000;97(25):13714–9.
195. Lu Q-L, et al. The status of exon skipping as a therapeutic approach to duchenne muscular dystrophy. Mol Ther. 2011;19(1):9–15.

196. Luu M, Cantatore-Francis JL, Glick SA. Prenatal diagnosis of genodermatoses: current scope and future capabilities. Int J Dermatol. 2010;49(4):353–61.
197. Sato M, Tanigawa M, Kikuchi N. Nonviral gene transfer to surface skin of mid-gestational murine embryos by intraamniotic injection and subsequent electroporation. Mol Reprod Dev. 2004;69(3):268–77.
198. Endoh M, et al. Fetal gene transfer by intrauterine injection with microbubble-enhanced ultrasound. Mol Ther. 2002;5(5 Pt 1):501–8.
199. Yoshizawa J, et al. Successful in utero gene transfer using a gene gun in midgestational mouse fetuses. J Pediatr Surg. 2004;39(1):81–4.
200. Endo M, et al. Efficient in vivo targeting of epidermal stem cells by early gestational intraamniotic injection of lentiviral vector driven by the keratin 5 promoter. Mol Ther. 2008;16(1):131–7.
201. Aiuti A, et al. Correction of ADA-SCID by stem cell gene therapy combined with nonmyeloablative conditioning. Science. 2002;296(5577):2410–3.
202. Gaspar HB, et al. Successful reconstitution of immunity in ADA-SCID by stem cell gene therapy following cessation of PEG-ADA and use of mild preconditioning. Mol Ther. 2006;14(4):505–13.
203. Hacein-Bey-Abina S, et al. Sustained correction of X-linked severe combined immunodeficiency by ex vivo gene therapy. N Engl J Med. 2002;346(16):1185–93.
204. Ott MG, et al. Correction of X-linked chronic granulomatous disease by gene therapy, augmented by insertional activation of MDS1-EVI1, PRDM16 or SETBP1. Nat Med. 2006;12(4):401–9.
205. Hacein-Bey-Abina S, et al. LMO2-associated clonal T cell proliferation in two patients after gene therapy for SCID-X1. Science. 2003;302(5644):415–9.
206. Hanawa H, et al. Efficient gene transfer into rhesus repopulating hematopoietic stem cells using a simian immunodeficiency virus-based lentiviral vector system. Blood. 2004;103(11):4062–9.
207. Hollis RP, et al. Stable gene transfer to human CD34(+) hematopoietic cells using the Sleeping Beauty transposon. Exp Hematol. 2006;34(10):1333–43.
208. Tabori U, et al. Detection of RAG mutations and prenatal diagnosis in families presenting with either T-B- severe combined immunodeficiency or Omenn's syndrome. Clin Genet. 2004;65(4):322–6.
209. Touzot F, et al. Gene therapy for inherited immunodeficiency. Expert Opin Biol Ther. 2014;14(6):789–98.
210. Williams ML, et al. Lentiviral expression of retinal guanylate cyclase-1 (RetGC1) restores vision in an avian model of childhood blindness. PLoS Med. 2006;3(6):e201.
211. Bedrosian JC, et al. In vivo delivery of recombinant viruses to the fetal murine cochlea: transduction characteristics and long-term effects on auditory function. Mol Ther. 2006;14(3):328–35.
212. Endo M, et al. Gene transfer to ocular stem cells by early gestational intraamniotic injection of lentiviral vector. Mol Ther. 2007;15(3):579–87.
213. Pinyon JL, et al. Close-field electroporation gene delivery using the cochlear implant electrode array enhances the bionic ear. Sci Transl Med. 2014;6(233):233ra54.
214. Acosta R, et al. Anticholinergic suppression of fetal rabbit upper gastrointestinal motility. J Matern Fetal Neonatal Med. 2002;11(3):153–7.
215. David AL, Peebles D. Gene therapy for the fetus: is there a future? Best Pract Res Clin Obstet Gynaecol. 2008;22(1):203–18.
216. Mehta V, et al. Long-term increase in uterine blood flow is achieved by local overexpression of VEGF-A(165) in the uterine arteries of pregnant sheep. Gene Ther. 2012;19(9):925–35.
217. RCOG. The role of emergency and elective interventional radiology in postpartum haemorrhage, vol. 6. London: RCOG; 2007. p. 2–3.
218. EVERREST, EVERREST. Available at: www.everrest-fp7.eu
219. Deprest J, Gratacos E, Nicolaides KH. Fetoscopic tracheal occlusion (FETO) for severe congenital diaphragmatic hernia: evolution of a technique and preliminary results. Ultrasound Obstet Gynecol. 2004;24(2):121–6.

220. Larson JE, Cohen JC. Improvement of pulmonary hypoplasia associated with congenital diaphragmatic hernia by in utero CFTR gene therapy. Am J Physiol Lung Cell Mol Physiol. 2006;291(1):L4–10.
221. Saada J, et al. Combining keratinocyte growth factor transfection into the airways and tracheal occlusion in a fetal sheep model of congenital diaphragmatic hernia. J Gene Med. 2010;12(5):413–22.
222. Muench MO. In utero transplantation: baby steps towards an effective therapy. Bone Marrow Transplant. 2005;35(6):537–47.
223. Flake AW, et al. Treatment of X-linked severe combined immunodeficiency by in utero transplantation of paternal bone marrow. N Engl J Med. 1996;335(24):1806–10.
224. Westgren M, et al. Prenatal T-cell reconstitution after in utero transplantation with fetal liver cells in a patient with X-linked severe combined immunodeficiency. Am J Obstet Gynecol. 2002;187(2):475–82.
225. Shaw SWS, et al. Autologous transplantation of amniotic fluid-derived mesenchymal stem cells into sheep fetuses. Cell Transplant. 2011;20(7):1015–31.
226. Orlandi F, et al. The risks of early cordocentesis (12-21 weeks): analysis of 500 procedures. Prenat Diagn. 1990;10(7):425–8.
227. Schoeberlein A, et al. In utero transplantation of autologous and allogeneic fetal liver stem cells in ovine fetuses. Am J Obstet Gynecol. 2004;191(3):1030–6.
228. De Coppi P, et al. Isolation of amniotic stem cell lines with potential for therapy. Nat Biotechnol. 2007;25(1):100–6.
229. Lee LK, et al. Placenta as a newly identified source of hematopoietic stem cells. Curr Opin Hematol. 2010;17(4):313–8.
230. Parolini O, et al. Concise review: isolation and characterization of cells from human term placenta: outcome of the first international Workshop on Placenta Derived Stem Cells. Stem Cells. 2008;26(2):300–11.
231. Parolini O, et al. Toward cell therapy using placenta-derived cells: disease mechanisms, cell biology, preclinical studies, and regulatory aspects at the round table. Stem Cells Dev. 2010;19(2):143–54.
232. Portmann-lanz CB, et al. Placental mesenchymal stem cells as potential autologous graft for pre- and perinatal neuroregeneration. Am J Obstet Gynecol. 2006;194(3):664–73.
233. Bollini S, et al. In vitro and in vivo cardiomyogenic differentiation of amniotic fluid stem cells. Stem Cell Rev. 2011;7(2):364–80.
234. Ditadi A, et al. Human and murine amniotic fluid c-Kit+Lin- cells display hematopoietic activity. Blood. 2009;113(17):3953–60.
235. Spinelli V, Guillot PV, De Coppi P. Induced pluripotent stem (iPS) cells from human fetal stem cells (hFSCs). Organogenesis. 2013;9(2):101–10.
236. Shaw SSWS, et al. Clinical applications of prenatal and postnatal therapy using stem cells retrieved from amniotic fluid. Curr Opin Obstet Gynecol. 2011;23(2):109–16.
237. EU, 2006. Committee for the Medicinal Products for Human Use (CHMP). Guideline on non-clinical testing for inadvertent germline transmission of gene transfer vectors. EMEA/273974/2005. 2006. European Medicines Agency.
238. Schneider CK, et al. Challenges with advanced therapy medicinal products and how to meet them. Nat Rev Drug Discov. 2010;9(3):195–201.
239. Crocker IP, Tansinda DM, Baker PN. Altered cell kinetics in cultured placental villous explants in pregnancies complicated by pre-eclampsia and intrauterine growth restriction. J Pathol. 2004;204(1):11–8.
240. Brownbill P, et al. Mechanisms of alphafetoprotein transfer in the perfused human placental cotyledon from uncomplicated pregnancy. J Clin Invest. 1995;96(5):2220–6.
241. Brownbill P, et al. Vasoactivity to and endogenous release of vascular endothelial growth factor in the in vitro perfused human placental lobule from pregnancies complicated by preeclampsia. Placenta. 2008;29(11):950–5.

Chapter 20
Transamniotic Stem Cell Therapy (TRASCET)

Beatrice Dionigi and Dario O. Fauza

Introduction

Transamniotic Stem Cell Therapy (TRASCET) is a novel therapeutic paradigm applicable to the management of different birth defects. It is based on the principle of harnessing and enhancing the natural biological role of select populations of stem cells that either naturally occur in the amniotic fluid, or are present therein in the setting of disease, for therapeutic benefit. It has only been described very recently and has yet to be attempted clinically as of this writing. At the same time, the experimental data available to date, along with the rationale behind it, substantiate the perspective of TRASCET eventually becoming a valid strategy for the treatment of a number of congenital anomalies. The appeal and practicality of simple intra-amniotic administration of select stem cells in large numbers as a means to boost their normal activity and provide significant therapeutic gain at minimal to no risk to the mother and fetus is self-evident. Such a minimally invasive approach would be relatively easily accessible to a majority of pregnant women. Although diverse applications of this therapeutic concept are conceivable and currently being investigated, experimental evidence of its usefulness is so far available solely for the management of neural tube and abdominal wall defects. Still, its potential reach as an original service-based model of on-demand individualized perinatal stem cell

B. Dionigi, M.D.
Department of Surgery, Brigham and Women's Hospital,
75 Francis St., Boston, MA 02115, USA
e-mail: bdionigi@partners.org

D.O. Fauza, M.D., Ph.D. (✉)
Department of Surgery, Boston Children's Hospital, Harvard Medical School,
300 Longwod Avenue, Fegan 3, Boston, MA 02115, USA
e-mail: Dario.Fauza@childrens.harvard.edu

D.O. Fauza, M. Bani (eds.), *Fetal Stem Cells in Regenerative Medicine*,
Stem Cell Biology and Regenerative Medicine,
DOI 10.1007/978-1-4939-3483-6_20

processing, while further validating amniotic cell banking as clinically relevant, justifies this separate discussion of TRASCET in this book.

To date, despite multiple experimental advances, much promise and unfortunately excessive publicity, the reality is that most new cell-based therapies have yet to deliver significant impact to patient care. Conspicuous exceptions are the therapies based on controlling and/or enhancing the biological role that given cells already normally have. Clear examples are the different variations of blood transfusions and of bone marrow transplantation. Those long established cell-based therapies have had unparalleled impacts in health care, to a large extent due to the fact that the cells involved are used to fulfill the very same roles that they already perform in nature. Therein lies much of the appeal of TRASCET, in that it is also based on the therapeutic augmentation of the normal biological activities of select cells, only in the prenatal period and in a unique environment. The data gathered thus far and the vastly successful precedents of analogous forms of therapies in the postnatal period (blood transfusion and bone marrow transplantation) underscore the expectation that TRASCET could possibly lead to significant impact on patient care. Another component of the appeal of this innovative approach is the perspective of providing significant therapeutic improvement to children with major congenital anomalies via wieldy autologous cell injections, which are essentially office-based procedures. This practicable methodology would be easily accessible to virtually all pregnant women carrying a baby with a birth defect and from a very early point in gestation, thus potentially maximizing impact.

Biological Role of Amniotic Mesenchymal Stem Cells

It was not until only a few years ago that a biological role of any cell present in the amniotic fluid was first described, specifically an activity of amniotic mesenchymal stem cells (afMSCs) in fetal wound healing [1]. In consecutive experiments in fetal lambs, that study showed that although not absolutely essential to the healing process, afMSCs do expedite wound closure and enhance its extracellular matrix profile (Figs. 20.1 and 20.2) [1].

Within the spectrum of reparative and regenerative processes, fetal wound healing is closer to the latter than to the former. When compared with healing at any stage of postnatal life, wound healing in the fetus involves significantly less inflammation and can be almost scarless, particularly early in gestation. The mechanisms behind the fetus' greatly enhanced capacity to heal wounds remain not fully understood [2]. Until that study, the focus of fetal wound healing research had been on the peculiarities of local molecular pathways and gene expression patterns, such as for example the widely described upregulation of hyaluronan, a major extracellular matrix component of the epidermis which controls keratinocyte proliferation and differentiation [3, 4]. That work showed that there was a hitherto overlooked cellular component to fetal wound healing. Such a finding was in accordance with the well-established fact that mesenchymal stem cells from other sources, most notably bone marrow, are known to home in to injured sites and help promote local repair in postnatal life [5–7].

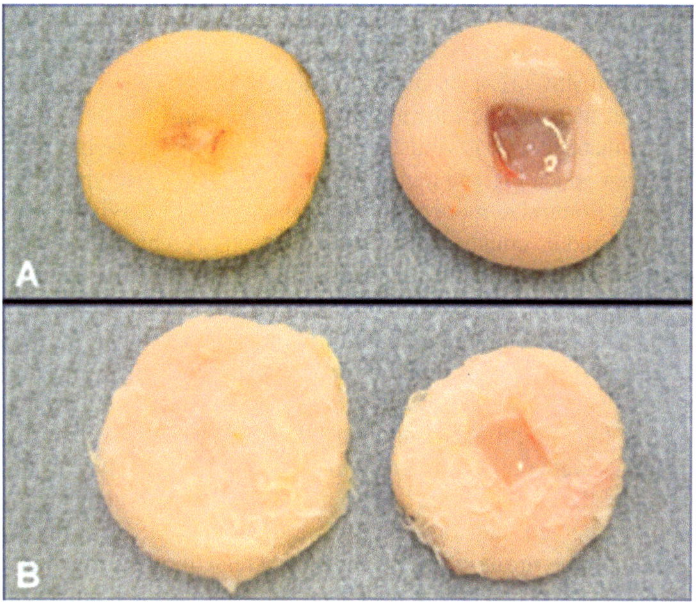

Fig. 20.1 Representative gross view of two sets of fetal lamb wounds—normal controls, on the *left*, and wounds deprived from exposure to amniotic cells on the *right*. Each set is from the same animal, on post-operative days 9 (**a**) and 20 (**b**), illustrating the evident differences in healing rate

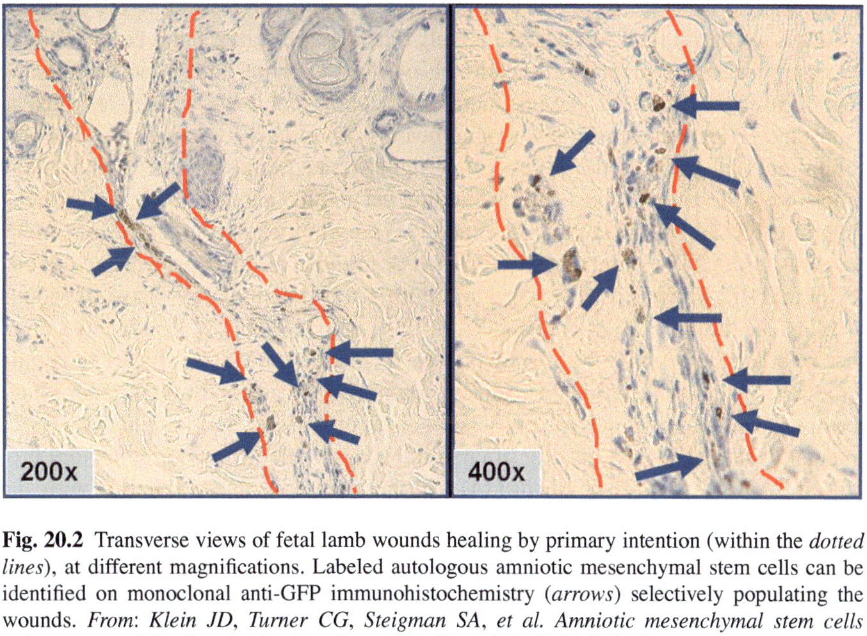

Fig. 20.2 Transverse views of fetal lamb wounds healing by primary intention (within the *dotted lines*), at different magnifications. Labeled autologous amniotic mesenchymal stem cells can be identified on monoclonal anti-GFP immunohistochemistry (*arrows*) selectively populating the wounds. *From: Klein JD, Turner CG, Steigman SA, et al. Amniotic mesenchymal stem cells enhance normal fetal wound healing. Stem Cells Dev 2011;20 (6):969–76*

Certainly, there is ample evidence that fetal wound healing is a complex process, not necessarily chiefly governed by an exogenous, even if autologous, cellular component such as afMSCs. For instance, marsupial fetuses exhibit improved wound repair even after they have left the marsupial pouch. Still, their healing post-pouch life is not as enhanced as it is whilst within the pouch and there is a temporal correlation between pouch life and transition into a scarring phenotype [8, 9]. Further, fetal fibroblasts synthesize more collagen, particularly of the type-III, and migrate faster than their adult counterparts, both of which are characteristics known to impact wound healing [3, 10]. A well-known drawback of the ovine model used to uncover the role of afMSCs in fetal wound healing is the fact that it is not conducive to in depth mechanistic analyses, due to the notorious unavailability of microarrays for pathway-specific investigations in large/domestic animals, including sheep. From a mechanistic perspective, however, that data at least allow for a few apposite speculations in light of current general knowledge on wound healing and afMSCs. Labeled afMSCs preferentially migrated to the areas of injury. Hyaluronic acid may have been one of the factors acting as a homing or chemotactic signal recognized by these cells. The default pathway of differentiation of afMSCs is the fibroblastic lineage. Previous leporine models have shown that fetal fibroblasts have an increased density of cell surface hyaluronic acid receptors compared to adult counterparts [11]. Elevated hyaluronic acid levels may make the fetal wound matrix more welcoming of a fibroblast influx, including possibly phenotypically related cells from the amniotic fluid, such as afMSCs. Indeed, in that study, the trend towards higher hyaluronic acid levels observed in the wounds in which afMSCs were prevented from engrafting may have been at least in part related to that very component of the experimental design, namely a semi-permeable membrane restricting the influx of afMSCs to certain fetal wounds. Relatively recently, Substance P has been shown to be upregulated in postnatal animal models of wound healing and to act systemically as a messenger of injury. The resultant mobilization of CD29+ cells in response to Substance P leads to accelerated wound healing, seemingly by stimulating cell proliferation, activation of the extracellular signal-related kinases (Erk) 1 and 2, and nuclear translocation of beta-catenin, adding yet another element to the wound healing puzzle [12]. In that study, Substance P expression was documented in all fetal wounds and afMSCs are robustly CD29+. A number of other growth factors and cytokines are known to be upregulated during the inflammatory phase of wound healing, including TGF-β isoforms, IL-8, IL-6, and TNF-α [13]. These and a number of other factors have been shown to be secreted by afMSCs, at least in certain culture conditions *in vitro* [14]. Indeed, such afMSC-preconditioned media have been used to augment dermal fibroblast migration *in vitro* and to enhance experimental postnatal wound healing *in vivo*, implying a role for paracrine factors produced by afMSCs in tissue repair [14]. All these previous data are in line with and add to the finding of a consequential role of afMSCs in fetal wound healing.

It has been previously reported that afMSCs proliferate more than twice as quickly in culture than bone marrow- and umbilical cord blood-derived mesenchymal stem cells of indistinguishable phenotype, when grown under identical conditions *in vitro* [15]. For example, a 3–5 mL aliquot of amniotic fluid obtainable during a routine diagnostic amniocentesis is all that would be needed for one to obtain several hundred million cells in 3–4 weeks' time [16, 17]. The mechanisms behind the remarkably enhanced self-renewal capacity of afMSCs have yet to be determined, yet such robust proliferation kinetics may be central to their role in enhancing fetal tissue repair. In like manner, the previously demonstrated unique matrix deposition patterns of afMSCs *in vitro* when compared with that of other mesenchymal stem cells may also be relevant to such role [15].

The finding that afMSCs play a role in fetal tissue repair has added a new dimension to fetal and general wound healing, as well as has lent biological support to the use of afMSCs in cell-based regenerative strategies, not only perinatally, but also later in life [16–24]. From a translational perspective, afMSCs are arguably more applicable in this setting than any other stem cell also because these are autologous fetal cells that can be procured from minute samples obtained by one of the least invasive of methods—a plain amniocentesis—which is often already indicated for diagnostic purposes in any mother carrying a fetus with a congenital anomaly, therefore placing no additional risk to the mother and fetus and thus rendering the TRASCET concept ethically unobjectionable.

In parallel to these putative advantages of using afMSCs as agents of TRASCET, general properties of MSCs are also relevant to this therapeutic principle. It is widely known that MSCs possess anti-inflammatory and immunomodulatory capabilities. Numerous experimental and even a few clinical studies have uncovered the benefits of MSCs from different sources, administered locally and/or systemically, in lung injury, articular damage, cardiac repair, inflammatory bowel disease, graft-versus-host-disease, and various other pathological processes [25, 26]. At the same time, despite a few isolated insights, much remains to be uncovered about the mechanisms underlying MSC homing and effects in tissue repair and inflammation [12].

Disease-Associated Amniotic Stem Cells

It has been recently shown that the amniotic fluid can harbor disease-specific, or disease-associated stem cells [27]. More specifically, it has been shown that neural stem cells (NSCs) can be isolated from the amniotic fluid in the setting of experimental neural tube defects—so-called amniotic fluid neural stem cells (afNSCs) [27]. While neural cells have long been known to exist in amniotic fluid in the presence of neural tube defects, and in fact often aid in the diagnosis of these anomalies, the existence of an undifferentiated, more primitive population of neural cells within the amniotic cavity had yet to be demonstrated until that report [28–32]. Previous data on the abundant presence of neural cells in human amniotic fluid in the setting of fetal neural tube defects would suggest that a comparable isolation of NSCs may well be practical clinically [32]. The fact that the amniotic fluid may be a practical source of (autologous) NSCs applicable to novel forms of therapies for spina bifida adds another dimension to the TRASCET concept. However, the role of NSCs in the setting of neural tube defects remains to be determined. Other recent reports at least suggest that this peculiar population of stem cells may have diagnostic value in the setting of congenital neural tube anomalies, in addition to a potential therapeutic one [33, 34]. Certainly, the perspective of other disease-associated amniotic stem cells being described, along with unique roles for such cells in select settings, is not to be discounted.

TRASCET for Neural Tube Defects

Some of the most common and morbid congenital (or acquired) perinatal diseases involve major central nervous system damage, typically with devastating long-term morbidity and significant mortality. Notable examples are congenital neural tube defects (NTD), which stem from the failure of the neural tube to close by the fourth week of embryonic development. These defects are commonly classified as open or closed based on the presence or absence of exposed neural tissue, and may involve any portion of the brain and/or spinal cord.Isolated neural tube defects are multifactorial in inheritance in the vast majority of cases [35]. At the same time, folic acid deficiency has long been known to be either a contributory, or a causative factor [36]. Folic acid supplementation during the first trimester of gestation can reduce the risk of NTDs by as much as 50–70 % [37, 38]. However, despite mandatory folate supplementation in all enriched cereal grain products, the incidence of spina bifida in the United States has remained relatively stable over the last several years, at 3–4 per 10,000 live births [39, 40].

Spina bifida is characterized by a midline vertebral defect, typically in the dorsal portion of the lumbosacral vertebrae. Depending on whether only the meningeal sac, or both the meningeal sac and the spinal cord protrude through the defect, it is referred to as meningocele, or myelomeningocele (MMC), respectively. Myelomeningocele leads to injury/loss of spinal cord tissue at and below the lesion. Common manifestations include paraplegia, urinary and fecal incontinence, sexual dysfunction, and secondary musculoskeletal deformities. Overall mortality of MMC at 5 years of age is approximately 14 %, reaching 35 % in children with hydrocephalus [41]. Morbidity rates are much higher, with the level of the lesion determining the type and severity of motor and sensory dysfunction. Spina bifida leads to spinal cord damage by both primary and secondary mechanisms. The former derives from the abnormal spinal cord development associated with the incomplete closure of the neural tube, while the latter is a result of spinal cord exposure to both the amniotic fluid (chemical insult) and local trauma (mechanical insult) within the uterus. Several studies point to the secondary mechanisms as being the most relevant clinically. More specifically, pathological analyses of human stillborns have shown that embryos or very young fetuses with MMC have little damage to the neural tube or spinal cord [42–45]. Fetal imaging studies have demonstrated that leg movement patterns tend to be normal in fetuses with MMC early on in gestation, only to deteriorate as pregnancy progresses [46, 47]. Further evidence of the impact of secondary damage to the spinal cord can be found in reports suggesting that delivery of MMC fetuses by cesarean section is associated with improved neurological outcome, when compared with vaginal birth [48, 49]. Animal models have also shown the importance of the secondary insult to the spinal cord in MMC [50–52]. These data, along with evidence of neurological improvement after experimental surgical prenatal closure of MMC, have served as the basis for surgical fetal MMC repair in humans.

Current treatment protocols for most forms of spina bifida are essentially supportive and aimed at minimizing additional central nervous system damage. This group of diseases is among the most morbid survivable congenital anomalies—so much so that it has prompted the first ever prospective randomized control trial of prenatal surgical repair of a non-life threatening anomaly, widely known as the Management of Myelomeningocele Study, or MOMS trial, a multicentric effort funded by the National Institutes of Health [53]. This sentinel undertaking has further validated multiple years of research pointing to the significant clinical impact of the secondary component of spinal cord damage in spina bifida and shown that, compared with the conventional postnatal coverage of the lesion, prenatal repair does offer some benefits, however only in a rather small subset of maternal-fetal units [53]. For example, only 183 mothers over well more than 1000 screened could be included in that study, for various reasons. Another limitation of surgical repair is the fact that it can only be safely performed no earlier than the second half of the second trimester of gestation. This is already quite late into the pathophysiological process in NTDs, which starts by the fourth week of embryonic development, when the neural tube is supposed to be closed. Like virtually all forms of fetal surgery, this type of intervention is not without significant maternal and fetal risks, most notably preterm labor and prematurity. Overall, it is fair to say that the benefits of both pre- and postnatal interventions remain relatively modest. Lifelong support, rehabilitation and variable degrees of institutionalization are typically necessary, if not the norm. A more effective therapeutic strategy should encompass accessibility to a larger proportion of patients, repair much earlier in gestation than what is viable via fetal surgical intervention, and feasibility through a minimally invasive approach, for example by inducing local regeneration. The TRASCET principle fulfills these requirements.

Amniotic Mesenchymal Stem Cells

In two studies published as of this writing, the TRASCET concept based on afMSCs has shown to have some impact in a rodent model of spina bifida [54, 55]. In the first experiment, time-dated pregnant Sprague-Dawley dams (n = 24) exposed to retinoic acid for the induction of fetal neural tube defects were divided into three groups. Group I had no further manipulations. Groups II and III received volume-matched intra-amniotic injections of either saline (group II) or a suspension of 2×10^6 cells/mL of afMSCs (group III) blindly in all fetuses (n = 202) on gestational day 17 (term = 21–22 days). Infused afMSCs consisted of syngeneic Lewis rat cells with identity confirmed by flow cytometry for CD29, CD44, CD45, CD73, and CD90 expressions, labeled with green fluorescent protein (GFP; 77–89 % positivity by FACS analysis). Animals were killed before term. Statistical comparisons were performed by generalized estimating equations, ANOVA, the Wald test, and Bonferroni comparisons, as appropriate ($P<0.05$). Results: A total of 165 fetuses were viable at euthanasia. Among fetuses with spina bifida (96/165; 58 %), there were no significant differences in the overall dimensions of the discernible defect across the groups ($P=0.19$). However, there was a statistically significant increase in the proportion of fetuses with variable degrees of coverage (some complete) of the defect by a thin, rudimentary skin, confirmed histologically (Figs. 20.3 and 20.4), in group III ($P<0.001$), with no differences between groups I and II ($P=0.98$). Donor afMSCs were identified in 83 % (33/40) of the fetuses in that group via immunohistochemistry for GFP, preferably engrafting in bone (Fig. 20.5). It could be concluded that afMSCs could induce partial or complete coverage of experimental spina bifida in the retinoic acid rodent model after concentrated intra-amniotic injection, seemingly via a paracrine effect, introducing the TRASCET concept as a potential option for the prenatal management of spina bifida.

A cell-based approach to the closure of experimental spina bifida through intra-amniotic injection had been previously described, albeit only in an *ex vivo* avian model (i.e. in eggs) and using either embryonic stem cells, or putative bone marrow stem cells [56–58]. The rodent study described above constituted the first report *in vivo*, in a mammal, and using afMSCs. From a translational perspective, the afMSC seems more appealing than bone marrow and embryonic stem cells. Autologous bone marrow cells would only be obtainable by much more invasive methods than a simple amniocentesis and the biological and ethical limitations of embryonic stem cells are many. There may also be differences as to the impact that that different (mesenchymal or not) stem cells have in this setting. As mentioned above, the fact that afMSCs have been shown to partake in fetal wound healing adds to their translational appeal, in that one would only be augmenting a role that they already play in nature.

Another experiment, subsequent to that initial report in rodent, was devoted to determining whether this therapeutic strategy could also have any impact on the Chiari-II malformation almost universally associated with spina bifida, also in the retinoic acid rodent model. Time-dated pregnant Sprague-Dawley dams ($n=11$) exposed to retinoic acid for the induction of fetal neural tube defects were divided into two groups: one ($n=6$) had no further manipulations and another ($n=5$) received volume-matched intra-amniotic injections of a suspension of 2×10^6 cells/mL of afMSCs blindly in all fetuses on gestational day 17 (term=21–22 days). Again, infused afMSCs again consisted of syngeneic Lewis rat cells with mesenchymal identity confirmed by flow cytometry, labeled with GFP. Animals were killed before term, when fetuses were divided into three groups: untreated controls with isolated spina bifida ($n=21$); isolated spina bifida treated with transamniotic delivery of afMSCs ($n=28$); and normal controls ($n=13$). Analyses included magnetic resonance imaging (MRI) with a high resolution (sub-millimeter) scanner and histology. The Chiari-II malformation was assessed on MRI by computer-generated

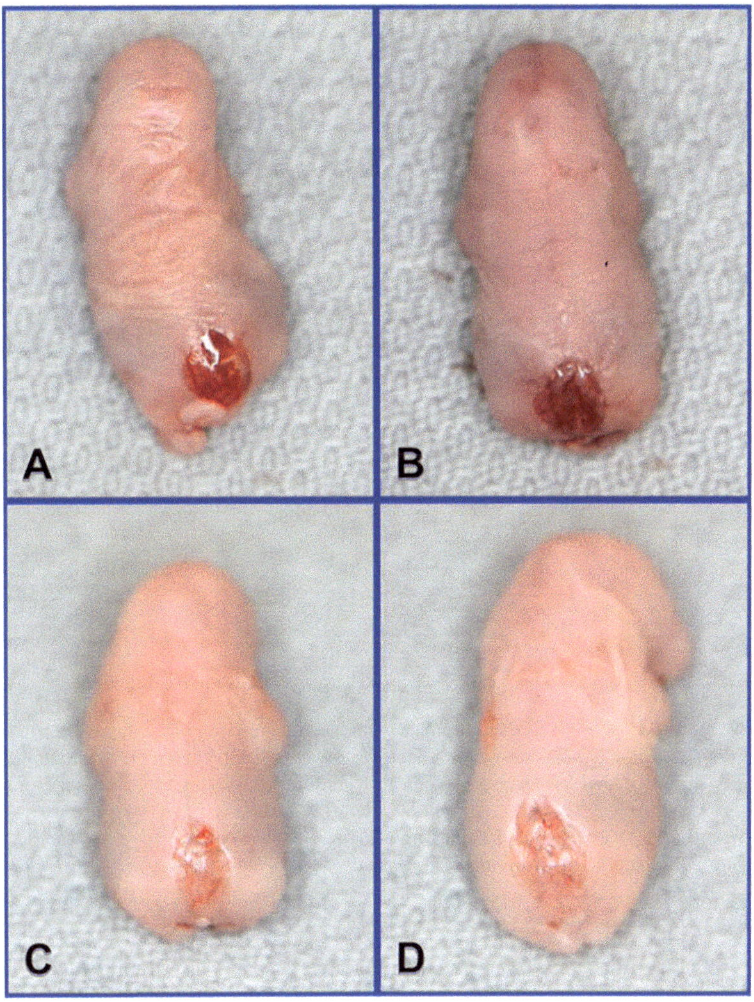

Fig. 20.3 Gross views of spina bifida defects in a rodent model at euthanasia (please refer to text). (**a**, **b**) Typical appearance from Groups I and II, in which a thin fibrous membrane could often be discerned covering the defect; (**c**, **d**) Appearance found in Group III, in which the defect is covered by a different looking membrane, shown to be rudimentary skin on histology (Fig. 20.4). *From: Dionigi B, Ahmed A, Brazzo J, 3rd, Connors JP, Zurakowski D and Fauza DO. Partial or complete coverage of experimental spina bifida by simple intra-amniotic injection of concentrated amniotic mesenchymal stem cells. J Pediatr Surg 2015;50(1):69–73*

specific angular and linear measurements of brainstem and cerebellar placement in relation to the basioccipital bone and the base of the skull, respectively (Fig.20.6). Statistical analyses were by Pearson chi-square, Fisher's exact test, and ANOVA with Bonferroni comparisons (2-tailed $P<0.05$). Results: As expected, there was a statistically significant increase in the proportion of fetuses with variable degrees of coverage of the spina bifida by a rudimentary skin confirmed histologically in the

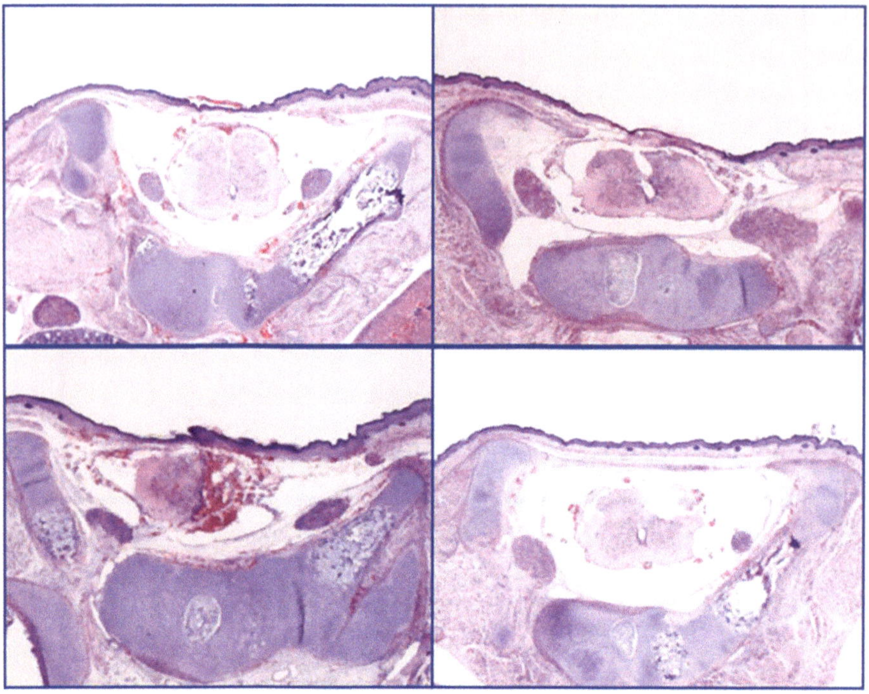

Fig. 20.4 Representative views of the spina bifida defect from Group III (please refer to text) on histology, showing the typical widely open vertebra arches, the spinal cord either apparently intact or variably deformed, and coverage of the defect by a rudimentary skin, with a paucity (or lack) of adnexa. H&E, 20× magnification. *From: Dionigi B, Ahmed A, Brazzo J, 3rd, Connors JP, Zurakowski D and Fauza DO. Partial or complete coverage of experimental spina bifida by simple intra-amniotic injection of concentrated amniotic mesenchymal stem cells. J Pediatr Surg 2015;50(1):69–73*

afMSC-treated group (P<0.001). Overall, there were statistically significant differences across the groups in linear and angular measurements of brainstem placement (P<0.001), with the untreated group displaying the highest caudal displacement. All pairwise comparisons of these parameters were statistically significant, with P=0.014 between treated and normal controls in angular brainstem (caudal) displacement and P<0.001 for all other angular and linear pairwise comparisons. Differences in cerebellar placement were also noted, albeit less pronounced, with P<0.001 overall and significance in most pairwise comparisons, except between treated and untreated groups (P=0.10). Donor afMSCs were identified in 71 % (20/28) of fetuses in the treated group via immunohistochemistry for GFP. It could be concluded that induced coverage of spina bifida by concentrated transamniotic delivery of amniotic mesenchymal stem cells does not completely reverse, however does minimize Chiari-II malformation in the retinoic acid rodent model. This led further support to afMSC-based TRASCET as a potential alternative or adjuvant for the prenatal management of spina bifida.

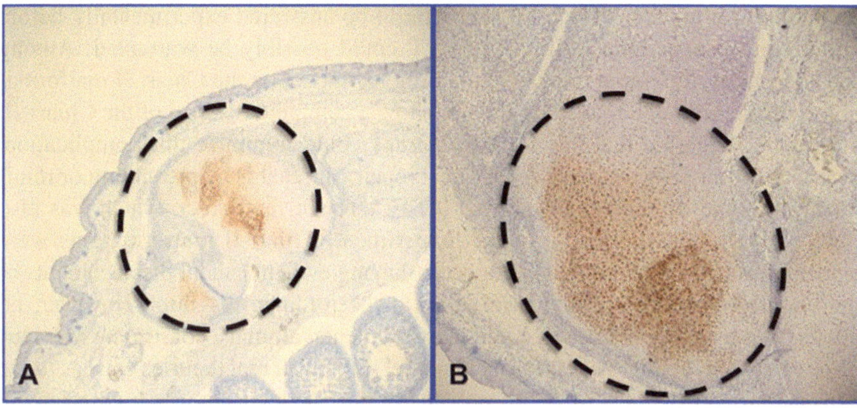

Fig. 20.5 Representative views of clusters of labeled donor amniotic mesenchymal stem cells identified on monoclonal anti-GFP immunohistochemistry (within the *dotted* perimeters) preferentially populating native bone in the vicinity of the spina bifida defect in a rodent model. (**a**) 20×; (**b**) 100× magnification. *From*: *Dionigi B, Ahmed A, Brazzo J, 3rd, Connors JP, Zurakowski D and Fauza DO. Partial or complete coverage of experimental spina bifida by simple intra-amniotic injection of concentrated amniotic mesenchymal stem cells. J Pediatr Surg 2015;50(1):69–73*

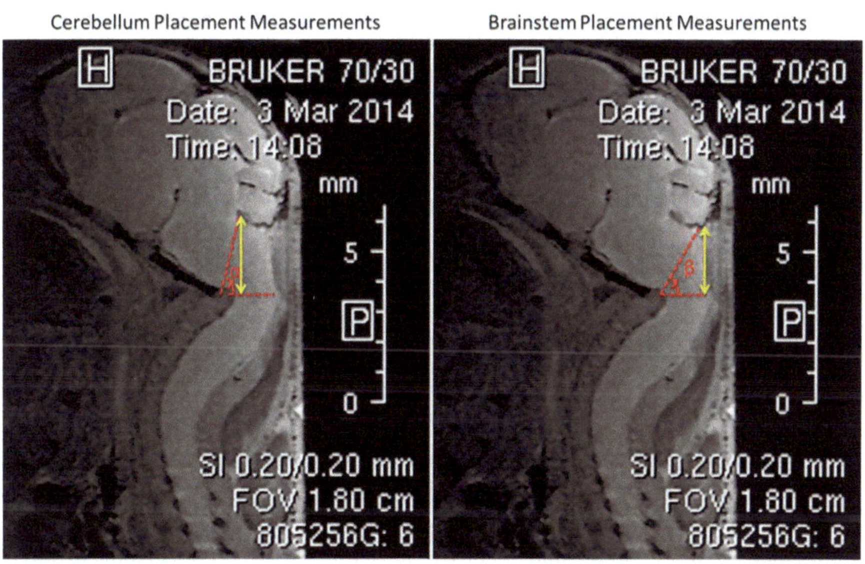

Fig. 20.6 The presence of a Chiari-II malformation was assessed in a rodent model of spina bifida on MRI scans by computer-generated specific linear (*double arrows*) and angular measurements of cerebellar (α) and brainstem (β) displacement in relation to the basioccipital bone and the base of the skull, respectively. *From*: *Dionigi B, Brazzo J, 3rd, Ahmed A, et al. Trans-amniotic stem cell therapy (TRASCET) minimizes Chiari-II malformation in experimental spina bifida. J Pediatr Surg 2015; in press*

Certainly, a number of questions remain to be answered experimentally before initial clinical experience with TRASCET could possibly be warranted. Among such questions, the importance of examining its impact on the Chiari-II malformation is clear. Although that study did not show a complete reversal of the Chiari-II, the results suggested that this form of therapy could minimize this complication, probably for the same reason why surgical repair can, that is by preventing or minimizing leakage of cerebrospinal fluid (CSF) through the defect, which was also covered, only non-surgically. In that experiment, Chiari-II malformations were evident in fetuses with myelomeningocele, having evident caudal displacements of the brainstem and cerebellum in relation to the foramen magnum. However, no evidence of hydrocephalus was noted in experimental animals, confirming previous observations involving the retinoic acid model [59]. Different theories exist as to the explanations for the lack of hydrocephalus in this model. Some have postulated a missed ventral rotation in the rodent hindbrain, allowing for CSF drainage even in presence of caudal displacement [60]. Others suggest that it is simply a time dependent matter, given the fairly short period of gestation in the rat, as well as the low rate and short term survival after birth. Time was also an inherent limiting factor in that experiment. Not only is the rat gestation quite short, it was feasible to perform the intra-amniotic injections with high survival rates on gestational day 17, which meant that there was less than a week between treatment and euthanasia. It is reasonable to speculate that the effects of the cell injections could have been more pronounced, should it have been possible to inject earlier and/or more than once. Perhaps, adding an additional control group receiving only saline injection would have enhanced that data. However, such a comparison had already been previously performed, with no difference in defect coverage between saline-injected and untreated groups [2]. In light of the fact that minimization of Chiari II is known to be directly related to the physical consequences of defect coverage, adding a saline-injected group in this second experiment would not have been unequivocally necessary.

Amniotic Neural Stem Cells

In light of the presence of afNSCs in the amniotic fluid in the setting of neural tube defects, as mentioned above, these stem cells have of course also become candidates for the TRASCET approach. The beneficial impact of a cell-based regenerative/reparative strategy may actually be maximized when applied to the developing, rather than the more mature spinal cord, a notion which has predicated the different fetal approaches to spina bifida repair. A recent study has shown that simple direct injection of expanded afNSCs into the amniotic cavity of rodent fetuses with spina bifida results in these cells selectively populating both the surface and deeper portions of the exposed neural tissue [61]. The fate of these donor cells *in vivo* after engraftment in the spinal cord and whether they lead to any meaningful degree of neural tissue repair or not, however, remains to be determined. Autologous afNSCs could conceivably be obtained by amniocenteses from a fetus with spina bifida, expanded *ex vivo* prenatally, and injected back into the amniotic fluid one or more times as a component of a broader therapeutic strategy to also include some form of closure of the defect, for example with afMSCs, at some point. Indeed, the perspective of using more than one cell type in TRASCET strategies for this disease, aiming at both neural repair and defect closure, is enticing.

TRASCET for Abdominal Wall Defects

Congenital abdominal wall defects are among the most common major structural congenital anomalies, with a combined incidence of approximately 1 in 2000 live births in the United States. Omphalocele and gastroschisis are by far the two most common forms of these diseases. Their etiologies are unknown. Omphalocele is typically the most common condition, although, more recently, the incidence of gastroschisis seems to be increasing, while the incidence of omphalocele has remained unchanged. Both omphalocele and gastroschisis involve an abnormal opening at the anterior aspect of the abdominal wall. In gastroschisis, the opening is almost always to the right of the umbilicus and quite small (usually less than 4 cm in diameter) when compared to the size/volume of the eviscerated organs. In omphalocele, the defect is central, at the site of the umbilical ring, and of highly variable dimensions. In omphalocele, the eviscerated contents are covered by a sac composed of peritoneum, Wharton's jelly, and amnion. In gastroschisis, the eviscerated content is totally exposed to the amniotic fluid/cavity, leading to significant local chemical and mechanical insult. Gastroschisis is thus associated with substantially more morbidity of the herniated structures than omphalocele is. Indeed, abnormalities of the gastrointestinal tract occur essentially in all cases of gastroschisis. The small and large intestines are almost invariably herniated, with other abdominal organs also possibly involved. The exposed bowel is grossly abnormal: edematous, inflamed, leathery, and foreshortened. Bowel loops are matted together, congested and/or ischemic, and coated with a thick peel. The mesentery is also thickened and short. Histologically, all layers of the bowel are considerably thickened, accompanied by varying degrees of amniotic peritonitis. More often than not, such morbidity is compounded by a time-dependent constriction of the abnormal abdominal opening, which compromises the blood supply to the eviscerated structures leading to congestion and/or ischemia and, in more severe cases, bowel strangulation and/or atresia. Incomplete fixation of the intestine leads to an increased tendency to develop midgut volvulus and the potential for infarction of large segments of bowel. Expectedly, functional consequences of such changes are the norm, with lack of peristalsis and a very slow return to normal gastrointestinal function practically always present. The period of dysfunction may last from weeks to several months. Variable degrees of prolonged intestinal transit, combined with decreased protein, carbohydrate, and fat absorptions in the neonatal period and beyond are the rule. It has long been established that the undue exposure to the amniotic fluid combined with local circulatory changes are the main mechanisms behind the significant gastrointestinal morbidity observed in gastroschisis (and in ruptured omphaloceles). It has also been well documented that the intensity of such local morbidity correlates with the functional/clinical manifestations.

The diagnosis of congenital abdominal wall defects is routinely made by ultrasound before birth, with many cases detected in the first trimester and even as early as at 10 weeks of gestation. The distinction between omphalocele and gastroschisis by fetal ultrasound is usually straightforward. In gastroschisis, the ultrasonographic appearance of the intestine correlates with clinical outcome. Despite many advances in fetal imaging, however, current prenatal management of these anomalies does not include any actual therapeutic intervention, with counseling on the choice of the type of delivery essentially constituting the sole intercession. Other than early delivery, a number of prenatal strategies have been proposed in order to minimize intestinal damage and improve intestinal function in gastroschisis. To date, either clinically or (mostly) only experimentally, these include amnio-exchange, amniotic fluid dilution, prenatal steroid administration, induced fetal diuresis, and even intrauterine repair of the defect, all with quite modest results [62–65]. Further, prenatal administration of steroids is not without risks, in that prolonged administration may lead to deterioration in maternal cardiovascular and metabolic status, as well as have deleterious effects on the fetus and the placenta.

A cell-based approach to the prenatal management of abdominal wall defects had yet to be described until a recent experimental report on afMSC-based TRASCET for gastroschisis in a surgical rodent model [66]. In that study, a gastroschisis was surgically created in 117 rat fetuses at 17–18 days of gestation (term = 21–22 days). Animals were then divided into three groups. One group (untreated; n = 62) had no further manipulations. Two groups received volume-matched intra-amniotic injections of either saline (n = 25) or a suspension of 2×10^6 cells/mL of afMSCs (n = 30) at the time of operation. Also in that study, infused afMSCs consisted of syngeneic rat cells with identity confirmed by flow cytometry for CD29, CD44, CD45, CD73, and CD90 expressions, albeit labeled with fluorescent cytoplasmic nanocrystals. Non-manipulated fetuses served as normal controls (NL). Animals were killed before term for analyses. Comprehensive computerized measurements of total and segmental (serosa, muscularis, and mucosa) intestinal wall thicknesses—long established surrogates for bowel damage in gastroschisis—were performed by two blinded observers. Statistical comparisons were by nested model ANOVA ($P < 0.05$). Overall survival was 25 %. Among survivors with gastroschisis, there were statistically significant decreases in total bowel wall, serosal, muscular, and mucosal thicknesses in the afMSC group vs. the untreated group (P = 0.001/0.035/0.001/0.005, respectively) and vs. the saline group (P = 0.003/0.05/<0.001/0.026, respectively) (Figs. 20.7 and 20.8). There were no differences between the afMSC group and NL, except for a significantly thicker muscular layer in the afMSC group (P = 0.014). There were no differences between the untreated and saline groups. Interestingly, unlike the findings in the spina bifida models of TRASCET, in that gastroschisis model there was not robust engraftment of nanocrystal-labeled afMSCs within the intestinal wall layers. Rather, donor cell engraftment seemed sparse, suggesting a

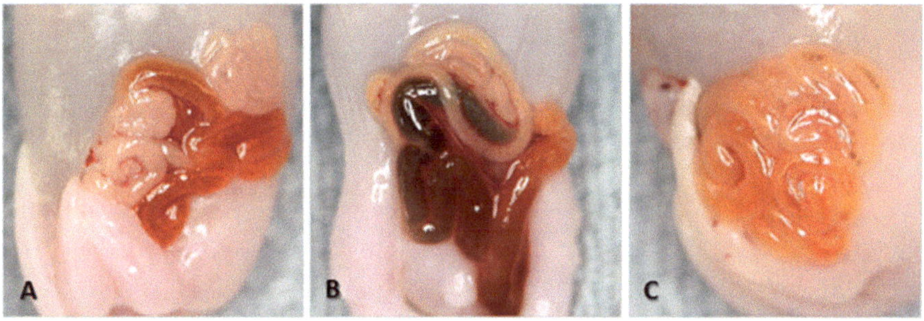

Fig. 20.7 Representative gross views of rat fetuses with gastroschisis at euthanasia. The exposed intestine in the (**a**) untreated and (**b**) saline groups (please refer to text) seemed heterogeneous, edematous, and thickened. The intestine in the (**c**) afMSC group appeared somewhat more homogenous and less affected macroscopically. *From: Feng C, Graham CD, Connors JP, et al. Transamniotic stem cell therapy (TRASCET) mitigates bowel damage in a model of gastroschisis. J Pediatr Surg 2015; in press*

paracrine effect. This finding was in accordance with that from a study on the effects of intra-peritoneal administration of comparable cells in a model of necrotizing enterocolitis [67]. It could be concluded from that first report that afMSCs mitigate bowel damage in experimental gastroschisis after concentrated intra-amniotic injection, thus suggesting that TRASCET could become a practical component of the management of gastroschisis. An in depth mechanistic analysis of how afMSCs may promote mitigation of intestinal damage in the setting of gastroschisis was beyond the scope of a first study such as that one, in that the very presence of such an effect would first have to be scrutinized. Certainly, an eventual understanding of such mechanisms will enhance the translational prospects for this particular application of TRASCET and beyond.

Regulatory Considerations and Future Perspectives

Unlike tissue engineering applications, the TRASCET principle does not necessarily involve the use of scaffolds, thus regulatory approval depends solely on the specific composition/processing and impact *in vivo* of the cell suspension. To date, no tumor or any other harmful effects have ever been identified in any of the multiple animal models having utilized native afMSCs therapeutically for well over a decade. These cells are not as primitive as embryonic stem cells and have been shown to be genetically and phenotypically stable during extensive cell processing under clinically acceptable FDA guidelines [16, 17]. The fact that they have proven effective in animal models in their native, undifferentiated state, without the need for any additional manipulation, after delivery as simple cell suspensions within the unique environment from which they are derived, further underlines the practicality and potential reach of the TRASCET approach.

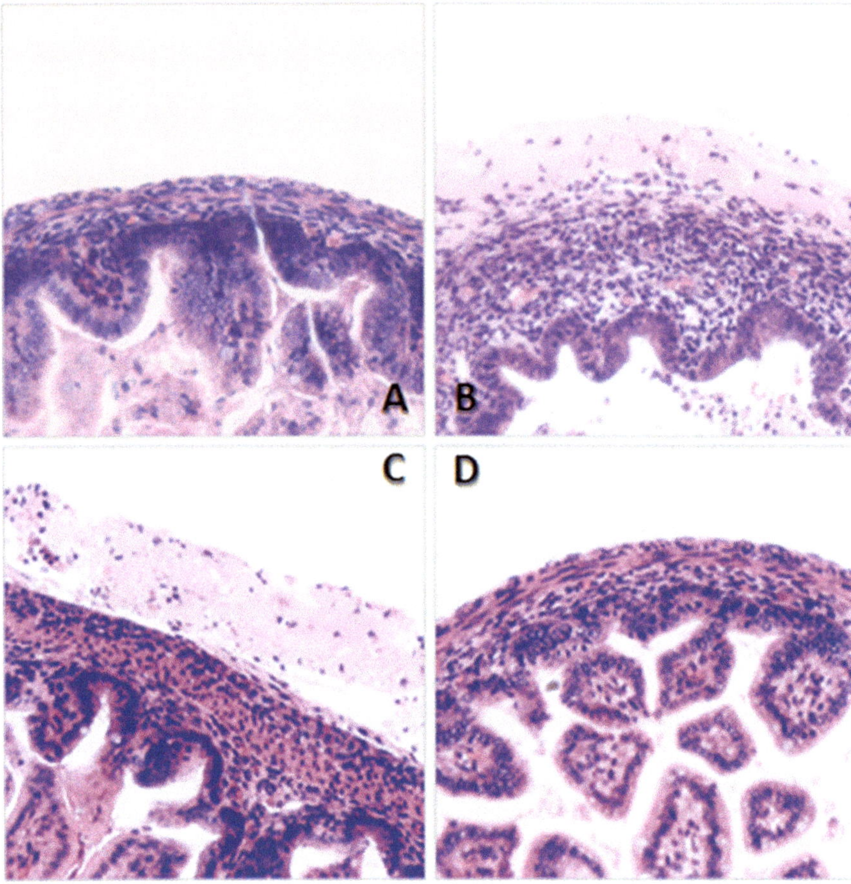

Fig. 20.8 Representative histological views of (**a**) normal rat fetus intestine and of the exposed intestine in (**b**) untreated, (**c**) saline, and (**d**) afMSC groups (please refer to text). The total bowel wall, serosal, muscular, and mucosal thicknesses appeared different in the afMSC group when compared with the untreated and saline groups, which were hardly distinguishable between each other. H&E, 200× magnification. *From: Feng C, Graham CD, Connors JP, et al. Trans-amniotic stem cell therapy (TRASCET) mitigates bowel damage in a model of gastroschisis. J Pediatr Surg 2015; in press*

Of course, the experimental efforts reported to date do not suffice as basis for eventual clinical trials of TRASCET. Further work in large animal models is paramount to clinical translation. Expectedly, the degree and pattern of therapeutic benefit in a large animal model (and in humans) will correlate more closely to variables such as volume and density of the cell suspension injected, timing and number of injections, and possibly also eventual pre-conditioning/targeted enhancements of the donor cells prior to delivery *in vivo*. Mechanistic insights, as well as data on the long-term fate of donor cells are needed. Still, the results reported to date, along with the fact that TRASCET is founded on the normal activity that select stem cells already perform in the amniotic fluid substantiate the expectation that TRASCET may become a practical, convenient, minimally invasive option, or an adjuvant to novel regenerative strategies for the prenatal treatment of different birth defects.

References

1. Klein JD, Turner CG, Steigman SA, Ahmed A, Zurakowski D, Eriksson E, et al. Amniotic mesenchymal stem cells enhance normal fetal wound healing. Stem Cells Dev. 2011;20(6):969–76.
2. Gurtner GC, Werner S, Barrandon Y, Longaker MT. Wound repair and regeneration. Nature. 2008;453(7193):314–21.
3. Tammi R, Pasonen-Seppanen S, Kolehmainen E, Tammi M. Hyaluronan synthase induction and hyaluronan accumulation in mouse epidermis following skin injury. J Invest Dermatol. 2005;124(5):898–905.
4. David-Raoudi M, Tranchepain F, Deschrevel B, Vincent JC, Bogdanowicz P, Boumediene K, et al. Differential effects of hyaluronan and its fragments on fibroblasts: relation to wound healing. Wound Repair Regen. 2008;16(2):274–87.
5. Wu Y, Wang J, Scott PG, Tredget EE. Bone marrow-derived stem cells in wound healing: a review. Wound Repair Regen. 2007;15 Suppl 1:S18–26.
6. Kwong FN, Harris MB. Recent developments in the biology of fracture repair. J Am Acad Orthop Surg. 2008;16(11):619–25.
7. Fu X, Li H. Mesenchymal stem cells and skin wound repair and regeneration: possibilities and questions. Cell Tissue Res. 2009;335(2):317–21.
8. Armstrong JR, Ferguson MW. Ontogeny of the skin and the transition from scar-free to scarring phenotype during wound healing in the pouch young of a marsupial, Monodelphis domestica. Dev Biol. 1995;169(1):242–60.
9. Morykwas MJ, Ditesheim JA, Ledbetter MS, Crook E, White WL, Jennings DA, et al. Monodelphis domesticus: a model for early developmental wound healing. Ann Plast Surg. 1991;27(4):327–31.
10. Gosiewska A, Yi CF, Brown LJ, Cullen B, Silcock D, Geesin JC. Differential expression and regulation of extracellular matrix-associated genes in fetal and neonatal fibroblasts. Wound Repair Regen. 2001;9(3):213–22.
11. Alaish SM, Yager D, Diegelmann RF, Cohen IK. Biology of fetal wound healing: hyaluronate receptor expression in fetal fibroblasts. J Pediatr Surg. 1994;29(8):1040–3.
12. Hong HS, Lee J, Lee E, Kwon YS, Ahn W, Jiang MH, et al. A new role of substance P as an injury-inducible messenger for mobilization of CD29(+) stromal-like cells. Nat Med. 2009;15(4):425–35.
13. Singer AJ, Clark RA. Cutaneous wound healing. N Engl J Med. 1999;341(10):738–46.
14. Yoon BS, Moon JH, Jun EK, Kim J, Maeng I, Kim JS, et al. Secretory profiles and wound healing effects of human amniotic fluid-derived mesenchymal stem cells. Stem Cells Dev. 2010;19:887.
15. Kunisaki SM, Fuchs JR, Steigman SA, Fauza DO. A comparative analysis of cartilage engineered from different perinatal mesenchymal progenitor cells. Tissue Eng. 2007;13(11):2633–44.
16. Kunisaki SM, Armant M, Kao GS, Stevenson K, Kim H, Fauza DO. Tissue engineering from human mesenchymal amniocytes: a prelude to clinical trials. J Pediatr Surg. 2007;42(6):974–9. discussion 9–80.
17. Steigman SA, Armant M, Bayer-Zwirello L, Kao GS, Silberstein L, Ritz J, et al. Preclinical regulatory validation of a 3-stage amniotic mesenchymal stem cell manufacturing protocol. J Pediatr Surg. 2008;43(6):1164–9.

18. Kaviani A, Perry TE, Dzakovic A, Jennings RW, Ziegler MM, Fauza DO. The amniotic fluid as a source of cells for fetal tissue engineering. J Pediatr Surg. 2001;36(11):1662–5.
19. Kaviani A, Guleserian K, Perry TE, Jennings RW, Ziegler MM, Fauza DO. Fetal tissue engineering from amniotic fluid. J Am Coll Surg. 2003;196(4):592–7.
20. Fuchs JR, Kaviani A, Oh JT, LaVan D, Udagawa T, Jennings RW, et al. Diaphragmatic reconstruction with autologous tendon engineered from mesenchymal amniocytes. J Pediatr Surg. 2004;39(6):834–8. discussion 8.
21. Kunisaki SM, Fuchs JR, Kaviani A, Oh JT, LaVan DA, Vacanti JP, et al. Diaphragmatic repair through fetal tissue engineering: a comparison between mesenchymal amniocyte- and myoblast-based constructs. J Pediatr Surg. 2006;41(1):34–9. discussion 9.
22. Kunisaki SM, Freedman DA, Fauza DO. Fetal tracheal reconstruction with cartilaginous grafts engineered from mesenchymal amniocytes. J Pediatr Surg. 2006;41(4):675–82.
23. Schmidt D, Achermann J, Odermatt B, Breymann C, Mol A, Genoni M, et al. Prenatally fabricated autologous human living heart valves based on amniotic fluid derived progenitor cells as single cell source. Circulation. 2007;116(11 Suppl):164–70.
24. Steigman SA, Ahmed A, Shanti RM, Tuan RS, Valim C, Fauza DO. Sternal repair with bone grafts engineered from amniotic mesenchymal stem cells. J Pediatr Surg. 2009;44(6):1120–6. discussion 6.
25. Barry F, Murphy M. Mesenchymal stem cells in joint disease and repair. Nat Rev Rheumatol. 2013;9(10):584–94.
26. Phinney DG, Prockop DJ. Concise review: mesenchymal stem/multipotent stromal cells: the state of transdifferentiation and modes of tissue repair--current views. Stem Cells. 2007;25(11):2896–902.
27. Turner CG, Klein JD, Wang J, Thakor D, Benedict D, Ahmed A, et al. The amniotic fluid as a source of neural stem cells in the setting of experimental neural tube defects. Stem Cells Dev. 2013;22(4):548–53.
28. Gosden C, Brock DJ, Eason P. The origin of the rapidly adhering cells found in amniotic fluids from foetuses with neural tube defects. Clin Genet. 1977;12(4):193–2U1.
29. Aula P, von Koskull H, Teramo K, Karjalainen O, Virtanen I, Lehto VP, et al. Glial origin of rapidly adhering amniotic fluid cells. Br Med J. 1980;281(6253):1456–7.
30. Greenebaum E, Mansukhani MM, Heller DS, Timor-Tristsch I. Open neural tube defects: immunocytochemical demonstration of neuroepithelial cells in amniotic fluid. Diagn Cytopathol. 1997;16(2):143–4.
31. Mendonca ED, Gutierrez CM, Peres LC. Brain tissue fragments in the amniotic fluid of rats with neural tube defect. Pathology. 2005;37(2):152–6.
32. Gosden CM, Brock DJ. Morphology of rapidly adhering amniotic-fluid cells as an aid to the diagnosis of neural-tube defects. Lancet. 1977;1(8018):919–22.
33. Pennington EC, Gray FL, Ahmed A, Zurakowski D, Fauza DO. Targeted quantitative amniotic cell profiling: a potential diagnostic tool in the prenatal management of neural tube defects. J Pediatr Surg. 2013;48(6):1205–10.
34. Pennington EC, Rialon KL, Dionigi B, Ahmed A, Zurakowski D, Fauza DO. The impact of gestational age on targeted amniotic cell profiling in experimental neural tube defects. Fetal Diagn Ther. 2015;37(1):65–9.
35. Main DM, Mennuti MT. Neural tube defects: issues in prenatal diagnosis and counselling. Obstet Gynecol. 1986;67(1):1–16.
36. Botto LD, Moore CA, Khoury MJ, Erickson JD. Neural-tube defects. N Engl J Med. 1999;341(20):1509–19.
37. MRC Vitamin Study Research Group. Prevention of neural tube defects: results of the Medical Research Council Vitamin Study. Lancet. 1991;338(8760):131–7.
38. Czeizel AE, Dudas I. Prevention of the first occurrence of neural-tube defects by periconceptional vitamin supplementation. N Engl J Med. 1992;327(26):1832–5.
39. Boulet SL, Yang Q, Mai C, Kirby RS, Collins JS, Robbins JM, et al. Trends in the postfortification prevalence of spina bifida and anencephaly in the United States. Birth Defects Res A Clin Mol Teratol. 2008;82(7):527–32.

40. Mohamed MA, Aly H. Birth region, race and sex may affect the prevalence of congenital diaphragmatic hernia, abdominal wall and neural tube defects among US newborns. J Perinatol. 2012;32(11):861–8.
41. Worley G, Schuster JM, Oakes WJ. Survival at 5 years of a cohort of newborn infants with myelomeningocele. Dev Med Child Neurol. 1996;38(9):816–22.
42. Patten BM. Embryological stages in the establishing of myeloschisis with spina bifida. Am J Anat. 1953;93(3):365–95.
43. Osaka K, Tanimura T, Hirayama A, Matsumoto S. Myelomeningocele before birth. J Neurosurg. 1978;49(5):711–24.
44. Hutchins GM, Meuli M, Meuli-Simmen C, Jordan MA, Heffez DS, Blakemore KJ. Acquired spinal cord injury in human fetuses with myelomeningocele. Pediatr Pathol Lab Med. 1996;16(5):701–12.
45. Meuli M, Meuli-Simmen C, Hutchins GM, Seller MJ, Harrison MR, Adzick NS. The spinal cord lesion in human fetuses with myelomeningocele: implications for fetal surgery. J Pediatr Surg. 1997;32(3):448–52.
46. Korenromp MJ, van Gool JD, Bruinese HW, Kriek R. Early fetal leg movements in myelomeningocele. Lancet. 1986;1(8486):917–8.
47. Sival DA, Begeer JH, Staal-Schreinemachers AL, Vos-Niel JM, Beekhuis JR, Prechtl HF. Perinatal motor behaviour and neurological outcome in spina bifida aperta. Early Hum Dev. 1997;50(1):27–37.
48. Luthy DA, Wardinsky T, Shurtleff DB, Hollenbach KA, Hickok DE, Nyberg DA, et al. Cesarean section before the onset of labor and subsequent motor function in infants with meningomyelocele diagnosed antenatally. N Engl J Med. 1991;324(10):662–6.
49. Shurtleff DB, Luthy DA, Nyberg DA, Benedetti TJ, Mack LA. Meningomyelocele: management in utero and post natum. Ciba Found Symp. 1994;181:270–80. discussion 80–6.
50. Meuli M, Meuli-Simmen C, Yingling CD, Hutchins GM, Hoffman KM, Harrison MR, et al. Creation of myelomeningocele in utero: a model of functional damage from spinal cord exposure in fetal sheep. J Pediatr Surg. 1995;30(7):1028–32. discussion 32–3.
51. Meuli M, Meuli-Simmen C, Hutchins GM, Yingling CD, Hoffman KM, Harrison MR, et al. In utero surgery rescues neurological function at birth in sheep with spina bifida. Nat Med. 1995;1(4):342–7.
52. Meuli M, Meuli-Simmen C, Yingling CD, Hutchins GM, Timmel GB, Harrison MR, et al. In utero repair of experimental myelomeningocele saves neurological function at birth. J Pediatr Surg. 1996;31(3):397–402.
53. Adzick NS, Thom EA, Spong CY, Brock 3rd JW, Burrows PK, Johnson MP, et al. A randomized trial of prenatal versus postnatal repair of myelomeningocele. N Engl J Med. 2011;364(11):993–1004.
54. Dionigi B, Ahmed A, Brazzo 3rd J, Connors JP, Zurakowski D, Fauza DO. Partial or complete coverage of experimental spina bifida by simple intra-amniotic injection of concentrated amniotic mesenchymal stem cells. J Pediatr Surg. 2015;50(1):69–73.
55. Dionigi B, Brazzo 3rd J, Ahmed A, Feng C, Wu Y, Zurakowski D, et al. Trans-amniotic stem cell therapy (TRASCET) minimizes Chiari-II malformation in experimental spina bifida. J Pediatr Surg. 2015;50:1037.
56. Lee DH, Kim EY, Park S, Phi JH, Kim SK, Cho BK, et al. Reclosure of surgically induced spinal open neural tube defects by the intraamniotic injection of human embryonic stem cells in chick embryos 24 hours after lesion induction. J Neurosurg. 2006;105(2 Suppl):127–33.
57. Lee DH, Park S, Kim EY, Kim SK, Chung YN, Cho BK, et al. Enhancement of re-closure capacity by the intra-amniotic injection of human embryonic stem cells in surgically induced spinal open neural tube defects in chick embryos. Neurosci Lett. 2004;364(2):98–100.
58. Lee DH, Phi JH, Kim SK, Cho BK, Kim SU, Wang KC. Enhanced reclosure of surgically induced spinal open neural tube defects in chick embryos by injecting human bone marrow stem cells into the amniotic cavity. Neurosurgery. 2010;67(1):129–35. discussion 35.
59. Danzer E, Schwarz U, Wehrli S, Radu A, Adzick NS, Flake AW. Retinoic acid induced myelomeningocele in fetal rats: characterization by histopathological analysis and magnetic resonance imaging. Exp Neurol. 2005;194(2):467–75.

60. Briner W, Lieske R. Arnold-Chiari-like malformation associated with a valproate model of spina bifida in the rat. Teratology. 1995;52(5):306–11.
61. Turner CG, Pennington EC, Gray FL, Ahmed A, Teng YD, Fauza DO. Intra-amniotic delivery of amniotic-derived neural stem cells in a syngeneic model of spina bifida. Fetal Diagn Ther. 2013;34(1):38–43.
62. Yu J, Gonzalez-Reyes S, Diez-Pardo JA, Tovar JA. Effects of prenatal dexamethasone on the intestine of rats with gastroschisis. J Pediatr Surg. 2003;38(7):1032–5.
63. Hakguder G, Olguner M, Gurel D, Akgur FM, Flake AW. Induction of fetal diuresis with intraamniotic furosemide injection reduces intestinal damage in a rat model of gastroschisis. Eur J Pediatr Surg. 2011;21(3):183–7.
64. Bittencourt DG, Barreto MW, Franca WM, Goncalves A, Pereira LA, Sbragia L. Impact of corticosteroid on intestinal injury in a gastroschisis rat model: morphometric analysis. J Pediatr Surg. 2006;41(3):547–53.
65. Till H, Muensterer O, Mueller M, Klis V, Klotz S, Metzger R, et al. Intrauterine repair of gastroschisis in fetal rabbits. Fetal Diagn Ther. 2003;18(5):297–300.
66. Feng C, Graham CD, Connors JP, Brazzo J, Pan AHS, Hamilton JR, et al. Trans-amniotic stem cell therapy (TRASCET) mitigates bowel damage in a model of gastroschisis. J Pediatr Surg. 2016
67. Zani A, Cananzi M, Fascetti-Leon F, Lauriti G, Smith VV, Bollini S, et al. Amniotic fluid stem cells improve survival and enhance repair of damaged intestine in necrotising enterocolitis via a COX-2 dependent mechanism. Gut. 2014;63(2):300–9.

Chapter 21
Stem Cells and Commercialization

Timothy Ken Mackey

Introduction

Stem cells have revolutionized the promise for cell-based therapies. A vast array of disease states are potentially amenable for stem cell use in treatment, providing hope for many patients who suffer from debilitating diseases and may not otherwise have access to treatment options. The potential promise of regenerative medicine is great: diseases potentially amenable to stem cell treatments include cancers, diabetes, cardiac disease, various muscular and neurological injuries, hematopoietic, immunologic, and a host of genetic disorders and countless other disease states [1].

Commercialization of traditional medication-based treatments have a well-established pathway. For example, in the United States, promising molecular entities enter and engage a systemic approach to evaluate the safety, grounds, and process of testing and evaluating the potential for use in humans. Through laboratory testing, investigational new drug application, and generally three phases of clinical trials testing with human subjects (Phase I with a small number of healthy volunteers for safety, toxicity and other profiling; Phase II with a larger number for patient effectiveness; and Phase III with a larger number of patients generally in multi-site healthcare settings to evaluate effectiveness and adverse events), drug regulatory authorities (DRAs) such as the US Food and Drug Administration (FDA) can assess whether such entities are viable under a safety and efficacy rubric to be marketed for use in its jurisdiction [2]. If approved, manufacturers can move to market and sell the product through standard commercialization efforts, and patients

T.K. Mackey, M.A.S., Ph.D. (✉)
Department of Anesthesiology and Division of Global Public Health, University of California, San Diego - School of Medicine, University of California, San Diego, 8950 Villa La Jolla Drive, La Jolla, CA 92037, USA
e-mail: tmackey@ucsd.edu; tmackey@ucsd.edu

D.O. Fauza, M. Bani (eds.), *Fetal Stem Cells in Regenerative Medicine*, Stem Cell Biology and Regenerative Medicine,
DOI 10.1007/978-1-4939-3483-6_21

can rely upon assessed research and peer review that has employed rigorous scientific and regulatory standards and oversight.

Although this process has been well established, the advent of new concepts and disruptive technologies—Internet-based technologies which can promote unproven treatments and therapeutic services directly to patients and enable the globalization of medical tourism—has created infrastructures that allow for illicit commercialization of unproven experimental materials. Importantly, this includes stem cells and its use as purported therapies, an issue that has come to the attention of the public, policymakers, public health agencies and drug safety regulators alike. Beyond the well publicized research fraud involving stem cell research [3], as well as clinical deceptions involving maternity care clinics and providers harvesting umbilical cord blood for illegal resale for use in the United States and other jurisdictions worldwide [4], the border-free nature of Internet trade and commerce creates the perfect storm that allows questionable claims, suspect providers, unethical research, and inappropriate bedside treatments that exploit desperate patients to be used and sold worldwide, virtually without oversight by the global community.

This confluence of factors places patients at risk for having their treatment hopes turned into significant financial loss due to fraudulent marketing and potential adverse events associated with poor quality and/or poor provider practices in manipulating and using stem cells. It also creates harm to patients who look to these purported treatments as a potential cure to their disease states and believe that these treatments are a better substitute for established, evidence-based efforts. This harm extends to financial fraud and abuse when patients and their families are subjected to illegitimate fees and charges that are largely paid for out of personal funding. It also harms translational researchers and legitimate manufacturers investing significant resources in establishing evidence-based science and uses for stem cells in regulated and accountable systems.

Medical and Transplant Tourism

An important infrastructure supporting illicit stem cell marketing and use is international medical tourism. Medical tourism is a general term signifying patients traveling outside their home country for medical treatment while assuming associated personal expense and risk [5–7]. Common procedures for medical tourism include cosmetic surgery, dentistry, cardiac surgery, and orthopedic surgery, though this market is rapidly expanding into a broad array of medical services. Importantly, key destinations for such services include countries such as Argentina, Brazil, Costa Rica, India, Malaysia, Mexico, Pakistan, Singapore, and Thailand [5]. These emerging and frontier market systems have uneven oversight and resources available for medical regulation compared with developed markets.

One area of medical tourism that is rapidly expanding, with concomitant regulatory and ethics concerns, is transplant tourism, including stem cell tourism [5].

As might be evident, transplant tourism is travel to a foreign country to obtain an organ *purchased* from unrelated donors. For solid organs, although the industry is clearly illicit and estimates hard to come by, reports indicate that, for example, approximately 5–10 % of all renal transplants in 2007 were from commercial living donors or vendors [8], while World Health Organization (WHO) estimates that during this same period 10 % of organ transplants performed worldwide involved transplant tourists [9]. This percentage has likely grown.

India, the Philippines, China, Egypt, Iraq, various South American countries, and Turkey have been reported as organ suppliers for transplant tourism patients globally, particularly for those patients coming from wealthy countries such as Australia, Israel, Japan, Taiwan, Saudi Arabia, and the United States [10]. Indeed, multiple globalized networks for organ trafficking supporting transplant tourism and travel for commercial organ transplantation have emerged in many developing countries [11–13]. Such illicit commercialization has been extant despite efforts by WHO and the World Health Assembly (WHA) attempting to address key concerns regarding exploitation of both patients and non-voluntary donors [14, 15]. For example, as early as 2004, the World Health Assembly adopted Resolution 57.18 that encouraged member states to safeguard the poorest and vulnerable groups from exploitation associated with transplant tourism, including the sale of tissues and organs. Further, in 2008 it attempted to reiterate concerns with transplant tourism, expressly noting the challenges in this illicit trade associated with organ trafficking, leading to a revision of the Guiding Principles on Human Cell, Tissue and Organ Transplantation and released those revised principles in 2009. Despite these efforts, transplant tourism continues to flourish.

Studies have noted potential concerns that have arisen from transplant tourism. For example, the donor and the procedures may lack documentation for future use and analysis; patients receive fewer immunosuppressive drugs than is current practice in developed countries such as the United States, and most patients do not receive antibiotic prophylaxis [5]. Importantly, the transplant tourism controversy provides a meaningful case study when examining the unregulated growth of the stem cell tourism and its potential risks to patient safety.

Direct-to-Consumer Advertising, The Internet, and Stem Cell Tourism

Unfortunately, the challenges of transplant tourism/organ trafficking are highlighted and exacerbated in the stem cell context. Stem cell tourism not only encompasses traditional high risk countries and practices, but it has also expanded globally through the rapid globalization of medical tourism, medical tourism-directed online marketing, promotion of questionable and often fraudulent claims, and unregulated direct-to-consumer advertising (DTCA). These risk factors represent the focal points of access and danger emanating from illicit stem cell commercialization.

Legal DTCA is permissible amongst developed nations only in the United States and New Zealand. However, its impact has been studied across borders, and has been shown to adversely impact patient safety and public health [16]. Importantly, however, the global reach of clearly illicit and unsubstantiated DTCA (such as advertising associated with illicit "no prescription" online pharmacies) has also been well documented, expanding in orders of magnitude via the Internet. This situation allows illicit stem cell vendors and clinics to join the plethora of potentially illicit, fraudulent, and harmful health information, services and products openly available over the Internet.

Online DTCA of all things medical now includes a wide range of medical products, and of course supports stem cell clinic marketing, sales, financing, infrastructure design, and clinical equipment. For example, medical products advertised online range from medications (lifestyle drugs, lifesaving drugs, drugs in shortage, withdrawn drugs, drugs approved but not yet released), vaccines on the WHO Essential Vaccines list, medical devices, implantable contraception, a whole host of medical tests, surgical/operating room equipment, and a wide array of other health-related goods and services [16–38]. These health products and services are often marketed and sold without healthcare professional oversight or the need for a prescriptions, and are freely advertised direct-to-the patient/consumer using search engine advertising, social media, and a combination of mediums that target underage and vulnerable groups [16–38].

These strategies have rapidly expanded and are amenable in the evolving stem cell treatment sphere, where a wide array of marketers and expanding claims are now engaged in stem cell DTCA via the Internet and associated technologies [39]. In the stem cell context, online DTCA allows vendors to present information on purported cures and treatments directly to the consumers outside of the physician-patient relationship. This medium also allows for multichannel marketing meeting consumers where they increasingly search for health information: on the Internet and social media.

Online stem cell DTCA is as extensive as it is inexpensive, using available webpages presence, social media, and direct solicitation, the unregulated ecommerce of health creates a conduit to a global stem cell tourist industry [19]. This follows pathways laid down by illicit Internet pharmacies, whose use of social media takes advantage of largely absent effective regulation, low cost of entry, and the growing popularity of social media channels among Internet users of all ages [40]. Indeed, the literature has reports that social media-based DTCA (that is not part of a separate fee-based Ad program) often requires no cost of entry and is largely unregulated for potential illegal content by service providers [19].

This confluence of Internet technology and the potential of stem cell innovation, has enabled and emboldened suspect stem cell providers and facilitators to both market to and recruit patients interactively using social media testimonials on YouTube, blogs, websites, and other online mediums [41, 44]. Traditional media outlets have also fueled such efforts by providing favorable coverage for stem cell "treatments" [41]. This situation exists despite the fact that stem cells have not been approved for any non-hematologic or immunologic indication by DRAs in research-based countries. Indeed, only one stem cell product has been approved by the US FDA in the hematologic context [42]. Further, aggressive online testimonials of stem cell treatment success have been put in doubt by investigative news reports following up with patients receiving stem cell treatments abroad, which found little to no actual improvement in patient conditions despite continuously advertised patient testimonial claims [43]. These unsubstantiated stem cell activities and misleading online advertisements create clear global patient safety risks.

Policy Issues: Marketing, Adverse Events, Quality

DTCA stem cell marketing and stem cell tourism has become more common in the USA and internationally—anywhere with access to the Internet—claiming cures or desired effects for virtually any disorder, condition, or disease [44]. Virtually all if not the vast majority of these claims are not evidence-based, and discount any key information associated with research, the experimental nature of stem cell therapy, clinical outcomes studies, or risk disclosure. Further, the entities engaged in these activities may be highly suspect and engaging in illicit unethical activities. Reports regarding these vendors have identified serious concerns, from clearly fraudulent activities and misleading advertising by stem cell clinics to trading in human fetuses to support their commercial activities [42]. There is a wide array of ethical issues needing attention in the stem cell tourism market unaddressed by suspect commercialization vendors who are not well regulated. These include exploitive and emotional marketing targeting parents to treat their children, treatment of the mentally incompetent, facilitation by some developed country scientists, challenges to registering stem cell activities, and other deeper, patient safety and professional ethical concerns [45–51].

Beyond suspect ethics, claims, and vendors, actual, known risks and adverse events are not disclosed in global stem cell DTCA promotion. Yet information regarding documented stem cell adverse patient safety events associated with unproven stem cell treatments has been reported in the literature and in media reports [52]. Importantly, these adverse events have occurred in a diverse set of developed and developing countries, indicating concerns regarding specific countries or facilities may not explain all occurrences and reifying the notion that stem cells are experimental as is their therapeutic use. Details are scant, but these poor clinical results include multiple patient deaths linked to stem cell use in China, death resulting from complications from stem cell treatment in the state of Florida, USA, death of a 18 month old baby in a German stem cell clinic, death from an experiential stem cell kidney treatment in Thailand, and the development of brain tumors in a pediatric patient injected with stem cells from at least two different donors in Russia [53–57]. In addition, beyond tragic deaths, potential medical risks include the possible transmission of infectious and genetic diseases to patients, uncontrolled cell/tumor growth, and other unreported negative outcomes have also occurred but are not disclosed [41, 58–64].

In addition to an environment of suspect claims and undisclosed risks, quality is of great concern. Fundamental questions regarding the consistency and quality of stem cell therapies, including concerns of whether advertised therapy actually contain the appropriate stem cell products and issues of sourcing and storage, are critically important to identify and address to ensure patient safety. As well, related issues of procurement of biological material used by vendors, including appropriate ethnically diverse and specific stem cell type [65], due to significantly different results across centers and poor clinical outcomes need integration for analysis [41, 62, 63, 66, 67]. This is particularly important as many clinical reviews have been critical of international use and reporting of benefits of stem cell treatments and experiments on the basis of their lack of rigor, methodology challenges, as well as reported side effects and outcomes [66, 68–70]. Other reviews have emphasized the need for substantive clinical investigations before allowing broad stem cell use [71] and attention to those "providers" that are rendering treatment [72].

Policy Concerns: Finances

From a regulatory and patient safety standpoint, the online stem cell marketplace encompasses an underground delivery system that is dangerous because of inappropriate use due to lack of scientific evidence and suspect quality of these products and services. These risks are exponentially higher when coupled with unrestricted DTCA marketing of stem cell treatments from locations worldwide, advertised using aggressive marketing techniques often specifically targeted to vulnerable patient populations.

However, these purported treatments also have tremendous financial downside risks for patients. Procedures are expensive: 2010 reports indicate an *average* of $47,000 per treatment, paid in cash since insurers consider these experimental treatments and do not pay for them [56]. Indeed, many stem cell clinic vendors actively solicit patients using financial programs to enable consumer payment of potentially expensive, but undisclosed medical costs. In fact, only one of the websites recently reviewed in a study that examined online stem cell clinic marketing tactics publicly disclosed estimated pricing of therapy options (ranging from US$6000–$22,000). Others simply stated that prices vary based on the scope of medical treatment and are disclosed after initial application/consultation [44].

Payment for possible treatment is accomplished by sites advertising acceptance of various payment methods common to e-commerce (e.g., credit cards, cash, wire transfer) and offering financing options or providing referrals to external third party financing companies [44]. This strategy may be necessary as stem cell treatments are likely underwritten completely out-of-pocket by patients (as they are likely considered experimental and not covered by a patient's insurance coverage), and are often bundled with non-medical costs (e.g., tests, transportation, lodging, interpretation services, etc.) that may increase the overall cost of treatment as well as increase potential profits for stem cell vendors. However, disclosure of this information on critical aspects of personal financial data in potentially unsecured online settings risks not only directly loss from an ineffective treatment, but also identity theft and fraud through use of credit card and bank information provided to these vendors.

Adding to the financial costs, emotional prices are paid by vulnerable patients attracted by testimonials of online DTCA "innovative" treatments by stem cell clinic websites [62, 66, 73–75]. This marketing often exploits the most ill and vulnerable in search of any treatment that may address their disease [76], similar to the Laetrile cancer treatment frauds that have occurred in the past [58, 64]. Although there has been closure of clinics engaged in potentially illegal stem cell activities, law enforcement efforts have been mostly ineffective. Indeed, some clinics have closed due to poor media coverage and resulting negative publicity, not direct regulatory or legal action [46, 58]. Others are discovered by investigative journalists, including a "60 min" investigation uncovering stem cell fraud by a physician with a revoked medical license who was selling and making false and misleading claims about stem cells, which were found to be of substandard quality [67]. Other national media outlets in the United States including NBC News and National Public Radio have also reported the use of the Internet in promoting suspect stem cell treatments [43, 77, 78].

The result of these forces is a global stem cell bazaar, with broad scope and coverage. Vendor employ a host of questionable procedures and providers are fueled by inexpensive and unregulated online marketing, permitting stem cell vendors from anywhere in the world to market unproven medical use of stem cells directly to prospective patients in a growing and largely unregulated *caveat emptor* e-commerce setting. In return, consumers pay high cash prices for unproven and experimental treatment marketed as legitimate care. The medical tourism infrastructures with online DTCA facilitates this trade, connecting these sellers to patients in any country who seek treatments unavailable or unapproved in their home countries, and allowing stem cell vendors to avoid regulatory or legal risk [6, 7].

Regulatory Efforts: Medical Practice versus Drug Manufacturing

It is well established that a variety of stakeholders believe that stem cell tourism and DTCA pharmaceutical marketing should be regulated [79–81]. Efforts to regulate Internet DTCA stem cell marketing have been attempted. However, there is regulatory conflict involving stem cell oversight. For example, in the USA, legal controversy exists as to whether the FDA has authority to regulate stem cell treatments as a drug/biologic as claimed by federal regulators, versus stem cell oversight being limited to only state medical board regulators as the practice of medicine, as claimed by vendors. This drug/biologic/tissue versus medical practice debate is a traditional division in many oversight regimes.

In attempting to exercise DRA authority over stem cell therapy, the USA FDA has sought termination of DTCA stem cell use by domestic providers through the court system by seeking injunctive relief to stop stem cell clinic activities. In the USA federal case of *United States v. Regenerative Sciences* [75], the FDA argued that forms and manipulation of stem cell therapy, including clinical harvesting, cell culture expansion, and reinfusion into patients, are subject to existing FDA regulation of drugs/biologics. Because of those activities, the FDA indicated this stem cell use must be approved as an Investigational New Drug, drug, or biologic, as well as manufactured using current good manufacturing practices (cGMP). Consequently, the vendor's online DTCA uses and marketing of stem cells represented drug misbranding and adulteration because they did not fulfill the mandates under these rules, nor do they fall within any recognized exception (such as only minimum manipulation for use in the same patient). Under this interpretation, DTCA stem cell marketing and purported therapy may promote unapproved or unregistered therapies/products in violation of DRA regulation [74, 75, 82–84].

In counterargument, stem cell clinics have claimed the FDA has no jurisdiction to regulate stem cell treatments involving autologous cell processing [75]. They argue that stem cell therapy simply constitutes the practice of medicine, a state-based oversight activity, and is therefore outside the scope and authority of the FDA [85]. They therefore have attempted to bypass extant biologic/drug rules and hence FDA regulatory oversight by indicating that any FDA attempt to control their operations is *ultra vires* (i.e., outside the scope of its authority). However, courts have disagreed, and ordered these stem cell vendors to cease operations and comply with FDA oversight rules though certain local jurisdictions may conversely allow stem cell practice to continue [84]. Stem cell clinics have indicated they will appeal. In response, they have also simply moved offshore to continue operations unfettered and outside of the purview of FDA activities [86].

Many countries and regions have taken proactive measures to specifically regulate stem cell therapy to avoid the issue of drugs versus medical practice, including the UK, the European Union, and Australia. For example, the UK has exemptions exclusively for short-term stem cell storage as noted in the UK Stem Cell Tool Kit, while Australia's Regulatory Framework on Biologicals directly defines and notes that human stem cells are biological products and must be regulated based on risk and the extent to which they are altered and/or manipulated [87].

However, the traditional dichotomy of drug/biologic oversight and practice of medicine exists in most jurisdictions and consequently, DRAs may find themselves embroiled in a similar legal argument. Like the USA, other jurisdictions also use independent entities such as Medical Councils to regulate medical practice separately from DRA oversight of drug/biologic safety. Because of the fact that most countries fall within this category, it is important for these countries to specifically regulate stem cell therapy utilizing DRA oversight, regardless of considerations of medical practice that may also do so, to ensure all marketing of stem cells and their use is regulated and patient safety oversight can be coordinated and effective.

The need for effective, proactive regulatory strategies against stem cell tourism is particularly important due to limited enforcement effectiveness. Enforcement of public health laws is critical to provide for accountability in this industry. Yet efforts have varied, with some countries such as Hungary, Costa Rica, and China reporting arrests and shut downs of clinics [6]. However, these enforcement actions are limited, and do not appear to have significant impact on operations or stem cell industry growth. For example, in China, with a vast network of private and publicly owned hospitals offering stem cell treatments, in 2009 announced new regulations in an attempt to better control the offering of unproven therapies (including stem cell therapies) [88]. However, questions regarding the enforceability of the new regulations and concerns regarding the impact of small fines for violations indicate the new rules are not acting as an effective deterrent [42, 43, 89] Subsequently, growth of China's stem cell industry appears to continue to expand and has an impact on other countries [6, 17, 42, 43, 86, 89].

Challenges of Self-Regulation

Some self-regulatory efforts have been made, but do not seem to have had substantive impact, as questionable stem cell treatment purveyors continue to increase in number [39]. Hence, uneven international self-regulation has not, so far, been effective.

Limited self-regulation activities includes the International Cellular Medicine Society (ICMS), an entity comprised of physicians and patients, some with close ties to the industry while also being co-founded by the medical director of a stem cell clinic involved in ongoing FDA litigation. ICMS offers its own accreditation process of stem cell providers and clinics [59, 60]. However, its physician-membership does not appear to involve credential verification; and its clinic membership does not require separate Institutional Review Board (IRB) approval because the Society sells its own IRB services per its own set of guidelines. ICMS therefore has significant potential for conflicts of interest, an issue that can mislead patients regarding the trustworthiness of DTCA-offered stem cell treatments [42].

Science-oriented organizations have also attempted to provide guidance in this setting. For example, efforts by the International Society for Stem Cell Research (ISSCR) to develop evidence-based and consensus guidelines for evaluation and oversight, informed consent, and increased transparency for stem cell use and treatment represent an important contribution in addressing necessary and appropriate global oversight of unproven stem cell therapies [90]. This was to include review and assessment of DTCA stem cell based claims. However, these efforts have been thwarted by stem cell vendor legal efforts threatening suit against ISSCR, including its efforts to provide patient guidance as to evidence-based availability to stem cell treatments. The independent, nonprofit organization of scientists and clinicians had attempted to launch a website which would have allowed online users to submit names of suspect stem cell clinics. ISSCR would then use this information and attempt to determine if reported stem cell providers had medical ethics oversight committees and if they were in compliance with regulatory agencies such as the

FDA and European Medicines Agency. However, due to the potential for lawsuits, ISSCR abandoned its plans for the website and instead engages in other forms of patient education/outreach [60].

However, ISSCR or similar binding standards provide an important basis that can begin to address the myriad of ethical and societal issues associated with DTCA-enabled stem cell sales, claims, and marketing. Enforcing such guidelines on scientists, clinical researchers, medical societies, healthcare providers, and information technology service providers while also educating patients/consumers across multiple jurisdictions could limit illicit stem cell use and marketing.

In addition, other voluntary guidance may be useful. For example, the Thai Medical Council has provided specific recommendations for stricter oversight of stem cell use and treatments by physicians in that country, which could represent a template or guide for international standards in emerging and developing countries regarding stem cell use. Most countries have the power to act against unprofessional actions of medical care providers and should take advantage of this authority in addressing unauthorized stem cell therapy [39, 91].

Potential Reform

At the present time, no governments, private sector approaches, nor institutional strategies have, in fact, provided any best practices to make illicit global sales and use of stem cells both accountable and ensure patient safety. Because efforts to regulate stem cells have been centered around geopolitical efforts necessarily focusing on local country jurisdiction, while the illicit sale and use of stem cells is a global concern due to borderless Internet, local regulation must combine with global oversight to garner any comprehensive, positive lessons to result. Hence, challenges regarding stem cell therapy, medical tourism, and DTCA stem cell marketing require formal, policy steps to curb misleading and fraudulent advertising to protect global patient safety. These should focus upon establishing clear and harmonized DRA authority, medical licensure and standards for practice, and potential global governance to address unproven stem cell activities across geopolitical borders.

Clarifying DRA Regulatory Authority

To protect patients against the harm and exploitation of potentially dangerous stem cell treatment, several avenues should be explored. First, recognizing the potential conflict that may occur across international settings in regulation of stem cells as drugs/biologics *versus* the practice of medicine, DRAs must be explicitly empowered to regulate all stem cell use as blood-derived products when such authority is not sufficiently clear. Explicit clarification of this authority for DRAs would ensure that offered stem cell therapies are subject to drug, biologic, cGMP, and current good tissue practice mandates explicitly required in other therapeutic treatments. This defined authority could also provide for clear rules for DRAs and regulated entities, enhance enforcement efforts, and limit potential legal challenges. Most importantly, rules expressly indicating that any stem cell therapy provider and/or advertiser is subject to DRA authority, and that the practice of medicine does not preempt DRA regulatory oversight power is needed. The current consistency across key jurisdictions such as Australia, the EU [92] as well as Canada, Japan, and Singapore, in stem cell regulation provides an established foundation to build this clarity empowering DRAs to directly regulate stem cell activities [93].

This regulation should augment existing interpretations that DRA's have jurisdiction over human cells, tissues, and cellular and tissue-based products, which are determined as processed or combined for metabolic purposes [59, 62]. In addition, successful approaches by DRAs can be shared with other countries to promote regulatory harmonization and discourage dangerous stem cell medical tourism.

DTCA Stem Cell Marketing Regulation and Disclosures

Beyond regulatory reform, misleading and fraudulent DTCA stem cell marketing should be identified and regulated as the experimental treatment it is. As a minimum standard, online disclosures should include affirmations and documentation that all stem-cell therapy providers adhere to IRB approval, patient informed consent, pre-/post-trial follow up, adverse event reporting, testing stem cell purity and potency, and screening and tracking of stem cell products [59]. All this information should be available online to prospective patients for inspection and public scrutiny.

Patients should also be made specifically aware of their rights and protections as research participants and educated that they are test subjects for experimental treatment and should not expect clinical benefit [94]. This is clearly not the case at the present time, where stem cell advertisers and providers have made claims that do not explain the experimental nature of these unapproved and unproven treatments.

DRAs should also use existing legal powers to target illegitimate online claims by stem cell marketers, including indictments against them for false and misleading statements and claimed treatment benefits [62]. For example, the FDA as a DRA should invoke its powers to bring *ex parte* hearings (i.e., abbreviated hearings with only one side present) for temporary restraining orders (TRO) against websites potentially violating the Food, Drug, and Cosmetic Act [16, 95]. Under a TRO, patient protections against suspect sites may be put into place until full hearings can occur. At these latter hearings, FDA should seek permanent injunctions against suspect marketing of these offending websites. If these websites and their owners are offshore, FDA should work with the Department of Justice and international partners to shut down these operations as they have in the past for illicit online pharmacies selling counterfeit medicines [96]. Other countries have similar public health powers [36, 97].

General consumer protection laws in each individual country that exist to protect consumers against false and misleading marketing should also be employed for DTCA stem cell vendors. In the USA, the Federal Trade Commission and individual states can bring suits for this kind of suspect advertising [98, 99]. Similarly empowering national laws include the Australian Competition and Consumer Act [63, 99] and the UK Consumer Protection from Unfair Trading Regulations [100]. These laws should be used to take down questionable DTCA stem cell purveyors and reinforce global proscriptions that prohibit DTCA.

Medical Licensure Oversight

Other methods to address inappropriate experimental stem cell use should include targeting practitioners engaging in this unproven/unsupported/unauthorized practice through existing national or state medical licensure oversight boards. Medical practice regulators should act quickly against healthcare providers promulgating experimental treatments as legitimate and well established that employ stem cells, and immediately suspend, then move permanently against their healthcare licenses. In the USA, state authorities have this power and should engage in licensure proceedings against unsupported and fraudulent stem cell therapy practitioners as is common for other unethical practices of medicine [66, 101].

Science as Standard

On the global level, cooperative means of governance should be engaged in the absence of clear, enforceable international legal regimes on stem cell use and advertising. The global community of researchers, policymakers, DRAs, pubic health agencies, patients safety advocates, and other civil society groups, should create a framework for preventing unauthorized stem cell experimentation and marketing in alignment with efforts to establish responsible ethical guidelines in stem cell research [46, 102]. These efforts should include standardizing stem cell clinic operations and developing substantive and independent accreditation that includes human research protections. Use of ISSCR materials and leadership with international support may be a good beginning with its established science, research, and patient base and extant created information and education materials.

These efforts are important, as poor quality and excess promises for stem cells harm legitimate research and clinical work in the field [103, 104]. It also confuses patients as to whether the purported therapies are 'science or swindle' [103]. The promise of stem cells and their future viability as a potentially ground-breaking treatment modality are highly dependent upon legitimate and considered allocation of resources to identifying and showing rigorously that such treatments work. Broken promises to patients by suspect stem cell users, adverse patient safety events, and other publicized failures by less robust providers who seek to capitalize on vulnerable patients and profit making opportunities may undermine legitimate efforts and dampen funding prospects for legitimate stem cell clinical research.

Beyond harmonized DRA regulations, educational activities, and coordinated enforcement efforts against errant stem cell purveyors, stem cell research and translational activities should be accelerated to bring legitimate science and promise to the stem cell treatment sphere. In this regard, other policy efforts aimed at promoting legitimate stem cell commercialization efforts should be expanded. For example, an accelerated process that holds great promise is the EU Advanced therapy medical products (ATMPs) pathway. The ATMP pathway, created by Regulation (EC) No. 1394/2007 [105], is relatively new, created under the EMA (European Medicines Agency) under the Committee for Advanced Therapies (CAT). The program aims to facilitate stem cell therapeutics work using a new, facilitated approach [106, 107].

The CAT EMA system reviews applications for marketing authorization for ATMP products, including stem cells. This dedicated function of the CAT creates efficiencies alone because of its clear mandate and scope of review. However, further, it addresses a key concern of many legitimate stem cell entities: they are generally small or medium size business entities (SME) with limited resources for extensive, full, regulatory guidance and engagement as seen for typical small molecule or biologic medicine. In the CAT EMA system, to avoid the high cost and generally unfamiliar regulatory procedures for drug marketing authorization, the CAT EMA created a certification procedure expressly for SMEs and ATMP development. Stem cell SMEs going through the CAT EMA certification process provide important recognition that the SME stem cell activities are viable, as well as a valuable progress point for potential investors and licensees, providing additional working capital for these legitimate entities [108]. The procedure is completely optional, free of any charges, and may be engaged at any time prior to marketing authorization application for the therapy [109]. Consequently, the CAT EMA ATMP process provides SMEs and other entities benefits from this exclusive pathway while also engaging European regulators early to promote good communications and (hopefully) rapid advancement in stem cell research and clinical applications. Other regions and countries such as Asia and Japan are also engaging in similarly innovative regulatory approaches to allow accountable, more facilitated entry into the regenerative medicine market [110]. These approaches should be assessed together in global forums to determine up to date, well established policy lessons for application more broadly.

In combination, a faster growing, accountable set of data and clinical information may emerge for legitimate commercialization of stem cell work while regulatory and law enforcement reforms can temper the massive fraudulent stem cell DTCA perpetrated worldwide. By focusing social resources on groups performing research responsibly, these latter groups will have a competitive advantage over illicit players, who have not undergone rigorous assessment and methodology review—and can be seen as such. This goal can benefit patients over the long run in combination with robust regulatory structures to combat online DTCA suspect stem cell marketing and treatments.

Conclusion

Stem cells have great potential to treat and cure a multitude of human disease. Yet like all systems of potential, illicit and unethical vendors have attempted to capture this market with outrageous claims, expensive treatments, and wide-ranging online DTCA marketing preempting legitimate stem cell commercialization. The direct impacts on patients as well as direct and indirect impacts on legitimate research and progress in the field are highly negative. Oversight, accountability, and innovation support for viable efforts in stem cell development are necessary to ensure that patient safety is maintained, patients are given the best opportunity for effective and proven treatment, and providers who render care using stem cells are qualified and ethically sound. Only through responsible and appropriately regulated commercialization supported by collective global action can the scientific validity and promise of stem cell therapy be realized.

References

1. NIH Center for Regenerative Medicine. Monthly abstracts of selected stem cell articles of interest. http://crm.nih.gov/resources/scientificUpdates.asp. Accessed Oct. 21, 2013.
2. FDA. How drugs are approved and developed. 2013. http://www.fda.gov/drugs/developmentapprovalprocess/howdrugsaredevelopedandapproved/. Accessed Oct. 18, 2013.
3. Murnaghan I. Stem cell scandal: a case study. Explore stem cells. 2014. http://www.explorestemcells.co.uk/stem-cell-scandal-case-study.html. Accessed Oct. 20, 2014.
4. U.S. Food and Drug Administration. FDA warns about stem cell claims. 2012. http://www.fda.gov/ForConsumers/ConsumerUpdates/ucm286155.htm. Accessed Oct. 20, 2014.
5. Lee CV, Balabian V. Chapter 2. Medical tourism. Centers for Disease Control and Prevention: The Pre-Travel Consultation – Counseling and Advice for Travelers. 2014. http://wwwnc.cdc.gov/travel/yellowbook/2014/chapter-2-the-pre-travel-consultation/medical-tourism. Accessed Oct. 18, 2013.
6. Einsiedel EF, Adamson H. Stem cell tourism and future stem cell tourist: policy and ethical implication. Dev World Bioeth. 2012;12(1):35–44.
7. Horowitz MD, Rosensweig JA, Jones CA. Medical tourism: globalization of the healthcare marketplace. MedGenMed. 2007;9(4):33.
8. Sajjad I, Baines LS, Patel P, Salifu MO, Jindal RM. Commercialization of kidney transplants: a systematic review of outcomes in recipients and donors. Am J Nephrol. 2008;28(5):744–54.
9. World Health Organization. WHO proposes global agenda on organ transplantation. 2007. http://www.who.int/mediacentre/news/releases/2007/pr12/en/index.html. Accessed Oct. 18, 2013.
10. Martin D. The Long road from the kidney bazaar: a commentary on Pakistan's progress towards self-sufficiency in organ transplantation. J Multidisc Inter Stud. 2011;8(2). https://epress.lib.uts.edu.au/journals/index.php/portal/article/view/1833. Accessed Oct. 18, 2013.
11. Shimazo Y. The state of the international organ trade: a provisional picture based on integration of available information. Bull World Health Org. 2007;85(12):955–62.
12. Budiani-Saberi D, Delmonico F. Organ trafficking and transplant tourism: a commentary on the global realities. Am J Transpl. 2008;8(5):925–9.
13. Budiani-Saberi D, Mostafa A. Care for commercial living donors: the experience of an NGO's outreach in Egypt. Transpl Int. 2010;24(4):317–23.

14. World Health Organization. Guiding principles on human organ transplantation. Geneva: World Health Organization. 2010. http://www.who.int/ethics/topics/transplantation_guiding_principles/en/index1.html. Accessed Oct. 18, 2013.
15. World Health Organization. Human organ and tissue transplantation. Geneva: World Health Organization. 2009 [updated 2012 Sep 18]. http://apps.who.int/gb/ebwha/pdf_files/A62/A62_15-en.pdf. Accessed Oct. 18, 2013.
16. Liang BA, Mackey T. Prevalence and global health implications of social media in direct-to-consumer drug advertising. J Med Internet Res. 2011;13(3):e64. doi:10.2196/jmir.1775.
17. Mackey TK, Liang BA, Strathdee S. Digital social media, youth and non-medical use of prescription drugs: the need for reform. J Med Internet Res. 2013;15(7):e143. doi:10.2196/jmir.2464.
18. Liang BA, Mackey TK, Archer-Hayes A, Shinn L. Illicit online marketing of lorcaserin before DEA scheduling. Obesity. 2013;21(5):861–4.
19. Mackey TK, Liang BA. Global reach of direct-to-consumer advertising using social media for illicit online drug sales. J Med Internet Res. 2013;15(5):e105. doi:10.2196/jmir.2610.
20. Liang BA, Mackey TK, Lovett KM. Illegal "No Prescription" internet access to narrow therapeutic index drugs. Clin Ther. 2013;35(5):694–700.
21. Mackey T, Liang BA. Oncology and the internet: regulatory failure and reform. J Oncol Pract. 2012;8(6):341–3.
22. Lovett KM, Liang BA, Mackey T. Online direct-to-consumer access to insulin: patient safety considerations and reform. J Diabetes Sci Technol. 2012;6(6):1503–6.
23. Liang BA, Mackey T, Lovett KM. Suspect online sellers and contraceptive access. Contraception. 2012;86(5):551–6.
24. Liang BA, Mackey T, Lovett KM. Emerging dangers from direct botulinum access and use. J Homeland Sec Emerg Manag. 2012;9(1):art.9.
25. Lovett KM, Mackey TK, Liang BA. Evaluating the evidence: direct-to-consumer screening tests advertised online. J Med Screen. 2012;19:141–53.
26. Liang BA, Mackey T. Sexual medicine: online risks to health-the problem of counterfeit drugs. Nat Rev Urol. 2012;9:480–2.
27. Liang BA, Lovett KM. Online elder abuse: addressing senior safety risks. Experience. 2012;22(2):17–22.
28. Lovett KM, Liang BA. Direct-to-consumer disease screening with finger-stick testing: online patient safety risks. Clin Chem. 2012;58(7):1091–3.
29. Mackey TM, Liang BA, Kubic T. Dangerous doses: fighting fraud in the global medicine supply chain. Foreign Affairs. 2012; May 14.
30. Lovett KM, Liang BA, Mackey T. Risks of online direct-to-consumer tumor markers for cancer screening. J Clin Oncol. 2012;30(13):1411–4. doi:10.1200/JCO.2011.37.8984.
31. Liang BA, Mackey T. Creating illicit laparoscopic appendectomy operating rooms using online sourcing. Am Surg. 2012;78(4):502–3.
32. Mackey T, Liang BA, Lovett KM. Waking up to the risks of online direct-to-consumer advertising of sleep apnea screening tests. J Clin Sleep Med. 2012;8(1):5–6.
33. Liang BA, Mackey T. Online availability and safety of drugs in shortage: a descriptive study of internet vendor characteristics. J Med Internet Res. 2012;14(1):e27. doi:10.2196/jmir.1999.
34. Liang BA, Mackey T. Vaccine shortages and suspect online pharmacy sellers. Vaccine. 2012;30(2):103–5. doi:10.1016/j.vaccine.2011.11.016.
35. Lovett KM, Liang BA. Risks of online advertisement of direct-to-consumer thermography for breast cancer screening. Nat Rev Cancer. 2011;11(12):827–8. doi:10.1038/nrc3170.
36. Lovett K, Liang BA. DTC cardiac screening and suspect risk evaluation. JAMA. 2011;305(24):2567–8.
37. Liang BA, Mackey T. Direct to consumer advertising with interactive internet media: global regulation and public health issues. JAMA. 2011;305(8):824–5.
38. Liang BA, Mackey T. Searching for safety: addressing search engine, web site, and provider accountability for illicit online drug sales. Am J Law Med. 2009;35(1):125–84.

39. Sipp D. Direct-to-consumer stem cell marketing and regulatory responses. Stem Cells Transl Med. 2013;2(9):638–40.
40. Brenner J. Pew internet: social networking (full detail). Pew Internet Monitoring and regulating offshore stem cell clinics. 2013. http://pewinternet.org/Commentary/2012/March/Pew-Internet-Social-Networking-full-detail.aspx. Accessed May 28, 2013.
41. Kiatpongsan S, Sipp D. Monitoring and regulating offshore stem cell clinics. Science. 2009;323:1564–5.
42. Food and Drug Administration. FDA warns about stem cell claims. 2013. http://www.fda.gov/ForConsumers/ConsumerUpdates/ucm286155.htm. Accessed Oct. 21, 2013.
43. Knox R. Offshore stem cell clinics sell hope, not science. NPR Website. 2010. http://www.npr.org/2010/07/26/128696529/offshore-stem-cell-clinics-sell-hope-not-science. Accessed May 28, 2013.
44. Mackey TK, Liang BA. Dangerous science: online promotion of unproven stem cell therapies and global health risks. J Comm Biotech. 2013;19:19–28.
45. Devereaux M, Loring JF. Growth of an industry: how U.S. scientists and clinicians have enabled stem cell tourism. Am J Bioeth. 2010;10(5):45–6.
46. Zarzeczny A, Caulfield T. Stem cell tourism and doctors' duties to minors--a view from Canada. Am J Bioeth. 2010;10(5):3–15.
47. Hauskeller C, Wilson-Kovacs D. Traveling across borders--the pitfalls of clinical trial regulation and stem cell exceptionalism. Am J Bioeth. 2010;10(5):38–40.
48. Chandler J. Stem cell tourism: doctors' duties to minors and other incompetent patients. Am J Bioeth. 2010;10(5):27–8.
49. Pandya SK. Stem cell transplantation in India: tall claims, questionable ethics. Indian J Med Ethics. 2008;5(1):15–7.
50. Chandrashekhar M. Stem cell quackery. Indian J Med Ethics. 2008;5(4):198.
51. Abraham SJ. Stem cell transplantation in India: tall claims, questionable ethics: a response. Indian J Med Ethics. 2008;5(2):84–5. discussion 85–6.
52. Liang BA, Mackey TK. Stem cells, dot-com. Sci Trans Med. 2012;4(151):151cm9.
53. Cyranoski D. Korean deaths spark inquiry. Nature. 2010;468(7323):485.
54. Amariglio N, Hirshberg A, Scheithauer BW, Cohen Y, Loewenthal R, Trakhtenbrot L, et al. Donor-derived brain tumor following neural stem cell transplantation in an ataxia telangiectasia patient. PLoS Med. 2009;6(2):e1000029.
55. Lyn TE. China stem cell therapies offer heartbreak for many. Reuters 2011. http://www.reuters.com/article/2011/09/21/us-stemcell-scams-idUSTRE78K18120110921. Accessed May 28, 2013.
56. Mendick R, Hall A. Europe's largest stem cell clinic shut down after death of baby. The Telegraph. 2011. http://www.telegraph.co.uk/news/worldnews/europe/germany/8500233/Europes-largest-stem-cell-clinic-shut-down-after-death-of-baby.html. Accessed May 28, 2013.
57. Fitzpatrick D, Griffin D. Florida suspends doctor accused of illegal stem cell therapy. CNN Health. 2012. http://www.cnn.com/2012/03/08/health/stem-cell-doctor-suspension. Accessed May 28, 2013.
58. Zarzeczny A, Rachul C, Nisbet M, Caulfield T. Stem cell clinics in the news. Nat Biotechnol. 2010;28:1243–6.
59. Halme DG, Kessler DA. FDA regulation of stem-cell-based therapies. N Engl J Med. 2006;355:1730–5.
60. Ledford H. Stem-cell scientists grapple with clinics. Nature. 2011;474:550.
61. Hyun I. Allowing innovative stem cell-based therapies outside of clinical trials: ethical and policy challenges. J Law Med Ethics. 2010;38:277–85.
62. Enserink M. Biomedicine: selling the stem cell dream. Science. 2006;313:160–3.
63. Sternberg S. Doctors offer unapproved stem cell therapies. USATODAY.com. 2011. yourlife.usatoday.com. http://yourlife.usatoday.com/health/story/2011/06/Doctors-offer-unapproved-stem-cell-therapies/48933666/1#.TgsY1_J56jg.email
64. Relman AS. Closing the books on Laetrile. N Engl J Med. 1982;306:236.

65. Fakunle ES, Loring JF. Ethnically diverse pluripotent stem cells for drug development. Trends Mol Med. 2012;18(12):709–16.
66. Lau D, Ogbogu O, Taylor B, Stafinski T, Menon D, Caulfield T. Stem cell clinics online: the direct-to-consumer portrayal of stem cell medicine. Cell Stem Cell. 2008;3:591–4.
67. Stem cell fraud: a 60 minutes investigation. cbsnews.com. 2012. http://www.cbsnews.com/8301-18560_162-57354695/stem-cell-fraud-a-60-minutes-investigation/. Accessed May 23, 2013.
68. Dobkin BH, Curt A, Guest J. Cellular transplants in China: observational study from the largest human experiment in chronic spinal cord injury. Neurorehabil Neural Repair. 2006;20:5–11.
69. Rubin JP. Commentary on: stem cell facelift: between reality and fiction. Aesthet Surg J. 2013;33(3):339–40.
70. Atiyeh BS, Ibrahim AE, Saad DA. Stem cell facelift: between reality and fiction. Aesthet Surg J. 2013;33(3):334–8.
71. Madjd Z, Gheytanchi E, Erfani E, Asadi-Lari M. Application of stem cells in targeted therapy of breast cancer: a systematic review. Asian Pac J Cancer Prev. 2013;14(5):2789–800.
72. Sipp D. Stem cell stratagems in alternative medicine. Regen Med. 2011;6(3):407–14.
73. United States v. Regenerative Sciences et al. Civil Action No. 1:10-cv-01327, Complaint (D.D.C. Aug. 6, 2010).
74. Lysaght T, Campbell AV. Regulating autologous adult stem cells: the FDA steps up. Cell Stem Cell. 2011;9:393–6.
75. Regenerative Sciences v. U.S. Food and Drug Administration, et al. Civil Action No. 1:10-cv-01055-RMC, Complaint for Declaratory and Injunctive Relief (D.D.C. June 22, 2010).
76. Pepper MS. Cell-based therapy-navigating troubled waters. S Afr Med J. 2010;100(5):286,288.
77. Caplan A. Stem cell clinics ripping off patients. NBC News. 2011. http://www.nbcnews.com/id/43760409/ns/health-health_care/t/stem-cell-clinics-ripping-patients-bullying-scientists/#.UaU76Ov3iyw. Accessed May 28, 2013.
78. CBSNews. 21st Century Snake Oil. CBSNews. 2010. http://www.cbsnews.com/stories/2010/09/09/60minutes/main6850496.shtml?tag=mncol;lst;3. Accessed May 28, 2013.
79. George B. Regulations and guidelines governing stem cell based products: clinical considerations. Perspect Clin Res. 2011;2(3):94–9.
80. Sipp D. Hope alone is not an outcome: why regulations make sense for the global stem cell industry. Am J Bioeth. 2013;10(5):33–4.
81. McMahon D, Thorsteinsdóttir H. Regulations are needed for stem cell tourism: insights from China. Am J Bioeth. 2010;10(5):34–6.
82. Licensing under the Quality and Safety Regulations. hta.gov.uk. http://www.hta.gov.uk/licensingandinspections/licensingunderthequalityandsafetyregulations.cfm. Accessed Oct. 19, 2013.
83. Biologicals framework Q&A: Part 1 - Regulation of biological. tga.gov.au. http://www.tga.gov.au/industry/biologicals-framework-qa-p1.htm. Accessed Oct. 19, 2013.
84. Cyranoski D. FDA challenges stem-cell clinic. Nature. 2010;466:909.
85. Aicher RH. Is the FDA regulating the practice of medicine? Aesthet Surg J. 2013;33(3):452–5.
86. Cyranoski D. FDA's claims over stem cells upheld. Nature. 2012;488(7409):14. http://www.nature.com/polopoly_fs/1.11082!/menu/main/topColumns/topLeftColumn/pdf/488014a.pdf.
87. Australian Medical Council, amc.org.au. http://www.amc.org.au/index.php/about. Accessed Oct. 19, 2013.
88. Cyranoski D. China's stem-cell rules go unheeded. Nature. 2012;484(7393):149–50.
89. The Economist. Stem cells in China: wild east or scientific feast? The Economist. 2010. http://www.economist.com/node/15268869. Accessed May 28, 2013.
90. International Society for Stem Cell Research. http://www.isscr.org. Accessed Oct. 21, 2013.

91. Rowe A, Gracia-Barbero AM. Regulation and licensing of physicians in the WHO European Region. Geneva: World Health Organization; 2005.
92. Ilic N, Savic S, Siegel E, Atkinson K, Tasic L. Examination of the regulatory frameworks applicable to biologic drugs (including stem cells and their progeny) in Europe, the U.S., and Australia: part I--a method of manual documentary analysis. Stem Cells Transl Med. 2012;1(12):898–908.
93. Kellathur SN, Lou HX. Cell and tissue therapy regulation: worldwide status and harmonization. Biologicals. 2012;40(3):222–4.
94. Protection of Human Subjects. 45 CFR Part 46 (2010).
95. Food and Drug Administration. Regulatory procedures manual - 5-5 – Administrative detention of devices. FDA Website. http://www.fda.gov/ICECI/ComplianceManuals/RegulatoryProceduresManual/ucm176980.htm. Accessed May 14, 2013.
96. Mackey TK, Liang BA. The global counterfeit drug trade: patient safety and public health risks. J Pharm Sci. 2011;100(11):4571–9.
97. Medicinal products for human use - quality of medicines and the fight against falsified medicines. Public health, European Commission. ec.europa.eu. http://ec.europa.eu/health/human-use/quality/index_en.htm. Accessed Oct. 21, 2013.
98. Federal Trade Commission Act. 15 U.S.C. §41 et seq.
99. Unfair Competition; False Advertising. Cal. Bus. & Prof. Code §§17200, 17500 (2010).
100. The Office of Fair Trading. Office of fair trading, Fleetbank House, 2-6 Salisbury Square, London, EC4Y 8JX; http://www.oft.gov.uk. Accessed May 21, 2013.
101. Liang BA. Benefits notwithstanding: discipline associated with efficacious medical treatment. J Clin Anesth. 2009;21:613–5.
102. Hyun I, Lindvall O, Ahrlund-Richter L, et al. New ISSCR guidelines underscore major principles for responsible translational stem cell research. Cell Stem Cell. 2008;3:607–9.
103. Beil L. Stem cell treatments overtake science. NY Times. 9 Sept. 2013.
104. Sipp D. The unregulated commercialization of stem cell treatments: a global perspective. Front Med. 2011;5(4):348–55.
105. Council of Europe. Regulation (EC) No 1394/2007 of the European Parliament and of the Council, of 13 November 2007, on advanced therapy medicinal products and amending Directive 2001/83/EC and Regulation (EC) No 726/2004. 2000. http://eur-lex.europa.eu/LexUriServ/LexUriServ.do?uri=OJ:L:2007:324:0121:0137:en:PDF. Accessed Oct. 23, 2013.
106. Ancans J. Cell therapy medicinal product regulatory framework in Europe and its application for MSC-based therapy development. Front Immunol. 2012;3:253.
107. Migliaccio G, Pintus C. Role of the EU framework in regulation of stem cell-based products. Adv Biochem Eng Biotechnol. 2013;130:287.
108. Berger A, Schüle S, Flory E. The certification of advanced therapy medicinal products. A quality label for product development in small and medium-sized enterprises [German]. Bundesgesundheitsblatt Gesundheitsforschung Gesundheitsschutz. 2011;54(7):816–21.
109. Voltz-Girolt C, Celis P, Boucaumont M, D'Apote L, Pinheiro MH, Papaluca-Amati M. The advanced therapy classification procedure. Overview of experience gained so far. [German]. Bundesgesundheitsblatt Gesundheitsforschung Gesundheitsschutz. 2011;54(7):811–5.
110. Bills passed on regenerative meds. Japan Times. 20 Nov. 2013; http://www.japantimes.co.jp/news/2013/11/20/national/bills-passed-on-regenerative-meds/. Accessed Nov. 21, 2013.

Index

D.O. Fauza, M. Bani (eds.), *Fetal Stem Cells in Regenerative Medicine*,
Stem Cell Biology and Regenerative Medicine,
DOI 10.1007/978-1-4939-3483-6

MIX
Papier aus verantwortungsvollen Quellen
Paper from responsible sources
FSC® C105338

If you have any concerns about our products,
you can contact us on
ProductSafety@springernature.com

In case Publisher is established outside the EU,
the EU authorized representative is:
Springer Nature Customer Service Center GmbH
Europaplatz 3, 69115 Heidelberg, Germany

Printed by Libri Plureos GmbH
in Hamburg, Germany